FRACTALS IN ROCK MECHANICS

GEOMECHANICS RESEARCH SERIES

1

Editor-in-chief:
MAREK A. KWAŚNIEWSKI
Faculty of Mining and Geology, Silesian Technical University, Gliwice

FRACTALS IN ROCK MECHANICS

HEPING XIE
China University of Mining and Technology, Xuzhou & International Centre for Material Physics, Academia Sinica, Shenyang

A.A. BALKEMA / ROTTERDAM / BROOKFIELD / 1993

Published by
A.A. Balkema, P.O. Box 1675, 3000 BR Rotterdam, Netherlands
A.A. Balkema Publishers, Old Post Road, Brookfield, VT 05036, USA

ISBN 90 5410 133 4

Printed in the Netherlands

Contents

Preface

This book is based on lectures on 'Fractals – Rock Mechanics' given by the author at the University of Utah and China University of Mining & Technology.

Important developments in the progress of the theory of rock mechanics during recent years are based on fractal geometry and damage mechanics. The concept of fractals introduced by Mandelbrot (1982) has proven to be a highly useful way to describe the statistics of naturally occurring geometries. Natural objects, from mountains and coastlines to clouds and forests, are found to have boundaries best described as self-similar fractals. Fluid flow through jointed (or fractured) rock masses and clusterings of earthquakes are found to follow fractal patterns in time and space. Fracturing in rocks at all scales, from the microscale (microcracks) to the continental scale (megafaults), can lead to fractal structures. The process of diagenesis and pore geometry of sedimentary rock can be quantitatively described by fractals, etc.

The book is mainly concerned with these developments, as related to fractal descriptions of fragmentations, damage, and fracture of rocks, rock burst, joint roughness, rock porosity and permeability, rock grain growth, rock and soil particles, shear slips, fluid flow through jointed rocks, faults, earthquake clustering, and so on. The prime concerns of the book are to give a simple account of the basic concepts, methods of fractal geometry, and their applications to rock mechanics, geology, and seismology, and also to discuss damage mechanics of rocks and its application to mining engineering.

This book contains some of the topics I found particularly interesting and useful in teaching and in our research. Many more ideas and interesting facts of fractals are found in the books of Mandelbrot (1982), Feder (1988), Falconer (1985, 1990) and in the rapidly growing research literature. I would like to thank all colleagues (Prof. J. Feder, Prof. A.H. Thompson, Prof. P.R. La-Pointe, Prof. Po-zen Wong, Prof. C.W. Lung, Prof. Lianke Dong, Prof. K. Falconer, Prof. Y. Termonia, Prof. L. Pietronero, Prof. H. Takayasu, Prof. D.J. Barr, Prof. R.E. Williford, Prof. M.F. Goodchild, Prof. Tianhu Hao, Prof. Zhenyu Tao and others) who sent their manuscripts and papers to me, supported and encouraged me in writing this book.

I am very grateful to Prof. W.G. Pariseau for giving me the opportunity to teach this course for graduate students at the University of Utah in 1991. The first draft of the book was improved while teaching. My advisors, Prof. Zhida Chen and Prof. W.G. Pariseau, and my friends, Jeffrey C. Johnson and James F. Kohler, gave me many penetrating comments and suggestions. Fractals in rock mechanics, our research project, was supported by the National Natural Science Foundation of China and China National Coal Corporation.

I wish to express my thanks to Ms. Diana Fox (Secretary, Department of Mining Engineering, University of Utah) for reviewing the draft and correcting the grammar of the book, and to Mr. Shengping Yan and Mr. Peng Zhao for typing the manuscript and other assistance. Especially, I would like to thank Dr. Marek A. Kwaśniewski, editor-in-chief of this series, for editing the manuscript and his many valuable remarks.

I am also grateful for the understanding and support provided by my wife, Dongxiao Li, while I worked on this volume.

Heping Xie
China University of Mining & Technology
Xuzhou, Jiangsu, People's Republic of China
November 1991

CHAPTER 1

Introduction

Geometry is a discipline in which the shapes of objects are studied. Classical geometry is used to investigate regular and smooth shapes (lines, curves, circles and so on). However, geometric shapes in classical geometry are approximate descriptions of objects in nature. For example, the rough surface of the earth is regarded as a smooth spherical or ellipsoid surface, and an upland is considered a trapezoidal slope (line) surface. Although these approximations do not prevent us from reaching conclusions that agree with practice, the need to obtain accurate descriptions of matter becomes critical to the development of science.

Mandelbrot (1975b, 1982) introduced a new geometry – fractal geometry – to describe irregular and nonsmooth phenomena in nature, and it appears that this new field of investigation will be more applicable to the study of natural phenomena than regular and smooth curves and surfaces of classical geometry.

The word 'fractal' is derived from the Latin fractus, meaning broken. The idea of 'fractal' originated during the mathematical crisis of 1875-1925 (Mandelbrot, 1982). At the beginning of 1870's, most mathematicians believed that Euclidean geometry provided an accurate description of shape properties in both material and abstract space. These dimensions, as derived from Euclid (300 B.C.), stated that, for plane geometry:

(1) A point has no part;
(2) A line is a breadthless length;
(3) The extremities of lines are points;
(4) A surface has length and breadth only, and
(5) The extremities of surfaces are lines.

For spatial geometry, the assumptions were that:

(1) A solid has length, breadth and depth, and
(2) The extremity of a solid is a surface.

Meantime, most mathematicians believed that any continuous function must be differentiable.

In 1875, Weierstrass introduced a continuous but undifferentiated function in the form

$$W_0(t) = (1 - w^2)^{-1/2} \sum_{n=1}^{\infty} w^n \exp(2\pi i b^n t),$$

where b is a real number larger than 1, and w is written either as $w = b^H$ with $H \in [0, 1]$ or as $w = b^{D-2}$ with $D \in [1, 2]$. This function does not suit the original concept because 'nowhere differentiable' means that a curve does not have a tangent at all points. Thus the question arises as to whether the geometrical expression of this function is a curve.

The Weierstrass function upset the belief that continuous functions must be differentiable and led to a great deal of skepticism concerning the definition of a curve. For these reasons, Jordan redefined the concept of a curve as follows: A curve is the image of a continuous injection $x = f(t)$, $y = g(t)$; $(t_0 < t < t_1)$. If $f(t_0) = f(t_1)$, $g(t_0) = g(t_1)$, this image is called a Jordan closed curve.

Although Jordan's definition of a curve eliminated some problems, it was too extensive, that is, only a continuum was required. In 1890, Peano found that the curve defined by Jordan would run over all points of a square; that is, the curve would result in continuous mapping of an interval on the whole of a square. This meant that the definition was still not satisfactory.

On the other hand, what did the length of a curve mean? The common formulation of the length of a curve was

$$L = \int_a^b (1 + y'^2)^{1/2}\,dx,$$

where the existence of a derivative of $y = f(x)$ was required.

Meantime, in 1904, Koch constructed a continuous, but nowhere differentiable closed curve which has an infinite perimeter, and by which a finite area is surrounded. Start with an equilateral triangle with side length $3S$. In the middle interval of its every side, a new equilateral triangle with the side length of S is made, and the base side of every new triangle is removed. Then, in the middle interval on each side S of this new figure, some newer equilateral triangles with the side length $S/3$ are made again. Analogically, their base sides are all erased. After infinitely repeating the same procedure, Koch's closed curve (Figure 2.6b) is constructed. Its perimeter length is successively $9S$, $12S$, $16S, \ldots$, finally approaching ∞, whereas the area surrounded by this curve becomes

$$A = 9\sqrt{3} \times S^2/4 + 3\sqrt{3} \times S^2/4 + \sqrt{3} \times S^2/3 + 4\sqrt{3} \times S^2/27 + \cdots$$

which is an infinite series. It is seen that the ratio of this series is $\frac{4}{9}$, then

$$A = 9\sqrt{3} \times S^2/4 + (\tfrac{3}{4})S^2\sqrt{3}/(1 - \tfrac{4}{9}) = 18\sqrt{3} \times S^2/5 = \tfrac{8}{5}A_0,$$

where A_0 is the initial area of the equilateral triangle with side length $3S$.

On the other hand, from set theory, Cantor proved an one-to-one correspondence between the points of a line and the points of a plane. Although this was not continuous correspondence from a line to a plane, it proved that dimension was not how many points existed in space, or the number of coordinates needed to fix a point in space. It indicated that the dimension of geometrical figures could change with one-to-one corresponding mapping, and the dimensional value could increase with single-valued and continuous mapping. This led to the lash of the concept of classical Euclidean dimension. Thus fractional dimension, similar dimension and fractal dimension were proposed successively. Finally a new geometry, fractal geometry, was born (Mandelbrot, 1975b, 1982, 1983). It is clear that fractal geometry is not a straight 'application' of 20th century mathematics, it is a new branch born belatedly from the crisis of mathematics (1875-1925).

The basic concept of fractal geometry is fractional dimension. Fractional dimension was proposed by Hausdorff (1919). But fractal geometry formed by generalization of this note is due to B. B. Mandelbrot. He proposed the idea of fractal geometry (early 1973) while holding a lecture in College de France. He believes that the concept of fractional dimension may be a powerful tool used to research physical phenomena, and fractal geometry can be applied to handle irregular shapes.

How do we understand the fractional dimension as opposed to an integer dimension for which we are well trained in classical geometry? An excellent example is a sponge model. A sponge cube looks three-dimensional in classical geometry, but it may be stressed in a plane because of a highly ordered structure of holes in it; thus, it also seems to be two-dimensional. This means that the integral dimensions such as 2 and 3 here only reflect apparent phenomena of objects. In fact, the dimension of the sponge cube may be in the range between 2 and 3 because of existence of a highly ordered structure of holes in it; that is, fractional dimension. This indicates that fractional dimension can reflect inherent properties of objects.

Since Mandelbrot's books (1975b, 1977, 1982, 1983) have been published, the concept of fractals has caught the imagination of scientists in many fields, and papers concerning fractals in various contexts now appear almost daily. Mandelbrot has discussed trees, rivers, lungs, water levels, turbulence, economics, word frequencies and many, many other topics by using his fractal geometrical concepts. In his books, he purposefully has no introduction and no conclusion in order to stress his belief that as more work is done in this field, his ideas will reveal further insight into the geometry of nature (Feder, 1988).

Fractals proved to be more surprising is, spurred on by developments in computer technology, their wide applications in computer art and in the motion picture industry to generate realistic but synthetic landscapes. Fractal 'planets' rising over the horizon of its moon, mountains, valleys and islands have been shown as beautiful cover photographs in journals and magazines. These illustrations are based on algorithms of fractals (Peitgen & Richter, 1986; Peitgen & Saupe, 1988). They look natural. Voss has described some of ideas he used in the generation of his spectacular pictures (Voss, 1985a, b). Some of the computer graphics involved in the creation of the beautiful fractal objects are described in the books of Barnsley (1988), Stevens (1989, 1990), and Edgar (1990).

As a series of discipline of studies, the intellectual beauty of fractals has led to a revolution in statistical physics over the last decade. 'Fractional dimension' has been studied for over 70 years, although the term 'fractal' was not coined until 1975 by Mandelbrot. A large number of physicists became involved around 1980 (Wong, 1988). Fractals are 'hybrids' between familiar shapes and typically have noninteger dimensions. The mass of a 'mass fractal' is proportional to its radius to a power between 1 and 3; this power is 3 for a solid ball, 2 for a plane and 1 for a line. Similarly, the area of a 'surface fractal' scales with the radius of the object with a noninteger power. This property of fractals can be used with great effectiveness in analyzing physical problems. There exists enough introductory and advanced material on fractals, which are particularly well suited for thermodynamics, statistical mechanics and applied mathematics studies at any level. Some striking books and proceedings have mainly described in detail the ideas and applications of fractals of physics, such as: Fischer & Smith (1985), Feder (1988), Pietronero & Tosatti (1986), Jullien & Rotet (1987), Stanley

& Ostrowsky (1986), Barnsley & Denko (1986), Banaver et al. (1987), Brien (1989), Vicsek (1989), Robert & Linda (1989), Scholz & Mandelbrot (1989), Falconer (1990), Herrmann & Roux (1990), and Cherbit (1991).

Fractal geometry is used mainly to study irregular shapes with self-similarity, irregular figures with self-inverse, the mapping with self-squaring, self-affine fractal sets and so on. But the dominant content of fractal geometry is fractals with self-similarity (linear fractal, in short).

Self-similarity means that any portion of the fractal that is magnified by an arbitrary factor looks the same as the original fractal; that is, scale invariance. The phenomenon with the behaviors of self-similarity and scale invariance has been observed in many fields, but especially in the field of rock mechanics in which we are very interested. It may be said that the complexity of million-year-old sedimentary rock is being unraveled by concepts of fractals (Wong, 1988). In the past, physicists paid little attention to rock, mainly because they were discouraged by its apparent complexity. They are well trained in working with idealized models, but when faced with a piece of rock, not only do they not know where to begin, but they may also question whether it is even possible to find interesting physics in a 'dirty' and uncontrolled system. In the last few years, however, physicists have gained a new insight into rock using such concepts as the random system of fractals, percolation and diffusion-limited growth, and damage evolution. Mechanical behavior of rock materials is so complicated that it may be constructed in a fractal dimensional space. In fact, an element of rock materials is very similar to a sponge cube with highly ordered structures of micropores. As discussed above, this element is a fractal dimensional structure. Obviously, deformation and fracture of the element, fluid flow through this element and its other mechanics properties all should behave fractal natures. These ideas have been gleaned from many experiments of rock materials and in situ data.

This book is concerned with the rock mechanics phenomena that generate fractal structures and a new mechanics of rock materials – damage mechanics of rocks. In order to provide a basis for various applications of fractals to rock mechanics and geological science, we start out in Chapter 2 with a discussion of general properties of fractals (mainly refer to Feder (1988), especially his some excellent descriptions of classical fractals). Related concepts of fractals (such as the relationship between perimeter and area, random fractals, multifractals and fractal growths) and examples combining these concepts (such as rock permeability, fracture of materials and so on) and our recent work on the theory of fractals are also discussed in Chapters 3 through 8. The theory of large deformation, the concepts of damage mechanics of rock and their applications to rock mechanics and mining engineering are introduced in Chapters 9 and 10, respectively. In the remaining chapters (Chapters 11-16), the applications of fractals to rock mechanics and geological science are discussed in detail.

The mathematical basis of fractal geometry is metric topology and measure theory. For this reason, I briefly introduce measure theory (Lebesque measure and Hausdorff measure) and metric topology (based on Falconer, 1985 and Edgar, 1990) in Section 2.1. It is probably the most difficult part of reading the book, but it may lead you to easily understand the basic definition of a fractal. The chapters and sections with more mathematics labeled by an asterisk (*) (Section 2.1, Chapters 7 and 8) are suitable for the reader interested in fractal theory. However, they are mostly independent of other parts of the book, so the reader interested only in the applications of fractals can neglect these chapters and sections.

Fractal geometry is a new attractive discipline. In particular, the rapid development of modern microtechniques and computer technology makes fractal measurements and fractal simulations possible and easy, which inevitably promotes the application of fractal geometry to many fields. The main purpose of this book is to introduce the concept of fractal geometry and its applications to rock mechanics and geological science to spark interest in this new subject. We believe that fractal geometry will provide a new powerful approach for investigation of complicated mechanics problems of rock materials.

CHAPTER 2

General properties of fractals

2.1 MEASURE AND DIMENSION*

2.1.1 *Basic measure theory*

Sets, sequences, series, and functions occur in every area of applied mathematics. Sets will be designated by capital letters, $A, B, A_1, C_2, \ldots$. Individual members of sets will be designated by lower case letters, $x, y, a_1, a_2, b, \ldots$. The unique set which contains no element is called the empty set and is denoted by $\varnothing$.

If x is (or is not) a member of A, we write $x \in A$ ($x \notin A$); then A is termed a subset of B, and written $A \subseteq B$, if $x \in A$ implies $x \in B$. We assume that all sets in a given discussion are subsets of some fixed 'Universal set' U.

The basic operations involving sets are:

(1) Complement	A^c:	$x \in A^c$	if $x \notin A$;	
(2) Union	$A \cup B$:	$x \in A \cup B$	if $x \in A$	or/and $x \in B$;
(3) Intersection	$A \cap B$:	$x \in A \cap B$	if $x \in A$	and $x \in B$;
(4) Difference	$A \backslash B$:	$x \in A \backslash B$	if $x \in A$	and $x \notin B$; and
(5) Symmetric difference	$A \triangle B$:	$x \in A \Delta B$	if $x \in A \backslash B$	or $x \in B \backslash A$.

We can picture these and subsequent definitions quite easily by means of so-called Venn diagrams (Figure 2.1). In Figure 2.1, the square regions represent the Universal set, U. By convention, all the members (elements) of a set are assumed to be different (distinct), so even if sets A and B have an element x in common, the set $A \cup B$ contains the number x represented only once.

Two sets, A and B, may or may not have members in common. If A and B have no member in common, they are called disjoint, and written $A \cap B = \varnothing$.

Sets differ also in the number of elements they contain. Set A is termed countable if its members can be placed in a one-to-one correspondence (mating) with a subset (possibly all) of the positive integers. Otherwise, the set is called uncountable.

A sequence of sets, $A_1, A_2, \ldots$, is termed monotone increasing or monotone decreasing if $A_1 \subseteq A_2 \subseteq A_3 \subseteq \ldots$ or $A_1 \supseteq A_2 \supseteq A_3 \supseteq \ldots$.

Let X be any set (We shall shortly take X to be n-dimensional Euclidean space R^n). A non-empty collection ϕ of subsets of X is termed a sigma-field (or σ-field) if ϕ is closed under complementation and under countable union. So if $E \in \phi$, then $X \backslash E \in \phi$,

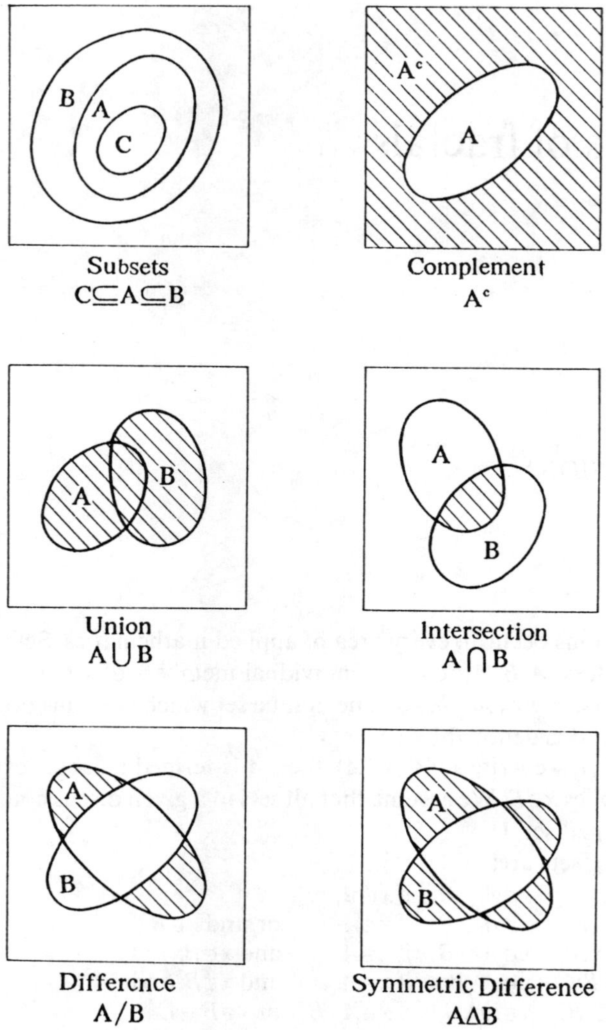

Figure 2.1. Venn diagrams.

and if $E_1, E_2, \ldots \in \phi$, then

$$\bigcup_{j=1}^{\infty} E_j \in \phi.$$

The lower and upper limits of a sequence of sets $\{E_j\}$ are defined as

$$\underline{\lim}_{j \to \infty} E_j = \bigcup_{k=1}^{\infty} \bigcap_{j=k}^{\infty} E_j$$

and

$$\overline{\lim}_{j \to \infty} E_j = \bigcap_{k=1}^{\infty} \bigcup_{j=k}^{\infty} E_j,$$

respectively. Thus, $\underline{\lim}\, E_j$ consists of those points lying in all but finitely many E_j, and $\overline{\lim}\, E_j$ consists of those points in infinitely many E_j. From the form of these definitions it is clear that if E_j lies in the σ-field ϕ for each j, then $\underline{\lim}\, E_j \in \phi$ and $\overline{\lim}\, E_j \in \phi$. If $\overline{\lim}\, E_j = \underline{\lim}\, E_j$ then we write $\lim E_j$ for the common value. This always happens if $\{E_j\}$ is either an increasing or a decreasing sequence of sets.

A measure, μ, is a function defined on some σ-field ϕ of subsets of X and taking values in the range $[0, \infty)$ such that

$$\mu(\phi) = 0 \tag{2.1}$$

and

$$\mu\left(\bigcup_{i=1}^{\infty} E_i\right) = \sum_{i=1}^{\infty} \mu(E_i); \tag{2.2}$$

for every countable sequence of disjoint sets $\{E_j\}$ in ϕ.

It follows from Equation (2.2) that μ is an increasing set function, that is, if $E \subset E'$ and $E, E' \in \phi$, then

$$\mu(E) \leq \mu(E').$$

Theorem 2.1: Continuity of measures. Let μ be a measure on a σ-field ϕ of subsets of X;

(a) If $A_1 \subset A_2 \subset \ldots$ is an increasing sequence of sets in ϕ, then

$$\mu\left(\lim_{j\to\infty} A_j\right) = \lim_{j\to\infty} \mu(A_j);$$

(b) If $B_1 \supset B_2 \supset \ldots$ is a decreasing sequence of sets in ϕ and $\mu(B_i) < \infty$, then

$$\mu\left(\lim_{j\to\infty} B_j\right) = \lim_{j\to\infty} \mu(B_j);$$

(c) for any sequence of sets $\{B_j\}$ in ϕ

$$\mu\left(\underline{\lim}_{j\to\infty} B_j\right) \leq \underline{\lim}_{j\to\infty} \mu(B_j).$$

Next we introduce outer measures which are essentially measures with property (2.2) weakened to subadditivity. Formally, an outer measure μ^* on a set X is a function defined on all subsets of X taking values in $[0, \infty)$ such that

$$\mu^*(\phi) = 0;$$

$$\mu^*(A) < \mu^*(A'), \quad \text{if } A \subset A';$$

and

$$\mu^*\left(\bigcup_{j=1}^{\infty} A_j\right) \leq \sum_{j=1}^{\infty} \mu^*(A_j) \quad \text{for any subsets } \{A_j\} \text{ of } X.$$

Outer measures are useful since there is always a σ-field of subsets on which they behave as measures.

A subset E of X is called μ^*-measurable if it decomposes every subset of X additively, that is, if

$$\mu^*(A) = \mu^*(A \cap E) + \mu^*(A \setminus E),$$

for every set $A \subset X$. Note that to show that set E is μ^*-measurable, it is enough to check that

$$\mu^*(A) \geq \mu^*(A \cap E) + \mu^*(A \setminus E).$$

It is trivial to verify that if $\mu^*(E) = 0$, then E is μ^*-measurable.

Let μ^* be an outer measure. The collection μ of μ^*-measurable sets forms a σ-field, and the restriction of μ^* to μ is a measure.

We say that the outer measure μ^* is regular if, for every set A, there is a μ^*-measurable set E containing A with $\mu^*(A) = \mu^*(E)$.

If μ^* is a regular outer measure and $\{A_j\}$ is any increasing sequence of sets, then

$$\lim_{j \to \infty} \mu^*(A_j) = \mu^*\left(\lim_{j \to \infty} A_j\right).$$

Let μ^* be a measure on the σ-field ϕ of subsets of X, $f: X \to R^n$ is said to be measurable if, for every real α the set,

$$f^{-1}((\alpha, \infty]) = \{x: x \in X, f(x) > \alpha\}$$

is in ϕ. A complex valued function $g + ih$ is measurable if g and h are both measurable.

What other functions are measurable? We may say certain combinations and limits of sequences of measurable functions are measurable. For example, let (f_n) be a sequence of measurable functions and define the function $\sup f_n$ by setting $(\sup f_n)(x) = \sup f_n(x)$ for each x; note that this function always exists as extended real valued functions are allowed. Then $(a > 0)$

$$\{x: \sup_n f_n(x) > a\} = \{x: f_n(x) > a \text{ for at least one } n\} = \bigcup_{n=1}^{\infty} \{x: f_n(x) > a\},$$

and since ϕ is closed with respect to countable unions, $\sup f_n$ is measurable. The following theorems summarize the position.

Theorem 2.2: Let μ^* be a measure on the σ-field ϕ of subsets of X, and suppose that $f, g: X \to R^n$ are measurable. Then the following are measurable:

(1) αf $(\alpha > 0)$;
(2) $|f|$;
(3) $f + g$ (with the proviso that $\pm(\infty) + (\mp\infty)$ should not occur);
(4) fg;
(5) $\max\{f, g\}$, $\min\{f, g\}$, (where $\max\{f, g\} = \max\{f(x), g(x)\}$ for $x \in X$);
(6) $f^+ = \max\{f, 0\}$, $f^- = \max\{-f, 0\}$.

Theorem 2.3: With μ as above, suppose that (f_n) is a sequence of measurable functions from X to R^n. Then the following are measurable:

(1) $\sup(f_n)$, $\inf(f_n)$;
(2) $\lim\sup(f_n)$, $\lim\inf(f_n)$;
(3) the pointwise limit $\lim(f_n)$, when it exists.

Let μ^* be a measure on a σ-field ϕ of subsets of X and suppose that $f:X \to R^n$ is measurable. Then,

(1) if $f \geq 0$, there is a monotone increasing sequence of non-negative simple functions which tends pointwise to f; and

(2) if f is bounded, there is a sequence of simple functions tending to f in the sup norm.

2.1.2 *Metric spaces (Falconer, 1985)*

One of the common ways to form a topological space is through the use of a distance function, also called a metric. For example, on R^n the standard distance,

$$d(\boldsymbol{x}, \boldsymbol{y}) = \left(\sum_{i=1}^{n} (x_i - y_i)^2 \right)^{1/2},$$

between $\boldsymbol{x} = (x_1, \ldots, x_n)$ and $\boldsymbol{y} = (y_1, \ldots, y_n)$ can be used to construct the open disks and from them the topology. The abstraction of this proceeds as follows.

Let X be a set. A metric on X is a function $d:X \times X \to R$ such that

(1) $d(x_1, x_2) = 0$, if $x_1 = x_2$/(definiteness);

(2) $d(x_1, x_2) = d(x_2, x_1)$/(symmetry); and

(3) $d(x_1, x_3) \leq d(x_1, x_2) + d(x_2, x_3)$/(triangle inequality).

A metric space is the pair (X, d); if there is no danger of confusion, just write X for (X, d).

Taking $x_1 = x_3$ in (3) shows that d is necessarily a non-negative function. It is proved in advanced calculus courses (and geometrically clear) that the standard distance on R^n satisfies (1)-(3). The topology determined by a metric is defined as follows: For $\varepsilon > 0$ and $x \in X$, the ε-disk about d is defined by $D_\varepsilon(x) = \{x' \in X \mid d(x', x) < \varepsilon\}$.

The collection of subsets of X that are unions of such disks is the metric topology of the metric space (X, d). Two metrics on a set are called equivalent if they induce the same metric topology.

Now let (X, d) be a metric space (for our purposes, X will usually be n-dimensional Euclidean space R^n). The sets belonging to the σ-field generated by the closed subsets of X are called the Borel sets of the space. The Borel sets include the open sets (as complements of closed sets), the F_σ-sets (that are countable unions of closed sets), and the G_σ-sets (countable intersections of open sets), etc.

An outer measure μ^* on X is termed a metric outer measure if

$$\mu^*(E \cup F) = \mu^*(E) + \mu^*(F),$$

where E and F are positively separated; that is, wherever

$$d(E, F) = \inf\{d(x, y): x \in E, y \in F\} > 0.$$

If μ^* is a metric outer measure, then the collection of μ^*-measurable sets includes the Borel sets.

2.1.3 *Lebesgue measure (Falconer, 1985)*

Let C be a coordinate block in R^n of the form

$$C = [a_1, b_1) \times [a_2, b_2) \times \cdots \times [a_n, b_n)$$

where $a_i < b_i$ for each i. Define the volume of C as

$$V(C) = (b_1 - a_1)(b_2 - a_2)\cdots(b_n - a_n)$$

in the obvious way. If $E \subset R^n$, let

$$L^n(E) = \inf \sum_i V(C_i), \tag{2.3}$$

where the infimum is taken over all coverings of E by a sequence $\{C_i\}$ of blocks. It is easy to see that L^n is an outer measure on R^n, known as the Lebesque n-dimensional outer measure. Further, $L^n(E)$ coincides with the volume of E if E is any block; this follows by approximating the sum in Equation (2.3) by a finite sum and then by subdividing E by the planes containing the faces of C_i. Since any block C_i may be decomposed into small subblocks, leaving the sum in Equation (2.3) unaltered, it is enough to take the infimum over δ-covers of E for any $\delta > 0$. Thus, L^n is a metric outer measure on R^n. It should be noted that the dimension in the Lebesque measure is an integer.

2.1.4 *Hausdorff measure (Falconer, 1985)*

If U is a non-empty subset of R^n, we define the diameter of U as $|U| = \sup\{|x-y| : x, y \in U\}$. If $E \subset \cup U_i$ and $0 < |U_i| \leqslant \delta$ for each i, we say that $\{U_i\}$ is a δ-cover of E. Let E be a subset of R^n and let D be a non-negative number. For $\delta > 0$ define

$$H^D_\delta(E) = \inf \sum_{i=1}^{\infty} |U_i|^D, \tag{2.4}$$

where the infimum is over all (countable) δ-covers $\{U_i\}$ of E. A trivial check establishes that H^D_δ is an outer measure on R^n (also called the Hausdorff outer measure).

To obtain the Hausdorff D-dimensional outer measure of E, we let $\delta \to 0$, thus

$$H^D(E) = \lim_{\delta \to 0} H^D_\delta(E) = \operatorname{Sup}_{\delta > 0} H^D_\delta(E); \tag{2.5}$$

the limit exists, but may be infinite. Since H^D_δ increases as δ decreases, H^D is easily seen to be an outer measure, but it is also a metric outer measure because if δ is less than the distance between the positively separated sets E and F, no set in a δ-cover of $E \cup F$ can intersect both E and F, so that

$$H^D_\delta(E \cup F) = H^D_\delta(E) + H^D_\delta(F),$$

leading to similar equality for H^D. The restriction of H^D to the σ-field of H^D-measurable sets is called the Hausdorff D-dimensional measure.

Note that an equivalent definition of the Hausdorff measure is obtained if the infimum in Equation (2.4) is taken over δ-covers of E by convex sets rather than by arbitrary sets, since any set lies in a convex set of the same diameter. Similarly, it is sometimes convenient to consider δ-cover of the open or closed sets. In each case, although a different value of H^D_δ may be obtained for $\delta > 0$, the value of the limit H^D is the same.

For any E and $\delta < 1$, it is clear that $H_\delta^D(E)$ is non-increasing, as D increases from 0 to ∞. Furthermore, if $D < t$, then

$$H_\delta^D(E) \geq \delta^{D-t} H_\delta^t(E),$$

which implies that if $H^t(E)$ is positive, then $H^D(E)$ is infinite. Thus, there is a critical value, dim E, called the Hausdorff dimension of E, such that

$$\begin{aligned} H^D(E) &= \infty \quad \text{if} \quad 0 \leq D < \dim E, \\ H^D(E) &= 0 \quad \text{if} \quad \dim E < D < \infty. \end{aligned} \tag{2.6}$$

If C is the cube of the unit side length in R^n then, by dividing C into K^n, subcubes of side $1/K$ in the obvious way, we see that if $\delta \geq K^{-1}n^{1/2}$ then $H_\delta^n(C) \leq K^n(K^{-1}n^{1/2})^n \leq n^{n/2}$, so that $H^n(C) < \infty$. Thus, if $D > n$, then $H^D(C) = 0$ and $H^D(R^n) = 0$, since R^n is expressible as the countable union of such cubes. It follows that $0 \leq \dim E \leq n$ for any $E \subset R^n$. It is also clear that if $E \subset E'$ then $\dim E < \dim E'$.

An H^D-measurable set, $E \subset R^n$, for which $0 < H^D(E) < \infty$ is termed a D-set.

Clearly, in contrast with the Lebesgue measure, the definitions of H^1 and L^1 on R^1 coincide. As might be expected, the outer measures L^n and H^n on R^n are related if $n > 1$, and, in fact, they differ only by a constant multiple. But it should be pointed out that the dimension in the Hausdorff measure is not required to be an integer.

2.1.5 *Hausdorff-Besicovitch dimension*

The definition of the Hausdorff measure may be generalized by replacing $|U_i|^D$ in Equation (2.4) by $h(|U_i|)$, where h is some positive function, increasing and continuous on the right (Mandelbrot, 1982).

Given a test (or 'gauge') function $h(\rho)$, where ρ is a prescribed radius to be defined continuously when the space is not Euclidean, a finite covering of the set E by balls of radii ρ_m can be said to have the measure $\Sigma h(\rho_m)$. One forms the infimum $\inf_{\rho_m < \rho} \Sigma h(\rho_m)$. As $\rho \to 0$, it has a limit

$$\lim_{\rho \to 0} \inf_{\rho_m < \rho} \Sigma h(\rho_m).$$

This limit defines the h-measure of R^n (the n-Euclidean dimension).

When $h_{(\rho)} = \tau_{(d)}\rho^d$, the h-measure is called d-dimensional, or the d-measure for short.

Given a set E which is two-dimensional, the h-measure may be evaluated $h_{(\rho)} = \pi\rho^2$. If one deals with a standard shape of unknown dimension, one will evaluate the measure for all test functions $h_{(\rho)} = \tau_{(d)}\rho^d$ with d an integer. If the length is infinite and the volume is zero, the shape can only be two-dimensional.

Besicovitch extended the core of this last conclusion to cases where d is not an integer and E is not of standard shape. He showed that for every set E there exists a real value D such that the d-measure H^d is infinite for $d < D$ and vanishes for $d > D$. This D is called the Hausdorff-Besicovitch dimension of E shown in Figure 2.2. In fact, the Hausdorff-Besicovitch dimension is same as the Hausdorff dimension defined by Equation (2.6), but the test function method discussed here leads us to easily determine the Hausdorff measure H^D. In the following chapters and sections, this method is widely used to measure fractal objects.

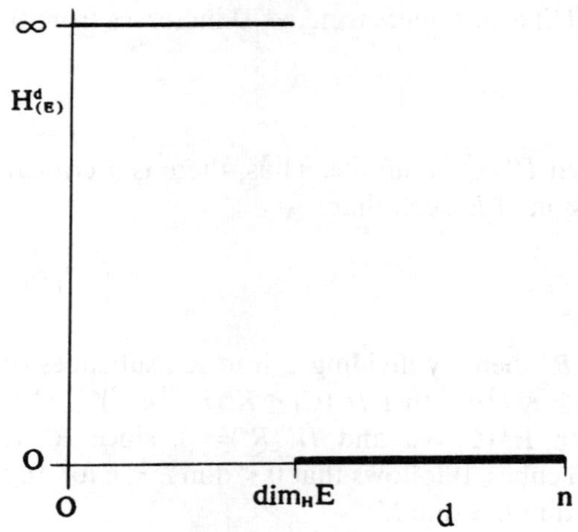

Figure 2.2. Graph of $H^d(E)$ against d for a set E. The Hausdorff-Bescovitch dimension is the value of d at which the 'jump' from ∞ to 0 occur (Falconer, 1990).

2.2 FRACTALS AND FRACTAL DIMENSION

Mandelbrot (1982) offered the following tentative definition of a fractal:

Definition 2.1: A fractal is by definition a set for which the Hausdorff-Besicovitch dimension D strictly exceeds the topological dimension D_T.

This definition requires a definition of the terms set, Hausdorff-Besicovitch dimension (D), and topological dimension (D_T) which is always an integer. In general, the topological dimension (D_T) of sets on R^n equals the Euclidean dimension (n), i.e., $D_T = n$. It should be noted that the above definition does not require any other conditions.

Mandelbrot (1986) proposed another definition of a fractal.

Definition 2.2: A fractal is a shape made of parts similar to the whole in some way.

It is clear that Definition 2.2 describes the essential feature – self-similarity in nature. It is only emphasized on the appearance of the matters. Whereas Definition 2.1 contains more content than Definition 2.2 because, in addition to Definition 2.2, it involves a kind of structure which satisfies the condition of $D > D_T$ without the feature of self-similarity.

Generally, one calls the Hausdorff-Besicovitch dimension fractal dimension. Hence, the measure of fractal sets should be the Hausdorff measure. For this reason, we can see that a fractal set (or fractal) must be required to satisfy two conditions: the first condition is that the measure of the set must be the Hausdorff measure, and the second one is $D > D_T$.

It should be pointed out that a neat and complete characterization of fractals is still lacking (Mandelbrot, 1987, shown in the book of Feder, 1988). At present, physicists really consider fractals as self-similarity. They think that Definition 2.1, although correct and precise, is too restrictive.

A rough definition (or explanation) of a fractal can be expressed as follows (Edgar, 1990):

Definition 2.3: A fractal set is a set that is more 'irregular' than the sets considered in classical geometry. No matter how much the set is magnified, smaller and smaller irregularities become visible.

I think that Definition 2.3 makes it easy to understand what a fractal is. Roughly speaking, fractal geometry is the geometry of irregular shapes. In fact, irregular geometric abstractions often fit the physical world better than regular arrangements or smooth curves and surfaces; as Mandelbrot (1982) wrote, 'Clouds are not spheres, mountains are not cones, coastlines are not circles, and bark is not smooth, nor does lightning travel in a straight line.' They may be fractals.

2.3 HOW TO MEASURE FRACTAL CURVES

The definition of a fractal set given in Section 2.2 may be abstract and difficult to clearly understand. However after you have measured the length of some geographical curves, you are able to capture inherent features of fractals and the basic concepts of fractal nature.

Geographical curves are so involved in their detail that their lengths are often infinite or, rather, undefinable. On the scale of a map, the deep fjords on seacoasts (or boundary) are clearly shown. The details encountered by moving one tip to another tip are more difficult to resolve. When sailing there, you find rocks, islands, bays, faults, and gorges such that each of their portions, in a statistical sense, can be considered a reduced-scale image of the whole. In that case, the way of describing the degree of complication of the geographical curves becomes a question that is difficult to answer. However, we could start to measure the length of geographical curves by following these simple experiments:

(1) Select any topographic curve (or map) and locate a reasonable length of a sinuous coastline;

(2) Take a pair of dividers and set them at a moderate interval, equivalent to, say, δ. We can term this distance, δ, the yardstick;

(3) Then, 'walk' the dividers along the selected feature so that each point falls on the line and count the number of steps this takes as 'N';

(4) Reduce the yardstick (δ) to one-half of the set in (2) and repeat (3). Continue to do this until the dividers are so close that measurements are impossible to perform, and at each step record 'δ', 'N' and, hence, the estimated length 'L' ($= N\delta$) (as shown in Figure 2.4a); and

(5) There are now several estimates of 'L'. Using logarithms (base 10) of both 'L' and 'δ' on both axis, the graph of these estimates of $\log_{10}$ 'L' on the ordinate with the yardstick $\log_{10}$ 'δ' on the abscissa can then be plotted.

It is quite probable that your measured result will look similar to Figure 2.3 (Feder, 1988), which shows how the measured length increases as the yardstick length, δ, is reduced. This log-log plot shows that the measured length of the geographical curves indicates no sign of reaching a fixed value as δ is reduced. In fact, the measured length is approximated nicely by the formula

$$L_{(\delta)} = L_0 \delta^{1-D}; \tag{2.7a}$$

where L_0 is a constant.

For an ordinary curve L_0 is expected to be true length of the curve, at least

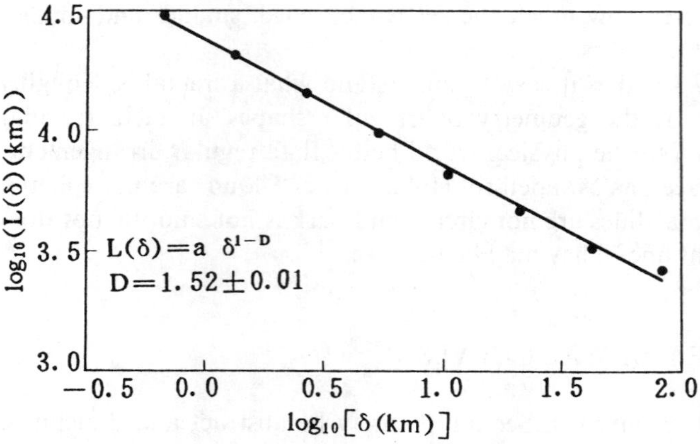

Figure 2.3. The measured length of the coastline of Norway as a function of the size δ of the δ × δ squares used to cover the coastline on the map. The straight line in this log-log plot corresponds to the relation $L_{(\delta)} = a \cdot \delta^{1-D}$, with $D \approx 1.52$ (Feder, 1988).

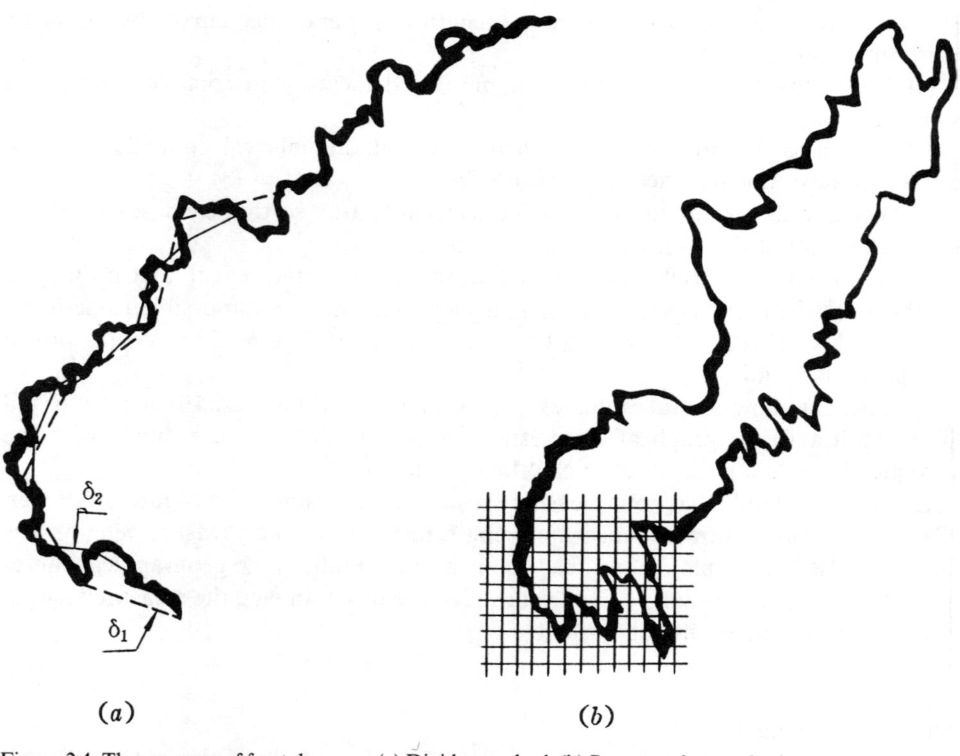

Figure 2.4. The measure of fractal curves. (a) Divider method; (b) Box covering method.

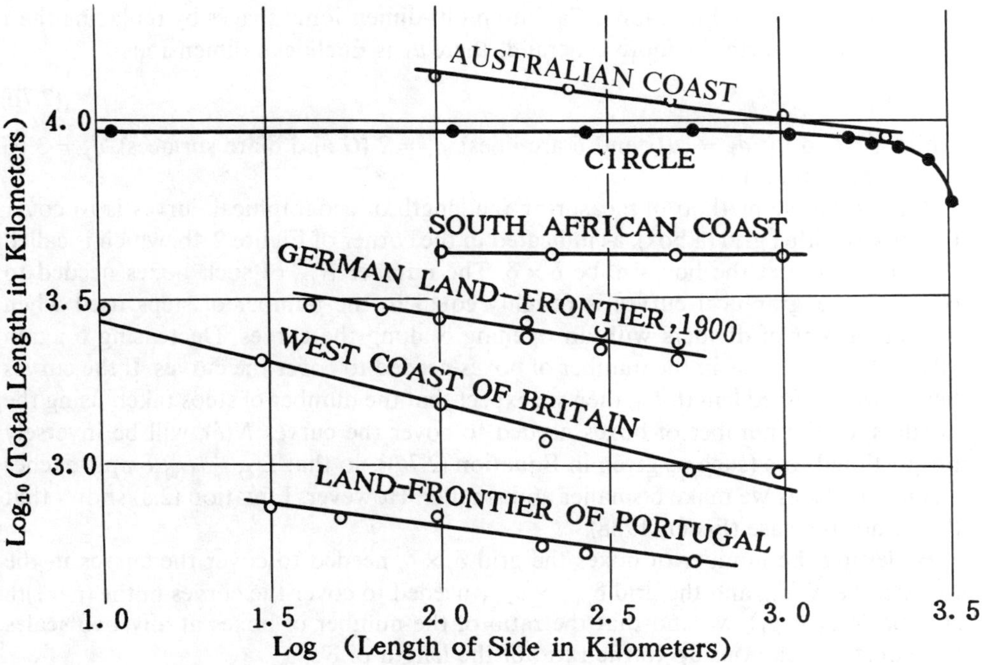

Figure 2.5. The length of coastlines as a function of yardstick length (Mandelbrot, 1967).

for a small enough δ, and exponent D should be equal to one. In fact, D for geographical curves is limited to the range $1 < D < 2$. Thus, Feder (1988) proposed that a fractal curve is defined as a curve whose length or perimeter is a function of the length of the measuring rod δ as shown in Equation (2.7a), where D is the fractal dimension. Feder (1988) measured the length of the coast of Norway, and found that $D = 1.52$. His results are reproduced in Figure 2.3. Mandelbrot (1967) analyzed data collected by Richardson, and Figure 2.5 is a reproduction of his figure showing the apparent length of various coastlines and boundaries. They all fall on straight lines in the log-log plot. The slope of the lines in this log-log plot is $1 - D$, where the fractal dimension D of the coastline of Britain is about 1.26.

From Equation (2.7a), we can also show the properties of fractal curves in mathematical structure as follows. In general, if $D > 1$ for fractal curves, then

$$\lim_{\delta \to 0} L_{(\delta)} = L_0 \lim_{\delta \to 0} \delta^{1-D} = \infty.$$

If δ does not approach zero, we have

$$(\partial/\partial\delta) L_{(\delta)} = L_0(1 - D)\delta^{-D} < 0. \qquad (2.8)$$

Equation (2.8) indicates in mathematical structure that $L_{(\delta)}$ increases as the yardstick length, δ is reduced.

Physically, the fractal dimension D measures the 'ruggedness' of geographical curves. Smaller D's represent smooth geographical curves, while large D's represent rugged geographical curves.

We can generalize Equation (2.7a) into multi-dimensional spaces by replacing the 1 in the exponent with the more general d_E (here d_E is Euclidean dimensions):

$$G_{(\delta)} = G_0 \delta^{d_E - D} \tag{2.7b}$$

which is valid for $d_E = 1$ (G and δ are lines), $d_E = 2$ (G and δ are surfaces), $d_E = 3$ (G and δ are volumes).

An alternative method of measuring the length of geographical curves is to cover the curves with a grid (a box), as indicated in the corner of Figure 2.4b, which is called box-covering. Let the box size be $\delta \times \delta$. The number $N_{(\delta)}$ of such boxes needed to cover the geographical curves is roughly equal to the number of steps used when walking a pair of dividers with an opening δ along the curves. Decreasing δ again gives a large increase in the number of boxes needed to cover the curves. If the curves have a well-defined length L_0, then we expect that the number of steps taken using the dividers, or the number of boxes needed to cover the curves $N(\delta)$, will be inversely proportional to δ (such as given in Equation (2.7a)), so that $L_{(\delta)} = N_{(\delta)} \cdot \delta$ approaches a constant L_0 as we make δ smaller and smaller. However, Equation (2.8) shows that this is not the case (Feder, 1988).

By letting the number of boxes the grid $\delta_i \times \delta_i$ needed to cover the curves in the i-th step be $N_i(\delta_i)$, and the grid $\delta_{i+1} \times \delta_{i+1}$ needed to cover the curves in the $(i+1)$th step be $N_{i+1}(\delta_{i+1})$, we find that the ratio of the number of boxes at any two scales is a constant relationship to the ratio of the length of δ's;

$$N_{i+1}/N_i = [\delta_i/\delta_{i+1}]^D. \tag{2.9}$$

Rearranging Equation (2.9) yields:

$$N_{i+1} = (N_i \delta_i^D)[\delta_{i+1}]^{-D} = a[\delta_{i+1}]^{-D}, \tag{2.10}$$

where $N_i \delta_i^D$ acts as base constant a against which the number of boxes N_{i+1} from any grid size δ_{i+1} may be predicted. Thus, Equation (2.10) can be easily generalized as

$$N = a\delta^{-D}; \tag{2.11}$$

with two scales of resolution. By rearranging Equation (2.9), D can be derived as

$$D = \log_{10}(N_{i+1}/N_i)/\log_{10}(\delta_i/\delta_{i+1}). \tag{2.12}$$

However, confirmation that a common fractal dimension D holds across a wider range of scales requires that the range of δ and corresponding N values be considered. Using the logarithms of Equation (2.11) yields

$$\log_{10} N = \log_{10} a - D \log_{10} \delta. \tag{2.13}$$

The range of δ and corresponding N values can be substituted into Equation (2.13), and the value of D is evaluated by using a simple linear regression procedure or the slope S of the lines in $\log N$-$\log \delta$ plot ($D = S$).

Similarly, from Equation (2.7a), we find that

$$\log_{10} L = \log_{10} L_0 + (1 - D) \log_{10} \delta. \tag{2.14}$$

The value of D is also evaluated from the slope S of the lines in $\log L$-$\log \delta$ plot (here $D = 1 - S$).

2.4 CLASSICAL FRACTALS

2.4.1 *The triadic Koch curve*

The triadic Koch curve (Figure 2.6) is one of the standard examples used to demonstrate that a curve has a fractal dimension $D > 1$.

Feder (1988) gave an excellent description of the construction of the Koch curve as follows. Start from a line segment of unit length. This starting form ($n = 0$) is called the initiator. The initiator is the 0-th approximation of the Koch curve. The construction of the Koch curve proceeds by a generator, removing the middle third of each segment, and replacing it by the other two sides of the equilateral triangle based on the removed segment, shown as the curve marked $n = 1$ in Figure 2.6a. Thus, we obtain the first approximation which is a curve of $N = 4$ line segments each of length $\frac{1}{3}$. The length of the $n = 1$ approximation becomes $L_{(1/3)} = \frac{4}{3}$. The next approximation is obtained by applying the same procedure to each of the segments in the $n = 1$ approximation, and so on. Thus, in the $n = 2$ approximation, we have a curve consisting of $N = 4^2$ segments each having length $\delta = 3^{-2} = (\frac{1}{3})^2$, and the length of this curve is $L_{(1/9)} = (\frac{4}{3})^2$. The n-th approximation comes by replacing

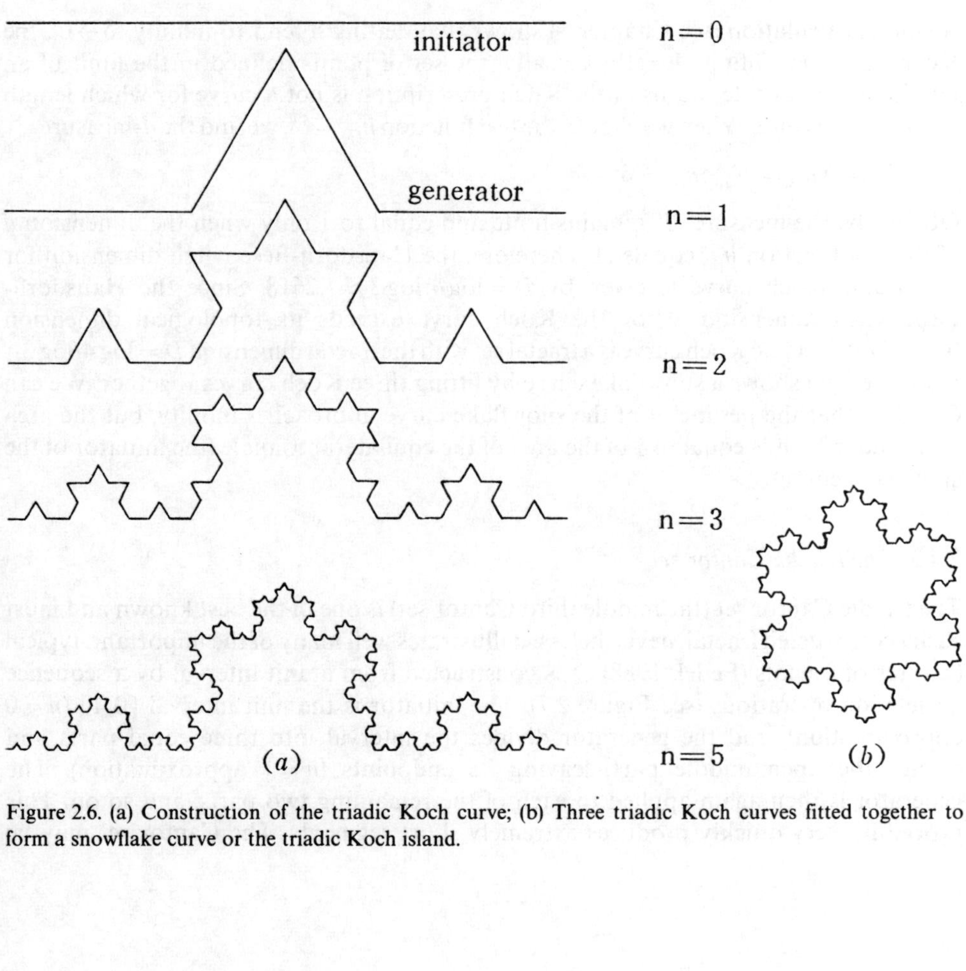

Figure 2.6. (a) Construction of the triadic Koch curve; (b) Three triadic Koch curves fitted together to form a snowflake curve or the triadic Koch island.

the middle third of each straight line segment of the $(n-1)$th approximation by the other two sides of the equilateral triangle, based on the removed segment. Such a curve is called a prefractal. The triadic Koch curve is this limit curve as $n \to \infty$. When n is so large that the $(n-1)$th approximation and n-th approximation differ only in fine detail, the length of the n-th approximation, is given by

$$L_{(\delta)} = (\tfrac{4}{3})^n.$$

The length of each of the small line segments is

$$\delta = (\tfrac{1}{3})^n.$$

From the above equation, the generation number n may be written in the form

$$n = -\log \delta / \log 3.$$

It is clear that the length of the n-th generation prefractal can be obtained as follows

$$L_{(\delta)} = (\tfrac{4}{3})^n = \exp\{-[\log \delta(\log 4 - \log 3)]/\log 3\} = \delta^{1-D},$$

where $D = \log 4/\log 3 = 1.2628$. This result is in agreement with that of Equation (2.7a). The number of segments is $N_{(\delta)} = 4^n = 4^{-\log \delta/\log 3}$ in n-th generation, and may be written as

$$N_{(\delta)} = \delta^{-D}. \tag{2.11}$$

A simple calculation (see Chapter 1) shows that, letting n tend to infinity ($\delta \to 0$), the Koch curve has infinite length. Visually, the set of points defined in the limit of an infinite number of iterations of the Koch prescription is not a curve for which length is a useful measure. When we choose the test function $h_{(\delta)} = \delta^d$, we find the d-measure

$$H^d = \Sigma h_{(\delta)} = N_{(\delta)} \cdot h_{(\delta)} = \delta^{-D}\delta^d.$$

Obviously, the measure H^d remains finite and equal to 1 only when the dimension d of the test function $h_{(\delta)}$ equals D. Therefore, the Hausdorff-Bescovitch dimension for the triadic Koch curve is given by $D = \log 4/\log 3 \approx 1.2618$. Since the Hausdorff-Besicovitch dimension D for the Koch curve exceeds its topological dimension ($D_T = 1$), the triadic Koch curve is a fractal set with the fractal dimension $D = \log 4/\log 3$.

Figure 2.6b shows a snowflake curve by fitting three Koch curves together. We can calculate that the perimeter of the snowflake curve approaches infinity, but the area surrounded by it is equal to $\frac{8}{5}$ of the area of the equilateral triangle (the initiator of the snowflake curve).

2.4.2 *The triadic Cantor set*

The triadic Cantor set (the middle third Cantor set) is one of the best known and most easily constructed fractal; nevertheless, it illustrates well many of the important, typical features of fractals (Fedel, 1988). It is constructed from a unit interval by a sequence of deletion operations (see Figure 2.7). The initiator is the unit interval $[0, 1]$ ($n = 0$ approximation), and the generator divides the interval into three equal parts and deletes the open middle part, leaving its endpoints ($n = 1$ approximation). The generator is then again applied to each of the remaining two parts, and so on. This procedure very quickly produces extremely short segments. The Cantor set may be

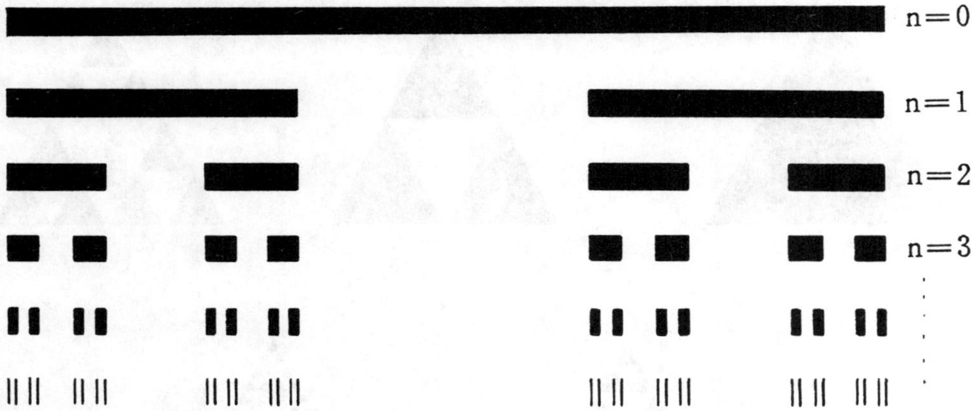

Figure 2.7. Construction of the triadic Cantor set. The initiator is the unit interval [0, 1]. The generator removes the open middle third. $D = \log 2/\log 3 = 0.6309$.

thought of as the limit of the sequence of sets of the n-th approximation as n tends to infinity, which is an infinite number of points scattered over the interval. This set is called 'Cantor dust' (Mandelbrot, 1982).

Now let us consider its Hausdorff-Bescovitch dimension. In the n-th approximation, 2^n intervals each of length 3^{-n} are remained. If we use line segments of length $\delta = (\frac{1}{3})^n$ to cover all of the segments generated in the n-th approximation, then $N = 2^n, \delta = (\frac{1}{3})^n$. The Hausdorff's measure of the Cantor set is given by

$$H^d = \sum_{n=1}^{\infty} \delta^d = 2^n(\tfrac{1}{3})^{nd} = \delta^{d-D}. \tag{2.15}$$

This measure diverges or tends to zero as δ is decreased, unless we choose $d = D = \log 2/\log 3 = 0.6309$. The topological dimension of the Cantor set (point set) is $D_T = 0$. Thus, the triadic Cantor set is a fractal set with a fractal dimension given by

$$D = \log 2/\log 3.$$

2.4.3 *The Sierpinski gasket and carpet*

As a plane analogue of the Cantor set, the Sierpinski gasket is obtained by repeatedly removing (or inverting) equilateral triangles from an initial equilateral triangle (initiator); see Figure 2.8. The generator eliminates a central triangle as shown. In each application of the generator, a filled triangle is replaced by $N = 3$ triangles that have been scaled down by the factor $r = \frac{1}{2}$. We can obtain that $N_{i+1}/N_i = 3$, $\delta_{i+1}/\delta_i = \frac{1}{2}$, and from Equation (2.12) its fractal dimension is $D = \log 3/\log 2 = 1.58$.

A related curve, the Sierpinski carpet, may be constructed using the same recursive procedures. At each stage, each remaining square is divided into nine smaller squares, of which eight are kept and the middle one discarded. The initiator is a square, and the generator is made of $N = 8$ squares. They are, respectively, the first and second figures as shown in Figure 2.9. Its fractal dimension is $D = \log 8/\log 3 = 1.89\ldots$

Figure 2.8. Construction of the triangular Sierpinski gasket. The fractal curve obtained in the limit of an infinite number of generations having the fractal dimension $D = \log 3/\log 2$.

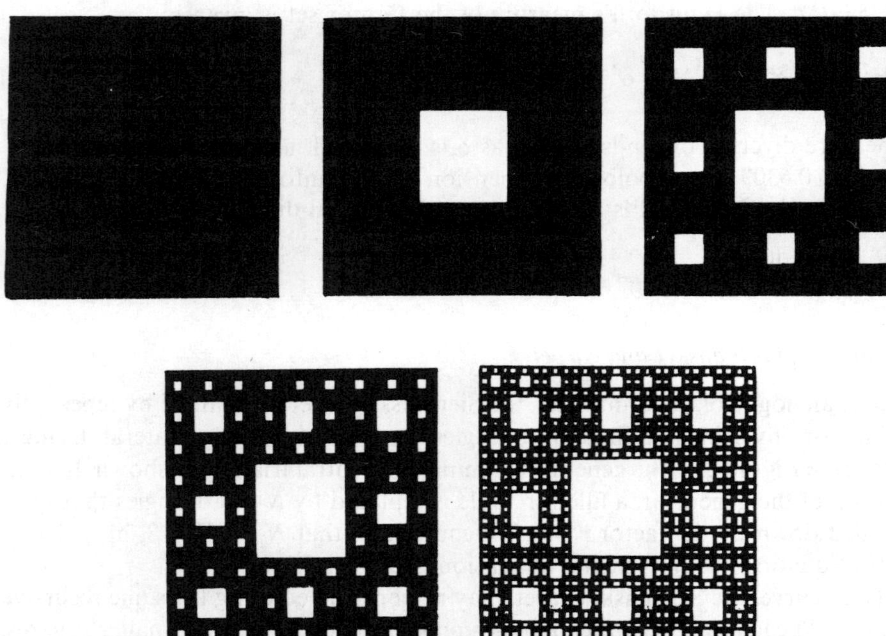

Figure 2.9. Construction of the Sierpinski carpet. The fractal curve obtained within the limit of an infinite number of generations has the fractal dimension $D = \log 8/\log 3 = 1.89\ldots$

(a)

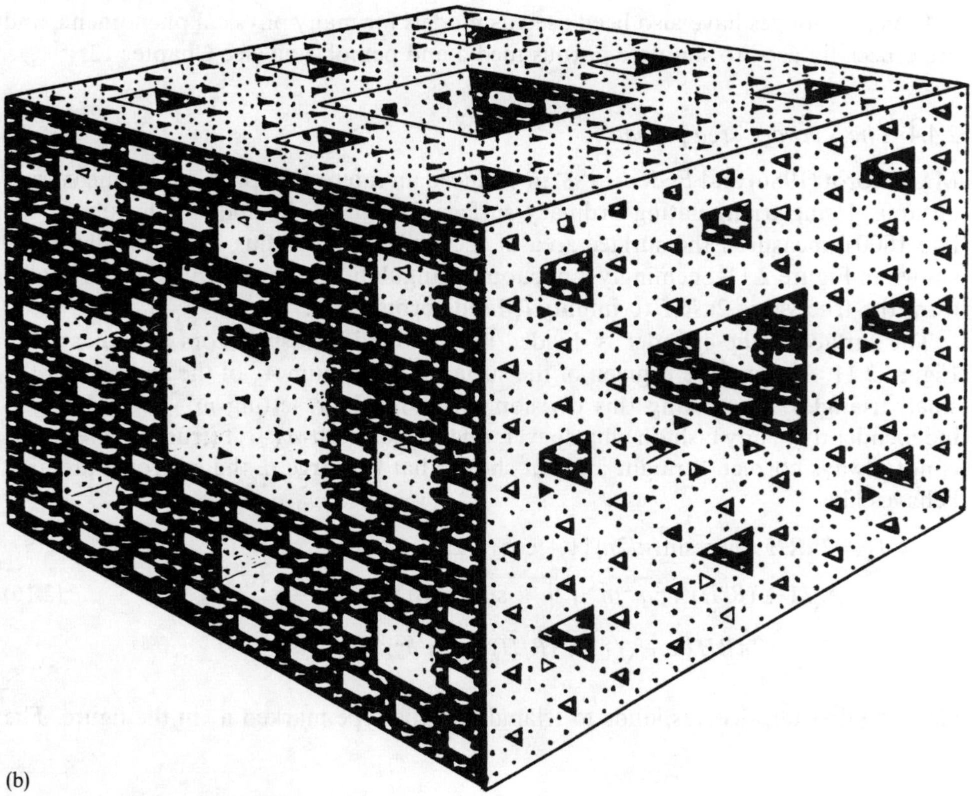

(b)

Figure 2.10. (a) The construction of Menger sponge. (b) Menger sponge $N_{i+1}/N_i = 20$; $\delta_{i+1}/\delta_i = \frac{1}{3}$; $D = \log 20/\log 3$.

The Sierpinski curves have been used as models for many physical phenomena. Gefen et al. (1980) reported the first systematic study of the critical phenomenon that occurs near phase transitions in spin systems carried by self-similar fractal lattices. In an interesting experiment, Gordon et al. (1986) measured the superconducting-to-normal phase transition temperature $T_c(H)$ as a function of the applied magnetic field H of a sample of an aluminum film with the structure of a tenth generation Sierpinski gasket prefractal. The phase boundary $T_c(H)$ is in excellent quantitative agreement with theoretical predictions (Feder, 1988).

A spatial formation of the Sierpinski carpet is the Menger sponge (see Figure 2.10b). The principle of the construction is only a spatial generalization of the plane Sierpinski carpet as shown in Figure 2.10a. The initiator is a cube, and the generator is made of $N = 20$ cubes. The construction starts from a unit cube from which seven cubes of side length $\frac{1}{3}$ are removed so that there remain, as first approximant, eight corner cubes joined by 12 bridging cubes (all of side length $\frac{1}{3}$). Analogous removal of seven cubes of length $\frac{1}{9}$, from each of the first approximant's 20 cubes, generates the second approximant, etc. This process of cutting out an increasing number of smaller and smaller cubes produces a surface supported by sponge-like structure of vanishing volume. The fractal dimension is $D = \log 20/\log 3 = 2.7268$.

Menger sponges have also been used as models for many physical phenomena, and are especially used to simulate porous media and percolation (see Chapter 12).

2.4.4 *The Schwarz area paradox*

Mandelbrot (1986) and Feder (1988) introduced an interesting problem for measuring an area. Using triangulating ordinary surface, Hermann Amandus Schwarz (in the late 1800s) measured the surface area of a cylinder with radius R and height H, as shown in Figure 2.11. Seemingly innocuous triangulation methods can yield any value from the true value $2\pi RH$ to infinity (Mandelbrot, 1986).

The surface of the cylinder is divided into m bands and n sectors as indicated in Figure 2.11. An estimate is given of the surface area as sum A_Δ of the areas of all the small triangles. By making this division finer and finer, letting $m \to \infty$ and $n \to \infty$ independently in any fashion whatsoever, one expects that $A_\Delta \to A$ (true area). But this is not always correct. Straight algebra shows that for large n, the approximate area behaves like

$$\begin{aligned} A_\Delta &= 2\pi RH \cdot 2n/\pi \cdot \sin(\pi/2n)\{1 + \cos(\pi/2n) \\ &\quad \cdot \sqrt{1 + (R/H)^2\pi^2 m^2/n^4 \cdot [2n/\pi \cdot \sin(\pi/2n)]^4}\} \\ &\xrightarrow[n\to\infty]{} 2\pi RH\{1 + \sqrt{[1 + (R/H)^2(\pi^2 m/n^2)^2]}\}. \end{aligned} \tag{2.16}$$

Here the first term corresponds to triangles of the type marked a_1 in the figure. The

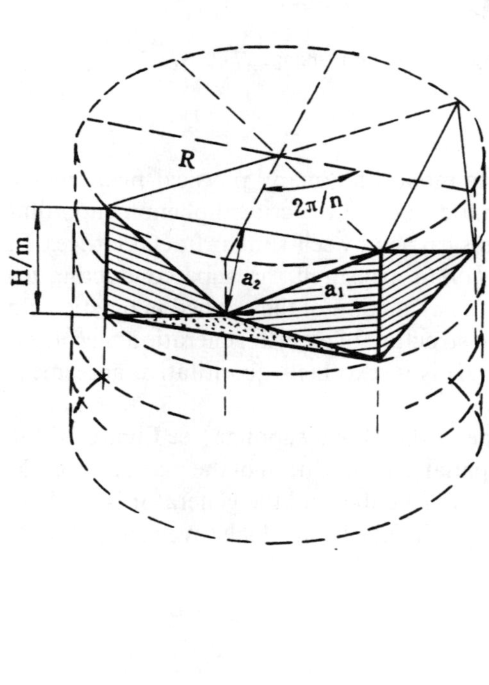

Figure 2.11. The vertical surface of a cylinder of radius R and height H is $2\pi RH$. The surface is approximated by a triangulation as illustrated (Feder, 1988).

second term with the square root corresponds to triangles of the type marked a_2. If $n \to \infty$, but $(m/n^2) \to 0$, the approximation converges to $2\pi RH$. However, if $n \to \infty$, and $m = \lambda n^2$ with λ a constant larger than zero, one can get any finite limit larger than $2\pi RH$ one wishes. In fact, by selecting $m = n^{\beta}$, with $\beta > 2$, then $A_{\Delta} \approx n^{\beta - 2}$. The approximation area then diverges as the individual triangles become smaller and smaller. It closely resembles a fractal. This is the Schwarz area paradox. It is easy to see that the approximate triangulation surface is more and more corrugated as the ratio m/n^2 increases and, in the limit the triangles of the type marked a_2 are practically perpendicular to the cylinder surface.

It should be noted that Schwarz area paradox is not a classical fractal, but mimics a fractal, it is a classical 'monster' surface. We can see that the Schwarz area paradox tells one how to select a good triangulation for measuring smooth surfaces, and how to estimate the area of a more complex or rough surface by means of the fractal method. For a detailed discussion about it see the paper of Mandelbrot (1986).

2.5 THE WEIERSTRASS FUNCTION

It is well known that fractal geometry is a new branch born belatedly by the crisis of mathematics caused by a continuous nondifferentiable function constructed by Karl Weierstrass (1857-1897). The crisis started in 1875 and lasted approximately till 1925.

The complex Weierstrass function is the sum of the series (Mandelbrot, 1982)

$$W_0(t) = (1 - W^2)^{-1/2} \sum_{n=0}^{\infty} W^n \exp(2\pi i b^n t). \tag{2.17}$$

Where b is a real number larger than 1, and W is written either as $W = b^H$, with $0 < H < 1$, or as $W = b^{D-2}$, with $1 < D < 2$. The real and imaginary parts of $W_0(t)$ are called Weierstrass cosine and sine functions, respectively. The function $W_0(t)$ is continuous, but nowhere differentiable, and is a fractal with a fractal dimension D.

Feder (1988) gives another form of the Weierstrass function from Mandelbrot's book (1982)

$$W(t) = \sum_{n=-\infty}^{\infty} (1 - e^{ib^n t}) e^{i\phi_n} / b^{(2-D)n}, \tag{2.18}$$

where ϕ_n is an arbitrary phase. Equation (2.18) is called the Weierstrass-Mandelbrot function. Berry & Lewis in 1980 discussed the Weierstrass-Mandelbrot cosine function

$$C(t) = \sum_{n=-\infty}^{\infty} (1 - \cos b^n t) / b^{(2-D)n}.$$

Feder (1988) has evaluated the function $C(t)$ for a few values of the parameters in the 'time' interval $0 \leq t \leq 1$. We reproduce the result of Feder (1988) as shown in Figure 2.12. The function $C(t)$ is lightly smooth for low values of D, but it fluctuates wildly as D increases toward 2, and can be scaled with a homogeneity equation given by

$$C(bt) = b^{2-D} C(t).$$

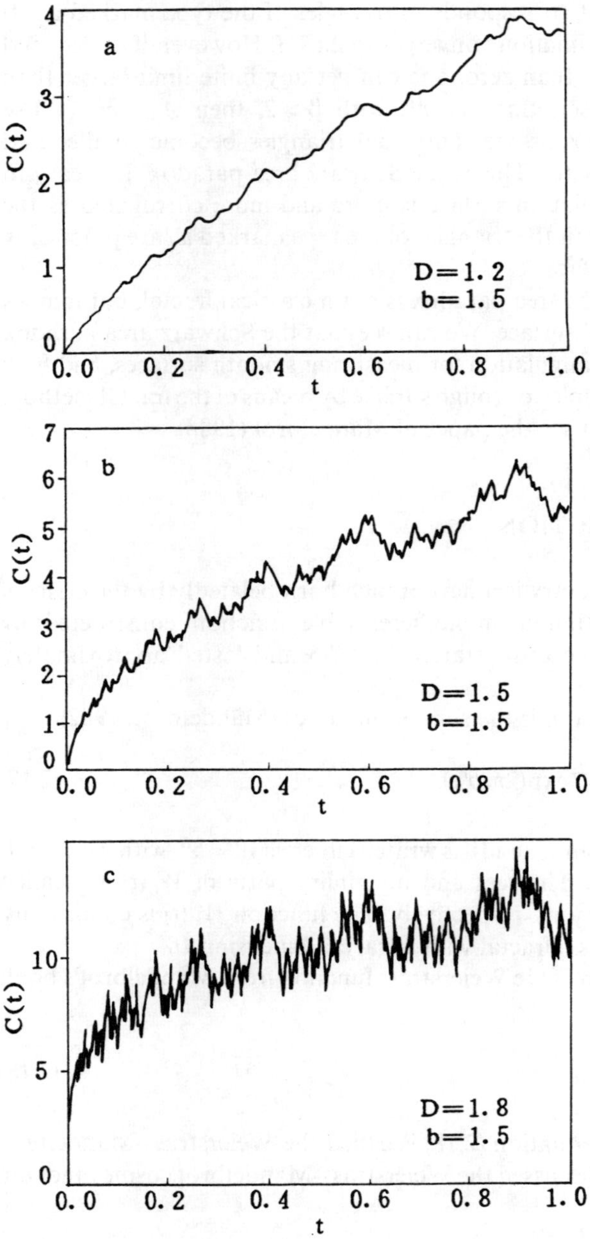

Figure 2.12. The Weierstrass-Mandelbrot fractal function $C(t)$, with $b = 1.5$; (a) $D = 1.2$, (b) $D = 1.5$, (c) $D = 1.8$ (Feder, 1988).

Another sine function discussed by Falconer (1990):

$$f(t) = \sum_{n=0}^{\infty} \left(\tfrac{3}{2}\right)^{-n/2} \sin\left[\left(\tfrac{3}{2}\right)^n t\right],$$

shows that infinite summation leads to the graph having a fine structure, rather than being a smooth curve to which classical calculus is applicable. The graph of $f(t)$ is reproduced in Figure 2.13.

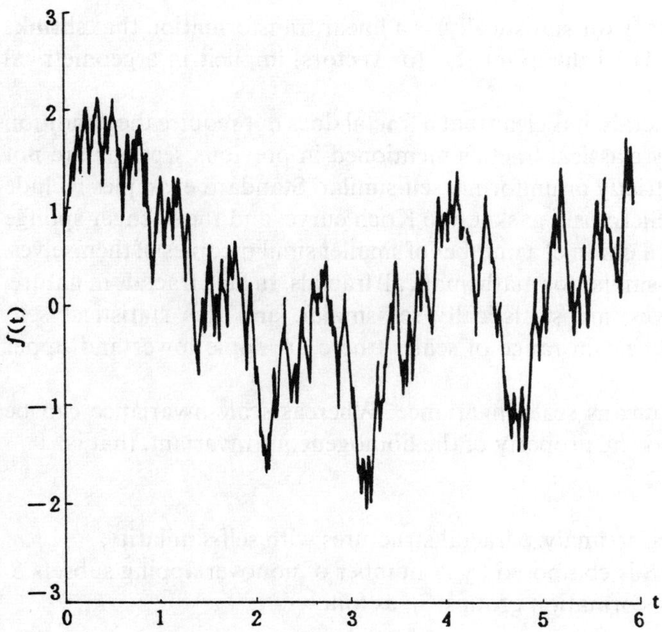

Figure 2.13. Graph of $f(t) = \sum_{n=0}^{\infty} (\frac{3}{2})^{-n/2} \sin[(\frac{3}{2})^n t]$ (Falconer, 1990).

2.6 SELF-SIMILARITY AND SELF-AFFINITY

From the idea of group theory, fractal geometry deals mainly with invariant properties and the scaling symmetry of continuous transformation groups. Suppose S_i $(i = 1, 2, \ldots, n)$ are nonoverlapping subsets of a set S, and S_i is obtained by transformation groups G_i from S:

$$S_i = G_i S,$$

where the transformation groups G_i could be associated with translation, rotation, stretching, shearing and similarity. If G_i is only a similar transformation group, that is,

$$G_i = K_i \begin{bmatrix} 1 & 0 & \cdots & 0 \\ 0 & 1 & & \\ \vdots & & \ddots & \\ 0 & & & 1 \end{bmatrix}_{n \times n},$$

where K_i $(i = 1, 2, \ldots, n) > 0$ are constants, when $K_i \neq K_j$ $(i \neq j)$, S is called nonuniform self-similar (Xie, 1989c); and when $K_i = K_j = r$ $(i, j = 1, 2, \ldots, n)$, S is called uniform self-similar or strictly self-similar, that is, G_i transforms the point set $X = (x_1, \ldots, x_i, \ldots, x_n)$ into the point set $X' = r(X) = (rx_1, rx_2, \ldots, rx_i, \ldots, rx_n)$, i.e.,

$$X' = G_i X = r(X),$$

and, hence, transforms S into $r(S)$, where r is also called the ratio of similarity. Thus, we say that the set S is a strictly self-similar fractal set.

The set S is statistically self-similar, if the equation $S_i = G_i S$ with similar transformation group G_i can be found in a statistical sense.

It is clear that self-similarity (or statistically) is a linear transformation that shrinks ($0 < r < 1$) or expands ($r > 1$) all the point sets (or vectors) implicit in a geometrical figure in the same ratio.

From the definition of fractals, it is clear that a fractal does not require the condition of self-similarity. But many classical fractals mentioned in previous sections are not only self-similar, but also strictly or uniformly self-similar. Standard examples include the triadic Cantor set, the Sierpinski gasket, the Koch curve, and the Menger sponge (see Figures 2.6-2.10), being a union of a number of smaller similar copies of themselves. So we call them strictly self-similar (or mathematical) fractals. In fact, fractals in nature, such as geographical curves, are statistically self-similar, and this statistical self-similarity only holds for a certain range of scales (there are some lower and upper cutoffs).

So-called self-similarity means scale-invariance. Whereas scale-invariance can be expressed in mathematics as the property of the homogeneous invariant, that is:

$$g(\lambda x) = \lambda^{\alpha} g(x).$$

This property is widely used to analyze fractal structures with self-similarity.

The set S is self-affine, if S is composed by N number of nonoverlapping subsets S_i of S obtained by affine transformation groups G_i, as follows:

$$G_i = \begin{bmatrix} a_{11} & a_{12} & \cdots & a_{1n} \\ \cdots & \cdots & \cdots & \cdots \\ a_{k1} & a_{k2} & \cdots & a_{kn} \\ \cdots & \cdots & \cdots & \cdots \\ a_{n1} & a_{n2} & \cdots & a_{nn} \end{bmatrix} + \boldsymbol{b}$$

where a_{kj} ($k, j = 1, 2, \ldots, n$) are constants, and $\boldsymbol{b}$ is a constant vector in R^n. Thus, an affine transformation group is a combination of a translation, rotation, dilation, and reflection.

The set S is diagonal self-affine, if S is composed by N number of nonoverlapping subsets S_i of S obtained by diagonal affine transformation groups G_i, as follows:

$$G_i = \begin{bmatrix} a_{11} & & & \\ & a_{22} & & 0 \\ & & \ddots & \\ & 0 & & a_{nn} \end{bmatrix} + \boldsymbol{b}$$

where a_{kk} ($k = 1, 2, \ldots, n$) are constants, and $a_{ii} \neq a_{jj}$ ($i, j = 1, 2, \ldots, n$).

Let a ratio vector $\boldsymbol{r} = (r_1, r_2, \ldots, r_n)$. Then a diagonal affine transformation group transforms a point set $\boldsymbol{X} = (x_1, x_2, \ldots, x_n)$ into a new point set $\boldsymbol{X}' = (r_1 x_1, r_2 x_2, \ldots, r_n x_n) = \boldsymbol{r}{:}\boldsymbol{X}$.

The set S is statistically diagonal self-affine, if S is composed by N number of nonoverlapping subsets S_i of S obtained statistically by diagonal affine transformation groups. In general, self-affine fractals in which we will be interested are all diagonal self-affine (Xie & Zhang, 1989).

Self-affine fractal sets form an important class of sets, which include self-similar sets as a particular case ($r_i = r_j = r$, $i = 1, 2, \ldots, n$). Obviously, affinity is a linear transfor-

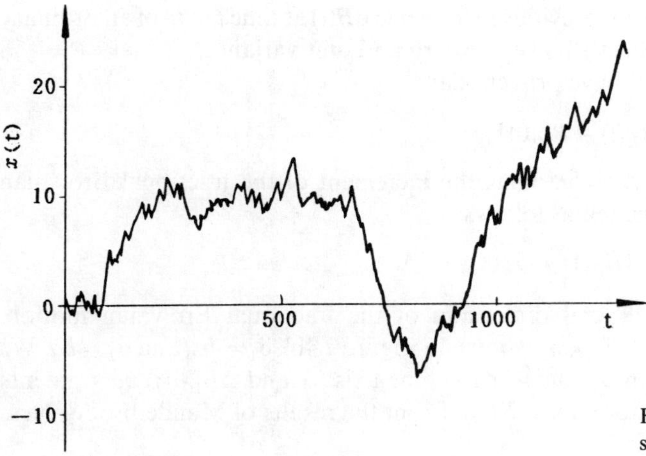

Figure 2.14. Graph of a Brownian sample function.

mation that shrinks or expands different vectors in a different ratio according to their directions.

One very special example of a self-affine fractal is the record of Wiener's scalar Brownian motion (1923), which is a random process with independent and stationary Gaussian increments (as shown in Figure 2.14).

Brownian motion (or the Wiener process) is given as a random process $X(t)$ having the following properties:

(1) with probability 1, $X(0) = 0$ (i.e., the process starts at the origin), and $X(t)$ is a continuous function of t;

(2) for any $t \geq 0$ and $h \geq 0$, the increment $X(t+h) - X(t)$ is normally distributed with zero mean and variance h; thus,

$$P_r((X(t+h) - X(t)) \leq x) = (2\pi h)^{-1/2} \int_{-\infty}^{x} \exp[-u^2/(2h)]\,du; \tag{2.19}$$

(3) If $0 \leq t_1 \leq t_2 \leq \ldots \leq t_{2m}$, the increments $X(t_2) - X(t_1), \ldots, X(t_{2m}) - X(t_{2m-1})$ are independent of each other.

Suppose that the position function of a Brownian particle is defined by

$$B(t) = X(t) - X(0) \sim \xi |t - t_0|^H; \quad (t \geq t_0),$$

or

$$B(t) \sim \xi t^H,$$

then, for any time t and t_0, we have $H = \frac{1}{2}$. The function $B(t)$ is continuous, but nondifferentiable ($X = 0$) (it is a fractal), where ξ is a random length determined by Gaussian distribution ($H = \frac{1}{2}$).

Mandelbrot & Ness (1968) generalized the value of $H = \frac{1}{2}$ to $H \in [0, 1]$, and defined the random function $B(t)$ (now denoted by $B_H(t)$, and called the fractional Brownian function), with zero mean roughly as follows:

$$B_H(t) = \{\int_{-\infty}^{t} (t_0 - t)^{H-1/2} dB(t)\} / \Gamma(H + \tfrac{1}{2}),$$

where $\Gamma(x)$ is the gamma function. This definition means that the value of the random function at time t depends on all previous increments $dB(t)$ at time $t < t_0$ of an ordinary Gaussian random process $B(t)$ with average zero and unit variance.

Mandelbrot & Ness (1968) have proven that

$$B_H(bt) - B_H(0) = b^H\{B_H(t) - B_H(0)\}.$$

In particular, let $t = 1$ and $\Delta t = bt$. Thus the increment of the fractional Brownian particle position can be rewritten as follows:

$$B_H(\Delta t) - B_H(0) = |\Delta t|^H\{B_H(1) - B_H(0)\} \sim |\Delta t|^H.$$

Now let us determine the fractal dimension of the fractional Brownian motion. Choosing the grid or box with $\delta_1 \times \delta_2$ (shown in Figure 2.4b), $\delta_1 = b\tau$, and $\delta_2 = ba$. We need $T/b\tau$ segments of length δ_1 to cover the time axis (t), and $\Delta B_H(b\tau)/ba$ segments of length δ_2 to cover the position axis ($X(t)$). From the results of Mandelbrot & Ness (1968):

$$\Delta B_H(b\tau) = b^H \Delta B_H(\tau),$$

we can obtain the number of grids or boxes needed to cover the record as follows:

$$N(b; a, \tau) = b^H \Delta B_H(\tau)/(ba) \times T/(b\tau) \sim b^{H-2}.$$

On the other hand, there is a relation between N and b (see Equation (2.11)) from the properties of a fractal curve:

$$N(b; a, \tau) \sim b^{-D}.$$

Comparing the above two equations, we have

$$D = 2 - H, \quad \text{for self-affine records.}$$

It is clear that the fractal dimension of the fractional Brownian motion (self-affine records) depends on the value of H. A detailed discussion about it can be seen in the book of Feder (1988) and Mandelbrot's paper (1986).

Voss (1985b) has generated many fractional Brownian curves, using his algorithm of successive random addition (as shown in Figure 2.15). His results indicate that: for $0.5 < H < 1$, the record exhibits rather clear trends with relatively little noise. In other words, the record is relatively smooth; but for $0 < H < 0.5$, the record appears very 'noisy' as shown in Figure 2.15.

Voss has also generalized an algorithm of the successive random addition to two and higher dimensions. He has specifically generated some fractal landscapes which are self-affine rather than self-similar because the required vertical and horizontal scalings are not necessarily equal. Visually, self-affinity implies that the landscapes possess no cues as to their scale. It is well known that the variogram of the fractional Brownian motion is a simple power function, i.e.,

$$V(t - t_0) = \langle [B_H(t) - B_H(0)]^2 \rangle \sim |t - t_0|^{2H},$$

the power being determined by a parameter H, $0 < H < 1$. By changing H, one can create a range of surfaces with different degrees of ruggedness from white noise ($H = 0$) to a plane ($H = 1$). Furthermore, contours of surfaces should be fractals, with the fractal dimension $D = 2 - H$, and the surface itself having the dimension $D = 3 - H$ (Mandel-

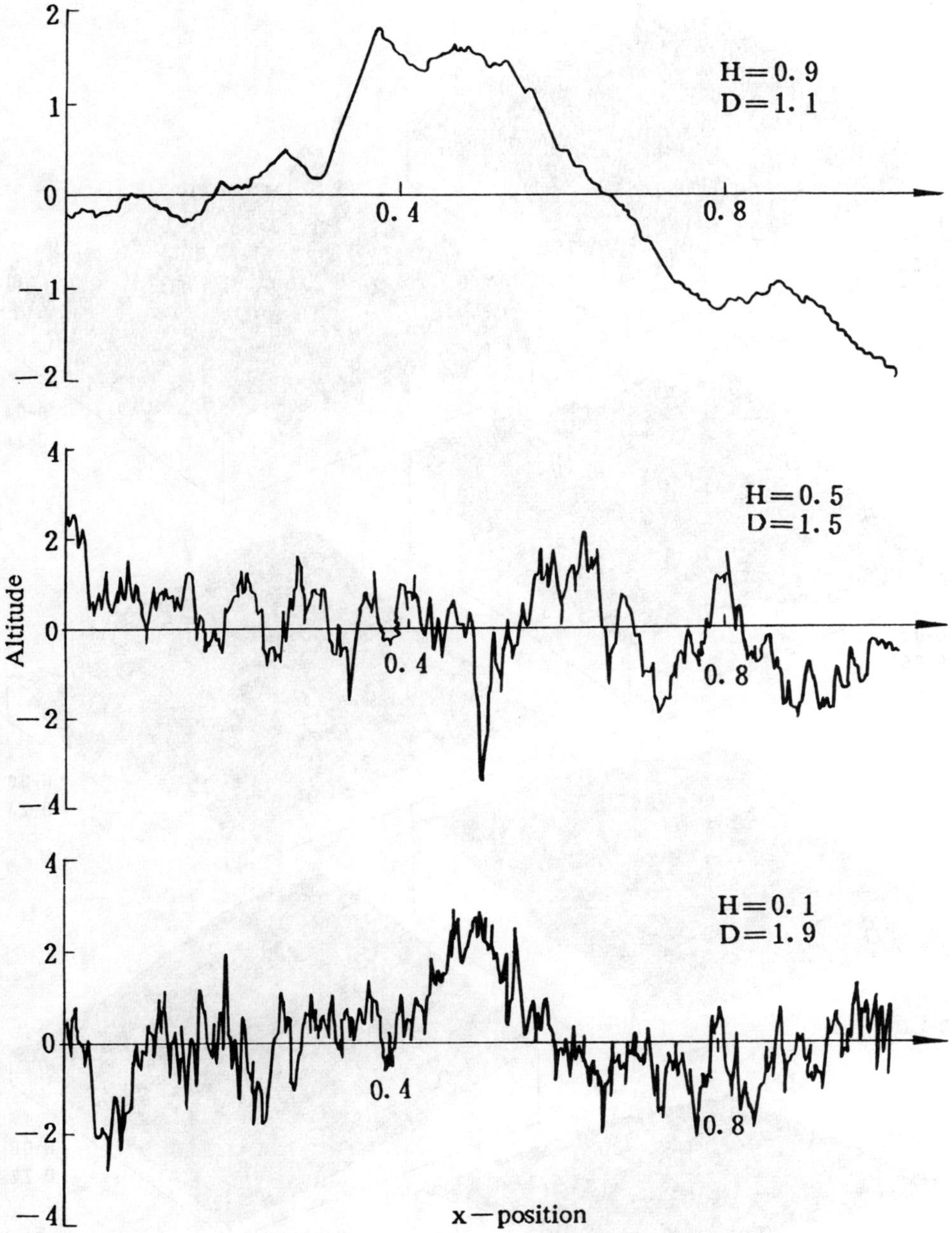

Figure 2.15. Fractional Brownian curves generated by Voss ($D = 2 - H$).

brot, 1982, p353). Goodchild (1988) and Goodchild & Mark (1987) have discussed fractal surfaces by fractional Brownian motion, and given self-affine surfaces generated by the fractional Brownian process with variable parameter H as shown in Figure 2.16, for modeling geographic phenomena.

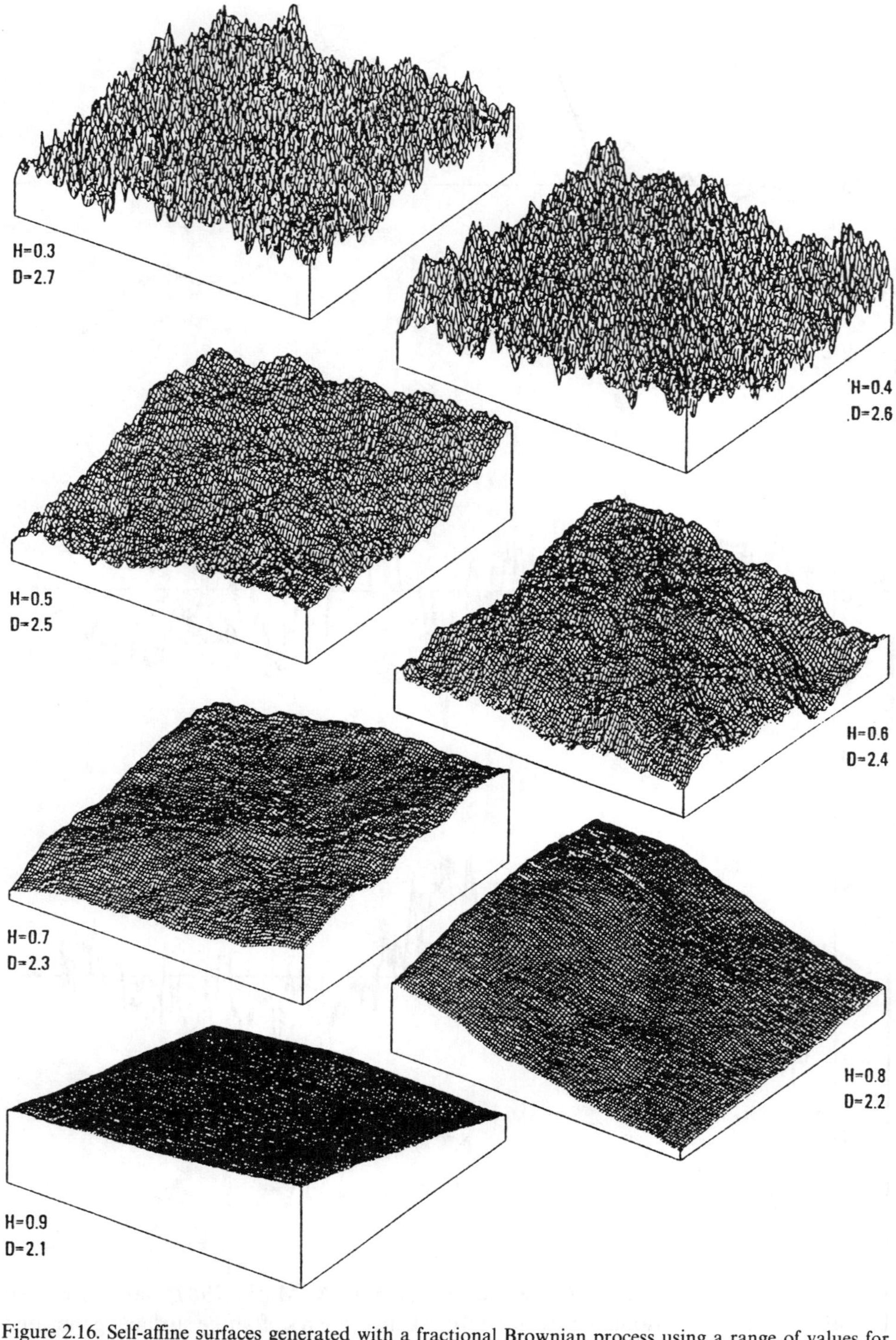

Figure 2.16. Self-affine surfaces generated with a fractional Brownian process using a range of values for the parameter *H* (Goodchild, 1988).

2.7 MEASUREMENTS OF FRACTAL DIMENSIONS

In order to characterize fractals occurring in many areas of physics (especially, fractal sets of strange attractors), one is very interested in the measurements of fractal dimensions. So far, there have been many different definitions of dimensions. They can be roughly classified into two groups, the first one being derived from purely geometrical requests, the second one being related to information theory. We introduce several common definitions of dimensions, as follows.

2.7.1 *The Hausdorff or Kolmogorov dimension*

Suppose $N(\delta)$ is the number of δ-spheres (sphere with diameter δ) needed to cover a set. The Hausdorff or Kolmogorov dimension of the set is defined by;

$$D_0 = \lim_{\delta \to 0} \log N(\delta)/\log(1/\delta). \tag{2.20}$$

D_0 is also called the capacity dimension. Using Equation (2.20), we can get $D_0 = 1.2618$ for the Koch curve, and $D_0 = 0.6309$ for the triadic Cantor set.

Kaplan & Koke (1979) have given an equilateral form of Equation (2.20) using the Lyapunov exponent;

$$D_0 = K + \left(\sum_{i=1}^{k} \lambda_i \right) \Big/ |\lambda_{k+1}|, \tag{2.21}$$

where λ_i $(i = 1, 2, \ldots, k)$ is the Lyapunov exponent, and k is a maximum which makes

$$\sum_{i=1}^{k} \lambda_i \geq 0.$$

2.7.2 *The information dimension*

From the definition of the capacity dimension, the number of δ-spheres is considered solely, but the number of points covered by each of the δ-spheres could not be analyzed. Thus, a definition of the information dimension is proposed:

$$D_1 = \lim_{\delta \to 0} \left[\sum_{i=1}^{N} P_i \log(P_i) \right] \Big/ \log \delta; \tag{2.22}$$

where P_i is the probability for a point of a long time series which falls in the δ-sphere i of the partitioning. The information dimension is a generalization of the capacity dimension. From Equation (2.22), let $P_i \equiv 1/N$, then Equation (2.22) becomes Equation (2.20).

2.7.3 *The correlation exponent dimension*

Grassberger & Procaccia (1983) have given the definition of the correlation exponent dimension:

$$D_2 = -\lim_{\delta \to 0} \log C(\delta)/\log \delta, \tag{2.23}$$

where the 'correlation integral' $C(\delta)$ is defined via a data series $x_1, x_2, \ldots, x_N$ as

$$C(\delta) = \lim_{N\to\infty} N^{-2}\{\text{number of pairs } (x_i, x_j) \text{ with } |x_i - x_j| < \delta\}$$

$$= \lim_{N\to\infty}\left[\sum_{i,j=1}^{N} \Theta(\delta - |x_i - x_j|)\right]\Big/ N^2; \tag{2.24}$$

where $\Theta(x)$ is the Heaviside function.

2.7.4 *The generalization dimension*

For any positive $q \neq 1$, and for a given partitioning, the order-q Kenyi entropy is defined as

$$K_q(\delta) = \left[\log \sum_{i=1}^{N(\delta)} (P_i)^q\right]\Big/(1-q). \tag{2.25}$$

The generalization dimension can be obtained by the definition of

$$D_q = -\lim_{\delta\to 0} K_q(\delta)/\log(\delta). \tag{2.26}$$

Obviously, D_q is equal to the capacity dimension D_0, the information dimension D_1, the correlation exponent dimension D_2, and so on, as q is equal to 0, 1, 2,..., respectively. In fact, up to a factor which for $\delta \to 0$ becomes independent of δ, one has

$$C(\delta) \approx \exp[-K_2(\delta)] = \sum_{i=1}^{N(\delta)} P_i^2,$$

so that

$$D_2 = \lim_{\delta\to 0} K_2(\delta)/\log(1/\delta).$$

From ordinary entropy

$$K(\delta) = -\sum_{i=1}^{N(\delta)} P_i \log P_i,$$

we have

$$K(\delta) = \lim_{q\to 1} K_q, \quad D_1 = \lim_{q\to 1} D_q, \quad \text{and} \quad D_0 = \lim_{q\to 0+} D_q.$$

We find that $D_q \geq D_p\,(p > q)$; i.e., $D_T \leq D_q \ldots \leq D_2 \leq D_1 \leq D_0$. For example, Grassberger (1983) has discussed the Henon map by means of the generalization dimensions and, given the results, fitting the interval $2^{-12} < \delta < 2^{-4}$, of $D_0 = 1.29 \pm 0.02$, $D_1 = 1.26 \pm 0.02$, $D_2 = 1.23 \pm 0.02$, and $D_3 = 1.20 \pm 0.02$. The fractal dimensions of the fractal sets can be estimated by the above definitions of dimensions, from Mandelbrot's definition of fractals, $D = D_q > D_T$. In particular, the different definitions of dimensions should be used to analyze complicated sets in nature, so that one can find a well-defined $D_q = D$ which can clearly characterize the physical properties of the system.

2.7.5 *The self-similarity dimension*

For a straight line, its Euclidean dimension is 1. Let a straight line have length X, and consist of $N = b$ equal segments. Each of the segments is the interval $(k-1)X/b \leq x \leq kX/b$, where k goes from 1 to b. Visually, each segment can be obtained from the whole by a ratio of similarity, $r_{(N)} = 1/b = 1/N$. Likewise, the Euclidean dimension of a plane is 2. We also divide a rectangle $0 \leq x < X, 0 \leq y < Y$ into $N = b^2$ parts. These parts are smaller rectangles, defined by the combined inequalities:

$$(k-1)X/b \leq x < kX/b$$

and

$$(h-1)Y/b \leq y < hY/b,$$

where k and h go from 1 to b. Each part can be obtained from the whole by a ratio of similarity

$$r_{(N)} = 1/b = 1/N^{1/2}.$$

For a rectangle parallelepiped, we can also obtain $r_{(N)} = 1/N^{1/3}$ by means of the same method.

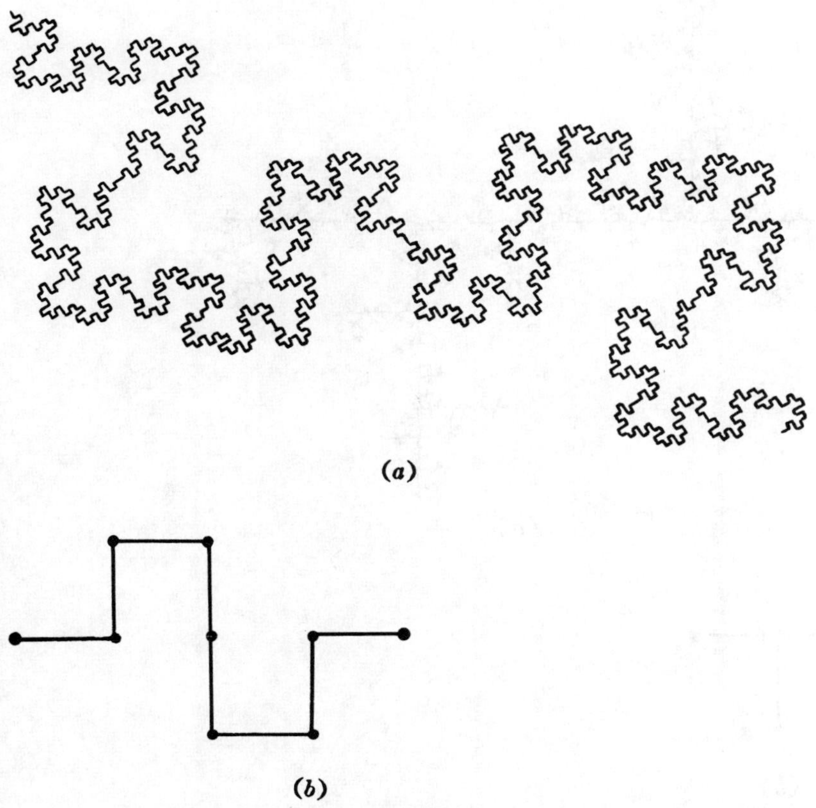

Figure 2.17. Minkowski sausage (also called the quadratic Koch island) (a); and its generator (b).

Similarly, all D-dimension parallelepipeds satisfy

$$r_{(N)} = 1/N^{1/D}. \tag{2.27}$$

Thus,

$$Nr^D = 1.$$

Equivalent alternative forms are

$$\log r_{(N)} = \log(1/N^{1/D}) = -(\log N)/D,$$

and

$$D = -\log N/\log r_{(N)} = \log N/\log(1/r). \tag{2.28}$$

For nonstandard shapes, such as fractal sets, the ratio of self-similarity has the same form as Equation (2.27), only if the shape is required to be self-similar, i.e., the whole may be split up into N parts which can be obtained from it by a ratio of similarity, r. Thus, Equation (2.28) can be used to estimate the fractal dimension of fractal sets with self-similarity. For a strictly self-similar fractal set, it is easy to determine N and the ratio of similarity, r from its generator. Provided that the generator of a strictly

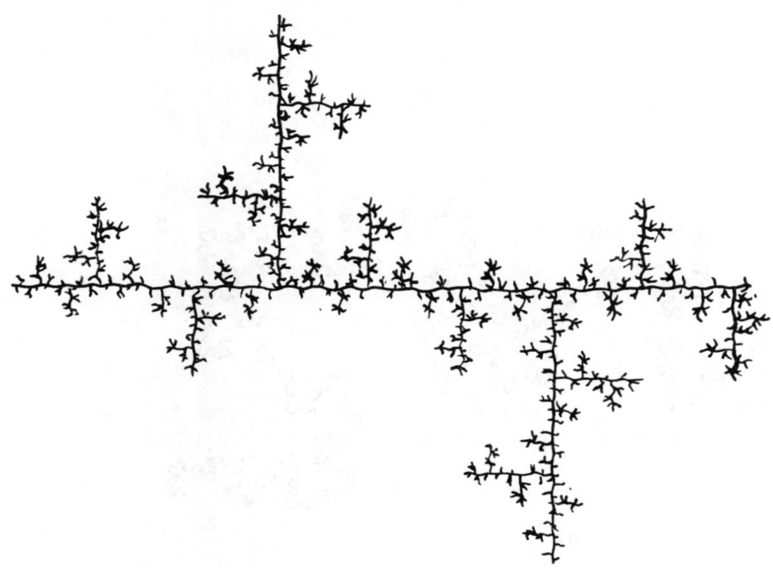

(a)

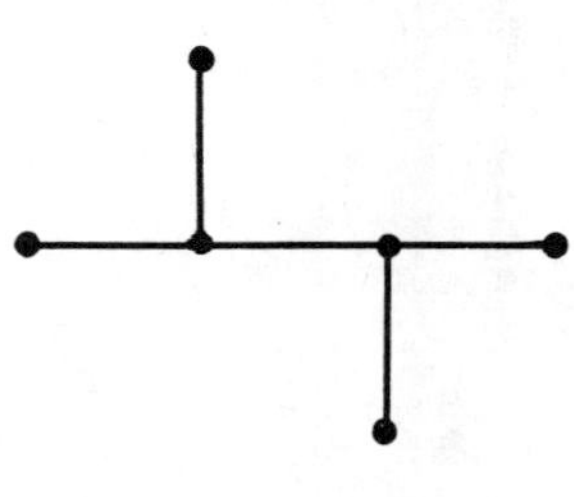

(b)

Figure 2.18. A tree-like fractal set (a); and its generator (b).

self-similar fractal set consists of a broken line connected by N straight lines with equal length, then, the ratio of the distance between the starting point and terminal point of the generator to the length of each of the straight lines is $1/r$, and the fractal dimension can be calculated by Equation (2.28). For instance, Figures 2.17-2.19 show some Mandelbrot-given curves (Mandelbrot, 1983) and generators of the curves. We can then determine quickly that: $N = 8$, $1/r = 4$, $D = \log 8/\log 4 = 1.5$ for a Minkowski Sausage (shown in Figure 2.17); $N = 5$, $1/r = 3$, $D = \log 5/\log 3 = 1.465$ for a tree-like fractal (shown in Figure 2.18); $N = 6$, $1/r = 3$, $D = \log 6/\log 3 = 1.63$ for Mandelbrot-given curve (shown in Figure 2.19). It is now interesting to return to the construction of the Koch curve. The generator of the Koch curve is shown in Figure 2.20. In Figure 2.20a, $N = 4$, $1/r = 3$, and $D = \log 4/\log 3 = 1.2618$. In Figure 2.20b, $N = 2$, $1/r = L_0/L = \sin\theta/\sin[(\pi - \theta)/2]$, and $D = \log 2/\{\log \sin\theta - \log \sin[(\pi - \theta)/2]\}$. The fractal dimension D will change with θ; $D = 2.0$ as $\theta = 90°$, and $D = 1.2618$ as $\theta = 120°$. In fact, the generator as $\theta = 120°$ (shown in Figure 2.20b) is an alternate of the triadic Koch curve.

We can say that fractal sets in nature are all statistically self-similar. It is not easy to determine their fractal dimensions, because it is impossible to get the number N and the ratio of similarity, r, directly from the process of construction. For such sets, the covering method, i.e., counting the number of boxes (squares, grids, or balls) with a characteristic linear dimension δ needed to cover the set, (see Equation (2.11)), is useful

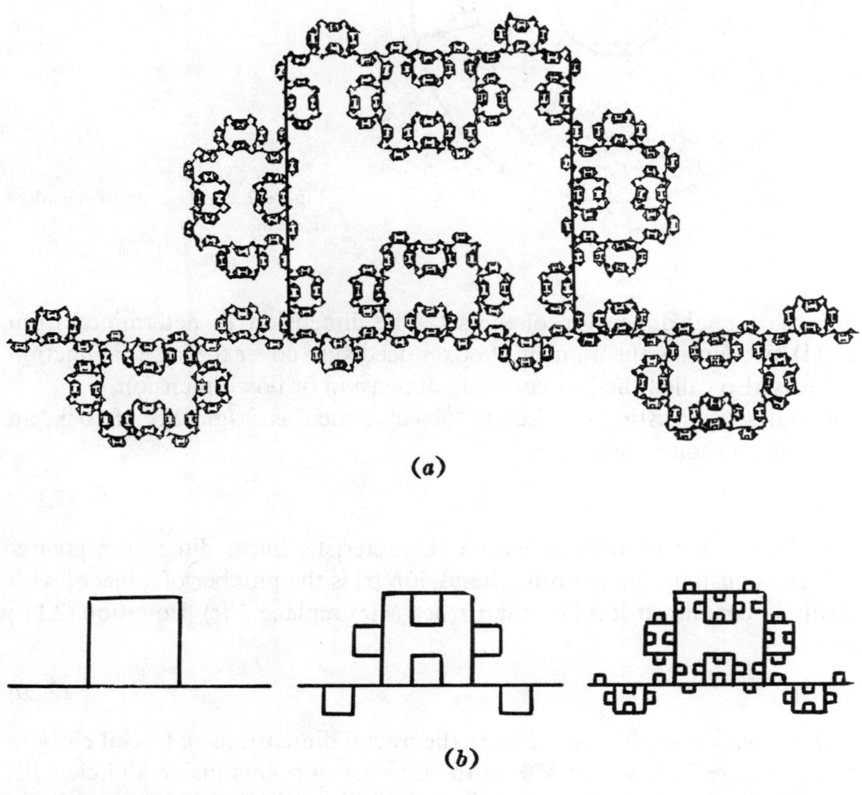

(a)

(b)

Figure 2.19. (a) Construction of a Mandelbrot-given curve without branches. This curve is obtained using a; (b) Generator with a single loop.

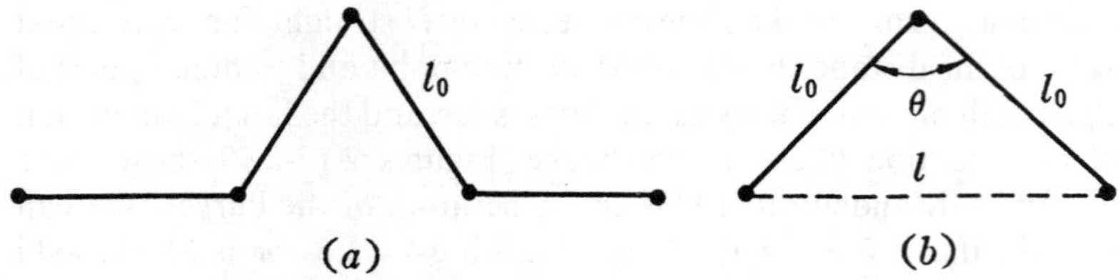

Figure 2.20. (a) The generator of the Koch curve; (b) A generalized generator of the Koch curve.

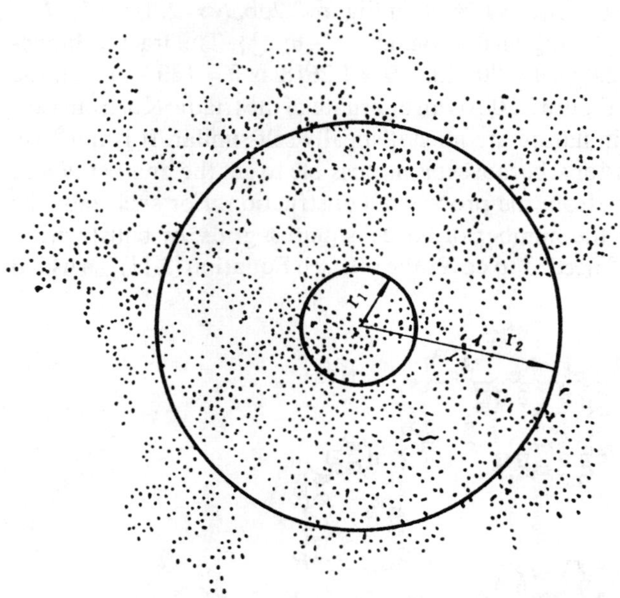

Figure 2.21. The number-radius relation.

in estimating the fractal dimension of the set. The dimension D, determined from Equation (2.11) by counting the number of boxes needed to cover the set as a function of the box size, is also called the box counting dimension or box dimension.

For the fractal set consisting of discrete 'objects', such as fragments of rock, an alternative box dimension is defined by

$$N(r) = Cr^{-D}; \tag{2.11'}$$

where $N(r)$ is the number of 'objects' with a characteristic linear dimension greater than r, and C is a constant. On the other hand, if $N(r)$ is the number of 'objects' with a characteristic linear dimension less than r, let $M(r)$ replace $N(r)$, Equation (2.11′) then becomes

$$M(r) = Cr^{D}. \tag{2.29}$$

Equation (2.29) is widely applied to measure the fractal dimensions of fractal clusters as illustrated in Figure 2.21, where $M(r)$ is the number of points inside a circle with radius r. For instance, we have discussed the fractal dimension of a galaxy with the distribution of large scales from Equation (2.29), where $D \in [0.86, 2.16]$ depends on the range of stars (see the paper of Xie, 1989c). Equation (2.29) is also called a number-radius relation, and the fractal dimension obtained from it is denoted by the cluster dimension (the application is shown in Section 14.10).

2.7.6 *The probability dimension*

In many cases, a combination of the probability and fractal dimension is needed Turcotte (1989). Let us consider the Cantor set illustrated in Figure 2.7. In each step of the construction, the middle third is removed from each of the remaining segments, so that $N = 2$, $r = \delta_{i+1}/\delta_i = \frac{1}{3}$. Thus, from Equation (2.28), $D = \log 2/\log 3$. The fractal dimension lies between 0 (the Euclidean dimension of the point set) and 1 (the Euclidean dimension of a line). The probability that a segment with length δ_i includes interval [0, 1] is $P\delta = 1$ when $\delta = 1$, $P\delta = \frac{2}{3}$ when $\delta = \frac{1}{3}$, and $P\delta = \frac{4}{9}$ when $\delta = \frac{1}{9}$. This is generalized to

$$P_{\delta i} = N_i \delta_i^1,$$

and substitution of Equation (2.11) gives

$$P_{\delta i} = \delta_i^{1-D}. \tag{2.30}$$

The plane form of the Cantor set is the Sierpinski carpet shown in Figure 2.9. The solid square of unit dimension has a square with dimension $\delta = \frac{1}{3}$ removed from its center. The process is repeated, we have $N = N_{i+1}/N_i = 8$, $r = \delta_{i+1}/\delta_i = \frac{1}{3}$, and $D = \log 8/\log 3 = 1.8928$. The fractal dimension lies between 1 (the Euclidean dimension of a line) and 2 (the Euclidean dimension of an area). The probability that a square with side length δ_i includes a solid is $P\delta = 1$ when $\delta = 1$, $P\delta = \frac{8}{9}$ when $\delta = \frac{1}{3}$, and $P\delta = \frac{68}{81}$ when $\delta = \frac{1}{9}$. Thus, we conclude that

$$P_{\delta i} = N_i \delta_i^2,$$

then

$$P_{\delta i} = \delta_i^{2-D}. \tag{2.31}$$

The space mimic of the Cantor set is the Menger sponge illustrated in Figure 2.10. The solid cube of unit length has a cube with side length $\delta = \frac{1}{3}$ deleted from its center. Using such recursive procedures, we can obtain $N = N_{i+1}/N_i = 20$, $r = \delta_{i+1}/\delta_i = \frac{1}{3}$, and $D = \log 20/\log 3 = 2.7268$. The probability that a cube with side length δ_i includes a solid is $P\delta = 1$ when $\delta = 1$, and $P\delta = \frac{20}{27}$ when $\delta = \frac{1}{3}, \ldots$. Similarly we have

$$P_{\delta i} = N_i \delta_i^3,$$

then

$$P_{\delta i} = \delta_i^{3-D}. \tag{2.32}$$

From Equations (2.30)-(2.32), a uniform formulation can be concluded as:

$$P_{\delta i} = \delta_i^{d_E - D}; \tag{2.33}$$

where d_E is the Euclidean dimension of a set. We can now use Equation (2.33) to obtain the fractal dimension of a set using probability. It should be noted that this relation can be applied to random processes, although it is derived for deterministic fractals.

CHAPTER 3

The perimeter-area relation of fractal sets and fractal surfaces

A relation between the circumferential length (perimeter) of a two-dimensional standard shape (such as a circle or square) and the area of a disc bounded by the standard shape can be obtained as follows:

$$(\text{Perimeter}) \propto (\text{area})^{1/2}. \tag{3.1}$$

For nonstandard shapes (such as clouds and inclusions etc.), the corresponding relation is generalized by Mandelbrot (1982):

$$(P_{(\delta)})^{1/D} \propto (A_{(\delta)})^{1/2}, \tag{3.2}$$

where $P_{(\delta)}$ is the perimeter of nonstandard shapes measured with a yardstick length of δ, and $A_{(\delta)}$ is the area measured in units of δ^2. Note that the dependence of $P_{(\delta)}$ upon δ is nonstandard, while the dependence of $A_{(\delta)}$ is standard (Mandelbrot, 1982). Equation (3.2) is the basic perimeter-area relation of fractal sets.

Mandelbrot (1982) and Feder (1988) show that the ratio between the perimeter and the area of fractal set defined by

$$\alpha_D(\delta) = [P(\delta)]^{1/D}/[A(\delta)]^{1/2}, \tag{3.3}$$

is only dependent on measure scales δ, independent of the shapes and sizes of fractal sets. For example, given two similar islands, the area and the length of coastline of each of islands are measured by using an area-dependent yardstick $\delta_i^* = \lambda\sqrt{A_{i(\delta)}}$ for the i-th island ($\lambda < 1$, and δ is a fixed yardstick), then

$$\alpha_D(\delta) = N_\lambda^{1/D}\lambda\delta^{(1-D)/D}, \tag{3.4}$$

where $\alpha_D(\delta)$ is called the shape factor, N_λ is the number of segments in the i-th approximation (such as to measure the length of a coastline) with a measure scale δ_i^* so that

$$P_i(\delta_i^*) = N_\lambda\delta_i^*. \tag{3.5}$$

Depending on the geometrical shapes, the parameters are N_λ and λ, but, for similarly-shaped sets, they are independent of the sizes of the sets.

3.1 THE PERIMETER-AREA RELATION FOR RAINS AND CLOUDS

One surprisely finds that the variety of cloud and rain size and shapes satisfy the perimeter-area relation. The fractal dimension, D, can be used to analyze the difference between clouds or rains, and processes of their change. The meteorologists expect, D to be an important parameter which characterizes the process of producing clouds and rains. In this section, we will mainly introduce Lovejoy's (after Feder, 1988) and Rys & Waldvogel's works on the fractal description of rains and clouds.

Lovejoy (1982) investigated the geometry of cloud and rain areas the very large size range from 1 to over $10^6\,km^2$, based on the data of radar observations from tropical Atlantic rain areas (with a rainrate above 2 mm/hr) and geostationary satellite infrared observations of cloud areas over the Indian Ocean. He found that the radar pictures of tropical rain areas at a resolution of 1 by 1 km and infrared pictures of Indian Ocean clouds from the geostationary operational environment satellite (GOES), sampled on a grid of 4.8 by 4.8 km, really behaved as the perimeter-area relation (Equation (3.2)). Taken separately, the radar and satellite data fall on parallel straight lines. When the satellite perimeters are multiplied by a correction factor, these two lines coincide as shown in Figure 3.1 (Lovejoy, 1982) and yield a least squares estimate of $D = 1.35$ with a correlation coefficient of 0.994. To explain why a correction factor is required to make the parallel straight lines coincide, Lovejoy explained that the radar data have a different spatial resolution from the satellite data and thus are more sensitive to the detailed structure of a perimeter; therefore, for a given area, the radar will always measure a longer perimeter than the satellite. Obviously, if the perimeter is a fractal with dimension D, the length of a perimeter measured at resolution δ is greater than the length of one measured at δ_0 by a factor $(\delta/\delta_0)^{1-D}$. The satellite perimeters were, therefore, multiplied by the factor $(1/4.8)^{-0.35} = 1.73$ to make them compatible with the radar data.

In Figure 3.1 the threshold defining cloud areas is somewhat arbitrarily set at -10°C for the cloud-top temperature. Physically, the rain and cloud fields are closely related in the tropics, since both occur in regions of convective updraft, which cause the moist, warm surface air to rise, cool by adiabatic expansion, and form clouds and rain in the resulting condensation procession. Many of these clouds, even at -20 or -30°C, are supercooled water rather than ice because of the relatively short time scale of the processes. Therefore, the cloud area described by the -10°C threshold contains regions of both cumulus and cirrus clouds.

Lovejoy's results as shown in Figure 3.1 have two striking aspects. The first is that the perimeter-area relation of cloud and rain areas hold over such a large range, from $1\,km^2$ to $1.2 \times 10^6\,km^2$, and there is an absence of any apparent bend or kink over a range of more than six orders of area magnitude. This is important because the evidence, to date, based largely on Fourier spectra of wind variations, has been inconclusive about the existence of length scales, i.e., there is no characteristic length scale, particularly in the range 1 to 1000 km. The second aspect is that the fractal dimension $D = 1.35$ is so close to the value $\frac{4}{3}$ that some fairly straightforward connection may exist because Kolmogorov's theory of isotropic, homogeneous turbulence in three-dimension (associated with the $-\frac{5}{3}$ power law energy spectrum) predicts $D = \frac{5}{3}$ and $\frac{4}{3}$ for isotherms and isobars, respectively (suppose $p \propto A^{D/2}$ for rain and cloud areas). An understanding of the physical origins of this value could be important.

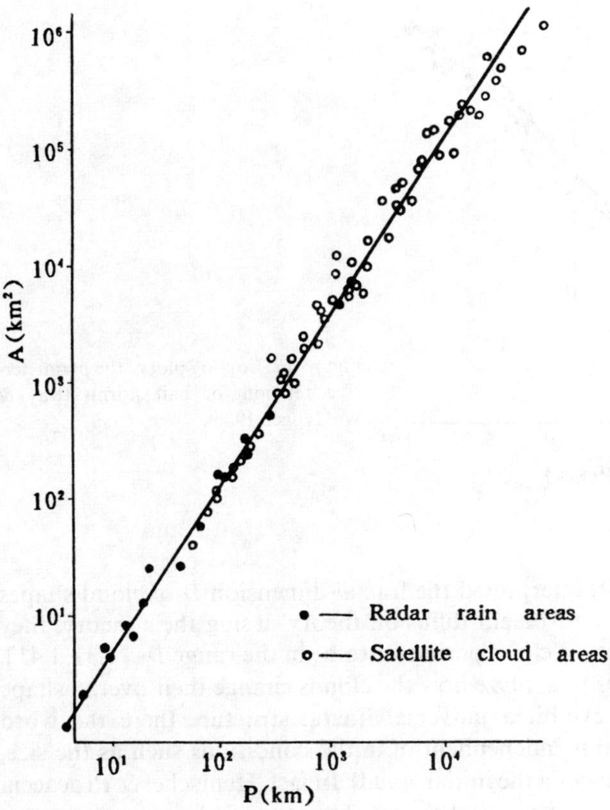

Figure 3.1. Area plotted against a perimeter of rain and cloud areas (Lovejoy, 1982).

Rys & Waldvogel (1986) investigated the fractal features of hail clouds. They show that severe convective clouds which occur in hail storms also behave like the perimeter-area relation as shown in Equation (3.2). In Figure 3.2, we reproduce their results, a log-log plot of the area A versus the perimeter P of 45 dBZ contours. Each point in the figure corresponds to a typical time (interval of one minute during the evolution of the hail storm). From similar plots of the contours of the dBZ values 55, 60 and 65, they concluded that smaller contours are smooth with $D = 1$, and larger contours are fractal structures with $D > 1$. On the other hand, Figure 3.2 clearly shows that there is a crossover at a critical value P_0 of the perimeter marked by the dotted line. This indicates the existence of a characteristic length $L = P_0/\pi = 3 \pm 1$ km. From the evaluation of 24 hail events, they have obtained

$$D = 1.36 \pm 0.1 \quad \text{for} \quad P > P_0;$$

and

$$D = 1.0 \pm 0.1 \quad \text{for} \quad P < P_0,$$

from the perimeter-area relation.

The fractal dimension of large contours $P > P_0$ is equal to the value of $D \approx 1.35$ obtained by Lovejoy.

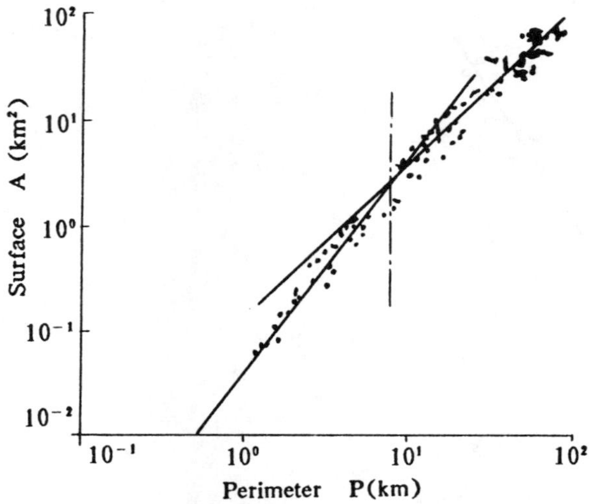

Figure 3.2. Log-log plot of the perimeter-area relation for hail storms (Ray & Waldvogel, 1986).

Hentschel & Procaccia (1984) interpreted the fractal dimension D of cloud shapes in the framework of the relative turbulent diffusion theory. Using their theory, they estimated the fractal dimension of a cloud perimeter to be in the range $D \in [1.37, 1.41]$. The main ideas of their theory is to analyze how the clouds change their overall shape in time, and why the clouds exhibit a universal fractal structure (here the word universal means that the fractal is independent of initial conditions such as the size, height, or other parameters specifying the initial cloud). In fact, Hentschel & Procaccia used only a simplified model to analyze cloud shapes. They specified some parameters such as temperature, pressure, water content, and droplet size, in addition to the turbulent velocity field, to characterize the state of the atmosphere. It is surprising that from such a simplified model of the atmosphere they are able to obtain the fractal dimension, $1.37 \leq D \leq 1.41$, in agreement with observation. It indicates that the phenomenon observed by experiments with fractal dimensions may be calculated from physical models of the system (Feder, 1988).

Note that a cloud in four-dimensional space is self-affine, not self-similar. However, surfaces of clouds in three-dimensional space may be self-similar (Feder, 1988). Voss (1985a, b) has discussed this distinction and has generated actual pictures of clouds similar to the paintings, using the fractal method.

Lovejoy & Schertzer (1985) proposed a self-similar fractal model of rain fields. They considered rain fields as the superposition of 'pulse', each pulse covering an area A, and obtained the increment in the rain rate $\Delta R = \pm A^H$ with $H = 0.6$. The resulting pictures look quite realistic.

3.2 THE FRACTAL DIMENSION OF RIVERS

Feder (1988) has discussed the perimeter-area relation of rivers. There exists a famous Horton's law in geography (Horton, 1945) as follows:

$$Y_u = KR^{\pm u}, \tag{3.6}$$

where u is the order of the streams, K is a constant, Y_u represents a characterized value such as the number, length, area and high difference of the stream order-u, and R is a proportional parameter related to this characterized value. This relation indicates that the number, length, area, and high difference of the basins, in a drainage system, vary with the stream order in the power-law relation. Hack (1957) studied rivers in Virginia and Maryland, as indicated in the drainage system illustrated in Figure 3.3. The length

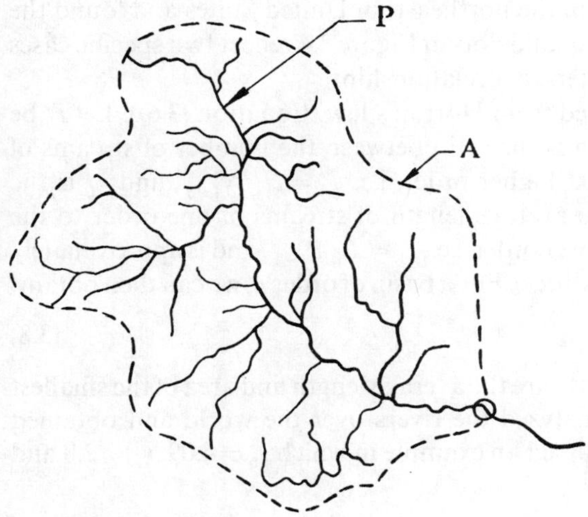

Figure 3.3. Drainage area A and the length P of the longest river above a given location (Feder, 1988).

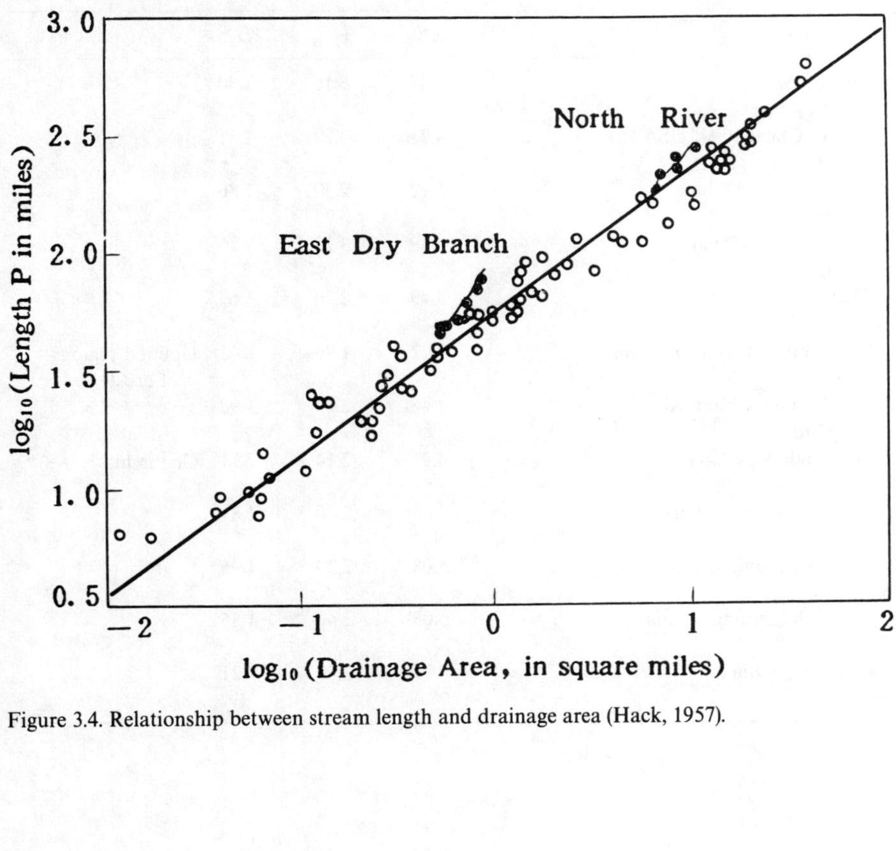

Figure 3.4. Relationship between stream length and drainage area (Hack, 1957).

P is measured along the longest stream to the drainage divide, and the area A is obtained from topographic maps and in a few cases from aerial photographs, by using a planimeter. He obtained the relationship between the length P and the area A from all the data as follows:

$$P = 1.4\sqrt{A^D}, \tag{3.7}$$

with $D = 1.2$, as shown in Figure 3.4. Hack also examined 400 similar measurements determined by Langbein et al. (1947) in the northeastern United States and found the same relationship (Equation (3.2)). The solid dots in Figure 3.4 reflect two specific cases of departure from the general perimeter-area relationship.

In fact, Equation (3.7) can be derived from Horton's law (Equation (3.6)). Let r_b be the bifurcation ratio of streams which is the ratio between the number of streams of one order and the number of the next higher order, i.e., $r_b = N_i/N_{i+1}$, and r_L is the length ratio (which is the ratio of the average length of streams of one order to the average length of streams of the next lower order, i.e., $r_L = P_i/P_{i-1}$, and is approximately a constant and independent of stream order). For a basin of order u, we can then obtain

$$P = P_0(A/A_0)^{D_u/2}; \tag{3.8}$$

with $D_u = \log r_b/\log r_L$. Where P_0 and A_0 are the average length and area of the smallest streams, respectively. Gau in 1982 analyzed the rivers over the world and obtained $D_u = 1.574$. Leopola & Langbein discussed an example in which $r_b = 4.09$, $r_L = 2.8$ and

Table 3.1. The fractal dimension of streams and rivers (Hong, 1988).

Region	r_b	r_L	D_u	
Allegheny river, USA (A.N. Strahler)	4.47	3.16	1.30	
Laurel branch of Cheat river, USA (R.E. Horton)	4.71	3.27	1.31	
Jiu Yuangou, Shanber in China (Chen Jicheng)	5.12	2.40	1.86	
She Jiagou, Shanber in China (Chen Jicheng)	4.39	3.22	1.26	
Gai river, Gansu, China (Chen Fahu)	3.79	2.20	1.69	
Streams of Hei Shuigou, Gansu, China (Chen Fahu)	5.22	1.96	2.45	Uplifted region of crust
Faults of Xian Shuihe, Chuan Xi, China (Zhang Daquan)	4.46	2.13	1.98	
Qian Fusi, Chuandong, China (Guo Yue)	4.97	2.14	2.11	Gentle hill
Chen Jiawan, Chuandong, China (Guo Yue)	4.88	2.40	1.81	
Yang Jiagou, Chuandong, China (Guo Yue)	5.08	2.29	1.96	
Niujia Caogou, Chuandong, China (Guo Yue)	5.05	2.40	1.85	
Wugui, Chuandong, China (Guo Yue)	3.94	3.09	1.21	

$D_u = 1.37$ by using a computer technique. Table 3.1 shows some results determined by Equation (3.8).

Mandelbrot (1982) and Feder (1988) have discussed the perimeter-area relationship of rivers in detail. The relationship (3.2) gives the fractal dimension of rivers and streams to be $D = 1.2$. Recently, Hjelmfelt (1988), Robert & Roy (1990) and Rosso et al. (1991) have also investigated the fractal nature of river length and catchment area measurements. Hjelmfelt (1988) found the fractal dimensions for length measurements to vary from 1.036 to 1.291, with an average value of 1.158. Specifically, Rosso et al. (1991) derived fractal dimension estimation from the Horton's laws of network composition. Their results provide a linkage between quantitative analysis of drainage network composition and scaling properties of river networks. However, river systems are not always self-similar and, therefore there are arguments about the length-area relation of rivers. Hack (1957) expressed the lack of self-similarity of rivers and streams as follows:

'It is obvious that on any stream the length increases downstream more rapidly than the 0.6 power of the area between all the major tributaries. There are stream reaches along which no large tributaries enter for long distances, and in these reaches the length must increase at a more rapid rate; this general principle applies to short as well as to long streams'.

Of course, we can not say that rivers and streams are not fractals because of the lack of this self-similarity, maybe some are fractals and others are not. However, how to define fractal dimensions for river systems still remains obscure and, visually, the geometry of the streams and rivers is required to further study them.

3.3 FRACTAL CHARACTERS OF FRACTURED SURFACES

The perimeter-area relation (Equation (3.2)) has been widely applied to quantitatively analyze fractured surfaces. The first research on fractal characters of fractured surfaces is due to Mandelbrot et al. (1984). They analyzed the roughness and irregularity of fracture surfaces of metal materials (300-Grade Maraging steel) by using the method of slit island analysis. In the slit island method, a fractured steel specimen is plated with electroless nickel and mounted in an epoxy matrix by vacuum impregnation in order to ensure edge retention. The specimen is then polished parallel to the plane of fracture so that 'islands' of steel surrounded by nickel are produced. In the microscope field of vision, the polished zones appear bright, so-called 'islands', and the unpolished zones appear dark, so-called 'lakes'. These islands (called slit islands by Mandelbrot et al.) consist of lakes within islands and islands within lakes. The roughness and irregularity of fracture surfaces are characterized either by lakes within islands or islands within lakes. The areas and perimeters of these islands can then be measured. The results obtained by Mandelbrot et al. (1984) are reproduced as shown in Figure 3.5. We can see that the perimeter-area relationship (Equation (3.2)) is well defined, with $D = 1.28$.

Mandelbrot et al. (1984) also used the method of fracture profile analyses based on the Fourier method to analyze these fractured surfaces. In this method, the fractured samples are sectioned perpendicular to the fracture surface to expose them in profile. As usual, the sample spectra of the profile exhibit wide oscillations. Most of them reflect

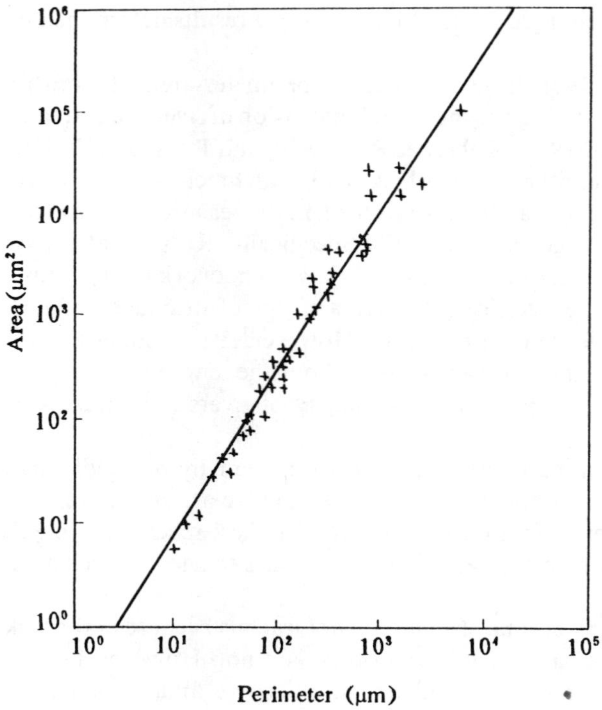

Figure 3.5. Fractal perimeter-area relstionship for slits, islands, using 300-Grade Maraging steel. Ruler = 1.5625 μm; fractal dimension increment $D = 1.28$ (Mandelbrot et al., 1984).

fundamental lengths of the microstructure and their accompanying higher harmonics, although some are statistically artificial products. Statistical fluctuations must be averaged out and the fluctuations, due to their microstructure, should be prevented from overly affecting the analysis. To achieve both goals, they averaged five spectra taken from serial sections and integrated them from high frequency to low. The integrated spectrum took the form $K^{-B'}$, with $B' = B - 1 = 6 - 2D$ (for details, see Chapter 12), which behaved as the fractal character of the fracture surface. The doubly logarithmic coordinate plots of the data reproduced in Figure 3.6 are indeed straight but over narrower ranges than the perimeter-area relationship plot. Thus, the fracture surface is again inferred to be a fractal with fractal dimension $D = 3 - B'/2$.

After heat-treating a series of identical 300-grade Maraging steel Charpy impact specimens at different temperatures, Mandelbrot et al. measured the impact energies dissipated for the fracture of samples and estimated the fractal dimensions of their fracture surfaces. In Figure 3.7 we reproduce their results, which show that the fractal dimensions, measured to be in the range $D \in [2.1, 2.28]$, are inversely proportional to the impact energy, i.e., the value of D decreases smoothly with an increase of the impact energy, we call this relationship a negative one.

Using the Mandelbrot's slit island method, we have analyzed fractured surfaces of rock materials and obtained similar results (Xie, 1987b) which will be discussed in detail in Chapter 14.

Pande et al. (1987) used the slit island method and the vertical section method, which involves taking a vertical section through the fractured surfaces and obtaining the

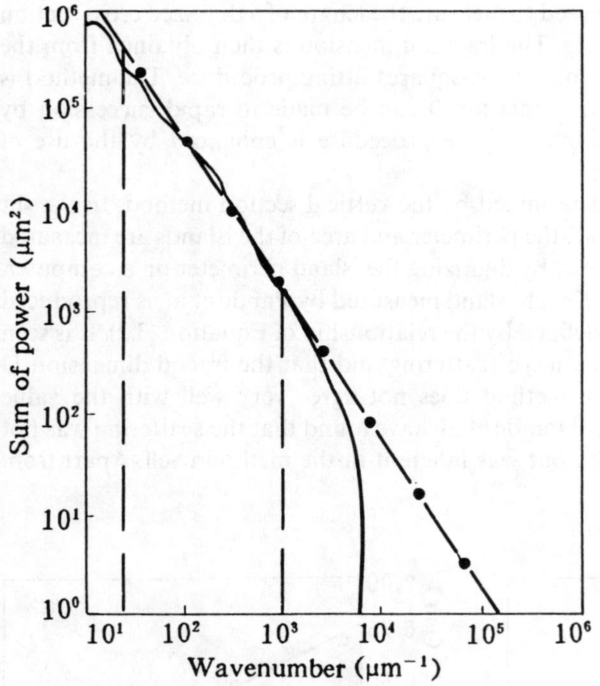

Figure 3.6. Cumulative spectrum for vertical cuts. 300-Grade Maraging steel, slope = 1.4818; fractal dimension D = 1.26 (Mandelbrot et al., 1984).

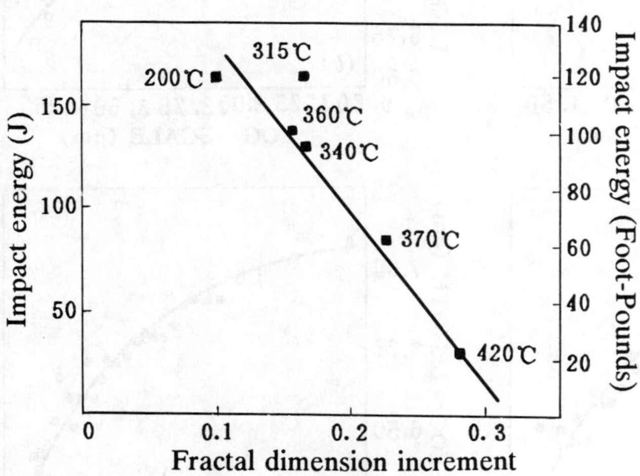

Figure 3.7. Impact energy versus fractal dimension, using 300-Grade Maraging steel at room temperature and Charpy impact (Mandelbrot et al., 1984).

fractal dimension of the profile by the relationship $L = L_0\delta^{(d_E - D)}$ or $D = d_E + (\Delta \log L/\Delta \log \delta)$, δ is the scale of measurement used to study the fractured surfaces of two titanium alloys.

In the vertical section method, the micrographs at a magnification of about 500 × were used to obtain fracture surface cross section profiles and analyzed by a computer program which can be used to measure the length of a digitized cross section profile using different scale lengths. The fractal dimension is then obtained from the slope of log L vs. log δ curves, using a least-squares fitting procedure. The method is quite simple, and several measurements for D can be made in rapid succession by further grinding. Statistical reliability of the procedure is enhanced by the use of random sections at several angles.

Figure 3.8 shows the results determined by the vertical section method. In the slit island method used by Pande et al., the perimeter and area of the islands are measured by using an electronic planimeter or by digitizing the island perimeter on a computer. The log(perimeter) vs. log(area) of each island measured by Pande et al. is reproduced in Figure 3.9, which can be well-defined by the relationship of Equation (3.2). It is seen that the data in Figure 3.8 have a large scattering and that the fractal dimension D measured by the vertical section method does not agree very well with the value obtained by the slit island method. Pande et al. have found that the scattering was not due to any error of measurement, but was inherent in the method itself. Apart from

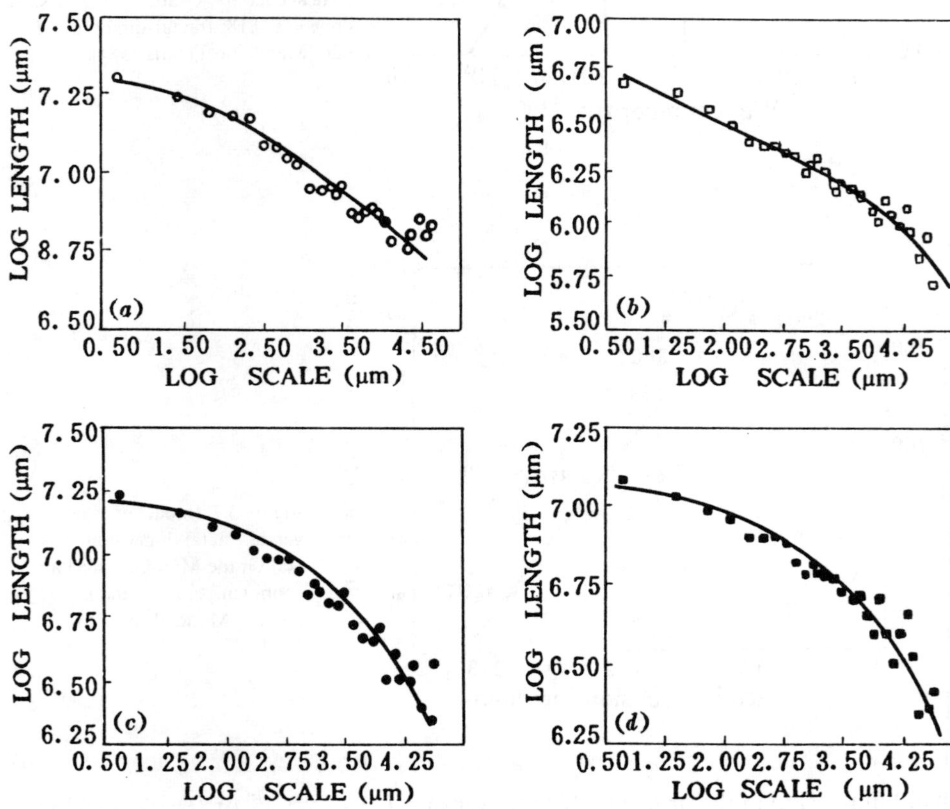

Figure 3.8. Plots of log(length) vs. log(scale) of four sections of the same fractured specimens of a titanium alloy (Ti-6Al-2Nb-1Ta-0.9Mo). (a) and (b) sections are basic surfaces, (c) and (d) sections are complementary surfaces with fractal dimensions 1.229 and 1.182, respectively. The continuous lines in (c) and (d) are arcs of circles (Pande et al., 1987).

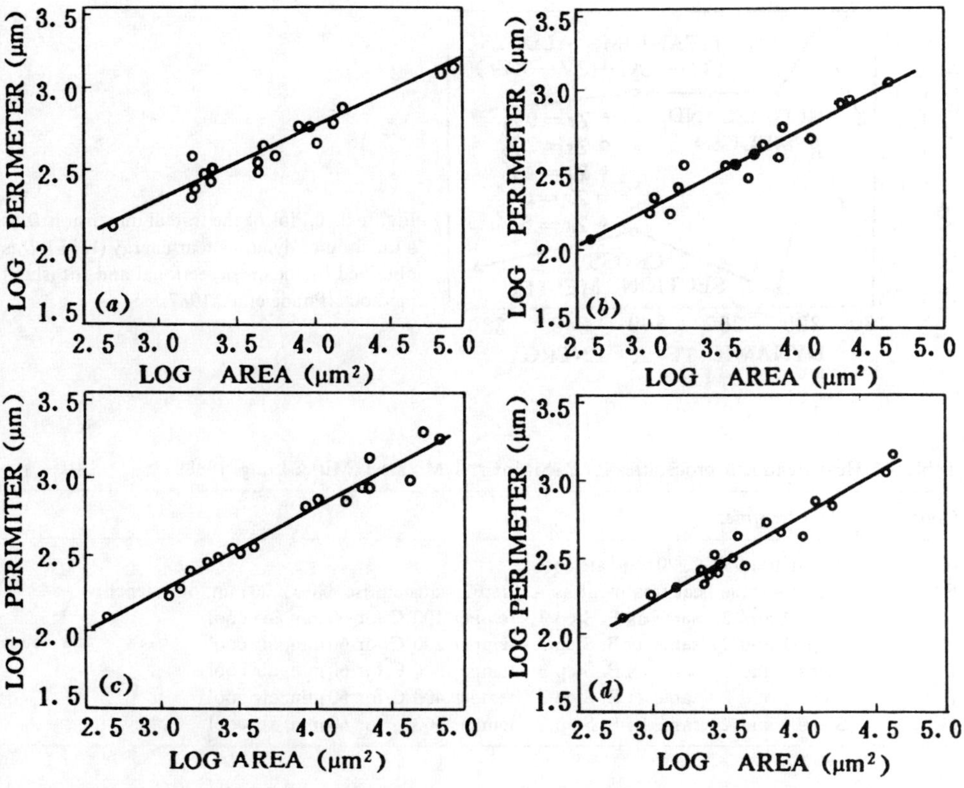

Figure 3.9. Plots of log(perimeter) vs. log(area) of the islands in the slit island method used for titanium alloy Ti-6Al-2V with varying amounts of Zr. (a) Zr = 0%; (b) Zr = 2%; (c) Zr = 4%; (d) Zr = 6%. The slope of a straight line drawn through the points can be used to obtain fractal dimensions. Notice the scattering of the data, which is not due to errors in measurement (Pande et al., 1987).

this, there are at least two other reasons for questioning the differences in results obtained by the two methods. An essential assumption of the vertical section method is that island shape (say ratio of length to breadth) is invariant with size. A systematic failure of this constraint will lead to spurious values for *D*. Another more fundamental reason is that roughness exhibited by an island contour cannot be expected to scale directly with the roughness of this surface. This difficulty arises because the observed surface is a section at grazing incidence to the fracture surface.

Using the titanium specimen with Zr, with the value of Zr ranging from 0 to 6% by weight, Pande et al. tried to ascertain whether any correlation exist between *D* and material properties. It would, thus, be interesting to see whether *D* correlates with the Zr content. On the other hand, the dynamic tear energy (DTE) (which is a measure of toughness), which decreases continuously with increasing Zr content, can be measured. Thus, Pande et al. plotted *D* as determined by the vertical section method and the slit island method as a function of DTE (shown in Figure 3.10, reproduction of their results). It is seen that both methods give a rough correlation between *D* and DTE. Further, the two correlations appear to be similar, although the magnitudes of *D* for

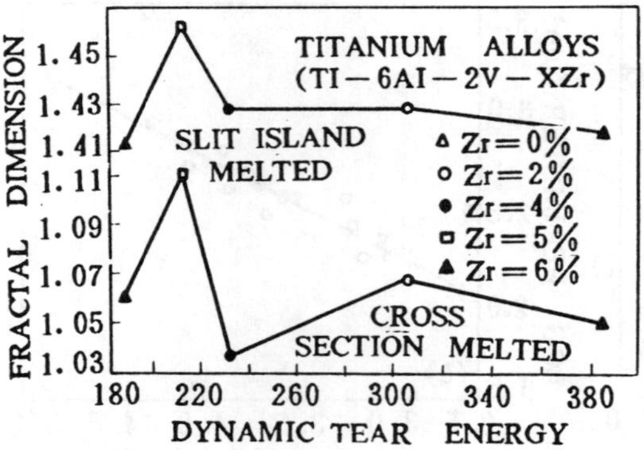

Figure 3.10.Plot of the fractal dimension D as a function of dynamic tear energy (DTE); D is obtained by the cross sectional and slit island methods (Pande et al., 1987).

Table 3.2. Heat treatment procedures for 24 $SiMnCrNi_2Mo$ steel (Mu & Lung, 1988).

Code	Heat treatment
A	Normalise 930°C, 20 min, air cool
B	Step 1 – same heat treatment as A. Step 2 – austenitise 890°C, 20 min, oil quench
C	Steps 1 and 2 – same as B. Step 3 – temper 100°C for 60 min, air cool
D	Steps 1 and 2 – same as B. Step 3 – temper 200°C for 60 min, air cool
E	Steps 1 and 2 – same as B. Step 3 – temper 300°C for 60 min, air cool
F	Steps 1 and 2 – same as B. Step 3 – temper 400°C for 60 min, air cool
G	Steps 1 and 2 – same as B. Step 3 – temper 500°C for 60 min, air cool

the same specimen obtained by the two methods are substantially different. Note that the correlation between D and DTE is still a negative one, the same as that determined by Mandelbrot et al. (1984) and Xie (1987b).

Mu & Lung (1988) measured fractal dimensions of fractured surfaces under plane strain conditions with the help of an image analysis technique for two high-strength steels by using the slit island method. The materials used by them are 24 $SiMnCrNi_2Mo$ and 30 $CrMnSiNi_2A$ steels. Table 3.2 shows the heat treatment procedures employed on the 24 $SiMnCrNi_2Mo$ steel. For 30 $CrMnSiNi_2A$ steel, the heat treatment was as follows: austenitise at 900°C for 30 min, transfer to a salt bath at 200°C for 60 min, temper at 280°C for 120 min, and then air cool. Mu & Lung carried out three-point bend (TPB) tests on the specimens at room temperature for 24 $SiMnCrNi_2Mo$ steel and at five different temperatures for 30 $CrMnSiNi_2A$ steel, and also measured the plane strain fracture toughness and the value of K_{1c} of the specimens. They then investigated fractal dimensions of the fractured surface of the specimens by using the slit island method. They also found that the relationship between perimeters and areas was very nearly a straight line when plotted on a double logarithm scale. We reproduce their results as shown in Figure 3.11. Meantime, they established the relationship between the logarithmic values of the fracture toughness K_{1c} and the fractal dimension D for the two steels, as illustrated in Figures 3.12 and 3.13. It is seen that the fractal dimensions D are approximately linear to the logarithm of the fracture toughness K_{1c}, but still with a negative slope.

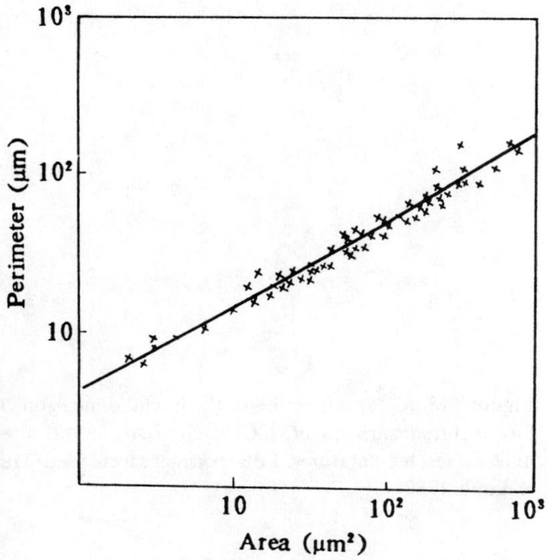

Figure 3.11. The fractal perimeter-area relationship for slit islands. The fractured surface is of $30CrMnSiNi_2A$ steel at −20°C. Ruler = 1.85 μm (Mu & Lung, 1988).

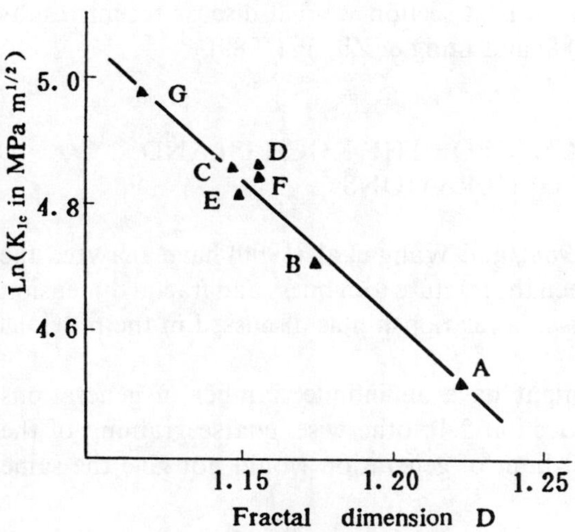

Figure 3.12. K_{1c} as a function of the fractal dimension D for fractured surfaces of $24SiMnCrNi_2Mo$ steel under seven different heat treatment conditions. A normalized, B normalized and quenched, C tempered at 100°C, D tempered at 200°C, E tempered at 300°C, F tempered at 400°C and G tempered at 500°C. TPB specimens were used (Mu & Lung, 1988).

From the above discussions, we can conclude that the perimeter-area relation, i.e., fractal method, can be widely applied to quantitative analyses of fractured surfaces of materials and become an important feature in the process of obtaining a better knowledge of the microstructural processes involved during the deformation and rupture of materials. However, the experimental results of the negative correlation between the fractal dimension D and material properties (K_{1c}, impact energy, and DTE) is difficult to explain if a high fractal dimension is supposed to be characteristic of a rough, as opposed to a smooth surface. Why does this occur? Is it caused by its physical mechanism or by geometrical scales? Recently, there has been a great deal of

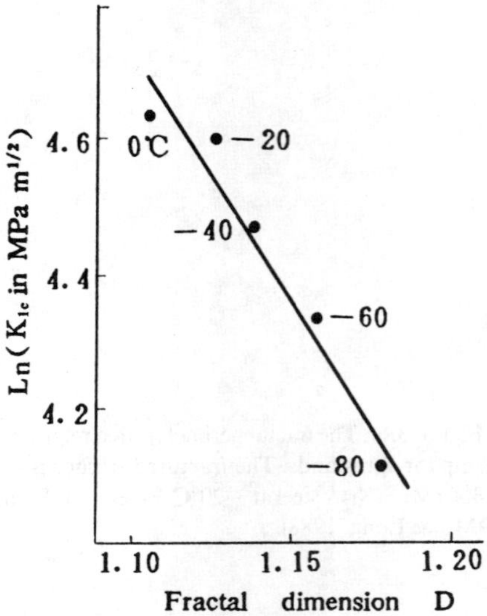

Figure 3.13. K_{1c} as a function of the fractal dimension D for fractured surfaces of $30CrMnSiNi_2A$ steel at five different test temperatures. TPB specimens were used (Mu & Lung, 1988).

interest in studying this phenomenon. In next section we will discuss recent results about it presented by Lung & Mu (1988) and Lung & Zhang (1989).

3.4 THE PERIMETER-AREA RELATION OF THE KOCH ISLAND WITH A FINITE NUMBER OF GENERATIONS

Lung & Zhang (1989), Lung & Mu (1988), and Wang et al. (1990) have analyzed the origin of the negative correlation between the fracture toughness and fractal dimension D as measured by using the perimeter-area relationship as discussed in the previous section.

The mathematical fractal model might have an infinite number of generations without upper and lower limits (see Section 2.4); otherwise, coarse graining of the initiator or fine graining of the lowest limit of generation would not give the same

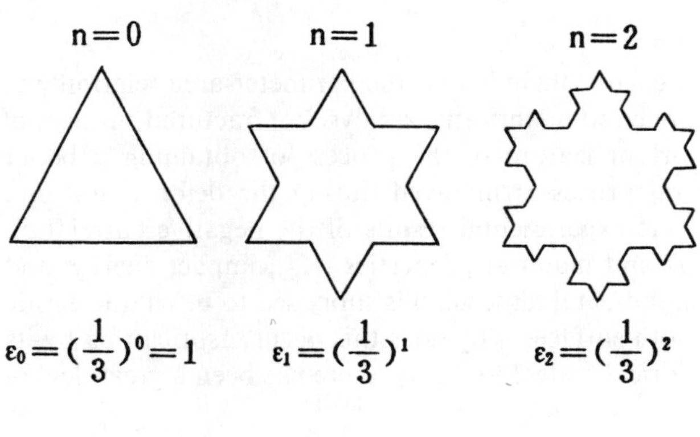

Figure 3.14. The triadic Koch island with an equilateral triangle initiator.

shape as the original one. However, realistic fractals in nature are always of a finite number of generations. As an example, the triadic Koch island (called the snowflake curve as shown in Figure 2.6b) is shown in Figure 3.14. The initiator is an equilateral triangle with side length $\delta_0 = 1$. The fractal dimension is $D = \log 4/\log 3 = 1.2618$.

Considering the perimeter-area (P-A) relationship,

$$[P(\delta)]^{1/D} = \alpha(\delta)[A(\delta)]^{1/2}, \tag{3.2}$$

of the triadic Koch island, the n-th approximation perimeter $P_n(\delta_n)$, and the area $A_n(\delta_n)$ (for details, see Section 2.4):

$$P_n/P_0 = (\tfrac{4}{3})^n$$

and

$$A_n/A_0 = 1 + 3 \times (\tfrac{1}{3})^2 + 3 \times 4 \times (\tfrac{1}{3})^4 + \cdots + 3 \times 4^{n-1} \times (\tfrac{1}{3})^{2n}.$$

Therefore,

$$D_n = D(n) = [(1-D)\log\delta + \log\alpha]/\{\log\alpha + [(1-D)/D]\log\delta + 0.5\log A_n(\delta)\}, \tag{3.9}$$

where P_0 and A_0 are the perimeter and area of the initiator, respectively. From Equation (3.9) and $\delta_n = (\tfrac{1}{3})^n$, it is seen that $D(n)$ drops to a minimum value as $n = 3$ at first and then goes up to a limited value of 1.2618 as $n \to \infty$ (Figure 3.15). Similarly, for the quadratic Koch island (Figure 2.17), the same conclusion has been obtained by Lung & Mu (1988) as shown in Figure 3.16.

Visually, the fractal dimension defined by the perimeter-area relationship depends on the step number of the generation, i.e., depends on the length of the yardstick. This means that the perimeter-area relationship may be expected to hold only in the limit of small yardsticks ($\delta \to 0$ as $n \to \infty$).

Note that if the construction of a fractal set will be terminated at some level n_{max}, such that no further detail structure is revealed by the refining ruler, then, for $n > n_{max}$, the perimeter $P_n/P_0(n > n_{max})$ is defined by its value at $n = n_{max}$, i.e.,

$$P_n/P_0(n > n_{max}) = P_n/P_0(n = n_{max}), \tag{3.10}$$

where P_0 corresponds to the perimeter of the initiator. Then P_n would hold a constant

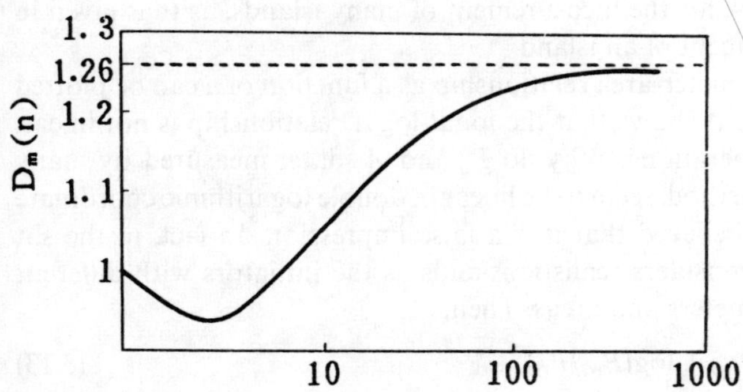

Figure 3.15. The relationship of measured fractal dimension $D_n(n)$ vs. the generation number n (Lung & Zhang, 1989).

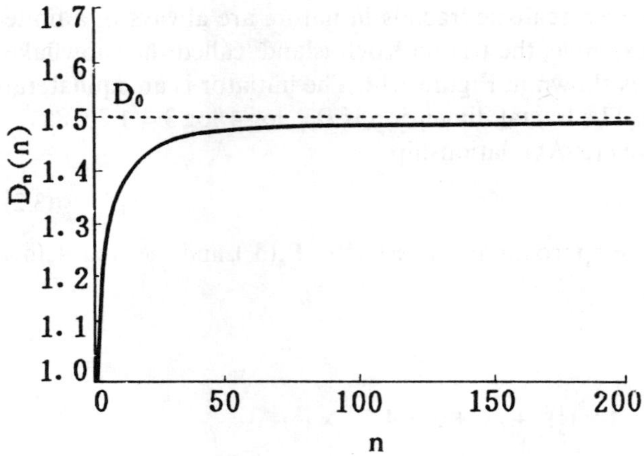

Figure 3.16. Fractal dimension $D_n(n)$ vs. the n curve for the quadratic ('quadric') Koch island (Lung & Wu, 1988).

value even under further refining of the ruler. This also means $D(n)$ would have

$$D(n > n_{max}) = D(n_{max}). \tag{3.11}$$

In all these physical situations, in fact, such a lower cutoff $P_n/P_0(n_{max})$ will be present, and an upper cutoff will also be present.

According to these relationships,

$$P_n^{1/D(n)} = \alpha_n A_n^{1/2}$$

and

$$2 \log P_n = D(n)(2 \log \alpha_n + \log A_n). \tag{3.12}$$

For the triadic Koch island, $\alpha_{n(\delta)}$ can be calculated by Equation (3.12) with P_n, A_n and $D_{(\delta)}$ from Equations (3.9-3.11). Using the computer simulation based on the regression analysis, Wang et al. (1990) have determined the change tendency of the fractal dimension D_n and the shape factor α_n with the number of steps of the construction, N, as shown in Figure 3.17. In Figure 3.17a, the curves 1 and 1′ represent the change tendency of the fractal dimension D_n and shape factor α_n with a different step number N of the construction. It is seen that D_n goes up to its true value as $n = 18$, and α_n approaches a constant as $n > 48$. Curves 2 and 3 in Figure 3.17a denote the convergence rate and means of D_n with different α_n values. Figure 3.17b shows the same change tendency of D_n and α_n with N, for the measurement of many islands, as that given in Figure 3.17a for the measurement of an island.

On the other hand, the perimeter-area relationship as a function of n can be plotted as illustrated in Figure 3.18a. It shows that the $\log P$-$\log A$ relationship is not linear, especially in the first few generations. Why do P_n and A_n data, measured by many authors using the slit island method, seem to be linear in double logarithmic coordinate plot? Lung & Zhang (1989) believed that it is a false impression. In fact, in the slit island method, one usually considers realistic islands as the initiators with different sizes, and measure their perimeters and areas. Then,

$$\begin{aligned}\log(P_{ni}/P_0) &= \log(P_{ni}/P_{0i}) + \log(P_{0i}/P_0) \\ &= \log[P_{ni}(n)] + \log r_i;\end{aligned} \tag{3.13}$$

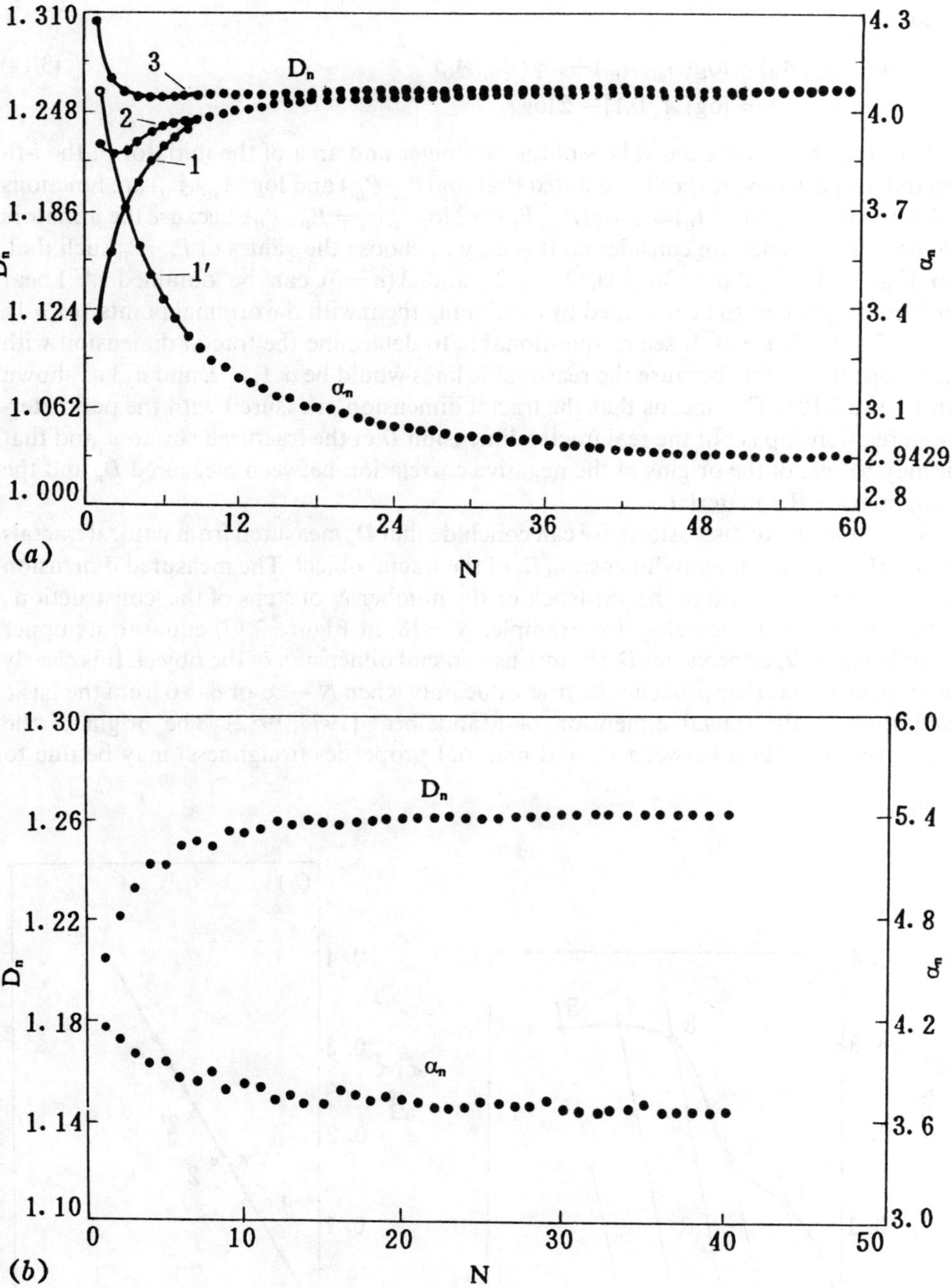

Figure 3.17. (a) The change tendency of fractal dimension D_n and shape factor α_n at the same island with the number N steps of construction; (b) The change tendency of fractal dimension D_n and shape factor α_n in the measurement of many islands with a different step number N of construction (Wang et al., 1990).

and

$$\log(A_{ni}/A_0) = \log(A_{ni}/A_{0i}) + \log(A_{0i}/A_0) = \log[A_{ni}(n)] + 2\log r_i, \tag{3.14}$$

where P_{0i} and A_{0i} are the values of the perimeter and area of the initiator of the i-th island, respectively. It should be noted that $\log(P_{ni}/P_{0i})$ and $\log(A_{ni}/A_{0i})$ are functions of n, and that $\log(A_{0i}/A_0) = 2\log(P_{0i}/P_0) = 2\log r_i, (r_i = P_{0i}/P_0)$, because the initiators of normal triangles are considered. It is easy to choose the values of P_{0i}/P_0 such that, in Figure 3.18b, data $1'(n = 1)$, $2'(n = 2)$, and $3'(n = 3)$ can be obtained. A linear relationship seems to be obtained by combining them with the original points $1(n = 1)$, $2(n = 2)$, and $3(n = 3)$. It seems questionable to determine the fractal dimension with the slope of this line because the reasonable lines would be $a_1 1$, $a_2 2$, and $a_3 3$ as shown in Figure 3.18a. This means that the fractal dimension, measured with the perimeter-area relationship is not the real fractal dimension D of the fractured surfaces, and that it may be one of the origins of the negative correlation between measured D_n and the toughness of the material.

From the above discussions, we can conclude that D_n measured from natural fractals is not the intrinsic fractal dimension, D, of the fractal object. The measured dimension depends on the length of the yardstick or the number N of steps of the 'construction', when N takes a large value (for example, $N = 18$, in Figure 3.17) equal to its upper cutoff length, D_n approaches D, the intrinsic fractal dimension of the object. It is clearly seen, in fact, that D approaches its true value only when $N \to \infty$ or $\delta \to 0$ from the basic definition of the fractal dimension of Mandelbrot (1977, 1982). The origin of the negative correlation between D_n and material properties (toughness) may be due to

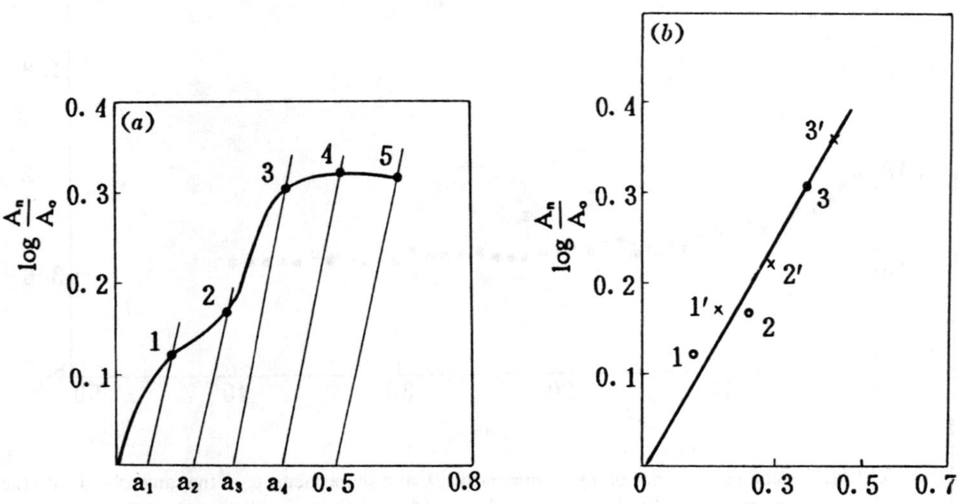

Figure 3.18. The perimeter-area relation as a function of n (a) Nonlinear Perimeter-area relation; (b) Linear perimeter-area relation. (Lung & Zhang, 1989).

fractal dimension measurement with a lower step number approximation of the fractal set.

Based on the above ideas, Lung & Wu (1988) measured D_n with different length yardsticks. For a sufficiently small yardstick length, a positive correlation between D_n and K_{1c} was observed (as shown in Figure 3.19). Figure 3.19 shows that there is a critical yardstick length, so that if the yardstick length is larger than the critical length, a negative correlation will be obtained, and if the yardstick length is less than the critical length, a positive one will be obtained. The critical length may be determined as the transition length between curves of negative and positive correlations.

Recently, Su et al. (1989) reported the result of fractal analysis of impact fracture of metal by using the slit island method with a small yardstick length $\delta = 0.7834$ μm. The metal used in their experiments was comprised of chemical components of C 0.1, Si 0.2-0.4, Mn 0.5-1.0, P 0.025, S 0.005, Cu 0.3-0.5, Ni 0.8-1.5, Cr 0.6-0.9, Mo 0.3-0.6, and V 0.04-0.08. The specimens were tempered at nine different temperatures (300-700°C), They obtained a positive correlation between the logarithms of impact toughness (IT) and fractal dimension D_n because the smaller yardstick length was used, as illustrated in Figure 3.20, showing that the logarithms of impact toughness increase monotonically with an increase of fractal dimension.

Another positive correlation between K_{1c} and the fractal dimension D_n has been reported by Mecholsky & Mackin (1988). They measured the fractal dimensions of the fractured surfaces of chert. A series of chert samples was received (sample U5) and heat-treated at 300°C (No. 3), 400°C (No. 4), and 500°C (F5), and their toughness, K_{1c}, was determined by Beauchamp & Purdy (1986). The toughness and dimensions are listed in Table 3.3, along with a microscopic and macroscopic description of the fractured surface for each treatment of the chert material. Notice in Table 3.3 that the fractal dimension decreases with heat treatment, consistent with the microscopic observations of the fracture surface roughness and also consistent with our theoretical analysis of fractal models of microfractures (Xie, 1987a, Xie & Chen, 1988c). Figure 3.21 shows a positive correlation between K_{1c} and the fractal dimension.

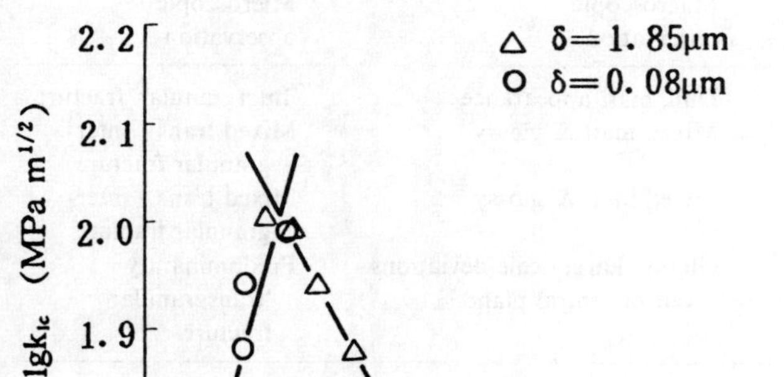

Figure 3.19. Log K_{1c} as a function of measured fractal dimension D_n for two different values of yardstick length δ (Lung & Wu, 1988).

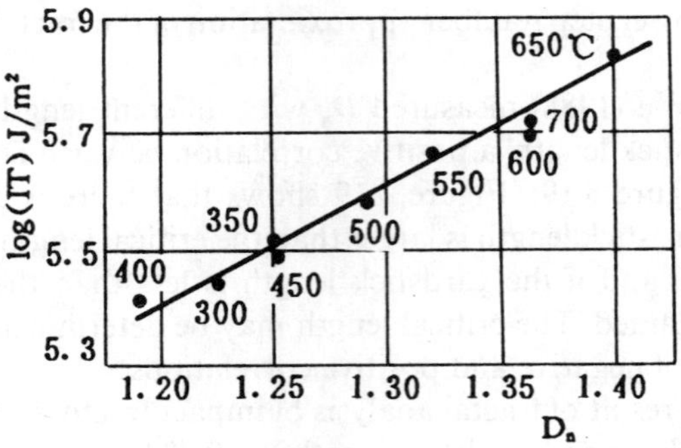

Figure 3.20. Log(IT) vs. fractal dimension D_n, specimens tempered at nine different temperatures (Su et al., 1989).

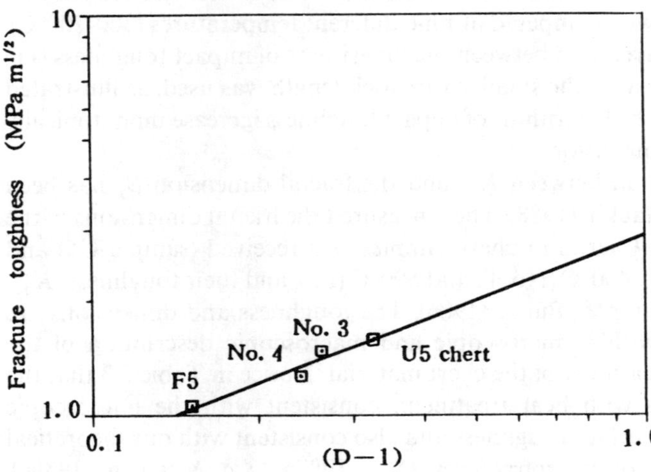

Figure 3.21. Fracture toughness as a function of the fractional part of the fractal dimension, $(D-1)$. The correlation coefficient is 0.96 (Mecholsky & Mackin, 1988).

Table 3.3. Toughness, fractal dimensions and fracture surface structure (Mecholsky & Mackin, 1988).

Sample	K_{1c} (MPa$m^{1/2}$)	Fractal dimension	Macroscopic observation	Microscopic observation
U5	1.55	1.32	Dull, matt appearance	'Intergranular' fracture
No. 3	1.46	1.26	Mixed matt & glossy	Mixed trans – inter-granular fracture
No. 4	1.25	1.24	Mixed matt & glossy	Mixed trans – inter-granular fracture
F5	1.05	1.15	Glossy. Large-scale deviations out of central plane	Predominantly 'transgranular' fracture

Lung & Zhang (1989), in another interesting study, have analyzed the origin of the negative correlation between D_n and K_{1c} by means of the micromechanism of fracture, rather than the slit island method of fractal dimension measurement. Suppose that new segments of microcracks of grain size were superimposed on the preceding larger

Table 3.4. The variation of fractal dimension of D_0 with the change of segment number N and the angle ϕ (Lung & Zhang, 1989).

ϕ (deg)	N							
	2	3	4	5	6	7	8	9
90	2.0000	1.3652	1.3333	1.2549	1.2398	1.2091	1.2000	1.1833
100	1.6247	1.2631	1.2380	1.1862	1.1747	1.1538	1.1470	1.1355
120	1.2619	1.1292	1.1158	1.0932	1.0873	1.0778	1.0743	1.0690
140	1.0986	1.0526	1.0470	1.0384	1.0360	1.0323	1.0308	1.0278

segment due to large inclusions. The fractal dimension may be calculated as

$$D = \begin{cases} \log N/\log[N\sin(\phi/2)]; & \text{as } \phi\in(2\sin^{-1}(1/N^{1/2}), \pi), \quad \text{if } N = \text{even}; \\ \log N/\log[1+(N^2-1)\sin^2(\phi/2)]^{1/2}; & \text{as } \phi\in(2\sin^{-1}[1/(N+1)^{1/2}], \pi), \quad \text{if } N = \text{odd}. \end{cases}$$

Here the angles between two adjacent segments are assumed to be equal. In this case, the fractal dimension D may decrease with an increase of the segment number N (Table 3.4).

Decreasing either the test temperature or the tempering temperature would produce materials with a high yield strength. High strength materials have a smaller critical crack length for propagation due to their low K_{1c}. It may induce a larger amount of smaller cracks (inclusions) to propagate. This makes the crack propagation between two small inclusions easier. Then, the segment number N decreases, and the fractal dimension of the fractured surface increases. This means that the fractal dimension of fractured surfaces increases with the decrease of fracture toughness. The correlation between D and K_{1c} is, therefore, negative. However, real fracture processes are quite complicated. A wide variety of mechanisms play relevant roles: grain boundaries, inclusions, second phase particles, etc. Material fractures are the total effect of many elementary processes. What process the main factor is may need to be studied further, depending on the co-action of interior and exterior conditions of the material.

It may be safe to say that, so far, we have not yet been able to clearly understand a negative correlation between fractal dimension D and material properties. However, after the discussion of the relationship between the material properties and fractal dimension, it can be said that whether the correlation between them is a negative or positive one, one at least has a more quantitative handle on the surface topography.

3.5 THE DEGREE OF STATISTICAL SELF-SIMILARITY OF FRACTURED SURFACES

In the previous two sections, we have discussed the fractal characters of fractured surfaces based on the perimeter-area relationship (Equation (3.2)). This relationship implies that the fracture surface behaves as a fractal, i.e., self-similarity. We say that a fracture surface, being a fractal, is based on the roughness and irregularity of the surface, rather than on the mathematical certification of self-similarity or statistical self-similarity of that surface. Thus, an obscure concept, all fracture surfaces have self-similarity or statistical self-similarity if they have roughness, is produced. On the other hand, fractal dimensions are estimated by using the slope of the relative curves of the double logarithms of perimeter and area data measured from the 'islands' of the fracture surface. If there is an outlier in the data, the slope measured by the theory of regression is certainly affected, hence the fractal dimension obtained from it will fall short of the realistic fractal dimension. Therefore, it is necessary to quantitatively describe how much statistical self-similarity there really is on a fracture surface, and the reliability of fractal dimensions measured from data of the log perimeter and log area. In this section we introduce some preliminary results in this respect (Zhang & Xie, 1991).

3.5.1 *Degree of statistical self-similarity*

In the fractal analysis of fracture surfaces, the coordinates of scatter points of the log-log coordinate system of perimeters and areas, determined by the vertical section or slit island method, can be defined as follows:

$$(x_1, y_1), (x_2, y_2), \ldots, (x_n, y_n). \tag{3.15}$$

Let $RXX = \sum_{i=1}^{n} (x_i - \bar{x})^2$, $RYY = \sum_{i=1}^{n} (y_i - \bar{y})^2$, $RXY = \sum_{i=1}^{n} (x_i - \bar{x})(y_i - \bar{y})$, then, the equation of the straight line is given by

$$\tilde{y} = \beta_0 + \beta_1 x, \tag{3.16}$$

where

$$\beta_1 = RXY/RXX \tag{3.17}$$

and

$$\beta_0 = \bar{y} - \beta_1 \bar{x}; \tag{3.18}$$

$\bar{x}$ and $\bar{y}$ are the average values of x_i and y_i $(i = 1, 2, \ldots, n)$, respectively. Hence, the fractal dimension $D = 1 - \beta_1$ in the vertical section method, and $D = 2\beta_1$ in the slit island method.

Given a set of data (3.15), a straight line can be determined by Equations (3.16-3.18). However, its reliability is dependent on the number of scatter points, the value distribution of the regression variable, and the cluster degree along a straight line of the scatter points. In general, the number of the scatter points and the value distribution of the regression variable X are controllable, but the cluster degree along the straight line of the scatter points is determined by the problem itself. Therefore, we must examine the cluster degree along the straight line and give a test criterion whether the fractal dimension D is accurately determined by the regression coefficient β_1 or not.

Letting $SS_{reg} = \sum_{i=1}^{n} (\tilde{y}_i - \bar{y})^2$, $RSS = \sum_{i=1}^{n} (y_i - \tilde{y}_i)^2$, we then have

$$RYY = SS_{reg} + RSS, \tag{3.19}$$

where RYY is called the total sum of squares. SS_{reg} is the sum of squares due to regression or the regression sum of squares, and RSS is the familiar residual sum of squares.

We define the sample correlation coefficient r_{xy} between x and y as follows:

$$r_{xy} = (SS_{reg}/RYY)^{1/2}. \tag{3.20}$$

Clearly, $r_{xy} \in [0, 1]$ and

$$r_{xy}^2 = SS_{reg}/RYY = 1 - RSS/RYY = \beta_1^2 RXX/RYY, \tag{3.21}$$

where

$$RSS = (1 - r_{xy}^2)RYY. \tag{3.22}$$

It is seen that the larger r_{xy} is, the smaller is RSS, and more favorably the scatter points coincide with the straight line obtained by regression. Thus, r_{xy} describes the degree of linear association between the variables x and y. We will now give the test criterion of the reliability by use of r_{xy}.

First, let the test linear hypotheses

$$y = \bar{\beta}_0 + \bar{\beta}_1 x + \varepsilon, \quad \varepsilon \sim N(0, \sigma^2);$$

i.e., the test hypotheses

$$H_0\colon \bar{\beta}_1 = 0.$$

According to the regression theory,

$$(\beta_1 - \bar{\beta}_1)[(n-2)RXX/RSS]^{1/2} \sim t(n-2); \tag{3.23}$$

where $t(\cdot)$ is the t distribution in the probability theory. If H_0 is true, $\bar{\beta}_1 = 0$, then

$$T = \beta_1[(n-2)RXX/RSS]^{1/2} \sim t(n-2). \tag{3.24}$$

If

$$|T| > t_{\alpha/2}(n-2), \tag{3.25}$$

then H_0 is untrue in the level of significance, α (in general, $\alpha = 0.05$), and this regression effect is notable, i.e., there is a good linear correlation between the data of x and y. Otherwise, this regression effect is not notable.

Since we use β_1 to estimate the fractal dimension D, β_1 is sensitive to the variation of the data, the method mentioned above is insufficient to test if a linear relationship exists between the variables x and y (for details, see the next section). Therefore, the test form (Equation (3.25)) should be modified by;

$$|T| > mt_{\alpha/2}(n-2), \tag{3.26}$$

where $m > 1$ is called the strengthening coefficient. After studying many numerical examples, we find that the experimental value of m equals 6 for $\alpha = 0.05$.

Using Equations (3.21) and (3.22), we have

$$|T| = \beta_1[(n-2)RXX/RSS]^{1/2} = [(n-2)RXX\beta_1^2/(1 - r_{xy}^2)RYY]^{1/2}$$
$$= [(n-2)r_{xy}^2/(1 - r_{xy}^2)]^{1/2};$$

then

$$r_{xy}^2 = T^2/[(n-2) + T^2].$$

Thus, an equivalent form of Equation (3.26) is

$$r_{xy} > mt_{\alpha/2}(n-2)/\{(n-2)+[mt_{\alpha/2}(n-2)]^2\}^{1/2}. \tag{3.27}$$

We define

$$r_\alpha(n-2) = mt_{\alpha/2}(n-2)/\{(n-2)+[mt_{\alpha/2}(n-2)]^2\}^{1/2}, \tag{3.28}$$

as the critical value of the correlation coefficient r_{xy} with $(n-2)$ degrees of freedom for the level of significance, α, and the strengthening coefficient m. If $r_{xy} > r_\alpha(n-2)$, there is a strong linear relationship between x_i and y_i. Thus, the fractal dimension can be accurately estimated by use of β_1, and there is a scale invariable range $[e^{\min\{x_i\}}, e^{\max\{x_i\}}]$ of the fractal set, (i.e., the lower cutoff is $e^{\min\{x_i\}}$ and the upper cutoff is $e^{\max\{x_i\}}$). Otherwise, these data are needed for continuous manipulation and analysis (a detailed) discussion is given in Section 3.5.3).

Visually r_{xy} is a measurement of the degree of the linear correlation between x_i and y_i. Where the degree of linear correlation between x_i and y_i behaves as the degree of statistical self-similarity of fractal sets. Therefore, we can give a definition of the degree of statistical self-similarity by use of r_{xy} and $r_\alpha(n-2)$ such that the statistical self-similarity is quantitatively described.

Definition 3.1: For a fractal set S, suppose that the correlation coefficient r_{xy} can be determined by n pairs of data (Equation (3.15)), and $r_\alpha(n-2)$ is the critical value of r_{xy}. The degree of statistical self-similarity of the fractal set S with the data (Equation (3.15)) for the level of significance α is then defined by

$$\rho_\alpha = e^{r_\alpha(n-2)(1-1/r_{xy})}, \tag{3.29}$$

and ρ_α is called the degree of statistical self-similarity for short.

It is clear that $r_{xy} = 1$ as $\rho_\alpha = 1$; i.e., y is a linear function of x, and the set S is self-similar. Since $r_{xy} \to 0$ as $\rho_\alpha \to 0$, i.e., there is no linear relationship between x_i and y_i, and the set S is not self-similar, this indicates that ρ_α can quantitatively described the degree of the statistical self-similarity of the set S.

3.5.2 *The fractal dimension estimate of fracture surfaces*

Figure 3.8 shows a large scattering in the data represented by Pande et al. (1987). Nevertheless, the degree of scattering in the data varies with the chosen unit of length of the abscissas and the ordinates. We replot the scatter points in Figure 3.8b by using the same length unit between abscissas and ordinates, as shown in Figure 3.22. Thus, by means of the scatter diagram, it is impossible to judge exactly whether the experimental data cluster along a straight line or not. Using the reliable test method given in Section 3.5.1, we have analyzed the data obtained from Figures 1-3 in the paper of Pande et al. (1987) and listed it in Table 3.5. From Table 3.5, we find that the statistic T is larger than $t_{\alpha/2}(n-2)$ for every group of data, and ρ_α can quantitatively characterize the degree of statistical self-similarity of every fracture surface. Although the distributions of the data given in Figures 3.8c and 3.8d look very similar, the quantitative analyses in Table 3.5 show a large difference. On the other hand, by the regression theory, a 95% confidence range can be determined by

$$\tilde{y} \pm t_{0.025}(n-2)[RSS/(n-2)\cdot(1+(1/n)(x-\bar{x})/RXX)]^{1/2}. \tag{3.30}$$

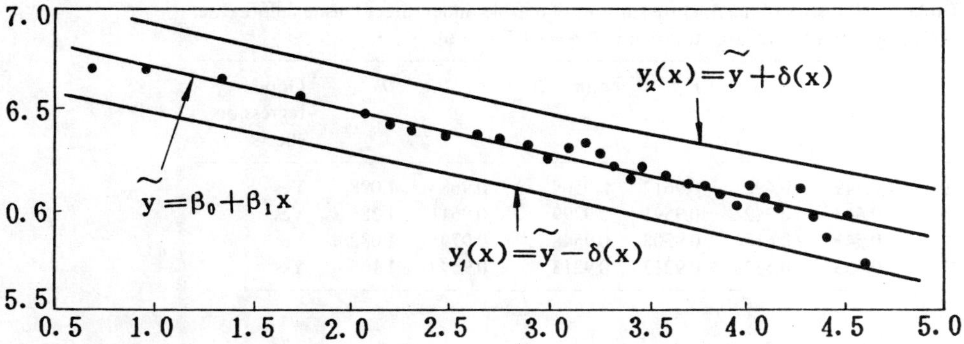

Figure 3.22. The straight line plot and 95% confidence range of the data in Figure 3.8b, by using the same length unit between abscissas and ordinates.

Table 3.5. The analysis results of the experimental data obtained by using the reliable test method ($\alpha = 0.05$, $m = 6$).

Group	n	β_1	β_0	T	$t_{\alpha/2}(n-2)$	r_{xy}	$r_\alpha(n-2)$	ρ_α	Good regression effects
1#	28	−0.135	7.402	11.40	2.0555	0.9129	0.9241	0.9156	No
2#	29	−0.210	6.909	14.56	2.0518	0.9419	0.9213	0.9448	Yes
3#	29	−0.203	7.455	6.73	2.0518	0.7913	0.9213	0.7842	No
4#	29	−0.185	7.347	11.75	2.0518	0.9173	0.9213	0.9203	No
5#	21	−0.029	5.412	5.22	2.0930	0.7678	0.9447	0.7515	No
6#	21	−0.023	5.372	9.51	2.0930	0.9089	0.9447	0.9097	No
7#	19	0.536	0.589	21.32	2.1089	0.9828	0.9508	0.9835	Yes
8#	19	0.545	0.569	15.39	2.1089	0.9659	0.9508	0.9670	Yes

Note: The data: groups 1#-4# correspond to Figures 3.8a-d, respectively; and groups 5#-6# correspond to Figure 3 in the paper of Pande et al. (1987); and groups 7#-8# correspond to Figures 3.9c and 3.9d, respectively.

Let $\delta(x) = t_{0.025}(n-2)\ [RSS/(n-2)\cdot(1+(1/n)(x-\bar{x})/RXX)]^{1/2}$, then a confidence range is the band region between the straight lines

$$y_1(x) = \tilde{y} - \delta(x) \quad \text{and} \quad y_2(x) = \tilde{y} + \delta(x);$$

in which the regression straight line is $\tilde{y} = \beta_0 + \beta_1 x$ as illustrated in Figure 3.22. For the data given in Figure 3.8b, we have

$$\delta(x) = 0.208\sqrt{[1.034 + 0.038(x - 3.278)^2]}.$$

From Figure 3.22, we can see that all scatter points lie within the confidence range. This means that the effect of the linear regression is notable for the data in Figure 3.8b. Therefore, the quantitative analysis and test method discussed here can eliminate the false impression produced from the apparent scatter diagram of the

Table 3.6. Statistics of the fracture surfaces of marble under direct tensile failure (Xie, 1987a) by using the reliable test method ($\alpha = 0.05$, $m = 6$).

n	β_1	β_0	r_{xy}	$r_\alpha(n-2)$	ρ_α	D	Good regression effects
19	0.549	0.492	0.9612	0.9508	0.9683	1.098	Yes
26	0.643	0.242	0.9591	0.9299	0.9611	1.286	Yes
19	0.541	0.617	0.9508	0.9548	0.9795	1.082	Yes
27	0.553	0.532	0.9213	0.9213	0.9542	1.106	Yes

experimental data, and become a simple and reliable method to measure fractal dimension.

Table 3.6 shows the quantitative analysis results of the data (measured by the slit island method) of fracture surfaces of marble under direct tensile failure by means of the reliable test method (Xie, 1987a). These results indicate that $\rho_\alpha > 0.95$; i.e., the fracture surfaces of marble have very good statistical self-similarity.

The detailed discussion about fractal analyses of fracture surfaces of rock materials will be introduced in Chapter 14.

3.5.3 *The analysis of residuals for the experimental data of the fracture surface*

In order to accurately enhance the measurement of fractal dimensions and search the origin of unnoticeable regression effects, we need to investigate the action of the experimental data on the fitting Equation (3.16) and find the outliers in the data, by use of residual analysis.

In Sections 3.5.1 and 3.5.2, we have used the test method to examine whether the premise conditions of the regression model for the measurement of fractal dimensions are satisfied or not. However, in the case when $r_{xy} < r_\alpha(n-2)$; i.e., the premise condition is not satisfied, what phenomenon will occur? Anscombe (1973) has given a simple example as shown in Figure 3.23. Four hypothetical data sets have the same statistics: $n = 11$, $\bar{x} = 9.0$, $\bar{y} = 7.5$, $\beta_0 = 3$, $\beta_1 = 0.5$, $RXX = 110.0$, $RYY = 41.35$, $RSS = 13.75$, $r_{xy} = 0.8165$, $r_{0.05}(9) = 0.9764$, and $r_{xy} < r_{0.05}(9)$. However, their scatter diagrams are mutually different. Residual analysis is a simple and effective method to overcome the occurrence of this kind of problems.

Let the residual of the i-th point be

$$e_i = y_i - \tilde{y}_i \quad (i = 1, 2, \ldots, n). \tag{3.31}$$

We define d_i, the standard residual of e_i, to be

$$d_i = e_i/\sigma_i;$$

where

$$\sigma_i = \{[1 + (1/n) + (x_i - \bar{x})^2/RXX](1 - r_{xy}^2)RYY/(n-2)\}^{1/2}. \tag{3.32}$$

A residual plot is the plot of d_i versus $\tilde{y}_i$, which has four kinds of fundamental forms

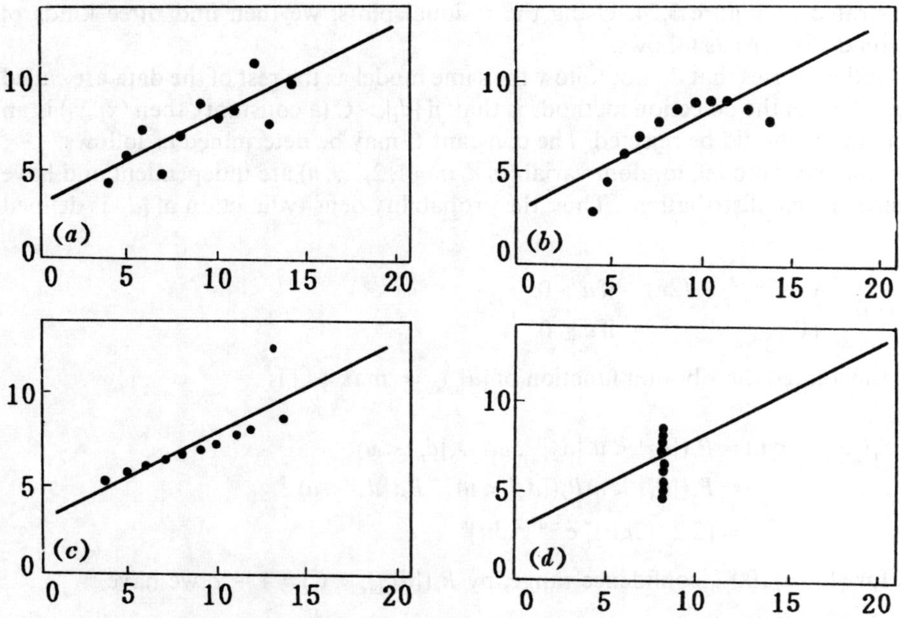

Figure 3.23. Four hypothetical data sets with the same statistics (Anscombe, 1973). (a) Data cluster along a straight line; (b) Data cluster along a curve; (c) Data with an outlier point; (d) Data with nonlinear scatter points.

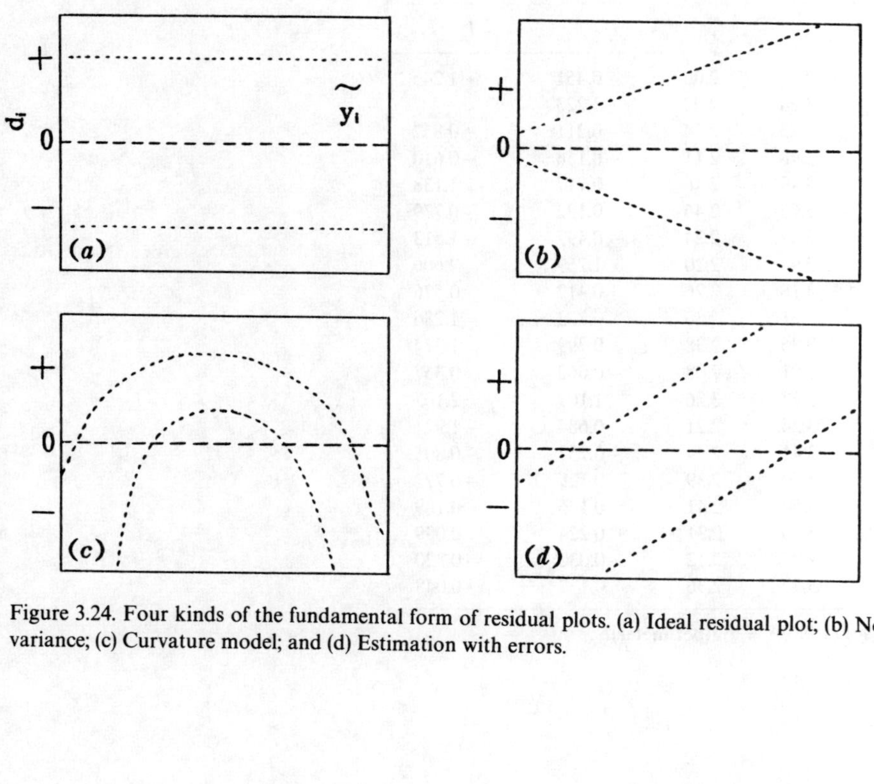

Figure 3.24. Four kinds of the fundamental form of residual plots. (a) Ideal residual plot; (b) Nonuniform variance; (c) Curvature model; and (d) Estimation with errors.

as illustrated in Figure 3.24. Using the residual plots, we then find three kinds of common deviations as follows:

(1) Outlier: Cases that do not follow the same model as the rest of the data are called outliers. One of the detection methods is that: if $|d_i| > C$ (a constant), then (x_i, y_i) is an outlier, and it should be rejected. The constant C may be determined as follows:

For the linear model, random variables d_i $(i = 1, 2, \ldots, n)$ are independent and have standard normal distributions. Thus, the probability density function of $|d_i|$ is defined by

$$f(u) = \begin{cases} 2e^{-u^2/2}/\sqrt{(2\pi)}, & \text{if } u > 0; \\ 0, & \text{if } u \leq 0. \end{cases}$$

Then, the jointed distribution function of $|d|_{\max} = \max\limits_{1 \leq i \leq n} |d_i|$ is

$$\begin{aligned} P_r(|d|_{\max} < u) &= P_r(|d_1| < u, |d_2| < u, \ldots, |d_n| < u) \\ &= P_r(|d_1| < u)P_r(|d_2| < u) \ldots P_r(|d_n| < u) \\ &= [2/\sqrt{(2\pi)} \cdot \textstyle\int_0^u e^{-u^2/2}\,du]^n. \end{aligned}$$

Thus, for $(1 - \alpha)100\%$ confidence range, by $P_r(|d|_{\max} < C) = 1 - \alpha$, we have

$$2/\sqrt{(2\pi)} \cdot \textstyle\int_0^c e^{-u^2/2}\,du = (1 - \alpha)^{1/n}$$

Table 3.7. A set of data obtained from the fracture surfaces of marble under direct tensile failure by the slit island method and their residuals (Xie, 1987a).

Point i	x_i	y_i	d_i	d_i^*
1	2.93	2.06	−0.451	−1.245
2	3.24	2.02	−3.223	—
3	3.23	2.24	−0.211	−0.852
4	2.98	2.11	−0.158	−0.610
5	3.49	2.46	0.717	1.138
6	3.63	2.46	−0.122	−0.779
7	3.27	2.24	−0.497	−1.513
8	2.94	2.20	1.259	2.606
9	3.18	2.26	0.412	0.576
10	3.21	2.30	0.733	1.286
11	3.35	2.38	0.799	1.374
12	2.91	2.08	−0.062	−0.357
13	3.27	3.26	1.107	2.102
14	3.24	2.21	−0.684	−1.921
15	3.41	2.34	−0.157	−0.805
16	3.50	2.39	−0.129	−0.772
17	3.50	2.41	0.125	−0.189
18	3.30	2.31	0.224	0.099
19	2.98	2.12	−0.030	−0.320
20	3.40	2.36	0.177	−0.048

Note: $x = \log(\text{area})$, $y = \log(\text{perimeter})$.

Hence,

$$1/\sqrt{(2\pi)}\cdot\int_{-\infty}^{c} e^{-u^2/2}\,du = [1+(1-\alpha)^{1/n}]/2, \tag{3.33}$$

or approximately

$$1/\sqrt{(2\pi)}\cdot\int_{-\infty}^{c} e^{-u^2/2}\,du = 1-\alpha/(2n). \tag{3.34}$$

Obviously, we can find the value of C from the table of the standard normal distribution function. In general, $C = 3$, for $n \geq 20$.

Note that we must be very careful to detect the outlier. An example will be introduced later.

(2) Heterogeneous variance: A band with the same width as illustrated in Figure 3.24a shows that the residuals are uniform and the variance of ε_i is a constant, while a band with a wedge of residual plots as shown in Figure 3.24b shows that the variance of ε_i is heterogeneous.

(3) The improper model: The model of linear regression is not available for the curvature residual plot, as given in Figure 3.24c. But, it may be useful to choose the model of nonlinear regression.

Moreover, the linear tendency of the residual plot as shown in Figure 3.24d means the estimation with errors.

A set of data obtained from the fracture surfaces of marble under direct tensile failure by the slit island method and their residuals are listed in Table 3.7. It is seen that there is only one point ($i = 2$) so that $|d_i| > 3$, and (x_2, y_2) is an outlier. The residual d_i^* listed in Table 3.7, represents the residuals after rejection of the outlier. The plot of residuals d_i^* versus $\tilde{y}_i$ is shown in Figure 3.25b and the statistics of these data without an outlier ($n = 19$) are listed in Table 3.6. In Figure 3.25a, the dotted line is the fitting straight line with the outlier ($n = 20$), and the solid line is the fitting straight line without the outlier ($n = 19$). This example shows that, for a scatter diagram of the original data, if the original data satisfy the condition of $r_{xy} > r_\alpha(n-2)$, then a better fitting effect is achieved by using the rejected outlier method, such that the accuracy of measurement

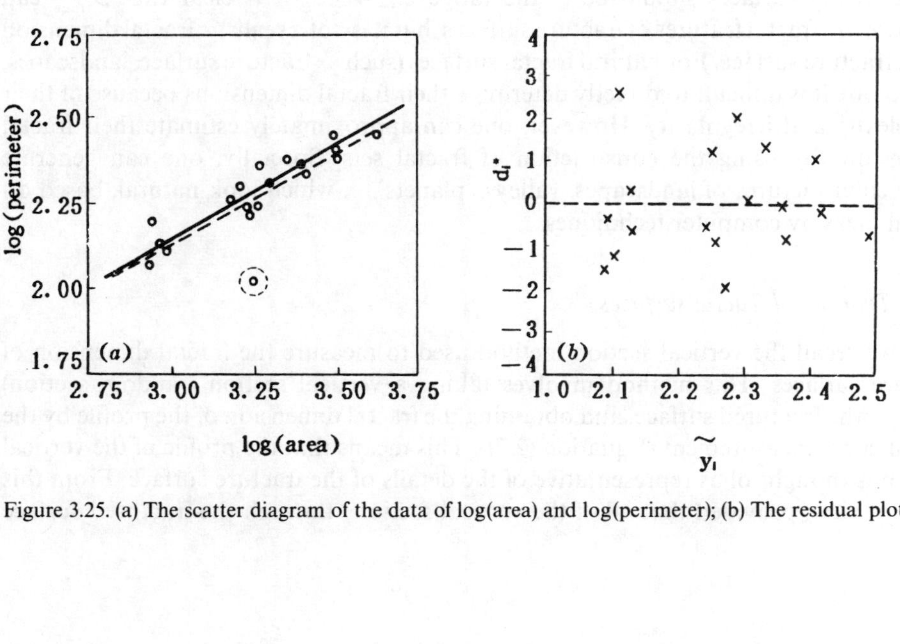

Figure 3.25. (a) The scatter diagram of the data of log(area) and log(perimeter); (b) The residual plot.

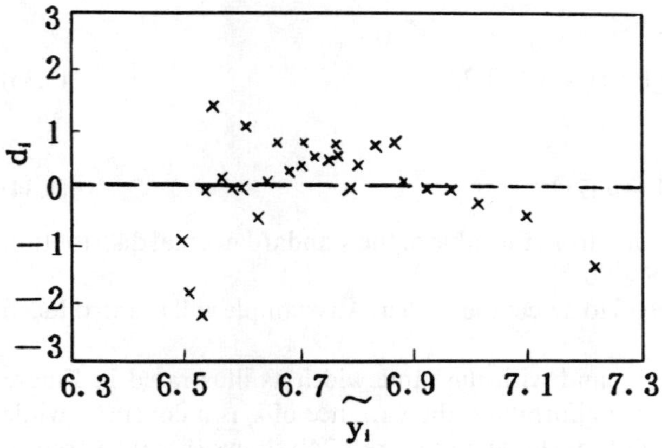

Figure 3.26. The residual plot of the curvature model, which is the same as that shown in Figure 3.24c.

of fractal dimensions is enhanced. But, when $r_{xy} < r_\alpha(n-2)$ for the original data, generally the new data after rejection of the outliers cannot still satisfy the condition $r_{xy} > r_\alpha(n-2)$. Figure 3.26 is the residual plot derived from the original data given in Figure 3.8d. It is seen that there is a nonlinear relationship between x_i and y_i, and it is impossible to satisfy the condition of $r_{xy} > r_\alpha(n-2)$. Therefore, this kind of data should be analyzed by using the nonlinear regression models.

3.6 THE FRACTAL SURFACES

The fracture surfaces may be fractal surfaces. However, in the fractal analyses of fracture surfaces discussed in the three previous sections, one uses the slit island (the perimeter-area relationship) and vertical section method, to measure the fractal dimensions of the surfaces, such that the two-dimensional problem (surface) is reduced to one dimension (line or boundary of the island). The fractal dimension determined by these methods is $D \in [1, 2]$, and we denote it by D_{trace}. In fact, the fractal dimension of the fractal surfaces should be in the range $2 \leq D \leq 3$. It is clear that D_{trace} can characterize fractal features of fracture surfaces, but it is not a realistic fractal dimension of the fracture surface. For natural fractal surfaces (such as fracture surface, landscapes, and so on), it is difficult to directly determine their fractal dimensions because of their complexity and irregularity. However, one can approximately estimate their fractal dimensions by using the construction of fractal sets. Specially, one can generate spectacular pictures of landscapes, valleys, 'planets'..., which look natural, based on fractal ideas by computer techniques.

3.6.1 *Translated fractal surfaces*

First we recall the vertical section method used to measure the fractal dimension of fracture surfaces. This method involves taking a vertical section (random section) through the fractured surface, and obtaining the fractal dimension of the profile by the fractal curve measurement (Equation (2.7)). This means that the profile of the vertical section is thought of as representative of the details of the fracture surface. From this point of view, provided that all of the vertical sections through the fractured surface

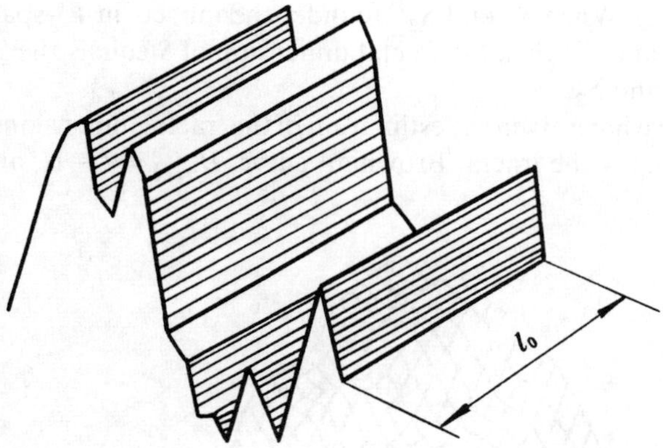

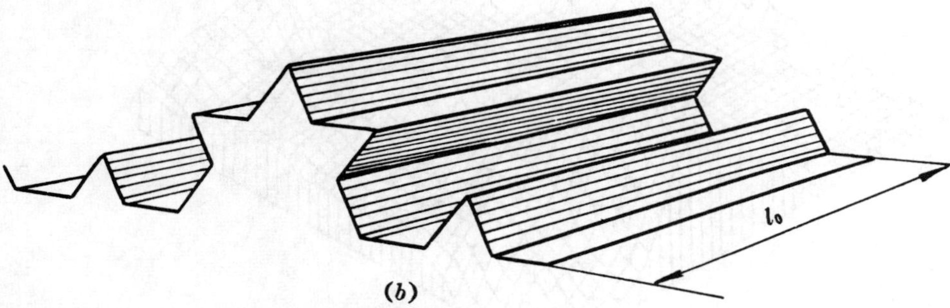

Figure 3.27. (a) The translated fractal surfaces; (b) A translated Koch fractal surface.

have the same profile, we can then easily construct a fractal surface (or fracture surface) by translating the profile along a direction perpendicular to the plane of the vertical section (or the profile) for a distance l_0, as shown in Figure 3.27a. In order to measure the area of such a surface, we would try to cover it with a rectangular of length l_0 and width δ. With the grids of small yardstick areas $\delta \times \delta$, we would need

$$N_{(\delta)} = (l_0/\delta)\cdot(a/\delta^{D_{\text{trace}}}) = al_0\delta^{-(1+D_{\text{trace}})}$$

to cover the surface. The Hausdorff's measure H^d is then given by

$$H^d = N_{(\delta)}\cdot\delta^d = al_0\delta^{d-(1+D_{\text{trace}})}.$$

This measure remains finite only when $d = D$ given by

$$D = 1 + D_{\text{trace}}. \qquad (3.35)$$

For example, the fractal dimension of the triadic Koch surface (shown in Figure 3.27b) is

$$D = 1 + 1.2618 = 2.2618.$$

Generally, there is an additive law of fractal dimensions for intersections of fractal sets

(Mandelbrot, 1982). When S_1 and S_2 are independent sets in R^n-space, and a set S is the product of S_1 and S_2, then the fractal dimension of S equals the sum of the fractal dimensions of S_1 and S_2.

In Section 2.6, we have given the estimation of the fractal dimensions of the fractional Brownian motion, for the fractal Brownian curve, $D_{\text{trace}} = 2 - H$, and for the fractal

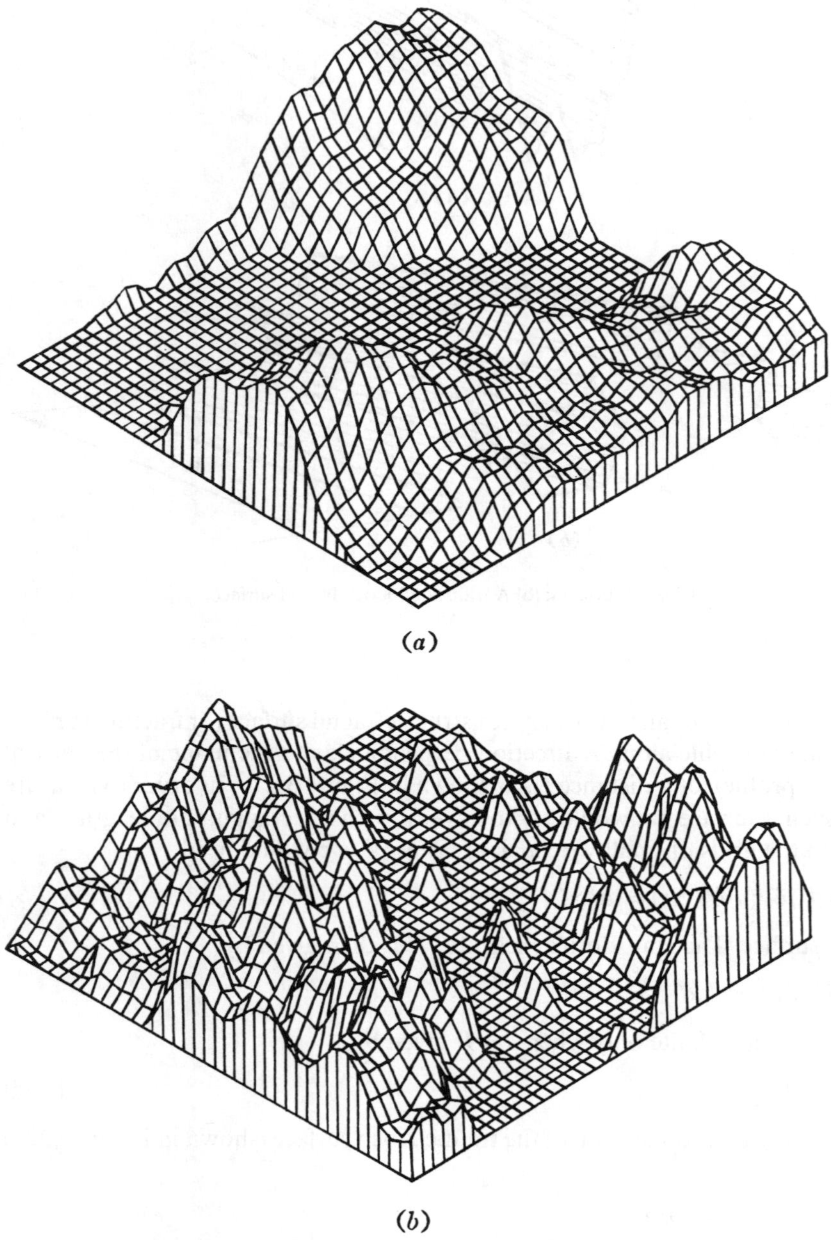

(*a*)

(*b*)

Figure 3.28. Fractional Brownian surfaces. (a) $H = 0.95$, $D = 2.05$; and (b) $H = 0.8$, $D = 2.20$ (Falconer, 1990).

Brownian surface, $D = 3 - H$. Indeed, $D = 1 + D_{\text{trace}}$, although the fractal Brownian surface is not a translated fractal surface.

3.6.2 *Brownian fractal surfaces*

A way of generating a more reasonable surface is to have a different height $Z(x, y)$, based on the translation of a fractal curve. There are many ways to define $Z(x, y)$ for generating more reasonable fractal surfaces. Falconer (1990) replaced the time variable t by the coordinate (x, y), so that the random variable $X(x, y)$ of the fractional Brownian motion may be thought of as the height $Z(x, y)$ of a surface at the point (x, y). Thus, he has effectively created fractional Brownian surfaces – or computer-generated landscapes.

For $0 < H < 1$, we define an index-H Brownian function $X(x, y): R^2 \rightarrow R$ to be a random function, such that:

(1) with probability 1, $X(0, 0) = 0$, and $X(x, y)$ is a continuous function of (x, y);

(2) for (x, y), $(h, k) \in R^2$ the height increments $X(x + h, y + k) - X(x, y)$ have the normal distribution with zero mean and variance $(h^2 + k^2)^H$. Thus,

$$P_r((X(x + h, y + k) - X(x, y)) \leq \Delta Z) \\ = (2\pi)^{-1/2}(h^2 + k^2)^{-\alpha/2} \int_{-\infty}^{\Delta z} \exp\{-r^2/[2(h^2 + k^2)^H]\}\, dr. \tag{3.36}$$

Some effort is required to demonstrate the existence of a process satisfying these conditions (i.e., the conditions of the random process of fractional Brownian motion). The correlation between the random variables $X(x, y)$ at different points is quite involved.

When $Z(x, y) = X(x, y)$ is used as the height of the generating fractal surface, some fractional Brownian surfaces that look like natural landscapes have been created by Falconer (1990, p251), as shown in Figure 3.28. It is seen that the graph obtained by intersecting $X(x, y)$ with any vertical section is that of a one-dimensional index-H Brownian function after adding a constant to ensure that $X(0) = 0$.

Goodchild (1988) and his coworkers have also created many landscapes by the fractional Brownian process that look natural (see Section 2.6).

Specifically, we should mention the work of Voss (1985) who has generated many fractional Brownian motions and fractional Brownian surfaces that look like natural landscapes by using a successive random addition algorithm. His successive random addition algorithm has the advantage that certain elevations (such as at the rim) may be fixed in advance. Thus one may use his algorithm to perform conditional simulations in which information on the elevations at some locations is given. In addition, this method can be iterated to any desired resolution.

3.6.3 *Feder-generated landscapes*

Feder (1988) has discussed in detail the generation of a fractal surface. He thought that a simple way of generating more reasonable surfaces is to add to the height $Z(x, y)$, which is generated by the translation of successive fractal curve layers of similar profiles obtained by rotating the first surface. Let $Z_D(x, y)$ be the surface profile generated by translating a fractal curve of dimension D in the (x, y) plane along the y-axis. We can then rotate this surface at an angle ϕ in the (x, y) plane to generate the surface

$Z_D(x, y|\phi)$, and finally we multiply the height by a factor h to define the surface $Z_D(x, y|h, \phi) = hZ_D(x, y|\phi)$. Thus, the height $Z(x, y)$ of generating fractal surfaces is defined by

$$Z(x, y) = \sum Z_D(x, y|h, \phi). \tag{3.37}$$

For a landscape, we only see a certain area – this defines an upper length scale L_{max}. In addition, we have a minimum length scale L_{min}, determined by the resolution of the eye or photographic film.

In order to generate surfaces that look like natural landscapes, Feder (1988) defined $Z(x, y)$ as

$$Z(x, y) = C_f \sin(fx), \tag{3.38}$$

where f is a spatial frequency, and C_f represents the amplitude. Note that the lowest frequency of interest is $f_{min} \propto 1/L_{max}$, since lower spatial frequencies simply appear to be constant additions to the height over the entire surface, and the highest spatial frequency of interest is $f_{max} \propto 1/L_{min}$, since finer detail cannot be resolved.

Feder tried to generate translation surfaces from superpositions of the oscillating function. First, he chose the spatial frequencies in the scaling form,

$$f_i = f_{j-1}^{\alpha} = f^{\alpha}, \quad (j = 1, 2, \ldots), \tag{3.39}$$

to define the frequency spectrum. This spectrum is discrete and defined by the exponent α and the lowest frequency f_0. To produce visible effects, the basic frequency f_0 must be larger than f_{min}. Increasing f_0 will result in pictures in which more of the landscape is seen in the observation window. To a large degree, the parameter α controls the type of landscape generated. Feder chose $\alpha \in [0.7, 1.4]$ here. Note that low values of α tend to generate alpine landscapes, whereas high values tend to generate rather smooth landscapes.

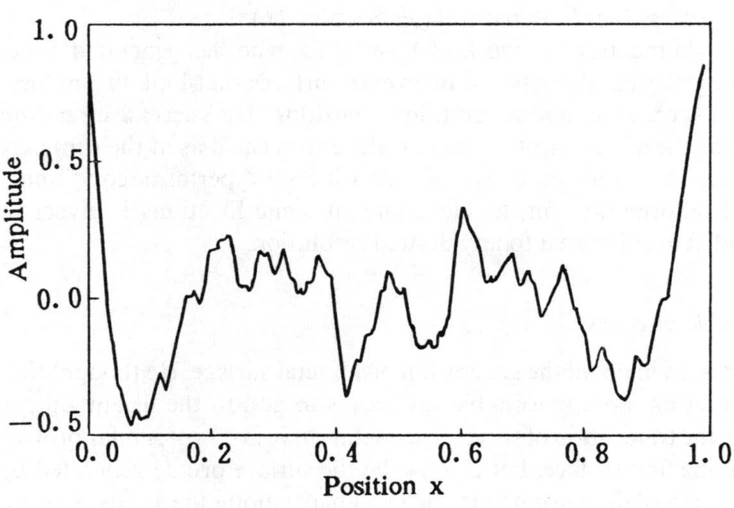

Figure 3.29. The profile $Z(x, 0|\psi = 0.616)$ (Feder, 1988).

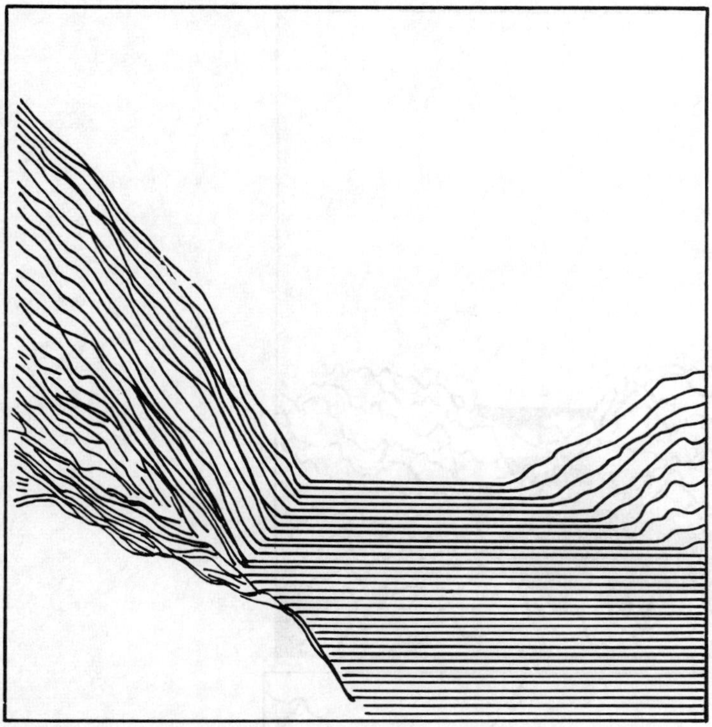

Figure 3.30. A landscape generated by 12 layers, where $\alpha = 1.1$ and $\beta = 1.1$ (Feder, 1988).

Figure 3.31. Landscape with respective and curvature, where $\alpha = 1.4$ and $\beta = 1.1$ (Feder, 1988).

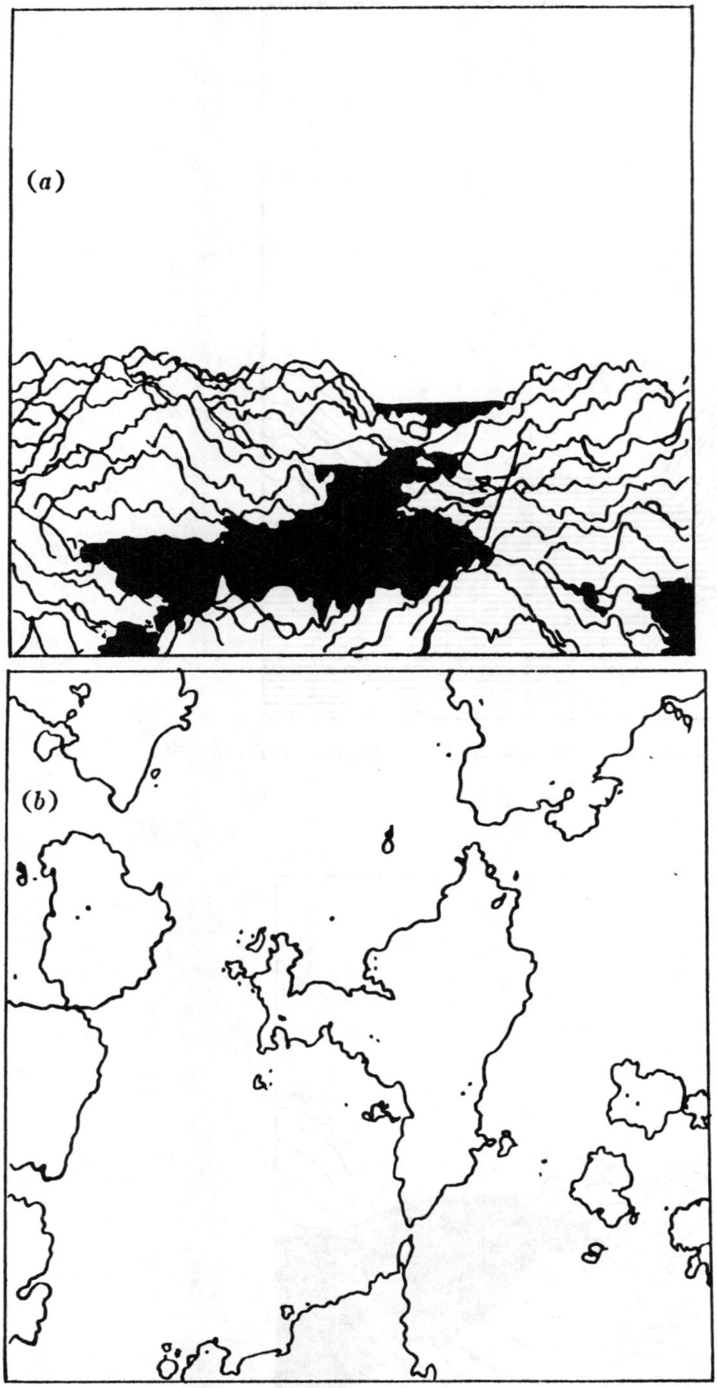

Figure 3.32. Fractal landscape of a coastline map, where $\alpha = 0.8$ and $\beta = 2.0$, the coastline fractal dimension $D = 1.15$. (a) Fractal landscape; (b) The coastline map. (Feder, 1988).

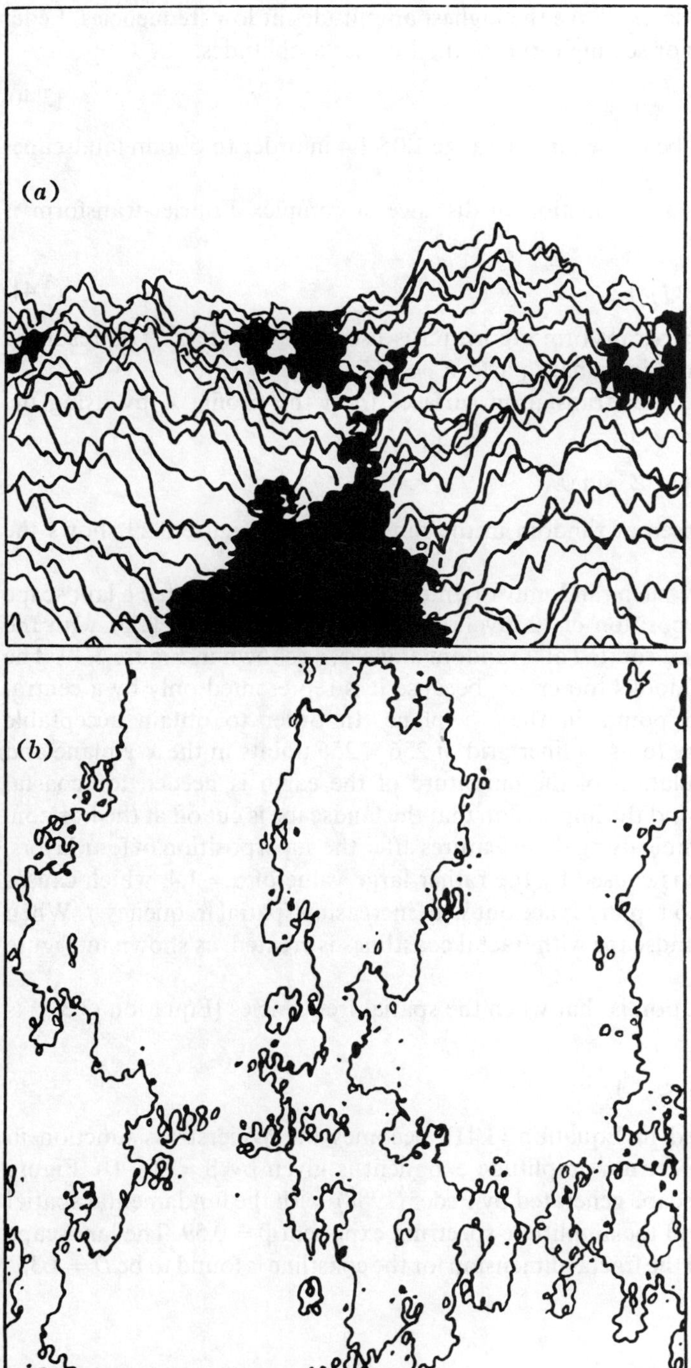

Figure 3.33. Weierstrass-Mandelbrot landscape; $\beta = 0.59$ and $b = 4.7$. The coastline fractal dimension $D = 1.35$. (a) The landscape; (b) The coastline map. (Feder, 1988).

Considering that landscapes have the highest amplitudes at low frequencies, Feder has chosen a self-similar or scaling form for the Fourier amplitudes,

$$C_f \propto f^{-\beta}. \tag{3.40}$$

The parameter β should be chosen in the range 1.05-1.4 in order to obtain landscapes that look 'natural'.

To obtain the height as a function of distance, a complex Fourier-transform is performed by

$$Z(x) = \Sigma C_{f_j} \exp(2\pi f_j x). \tag{3.41}$$

The standard fast Fourier-transform algorithm is used to evaluate Z. The resulting surface profile has real and imaginary parts: $Z' = RZ$ and $Z'' = JZ$.

Thus Feder generated real translation surfaces from the profile Z by using the expression

$$Z(x, y \,|\, \psi) = Z' \cos\psi + Z'' \sin\psi.$$

where ψ is an angle chosen at random in the range 0 to 2π. Figure 3.29 shows the profile $Z(x, 0 \,|\, \psi = 0.616)$.

By the addition of a few such randomly oriented layers, Feder generated a landscape resulting from the superposition of 12 layers of height profiles $Z(x, y \,|\, \psi, \phi)$, with the random phase angle ψ and rotated at a random angle ϕ, as shown in Figure 3.30. The landscape in Figure 3.30 looks too crude, because it is represented only by a central portion of 64×64 grid points in the x-y plane. In order to obtain acceptable landscapes, it is necessary to use a finer grid of 256×256 points in the x-y plane and add perspective. A simulation of the curvature of the earth is needed for coastal landscapes in order to avoid the impression that the landscape is cut off at the horizon. Figure 3.31 gives an example using these features after the superposition of four layers. The smooth appearance is caused by the rather large value of $\alpha = 1.4$, which causes the frequency spectrum to rapidly space out with increasing spatial frequency f. When $\alpha = 0.8$, a rather rough landscape with fractal coastlines is created, as shown in Figure 3.32.

An interesting observation is that when the spatial frequencies (Equation (3.39)) is replaced by

$$f_j = b f_{j-1}, \quad (j = 1, 2, \ldots)$$

with $f_0 = 1$, Z as defined in Equation (3.41), becomes the Weierstrass function in Equation (2.18), if the spectrum amplitude exponent is given by $\beta = (2 - D)$. Figure 3.33 shows a fractal landscape generated by Feder (1988), with the fundamental spatial frequency $f_0 = 4.7 = b$ and the amplitude spectrum exponent $\beta = 0.59$. The landscape looks quite irregular, and the fractal dimension for the coastline is found to be $D = 1.35$.

CHAPTER 4

Random fractals

Fractal sets in nature are not only statistically self-similar, but are also always random. It may be said that, random fractal sets are certainly statistically self-similar where enlargements of small parts have the same statistical distribution as the whole set. Therefore, we should use the language of the probability theory to precisely describe the fractal sets involving infinite random processes.

4.1 RANDOM CANTOR SETS

A specific statistically self-similar construction is the randomized linear dusts presented by Mandelbrot (1982, Chapter 23) and Chiles (1988).

In Section 2.4, we used the regular operation to construct many classical fractal sets (the triadic Koch curve, the triadic Cantor dust, and so on). We will now use the irregular operation, in which we may choose one generator at random each time to create random Cantor sets (Chiles, 1988).

4.1.1 *Random linear dust: Model 1*

As in the construction of the triadic Cantor Set, let us start from interval $[0, 1]$. The irregular operation at each step of the construction is defined by the fact that each remaining segment is subdivided randomly into b parts, and the number $c < b$ of the parts will be kept at random, as shown in Figure 4.1a.

The choice of the c subintervals is made by tossing with equiprobability among the possible combinations between b and c. The fractal dimension can be estimated by $\log c/\log b$, but b and c must be integer numbers. Note that the fractal dimension in this model is a constant.

4.1.2 *Random linear dust: Model 2*

Let p be the probability that the number c of subintervals will be kept at random. The irregular operation at the i-th step of the construction is defined by each remaining segment being subdivided randomly into b parts, and the number $c = pb$ in the range

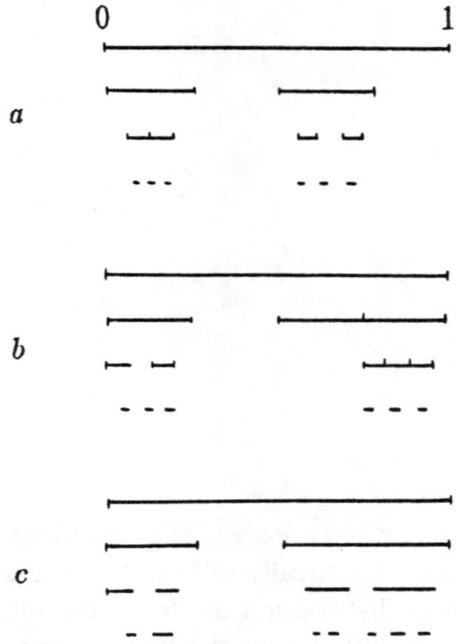

Figure 4.1. Random Cantor dusts. (a) Model 1; (b) Model 2; (c) An intermediate model (Chiles, 1988).

from 0 to b of the parts will be kept at random, as shown in Figure 4.1b. This model has the following properties;

(a) if $c = pb \leq 1$, the remains will be an empty set of points, and $D = 0$;

(b) if $c = pb > 1$, the remains will be random dust with a nonzero probability of extending to an infinite number of generations, where

$$D = \log c/\log b = 1 + \log p/\log b. \tag{4.1}$$

4.1.3 *An intermediate model*

Model 1 has the advantage of always yielding random dust; then, at each stage, c subintervals are selected. Model 2 has the advantage of allowing any D value but can lead to an empty set. In order to best simulate the fracture network, when an interval contains one or more points, at least one of the b subintervals need to contain a point. Chiles (1988) has introduced an intermediate model where survivals are tossed, each with a probability p, however, at least one needs to survive (as shown in Figure 4.1c). This is simple to achieve: only an independent toss has to be made for each subinterval (as shown in Model 2) and, in the case where no subinterval is selected, these tosses must be cancelled and the toss repeated. At each elementary operation, the number c of the subinterval has the expectation that

$$E(c) = pb/[1 - (1 - p)^b],$$

and

$$D = \log E(c)/\log b = 1 + \log\{p/[1 - (1 - p)^b]\}/\log b. \tag{4.2}$$

If b is a constant, one has only to adapt the probability p with the scale: starting from a segment of length L, at the i-th step subintervals, let length $r_i = L/b^i$, and thus, the probability p_i can be chosen in such a way as to give the correct local dimension $D(r_i)$.

Specifically Chiles has applied this model to two- or three-dimensional simulation of the fracture network. He considered a simple case of fractures or traces with the same orientation (so the fractures are separated into several sets corresponding to the various main orientations). If traces are discretized by their intersections with equispaced lines orthogonal to the fracture traces (or in the three-dimensional case by using a network of lines orthogonal to the fractures), the fractures will simulate correlated linear dusts, when a point is on a given line, a certain probability exists that the corresponding point occurs on neighboring lines. This probability is linked to the trace length distribution. Two parallel lines a distance h apart are used to intersect the fracture traces, as shown in Figure 4.2. Let $f(L)$ be the probability density of traces of length L. A simple calculation (Chiles, 1988) shows that the probability for a trace intersecting the first line to intersect the second one is

$$t = (1/E(L)) \cdot \int_h^\infty (L-h) f(L)\, \mathrm{d}L. \tag{4.3}$$

On the other hand, when simulating on one line up to n iterations the selection probability of a given final subinterval is

$$P = \prod_{i=1}^{n} P_i / [1-(1-P_i)^b], \tag{4.4}$$

where P_i ($i = 1, 2, \dots, n$) are values associated with the local dimensions $D(r_i)$. Among the selected subintervals, if a proportion t must be on both lines, the coefficients α_i may be defined such that;

$$\prod_{i=1}^{n} \alpha_i P_i / [1-(1-\alpha_i P_i)^b] = tP. \tag{4.5}$$

In order to simulate on both lines with the correct correlation, one has only to select a subinterval of the first line at each step and the corresponding subinterval of the second line with the following procedure:

(1) Select a random value u with a uniform distribution on $[0, 1]$;

(2) If $u \leq \alpha_i P_i$ the subintervals of both the lines are selected;

If $\alpha_i P_i < u < P_i$ the subinterval of line 1 is selected and another selection is made for the other line;

If $u > P_i$ the subinterval of line 1 is not selected, and another selection is made for line 2;

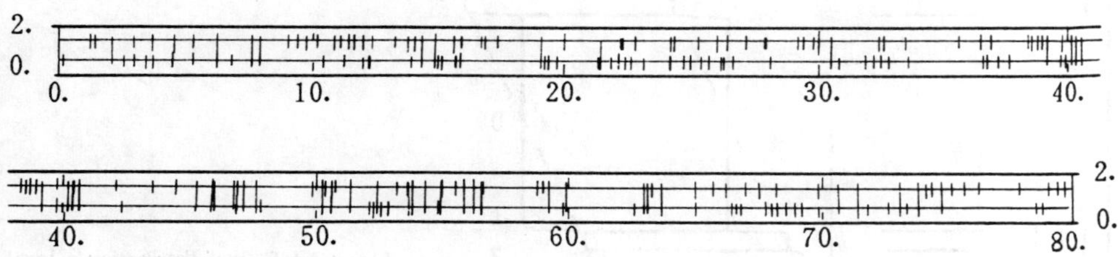

Figure 4.2. Fractal simulation of a set of subvertical fractures on a 80 m × 2 m drift wall. Only the intersections with two horizontal lines are simulated. The distance between these two lines corresponds to the mean trace length. The figured trace lengths are conventional (Chiles, 1988).

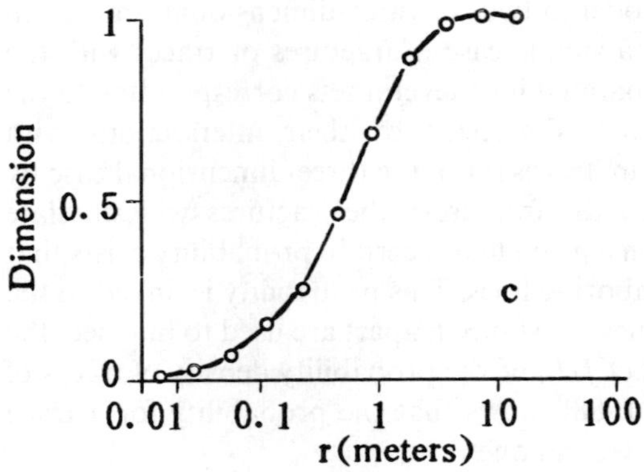

Figure 4.3. The graph of the local similarity dimension of the fractures in one-dimensional surveying along a 80 m section of drift wall (Chiles, 1988).

(3) The second selection for line 2 (when $u > \alpha_i P_i$) is made with a random number u' with a uniform distribution on $[\alpha_i P_i, 1]$; the subinterval is selected if $u' \leq P_i$.

Chiles has analyzed the fracture traces as shown in Figure 4.2 by this procedure and determined the local similarity dimension with $b = 2$ for successive subintervals, i.e., by subdividing each interval into two subintervals, as illustrated in Figure 4.3.

Falconer (1990) has given another construction and calculated related statistics. In fact, if we know the rule of the construction of the random fractal set, we can generate many of random Cantor sets, random Koch curves, random Sierpinski gaskets, and so on, and apply them in the simulation of engineering problems. Recently, Hornbogen (1989) has introduced the application of the random Cantor dusts to the simulation of the local shear field of materials. It is well known that the onset of plastic deformation at the yield stress is due to the formation of dislocation and to their tendency to form pile ups or tangles in the interior or steps in the surface. This is usually regarded as a random process. Slip line structures are characterized by the spacing S_s and height h. A macroscopic plastic shear r_p results from the sum of microshears (h/S_s) in all slip lines per unit volume (as shown in Figure 4.4),

$$\tan r_p = \Sigma(h/S_s). \tag{4.6}$$

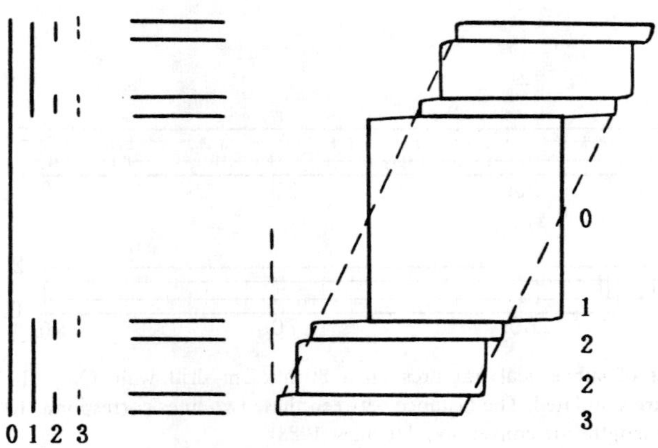

Figure 4.4. Fractal distribution of local shear (spacing of slip steps and amount of shear), $N = 2$, $1/r = 4$, and $D = 0.5$ (Hornbogen, 1989).

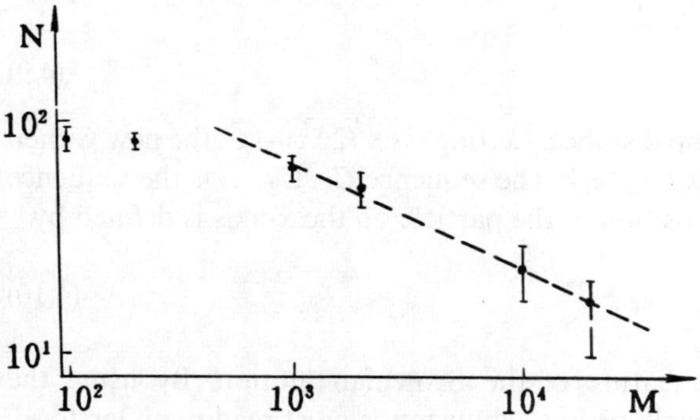

Figure 4.5. Fractal analysis of slip in C_d (99.99$_{at}$%); single crystal orientated for single slip and deformed at room temperature, the log-log plot of slip step density or spacings, and the fractal dimension $D = 0.49 \pm 0.03$ (Hornbogen, 1989).

One can measure the distribution of the spacing S_s or the number of slip steps found at a certain magnification M. If the distribution of local shear were self-similar and, consequently, fractal, it should resemble a random Cantor dust with $D < 1$. In an early investigation in this field, C_d single crystals were deformed under uniaxial tension. The microscopic investigation was conducted in the range $100 < M < 20\,000$. The number of slip steps counted and plotted as a function of M led to a fractal dimension of $D = 0.49$ (reproduced in Figure 4.5). Subsequent work on C_u single crystals led to a similar result. The value of the fractal dimension is in the range $0.4 < D < 0.7$. Clearly, the structure of the random Cantor sets may be useful for the simulation of the slip line system of materials with single crystals. However, it must be stated that fractal analyses of slip line patterns is a rather unexplored field, because the slip distribution does not depend only on the orientation and amount of strain, but also on the stacking fault energy and state of precipitation. A more reasonable simulation of slip lines in different systems should be realized by a different type of random fractal structure models.

4.2 RANDOM WALK IN ONE-DIMENSION

The so-called random walk in one-dimension occurs when the 'particle' moves on a line (x-axis) by jumping a step length of $+C$ or $-C$ every τ seconds. In the diffusion process, C may be regarded as some microscopic length (such as the particle diameter), and τ is said to be a microscopic time – the collision time.

Consider the step-length C to have a Gaussian or normal probability distribution;

$$P_r(C,\tau) = 1/\sqrt{(4\pi\mathfrak{C}\tau)}\exp[-C^2/(4\mathfrak{C}\tau)]. \tag{4.7}$$

Then, the random walk process is that, at time interval τ, a step-length C is randomly chosen so that the probability of finding C in the range $[C, C+dC]$ is $P_r(C,\tau)dC$. A sequence of such steps $\{C_i\}$ is a set of independent Gaussian random variables. The variance of this process is

$$\langle C^2 \rangle = \int_{-\infty}^{+\infty} C^2 P_r(C,\tau)dC = 2\mathfrak{C}\tau \tag{4.8}$$

where $\mathfrak{C}$ is the diffusion coefficient, and

$$\mathfrak{C} = (\tfrac{1}{2}\tau)\langle C^2 \rangle, \tag{4.9}$$

where $\langle C^2 \rangle$ is the mean square jump distance. Letting $C/\sqrt{(2\mathfrak{C}\tau)} \rightarrow C$, the new C then has a zero mean and a variance of $\langle C^2 \rangle = 1$. The sequence $C_1, C_2, \ldots$ is the sequence of steps of the random walk. The position of the particle on the x-axis is defined by

$$X(t = n\tau) = \sum_{i=1}^{n} C_i. \tag{4.10}$$

Note that the random walk is a feature of the Brownian motion. By using the Brownian motion, or the random walk process, diffusion is most readily understood. Nevertheless, we observe the Brownian motion only at intervals $b\tau$, rather than with infinite resolution, where b is an arbitrary constant. Thus, Equation (4.7) should be generalized by

$$P_r(C, b\tau) = 1/\sqrt{(4\pi\mathfrak{C}b\tau)}\cdot\exp[-C^2/(4\mathfrak{C}b\tau)]. \tag{4.11}$$

Similarly, we have a zero mean ($\langle C \rangle = 0$) and

$$\langle C^2 \rangle = 2\mathfrak{C}t, \quad \text{(here } t = b\tau). \tag{4.12}$$

This means that the record of the Brownian motion looks 'the same' under a change of resolution (comparing Equation (4.8) with Equation (4.12)), which is called a scale invariance or symmetry of the Brownian record.

In Equation (4.7), if we respectively replace $C^* = b^{1/2}C$ and $\tau^* = b\tau$, then Equation (4.7) becomes

$$P_r(C^* = b^{1/2}C, \tau^* = b\tau) = b^{1/2}P_r(C, \tau). \tag{4.13}$$

Where the prefactor b ensures that the probability density is properly normalized:

$$\int_{-\infty}^{+\infty} P_r(C^*, \tau^*)\mathrm{d}C^* = \int_{-\infty}^{+\infty} P_r(C, \tau)\mathrm{d}C = 1.$$

Equation (4.13) shows that the Brownian random process is invariant in distribution under a transformation that changes the time scale by b and the length scale by $b^{1/2}$. Clearly, the one-dimensional random walk or Brownian random process is a self-affine fractal. The position function and fractal dimensions of the Brownian random process have been discussed in Section 2.6.

An interesting application of an one-dimensional random walk to the fracture of materials has been considered by Takayasu (1982). He thought that the crack profile would be considered as a random walk in a plane. Takayasu's analysis of an inertial random walk on the x-axis shows that its fractal dimension in the (x, t)-plane approaches the value of 2 with an increasing measurement scale.

4.3 FRACTAL PERCOLATION

Percolation theory originates from the first paper of Broadbent & Hammersley (1957). This theory satisfactorily describes many systems exhibiting a purely geometric phase transition. Examples include: the conductor–insulator transition in mixtures of conducting and insulating particles, the spread of water displacing oil in porous rocks, the propagation and interconnection of cracks and fractures in rocks and engineering materials, etc. The theory is used to describe the elasticity of polymer gels, hopping conduction in doped semiconductors, Anderson localization in disordered systems, and many other applications. Presently, there is a vast amount of literature on percolation processes. A very nice introduction, written by Stauffer (1985) is available. Aharony (1986), Sokolov (1986), and Aharony & Stauffer (1987) have given concise presentations and discussed several important applications. Recently, Feder (1988), Wong (1988), Hornbogen (1989), and Falconer (1990) gave many essential details, and some fractal structures of the percolation process.

We will now consider the simplest lattice problems in the percolation theory for introduction of the basic idea of the percolation theory. A lattice can usually be regarded as a set of sites and bonds. A given site can be labeled (for example, it can be of black color) with a probability p. A set of black sites connected to one another will be called a black cluster. Clearly, if $p = 0$, the system has no black clusters, whereas for $p \ll 1$ the black clusters mainly denote sets with small number of sites. On the other hand, if $p = 1$, then all sites are black. This means that for $1 - x \ll 1$, there is an infinite black cluster in the system. There should be a critical value p_c at which a transition takes place from one regime to the other, i.e., when an infinite cluster first appears. This process is shown in Figure 4.6. A similar formulation can be used to deal with the problem of bonds, in which we regard a bond as being labeled with a probability p. Here we define a cluster as a set of sites linked by labeled bonds.

The formulation of an infinite cluster represents a phase transition. The order parameter of this transition is the power of an infinite cluster, defined as the probability that a site belongs to such a cluster. The critical behavior of this quantity is described by

$$P_\infty \propto (p - p_c)^\beta \quad \text{for } p > p_c \quad \text{and} \quad p \to p_c; \tag{4.14}$$

Let $\tau = (p - p_c)/p_c$ be the dimensionless deviation of the concentration from the critical

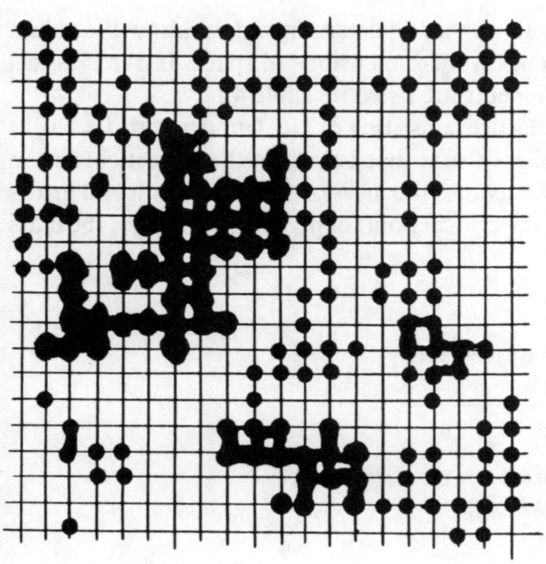

Figure 4.6. The percolation cluster in a lattice.

value. The average number of sites in a finite cluster behaves as follows in the limit as $\tau \to 0$:

$$S \propto |\tau|^{-\gamma} \tag{4.15}$$

We introduce a generalized specific heat c

$$c \propto |\tau|^{-\alpha} \tag{4.16}$$

where

$$c = (\partial^2 F/\partial \tau^2)|_{h=0}; \quad F(P, h) = \Sigma n_S e^{-Sh}; \tag{4.17}$$

and S is the number of sites, n_S is the fraction of clusters with this number of sites, and h is the field quantity with a specific meaning in the percolation theory.

We regard 'the correlation length' C as the characteristic spatial scale of the system. Assume that C is the characteristic size of a black cluster when $p < p_c$ and the characteristic size of the voids in the cluster when $p > p_c$. Thus, the critical behavior of this quantity is obtained by

$$C \propto |\tau|^{-\mu}, \tag{4.18}$$

if $h \neq 0$, a further exponent δ is introduced by

$$P_\infty(\tau = 0, h) = h^{1/\delta}. \tag{4.19}$$

All the critical exponents introduced above (so called thermodynamic) are linked by the useful scaling relationships:

$$\alpha = 2 - d_E \mu = 2 - 2\beta - \gamma,$$
$$\delta = \gamma/\beta - 1,$$
$$\eta = 2 - \gamma/\mu,$$

and

$$d_E \mu = 2\beta + \gamma. \tag{4.20}$$

The values of all the exponents given above can be determined if we know the values of two of them, and they are dependent on the dimensions of space. In the classical range $d_E \leq 6$, the values for all the exponents are listed in Table 4.1.

Let us consider again the quadratic lattice consisting of $L \times L$ nodes with $L = 20$, as shown in Figure 4.6. $M(L)$ is the number of sites that belong to the largest cluster on the $L \times L$ lattice. We can find that $M(L)$ denotes the mass of the percolation and grows with the size L of the lattice. In fact, at the critical point of percolation, $p = p_c$, the mass

Table 4.1. Principal critical exponents in percolation theory (Sokolov, 1986).

d_E	2	3	4	5	6
μ	4/3	0.88	0.7	0.6	1/2
β	5/36	0.40	0.5	0.7	1
t	1.29	1.7	2.4	2.7	3

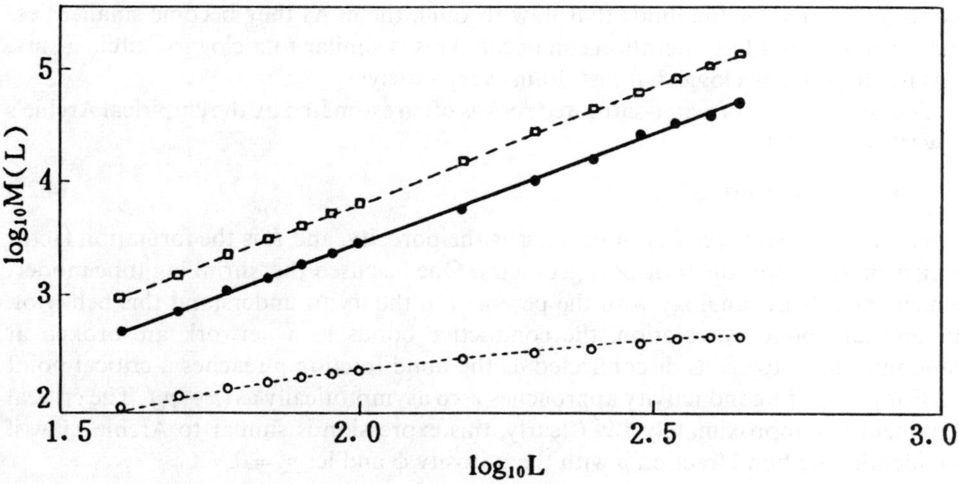

Figure 4.7. The mass of the largest cluster as a function of the linear dimension L of the quadratic lattice. The filled circles are for $p = p_c = 0.0593$. The solid line is $M(L) = AL^D$ with $D = 1.89$. For $p = 0.65$ (open boxes) a fit (dashed line) gives $D = 2.03$. The results for $p = 0.5$, i.e., below p_c, have been fitted to the form $M(L) = A + B\ln L$, shown as the dotted line through the open circles (Feder, 1988).

of the percolation increases with L as the power-law:

$$M(L) \propto L^D, \quad \text{near the critical point } p = p_c. \tag{4.21}$$

Feder (1988) has given the results of simulations on the square lattice, as illustrated in Figure 4.7. These results show that the percolation cluster at the critical point is fractal, or statistically self-similar, with fractal dimension D. Note that D is also called the cluster dimension (or mass dimension). $M(L)$ can be regarded as the mass of the cluster that consists of black balls or circles (black sites) with a diameter less than L (the same as Equation (2.29)).

The relationship between the critical exponents and the cluster dimension D have been obtained as follows (Feder, 1988; Sokolov, 1986; and others):

$$D = d_E - \beta/\mu. \tag{4.22}$$

One of the applications for the percolation theory is the problem of electrical conductivity of random networks (conductivity of mixtures of conducting and insulating particles, Sokolov, 1986). In the simplest case, such a system can be modeled in the bond problem: each whole bond has the resistance $r = 1$, whereas each broken bond has the resistance $r = \infty$. The resistance of a large part of such a network located between two perfectly conducting plates behaves as follows:

$$R = \begin{cases} \infty, & \tau \leq 0 \\ R_0\tau^{-t}, & \tau > 0 \end{cases}.$$

The values of t are listed in Table 4.1.

Wong (1988) has applied the percolation theory to describe porosity and conductivity of rock. He thinks that rocks have a zero percolation threshold. The reason is that the pores are altered by the fluids that flow through them. As they become smaller, less fluid can flow and less alteration can occur. This is similar to a clogged kitchen sink, it is never perfectly clogged, it just drains very slowly!

The conductivity of water-saturated rock is often estimated by the empirical Archie's law (Wong, 1988):

$$\sigma_r = \sigma_w/F \propto a\sigma_w\phi^{-m}, \tag{4.23}$$

where σ_w is the conductivity of water, ϕ is the porosity, and F is the formation factor which characterizes the transport geometry. One has used the 'shrinking tube model', which exploits the analogy with the percolation theory to understand this behavior. In ordinary bond percolation, the conductive bonds in a network are broken at random. The network is disconnected as the bond fraction p reaches a critical point p_c. For $p \geq p_c$, the conductivity approaches zero asymptotically as $(p - p_c)^t$. The critical exponent t is approximately 1.9. Clearly, this expression is similar to Archie's law if we identify the bond fraction p with the porosity ϕ and let $p_c = 0$.

Ambegaokar et al. (1971) showed in their work on electron hopping in amorphous semiconductors that transport in a random system with a broad distribution of conductances is dominated by conductances with magnitudes greater than some characteristic value g_c. The characteristic conductance g_c is the largest conductance such that the set of conductances $\{g|g > g_c\}$ forms an infinite connected cluster. Transport in such a system reduces to a percolation problem with a threshold value g_c. After assigning all local conductances with values $g \geq g_c$, and setting all conductances with values $g < g_c$ to zero, Kirkpatrik (1979) and Shante (1977) have obtained a trial solution for the sample conductance of the form

$$g = ag_c[P(g_c) - p_c]^t, \tag{4.24}$$

where $P(g_c)$ denotes the probability for a given conductance to be greater than or equal to g_c, and a is an appropriate constant. In three dimensions the percolation conductivity exponent t is about 1.9.

Katz & Thompson (1986) have applied this same percolation scheme to the permeability of porous rock. Determining the permeability of a porous rock reduces to solving the fluid-flow problem for a random lattice composed of elements of widely varying hydraulic conductances. The permeability of a single straight pipe with diameter d embedded in a medium of cross-sectional area A is about $(\frac{1}{32})d^2\phi$, where $\phi = \pi d^2/4A$. On the other hand, since the local hydraulic conductance is determined by the length, L, associated with a pipe of diameter, d, and length, L, the threshold conductance g_c defines a characteristic length L_c. Moreover, if the pores are filled with an electrically conducting fluid and an electric potential is applied across the sample, L_c defines the threshold in the electrical-conductance case. Thus, Equation (4.24) can be rewritten as a function of the length parameter L:

$$g(L) = \phi g_c(L)[p(L) - p_c]^t. \tag{4.25}$$

Here, the porosity ϕ ensures a proper normalization of the fluid or the electric-charge density. For appropriate choices of the function $p(L)$, the conductance $g(L)$ assumes a maximum value for some $L_{max} \leq L_c$. In general, L^h_{max} for the hydraulic conductance is different from L^e_{max} for the electrical conductance, because the conduction pathways have different weights for the two cases. Therefore, the electrical conductivity σ and the permeability K are defined by

$$\sigma = \alpha\phi[P(L^e_{\max}) - p_c]^t \quad \text{and} \quad K = \beta\phi(L^h_{\max})^2[p(L^h_{\max}) - p_c]^t. \tag{4.26}$$

To interpret the constants α and β, Thompson et al. (1987) assumed that the local rock conductivity is σ_0, the conductivity of the brine saturating the pore space, and that the local pore geometry is cylindrical. These assumptions imply that $\alpha = \sigma_0$ and $\beta = \frac{1}{32}$. Thus, for very broad pore-size distributions, Thompson et al. have determined $L^e_{\max}$ and $L^h_{\max}$ by the forms

$$\begin{aligned} L^e_{\max} &= 0.34L_c, \\ L^h_{\max} &= 0.61L_c, \end{aligned} \tag{4.27}$$

and

$$K = (\tfrac{1}{226})L_c^2(\sigma/\sigma_0). \tag{4.28}$$

In the limit where $L^e_{\max} = L^h_{\max}$, the constant in Equation (4.28) becomes $(\frac{1}{32})$. The magnitude of the constant in Equation (4.28) directly reflects the different weighting of the flow paths for the hydraulic and electrical conductance problems.

Figure 4.8 (Thompson et al., 1987) shows the gas permeabilities extrapolated to the infinite pressure sandstone samples and a few porous carbonate rocks against the permeabilities calculated from Equation (4.28). The gas permeabilities extrapolated to infinite hydrostatic pressure are customarily assumed to present the geometrically determined absolute permeabilities. The precision and accuracy of these measurements are $\pm 2\%$ for the more permeable samples. The permeabilities predicted by Equation

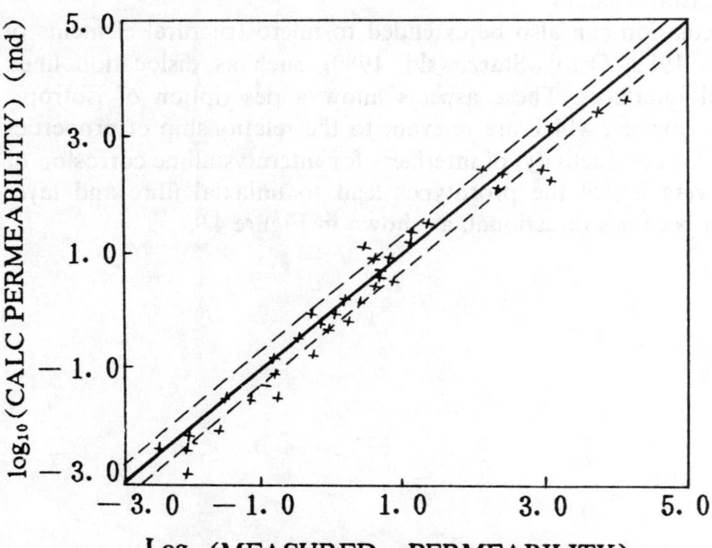

Figure 4.8. The logarithm of the permeability for a set of sandstones against the logarithm of the measured permeability. The solid line is the prediction with no adjustable parameters, while the dashed lines mark a factor of two deviations (Thompson et al., 1987).

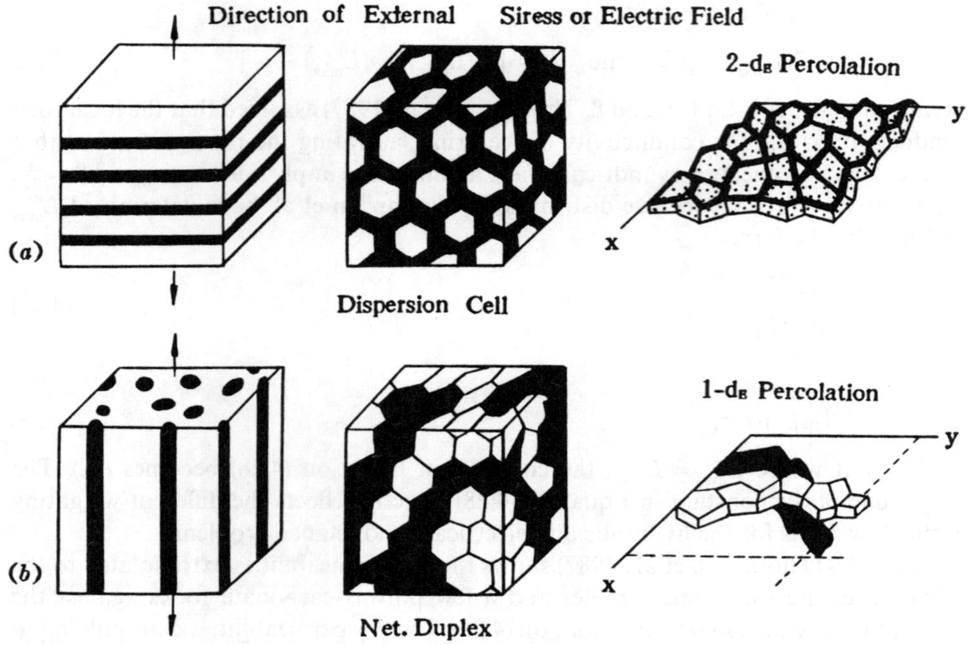

Figure 4.9. Isotropic and anisotropic two-phase microstructures and dimension of percolation: $d_E = 1$ in fibre, net, and duplex structure; $d_E = 2$ in layer, dispersion, and cell structure (Hornbogen, 1989).

(4.28), originating from the percolation theory as illustrated in Figure 4.8, is in agreement with the measured values.

The concept of percolation can also be extended to microstructural elements of materials (Hornbogen, 1989; Ostoja-Starzewski, 1989), such as dislocation lines, grain boundaries, and interfaces. These aspects allow a description of isotropic microstructures by parameters, which are relevant to the relationship of properties (percolation of phases for conductivity) of interfaces for intercrystalline corrosion or fracture. Anisotropic versions of the prototypes lead to uniaxial fibre and layer structures. Percolation becomes directional, as shown in Figure 4.9.

CHAPTER 5

Fractal growth

Many natural objects apparently grow in a fractal form. The study of the phenomenon of fractal growth is one of the main interests in fractal physics. Most of the activity in this field has been based on computer simulations and experimental results. One tries to describe the physical process of fractal structure growth from these activities, such that the physical features of the fractal structure from its initiation to growth can be captured.

The prototype fractal growth models are diffusion-limited aggregation (DLA), viscous fingering, and the more general dielectric breakdown model (DBM). These models have been extensively studied in view of their intrinsic theoretical interest, and they are believed to capture the essential features of pattern formation in seemingly different physical phenomena, especially rock mechanics phenomenon, in which we are very interested. In the future, it may be possible to account for the principal features of rock microgeometry by these fractal growth models.

5.1 DIELECTRIC BREAKDOWN MODEL

Niemeyer et al. (1984) have studied the planar discharge phenomenon of the SF_6 gas between two parallel glass plates and simulated the complex ramified structure of this discharge by computer techniques. The pattern thus generated by dielectric breakdown is called DBM.

Let us consider Figure 5.1 in which a schematic lattice structure in two-dimensional (planar) geometry is presented. The center of the lattice is regarded as the first electrode with the same potential; say $V = 0$. A circle far away at the center represents the other electrode; say $V = 1$. Here V is the electrical potential. Niemeyer et al. thought that the breakdown started from the center and occurred randomly; thus, the plasma channels in this pattern discharge became ideal conductors and $V = 0$ everywhere. They also considered that this pattern discharge was the process of stable electrical conductivity. Thus, the pattern discharge can be described by means of the Laplace equation,

$$\left.\begin{aligned} \nabla^2 V &= 0; \\ V(0,0) &= 0; \\ V|_{\text{circle}} &= 1. \end{aligned}\right\} \tag{5.1}$$

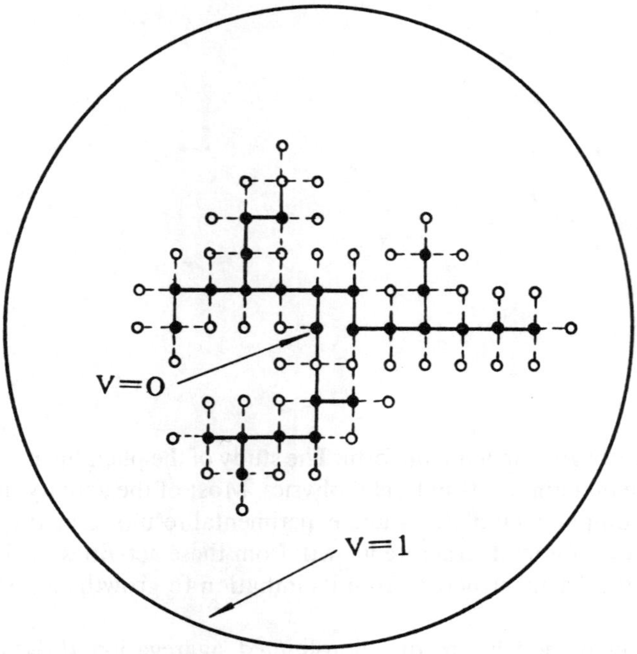

Figure 5.1. Illustration of the stochastic model introduced to simulate dielectric breakdown of a lattice. The central point represents one of the electrodes, while the other electrode is modeled as a circle at large enough distance. The discharge pattern is indicated by the black dots connected with thick lines, and it is considered equipotential ($V = 0$). The dashed bonds indicate all the possible growth processes. The probability for each of these processes is proportional to the local electric field (Pietronero et al., 1986).

This problem can be solved by discretizing the Laplace equation on the lattice and using iterative numerical methods (Pietronero et al. 1986),

$$(V_{i+1,k} - 2V_{i,k} + V_{i-1,k}) + (V_{i,k+1} - 2V_{i,k} + V_{i,k-1}) = 0, \tag{5.2}$$

where i and k are the labels of sites at the x-direction and y-direction, respectively.

In Figure 5.1, the black point represents the breakdown site, and the white point represents the unbreakdown site. Two black points are connected by a solid bond, which represents a conductor, whereas the dashed bonds connecting a black point to a white one represent insulators.

An important step in the computer simulation of this process is to determine the propagation rule. Since initial patterns are random, we can predict that there has to be a random relationship between the local field (i.e., breakdown sites) and its growth. Therefore, the propagation rule can be described by breakdown probability, which is determined by a power function of the differences in electric potentials between the nearest points, i.e.,

$$P_r = (V_{i+1,k} - V_{i,k})^m, \quad \text{or} \quad P_r = (V_{i,k+1} - V_{i,k})^m. \tag{5.3}$$

Now, the stochastic model is completely defined. The computer simulation procedure consists of the following steps (Pietronero et al., 1986):

(1) For a given pattern (black dots), solve the Laplace equations with the appropriate boundary conditions. This gives the local field corresponding to the bonds that are candidates for growth;

(2) Construct the probability field as given by Equation (5.3) and select one single bond by the Monte Carlo procedure based on these probabilities;

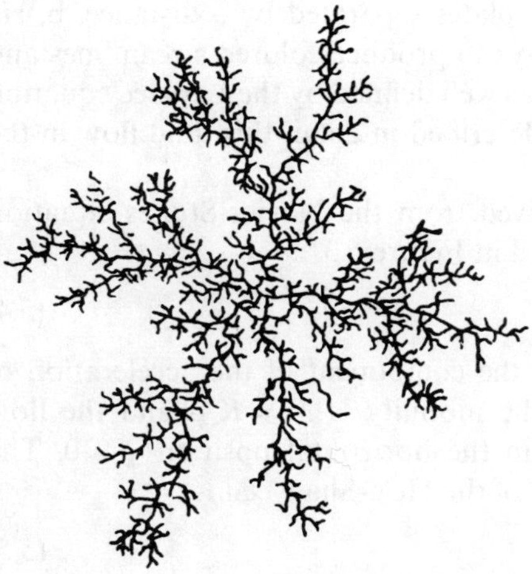

Figure 5.2. A space filling structure for $m=1$ (in Equation (5.3)), the fractal dimension $D \approx 1.7$ (Satpathy, 1986).

Table 5.1. The correlation between fractal dimensions and m values (Satpathy, 1986).

Dimension of the field	Dimension of the breakdown graph	$m=0$	$m=1$	$m=2$	$m=3$	$m=4$
2	2	2	1.75	~1.6		
3	2	2	1.96	1.82	1.78	1.69
3	2	3	2.48	2.11	1.96	1.75

(3) The selected bond is added to the equipotential pattern (set of black dots) and once again starts with (1), solving the Laplace equation with this new boundary condition.

For $m=0$ (in Equation (5.3)), the effect of the local field is irrelevant, and the probability is identical for all the bonds that are candidates for growth. The graph of the pattern discharge gives rise to a space filling structure; its fractal dimension is equal to the Euclidean dimension: $D=d_E=2$. More interesting is the case where $m=1$ that gives rise to Figure 5.2. This corresponds to a fractal dimension $D \approx 1.7$. When $m=2,3,4$, the fractal dimension D decreases with an increase of m. Table 5.1 shows the correlation between the fractal dimension D of the pattern discharge and m values (obtained by Satpathy, 1986).

On studying DBM, we can see that a suitable combination of the Laplace equation and a stochastic field generates stochastic fractal structures. This shows how fractal structures can be generated by a well-defined physical model.

5.2 VISCOUS FINGERING

The theory of viscous fingering originated from fluid flow in the Hele-Shaw cell. Hele-Shaw (1898) studied the flow of water around various objects placed in the

Hele-Shaw cell, consisting of two transparent plates separated by a distance, b. He examined the flow streamlines by injecting a dye to produce colored streamlines and found that the fluid flow in a Hele-Shaw cell was well defined by the Laplace equation for low Reynolds numbers. Feder (1988) has described in detail this fluid flow in the Hele-Shaw cell.

The equation for the flow velocity $\boldsymbol{U}$, derived from the Navier-Stokes equation governing flow in the Hele-Shaw cell illustrated in Figure 5.3, is

$$\boldsymbol{U} = -(K/\mu)\nabla(p + \rho g z) = -M\nabla\phi, \tag{5.4}$$

where p is the pressure, ρ the density, and g the component of the acceleration of gravity along the z-coordinate of the cell. The mobility is $M = K/\mu$ and the flow potential is $\phi = (p + \rho g z)$. For a cell placed in the horizontal position, $g = 0$. The viscosity of the fluid is μ, and the permeability of the Hele-Shaw cell is

$$K = b^2/12. \tag{5.5}$$

Note that the velocity in Equation (5.4) is the average velocity over the thickness of the cell. For incompressible fluids, the equation of continuity gives

$$\nabla \cdot \boldsymbol{U} = -\nabla^2(p + \rho g z) = -\nabla^2\phi = 0. \tag{5.6}$$

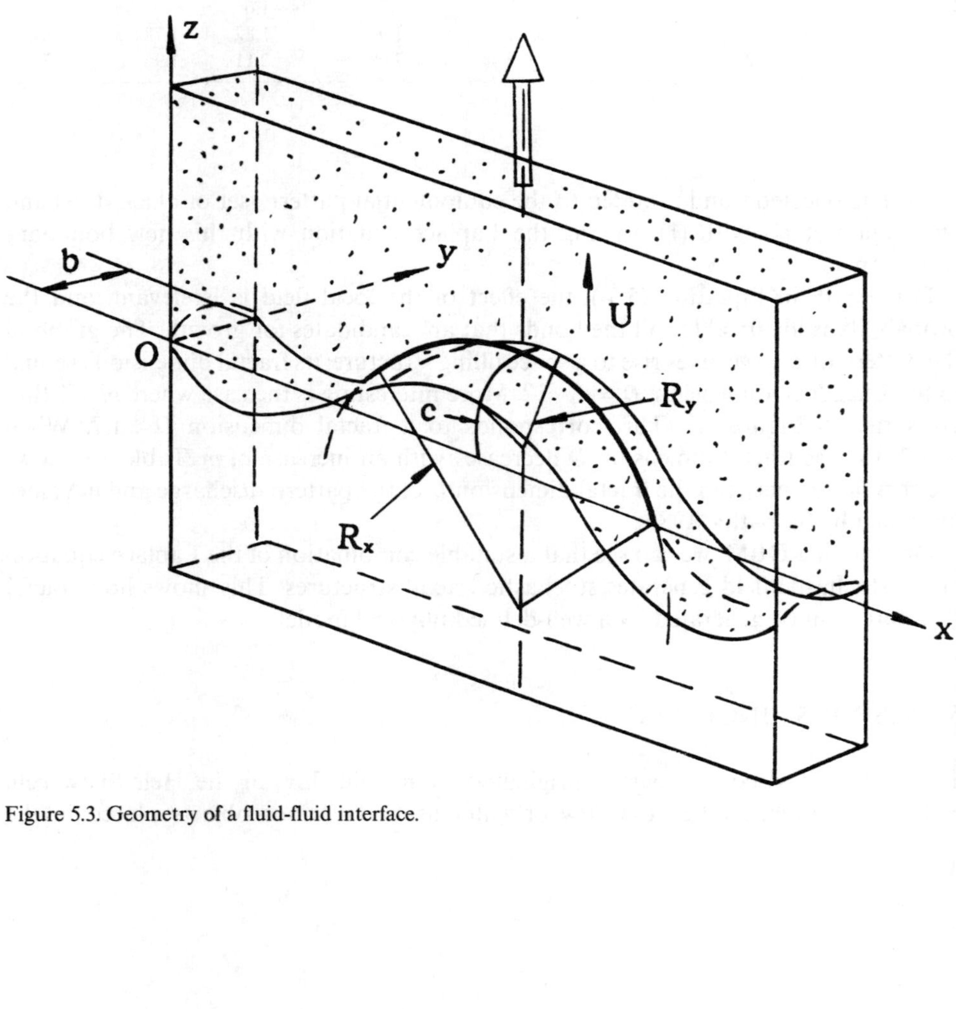

Figure 5.3. Geometry of a fluid-fluid interface.

This is the Laplace equation, and it is characteristic of potential problems. Therefore, the flows controlled by Equation (5.6) are called potential flows.

Feder (1988) discussed the situation illustrated in Figure 5.3, where a fluid (index 1) displaced another fluid (2). The interface between the two fluids is controlled by capillary forces when the fluids are at rest and there is a pressure difference between the two fluids.

$$(p_1 - p_2) = \sigma[(1/R_x) + (1/R_y)], \tag{5.7}$$

where σ is the interfacial tension between the two fluids. The two principal radii of curvature R_x and R_y describe the interface locally as indicated in Figure 5.3. The radii of curvature are defined to be positive if they have their centers in fluid (1). The radius of curvature R_y is controlled by the contact angle, θ, describing how the two-fluid interface contacts the plates that define the cell geometry. Typically, one finds that $R_y \propto b/2$, and it is also assumed that $R_x \gg R_y$. When the fluids are at rest and (2) is the wetting fluid, we have $p_1 > p_2$.

If we consider the case in which fluid (1) is injected at a constant rate U at $z = -\infty$ and fluid (2) is withdrawn at the same rate at $z = \infty$, then the interface between the two fluids will move with a velocity $\boldsymbol{U} = (0, 0, U)$ along the z-axis. However, the interface is unstable if the viscosity of the driving fluid is smaller than the viscosity of the fluid being driven. This instability is called viscous fingering (Engelberts & Klikenberg, 1951).

The physics of viscous fingering lies in the dynamics of the moving boundary. Assume that a pressure difference $\Delta P = P_1(z=0) - P_2(z=L)$ is maintained over the length L of a finite Hele-Shaw cell where air displaces a high-viscosity fluid. Clearly, the pressure in the air is constant and equal to the input pressure $P_1(z=0)$ when its viscosity is ignored. Next, assume that a finger gets ahead of the rest of the displacement front; then the pressure at the finger tip is also $P_1(z=0)$. The largest pressure gradient in the high-viscosity fluid is, therefore, at the finger tip and is defined by

$$\nabla P = \{P_1(z=0) - P_2(z=L)\}/(L-z);$$

where z is the position of the finger tip. This is clearly an unstable situation.

In Figure 5.3, assume that the straight interface is perturbed by a sinusoidal displacement. The position of the interface can then be obtained from the real part of the form

$$\varsigma = \varepsilon \exp[2\pi r t + i(2\pi x/\lambda)], \tag{5.8}$$

where λ is the wavelength of the perturbation, r the growth rate of the perturbation, and ε an infinitesimal amplitude of the perturbation.

Suppose λ_c is a critical wavelength. The solution of Equations (5.3-5.6), including only the linear terms in ς, gives the result that the front is unstable with respect to perturbations having a wavelength $\lambda > \lambda_c$, where λ_c is determined to be:

$$\lambda_c = 2\pi\{\sigma/[(\mu_2/K_2 - \mu_1/K_1)(U - U_c)]\}^{1/2}. \tag{5.9}$$

Here the critical velocity U_c is defined by

$$U_c = g(\rho_1 - \rho_2)/[(\mu_2/K_2) - (\mu_1/K_1)]. \tag{5.10}$$

Clearly, for $\rho_2 > \rho_1$, we have $U_c < 0$ for $\mu_2 > \mu_1$, so that the system is unstable even at $U = 0$. The interface is especially unstable at any velocity in the absence of gravity effects ($g = 0$).

All wavelengths $\lambda > \lambda_c$ are unstable; however, perturbations with a wavelength λ_m given by

$$\lambda_m = \sqrt{(3)}\lambda_c$$

have the largest growth rate and will dominate the dynamics of the front. Therefore, λ_m is always expected to be a characteristic period of viscous fingers. Using expression

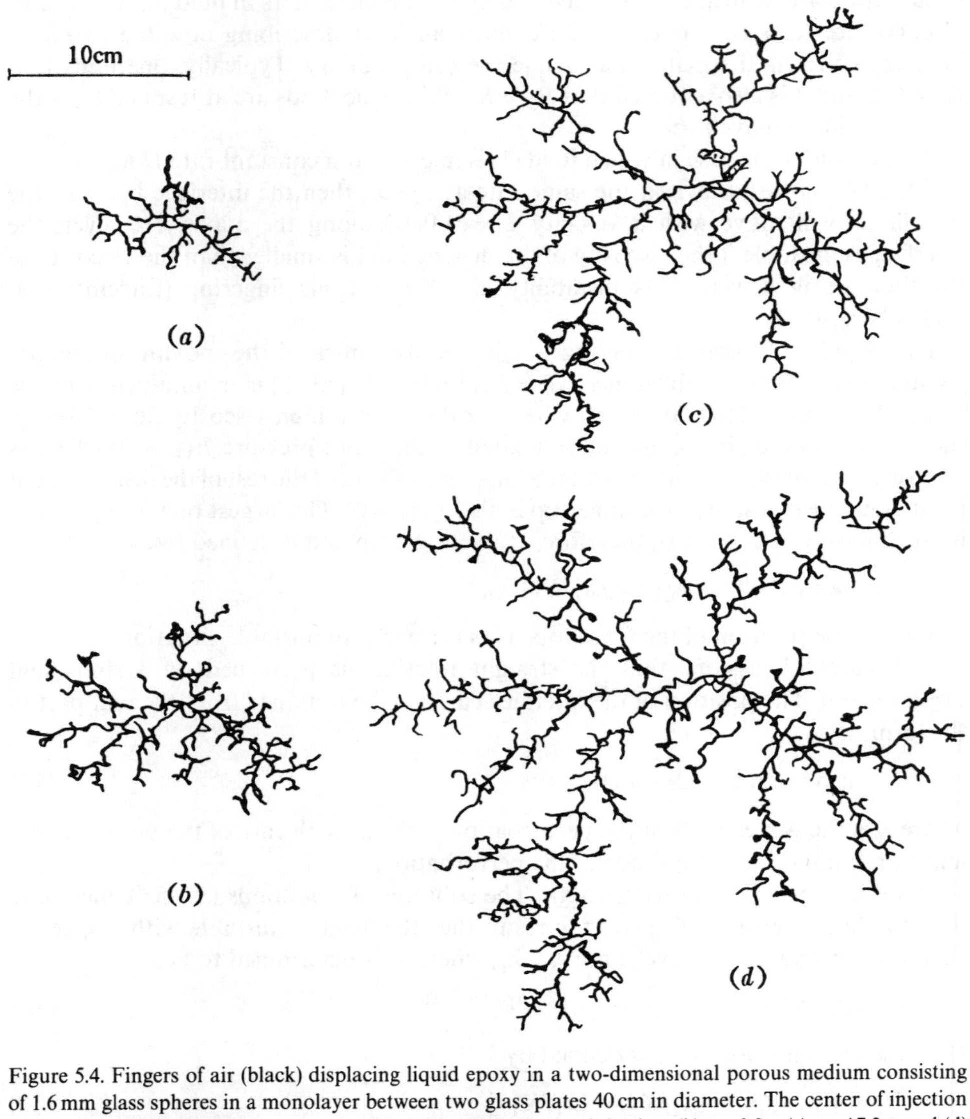

Figure 5.4. Fingers of air (black) displacing liquid epoxy in a two-dimensional porous medium consisting of 1.6 mm glass spheres in a monolayer between two glass plates 40 cm in diameter. The center of injection is near the middle of the structure. (a) $t = 2$ s after the beginning injection; (b) $t = 3.9$ s; (c) $t = 17.2$ s; and (d) $t = 19.1$ s. The capillary number is $N_{ca} = 0.04$ (Maløy et al., 1985a, b).

in Equation (5.4) for the permeability and supposing that the viscosity of the driving fluid is negligible ($\mu_1 \ll \mu_2$), as is the case when glycerol is displaced by air, one finds that a period of the expecting finger for a horizontal cell is defined by

$$\lambda_m = \pi b\sqrt{(\sigma/(U\mu))} = \pi b/\sqrt{N_{\text{ca}}}, \tag{5.11}$$

where N_{ca} is the capillary number given by

$$N_{\text{ca}} = U\mu/\sigma, \tag{5.12}$$

which measures the ratio of viscous to capillary forces.

As K is the actual permeability of a porous medium, Equations (5.3)-(5.5), representing the flow in the Hele-Shaw cells, can be used to simulate the flow of a fluid in the porous medium. Maløy et al. (1985a, b) have studied viscous fingers in two-dimensional porous media consisting of a randomly-packed single layer of glass spheres glued between two glass plates. They displaced a high viscosity fluid (epoxy) by air at a moderately high capillary number, $N_{\text{ca}} = 0.04$. Their results are reproduced in Figure 5.4. The fingering structures have been analyzed by Maløy et al. in terms of the covering method (i.e., the number-radius relationship – cluster dimension). The pictures were digitized and each pixel that contained a part of the air (black in Figure 5.4) was counted as a ball, and the distance r_i from the center of injection was measured. The number $N(r) = N(r_i < r)$ of pixels containing air inside the radius r from the center of injection was determined to be of the form

$$N(r) = N_0(r/R_g)^D f(r/R_g), \tag{5.13}$$

where N_0 is the total number of pixels containing air, and $R_g = (N_0^{-1}\Sigma r_i^2)^{1/2}$ represents the radius of gyration for the fingering structure. The experimental results in Figure 5.4 are best matched by choosing $D = 1.62 \pm 0.04$, which is the fractal dimension of the viscous fingering in two-dimensional porous media.

The theory of viscous fingering was developed and independently compared to experiments by Saffman & Tayloy (1958) and Chuoke et al. (1959). Engelberts & Klikenberg (1951) observed the fingers in the pore volume. Many recent theoretical and experimental results in this field have been published, e.g., those of Jensen et al. (1987), Homsy (1987), Feder (1988), and Wong (1988).

5.3 DLA MODELS

DLA – the diffusion-limited aggregation models were first proposed by Witten & Sander in 1981. Witten & Sander (1983) have further studied the correlation between DLA and the diffusion equation. DLA models consist of particle-particle and cluster-cluster models. The particle-particle model represents a process where particles are left to wander randomly until they reach the 'surface' of the cluster in which they come to rest, thus making the surface grow one step at the point of attachment (as shown in Figure 5.5). The cluster-cluster model is the process in which clusters walk at random or come together to form a larger cluster.

The so-called diffusion-limited aggregation process, as the term suggests, is required to fit in the diffusion process which can be described by the Fick diffusion law

$$j = -\mathfrak{C}\nabla u, \tag{5.14}$$

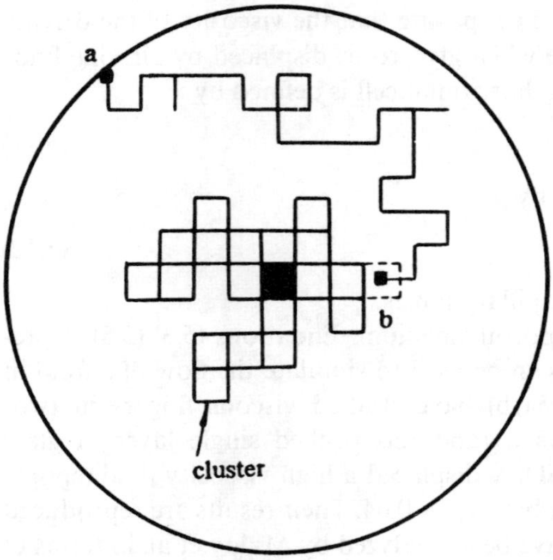

Figure 5.5. The diffusion-limited aggregation (DLA) model (particle-particle model). A particle is released from a random point a on the circle and performs a random walk until it either leaves the circle or reaches a square b next to one that has already been shaded, in which case this square is also shaded (Falconer, 1990).

and

$$\partial u/\partial t = -\nabla \cdot j = \mathfrak{C}\nabla^2 u, \tag{5.15}$$

where j is the diffusion flow and u the concentration of walking particles (or clusters). $\mathfrak{C}$ represents the diffusion coefficient, in general given by the Einstein relationship $\mathfrak{C} = 0.5\Gamma a^2$ for particles that take steps of length a in random directions at a rate Γ.

In the DLA model, u denotes the probability of particles sticking to the aggregate. Without the center of the aggregate, u is the uniform distribution, but with the center of the aggregate, u is the nonuniform distribution, i.e., there is a field of the probability gradient. Suppose that the lattice is so large, and the center cluster is so small, that the probability is regarded as independent of time change. This is a steady diffusion process. Thus, Equation (5.15) becomes

$$\nabla^2 u = 0; \quad \text{or} \quad j = 0. \tag{5.16}$$

Here we consider a two-dimensional geometry, with the boundary conditions $u = 0$ for all the points on the boundary of particles or particle clusters, and $u = 1$ on far away points. Thus, a mathematical description of DLA models can be given by

$$\begin{aligned} \nabla^2 u &= 0; \\ u|_{\text{boundary}} &= 0; \\ u|_{\text{far away}} &= 1. \end{aligned} \tag{5.17}$$

This problem, being the same as the DBM problem, can be solved by discretizing the Laplace equation on the lattice and using iterative numerical methods. The difference equation and boundary conditions can be written as

$$\begin{aligned} (u_{i+1,k} - 2u_{i,k} + u_{i-1,k}) + (u_{i,k+1} - 2u_{i,k} + u_{i,k-1}) &= 0; \\ u|_{\text{boundary}} &= 0; \\ u|_{\text{far away}} &= 1. \end{aligned} \tag{5.18}$$

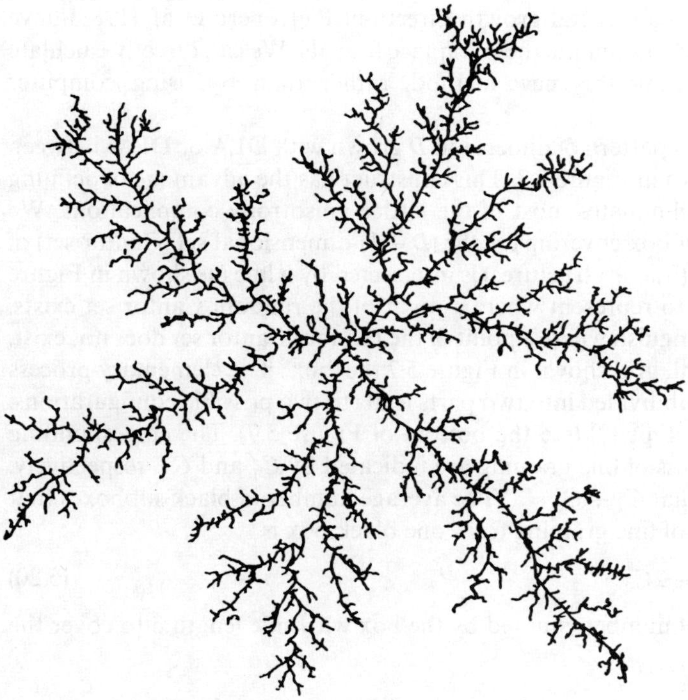

Figure 5.6. A computer realization of diffusion-limited aggregation. The square was divided into a 700 × 700 mesh, from which 16 000 squares were selected using the described method (Falconer, 1990).

A local smooth boundary grows with a velocity given by

$$V_n = \mathfrak{C}\boldsymbol{n}\cdot\nabla u|_{\text{boundary}}, \tag{5.19}$$

in a direction perpendicular to the 'surface' with normal vector $\boldsymbol{n}$ (Witten & Sander, 1983). Visually, it is the same as that in DBM for $m = 1$. Witten & Sander (1983) have shown their results by computer simulation, with a fractal dimension $D = 1.66$.

Falconer (1990) has given a computer realization of DLA, as illustrated in Figure 5.6, with a fractal dimension $D = 1.70$.

5.4 LAPLACE FRACTALS

From discussions in the three previous sections, we find that the mathematical structures of DLA, DBM, and viscous fingers all consist of the Laplace equation, boundary conditions, and the given growth rate ruler. Whereas the fractal growth processes are all determined by numerical simulation of this mathematical structure, the scale-invariance of the pictures of the fractal growth is due to the scale-invariance of the Laplace equation and consistency of the growth probability. These are called Laplace fractals.

Based on the invariance of fractal growth direction, Pietronero et al. (1988) have proposed a new approach to simulate these Laplace fractals. We can directly calculate fractal dimensions by using this new method, rather than by using computer simulation.

Let us consider a frozen pattern of dimension D grown with DLA or DBM between two parallel lines as shown in Figure 5.7. This construct has the advantage of defining a growth direction and eliminates most of the lattice anisotropy complications. We now consider a process of box covering for the $(D-1)$-dimensional set (Cantor set) of points where the pattern (fractal structures) is intersected by a line (as shown in Figure 5.7). We use a black dot to represent where a point of the random Cantor set exists, and the white one to distinguish where a point of the random Cantor set does not exist. Pietronero called the small grid shown in Figure 5.7 the 'box'. The elementary process by which a black box is subdivided into two parts leads to two possible configurations, indicated as type (1) and type (2) (see the bottom of Figure 5.7). The corresponding probabilities in this process of fine graining are indicated by C_1 and C_2, respectively. Normalization requires that $C_1 + C_2 = 1$. The average number of black subboxes that appears at the next level of fine graining from one black box is

$$\langle n \rangle = \Sigma n_i C_i = C_1 + 2C_2. \tag{5.20}$$

In box covering, the total number counted by the box with side length δ to cover the

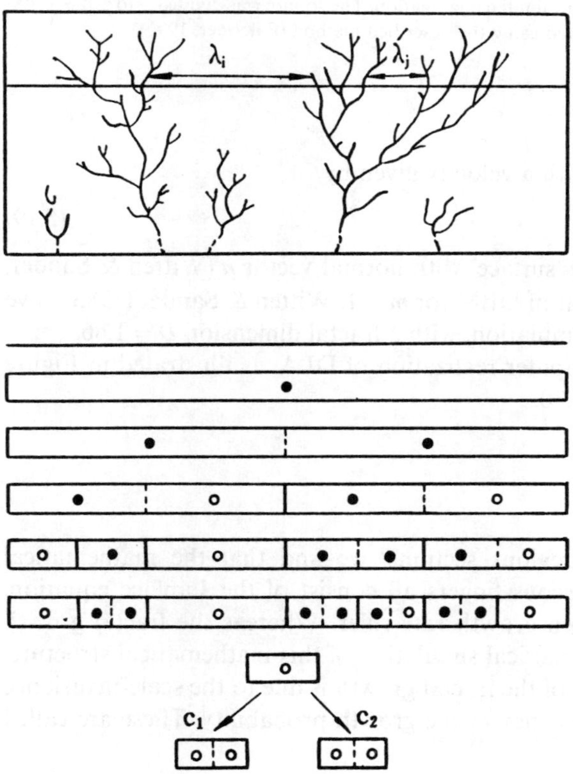

Figure 5.7. Schematic picture of the process of box covering for a set of points given by the intersection of the fractal structure with a line. Bottom: elementary process of fine graining for a box (black) that contains some elements of the set (Pietronero et al., 1988).

fractal structure with fractal dimension D is defined by

$$N_{b(\delta)} \propto \delta^{-D}.$$

In each step in this process, $N_{b(\delta)}$ adds the number of $\langle n \rangle$ as $\delta \rightarrow \delta/2$. It is clear that the fractal dimension of the whole structure is related to the values of C_1 and C_2 by

$$D = 1 + \log\langle n \rangle / \log 2 = 1 + \log(C_1 + 2C_2)/\log 2. \tag{5.21}$$

Given a certain level of fine graining, the distribution (C_1, C_2) corresponds to the relative density of newly-formed pairs (characterized by a dashed line) of types (1) and (2), respectively. However, this description is not complete, because two adjacent branches are indicated; for example, by λ_i and λ_j in Figure 5.7. The probability of occurrence $P_r(\lambda)$ of a certain value of λ is also related to (C_1, C_2). Letting b to be the lowest scale and $\lambda_n \neq b^n$ $(n = 1, 2, \ldots;\ \lambda_0 \sim 0)$, it can be shown by a nontrivial combinatorial analysis that

$$P_r(\lambda_0 \sim 0) = C_2/(1 + C_1/2), \quad (n = 0); \tag{5.22}$$

$$P_r(\lambda_n \sim b^n) = (\tfrac{3}{2})[C_1 C_2/(3 - C_2)](0.5 + 0.5C_1)^{(n-1)}, \quad (n = 1, 2, 3, \ldots). \tag{5.23}$$

Note that the λ distribution requires an analysis of all the steps of the fine graining procedure.

Clearly, all the properties of the system are related to the distribution (C_1, C_2). Because of the scale invariance of the Laplace equation, the same distribution (C_1, C_2) is found if we consider the growth process at different scales. Pietronero et al. (1988) have shown a detailed description of the way to compute this distribution. Their research results have also shown that the scale-invariance of the Laplace equation implies that the structure is self-similar under both growth and scale transformations.

5.5 LAPLACE FRACTALS AND FRACTURE OF MATERIALS

A wide variety of patterns derived from different growth processes has been shown to have a fractal structure. Louis et al. (1986) and Takayasu (1986) have studied fracture patterns in materials by means of Laplace fractals. Their results show that fracture patterns in materials also have Laplace fractal characters.

In DLA and DBM models, random patterns grow stochastically with a probability closely related to the solution of the Laplace equation. Contrary to those stochastic models, Takayasu (1986) has proposed a deterministic model of electric breakdown. Assume that a constant voltage V is applied to a resistor. When V is smaller than a certain critical voltage V_c, the conductivity of the resistor is G. Once V exceeds V_c, then the resistor is regarded as broken, and the conductivity is enhanced to λG $(\lambda \gg 1)$. This parameter λ represents the ratio of increase of the conductivity and, in an ideal case, $\lambda = \infty$. After breakdown, the resistor keeps the enhanced conductivity, regardless of the magnitude of V (as shown in Figure 5.8). Next consider a square network of resistors as indicated in Figure 5.9. In this network, the electric potential u satisfies the following equation for any lattice point (i, j) (its Laplace equation is defined by: $\nabla[G\nabla V] = 0$):

$$\sum_{k=1}^{4} G_k(i,j)\{u_k(i,j) - u(i,j)\} = 0, \tag{5.24}$$

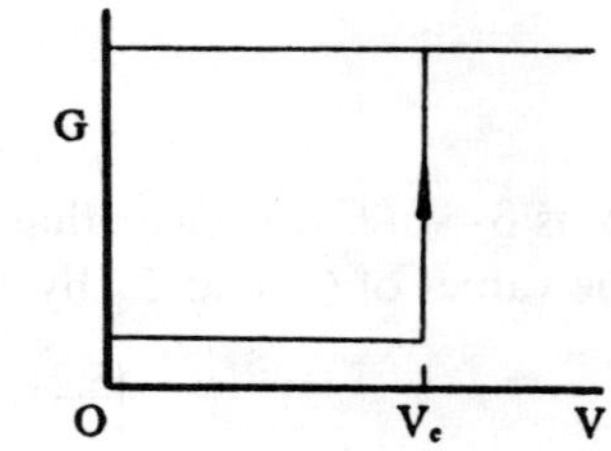

Figure 5.8. The response of conductivity G with respect to voltage V (Takayasu, 1986).

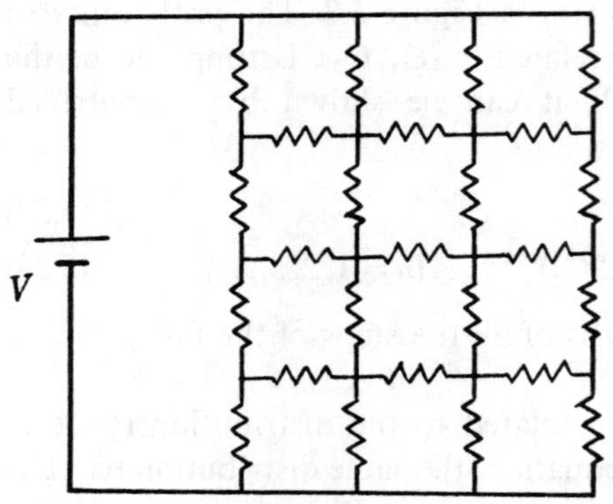

Figure 5.9. Network of 4×4 electric breakdown pattern (Takayasu, 1985).

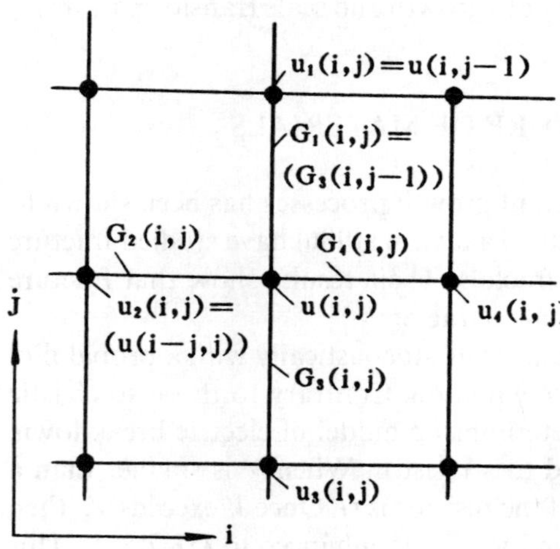

Figure 5.10. The definition of $u_k(i,j)$ and $G_k(i,j)$, where the suffix k denotes up, left, down, and right for $k = 1, 2, 3, 4$ (Takayasu, 1986).

where the subscript k represents the direction and $G_k(i,j)$ indicates the conductivity of the corresponding resistor as illustrated in Figure 5.10. If a lattice point is located on a boundary, we have to use $G_k = 0$ for a missing resistor. Once $\{G\}$ and the boundary conditions of $\{u\}$ are given, Equation (5.24) for all combinations of (i,j) constitutes a set of linear equations of $\{u\}$, which can be solved by a numerical method.

The time evolution of this system can be simulated by the following procedure:

(1) Given $\{G\}$ and a boundary condition of $\{u\}$;

(2) Solve $\{u\}$ by Equation (5.24);

(3) Check every resistor (except the ones already broken), if the broken condition, $|u_k(i,j) - u(i,j)| > V_c$ is satisfied then let $G_k(i,j) = \lambda G_k(i,j)$; and

(4) Stop if no resistor has been newly broken in the preceding procedure. Otherwise, return to (2) and continue the entire routine.

An example of evolution on a 10 × 10 network was given by Takayasu (as shown in Figure 5.11). Here, the boundary condition is defined by

$$u(i,0) = 0 \quad \text{and} \quad u(i,10) = C, \quad (i = 1,2,\ldots,10); \tag{5.25}$$

where C is chosen to be sufficiently large so that at least one resistor breaks, and the initial value of $\{G\}$ is randomly given as

$$G_k(i,j) = G' + G^*Z, \tag{5.26}$$

where G' and G^* are constants, and Z is a random number uniformly distributed on [0, 1]. Although the time step in this model is not in real time, it may be quite an interesting problem to quantitatively estimate the growth rate of the breakdown by examining 32 × 32 networks. Takayasu has found the exponential growth of the form

$$N_{(T)} \propto \exp(bT), \quad b = 0.85 \pm 0.03; \tag{5.27}$$

where $N_{(T)}$ indicates the average number of broken resistors at the time step T. This relationship is valid for a time step less than the critical percolation time.

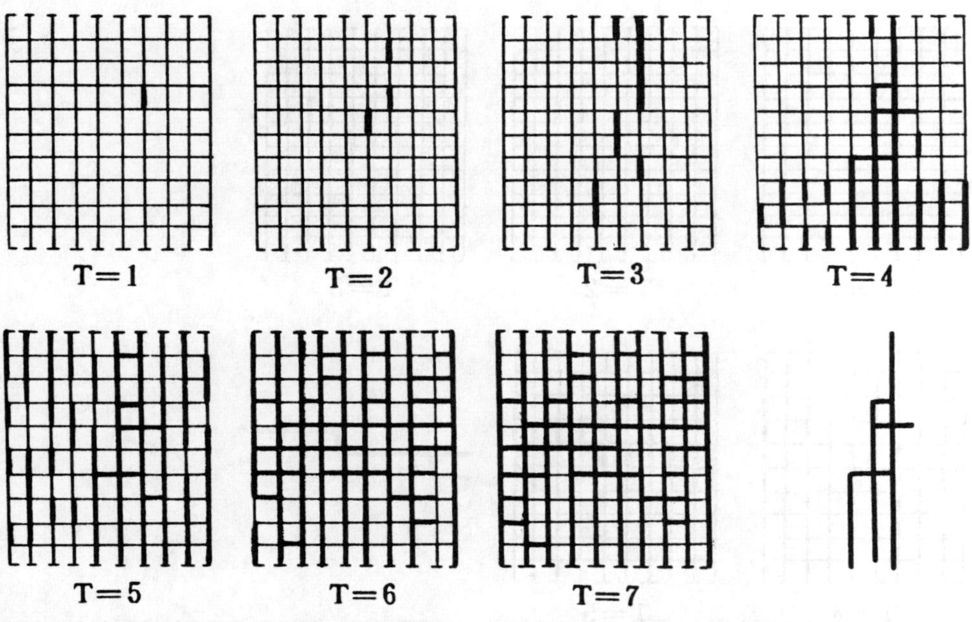

Figure 5.11. An evolution of the electric breakdown pattern on a 10 × 10 network. The last figure shows the pattern of the percolation cluster at $T=4$. The parameters are $G' = 2.0$, $G^* = 2.0$, $V_c = 1.0$, and $\lambda = 100$ (Takayasu, 1986).

Assume a situation in which one end of a thin, brittle stick is fixed while the other end is free. If the displacement of the free end d is less than a critical value d_c (the displacement at failure), then its modulus of rigidity, E, is a constant. Once d exceeds d_c, the stick may be broken and E may suddenly be reduced to a very small value, $\alpha E (0 \leq \alpha \leq 1)$, as shown in Figure 5.12.

Considering this elastic-brittle material (a plane square net consisting of such brittle sticks) and assuming the case wherein displacements at its lattice points are all perpendicular to the plane (anti-plane shear problem), Takayasu has simulated this brittle fracture process from Equation (5.24) in which $\{u\}$ and $\{G\}$ denote the displacements and the moduli of rigidity, respectively. Note that the boundary condition (Equation (5.25)) and the initial condition (Equation (5.26)) are also available in this case. This boundary condition corresponds to the physical situation such that the top edge of the brittle square plate is pulled up, while the bottom edge is fixed. Figure 5.13 shows the evolution of fracture under the same conditions with the case

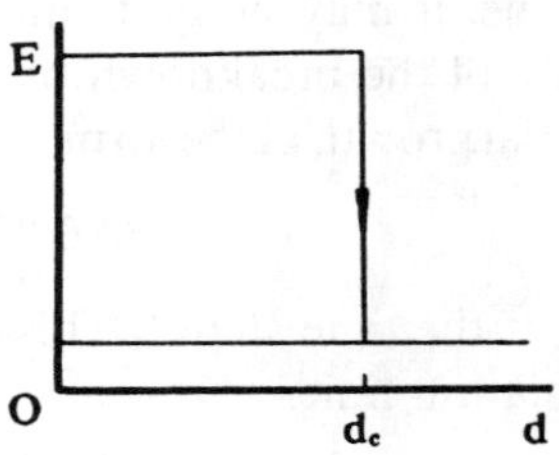

Figure 5.12. Response of the modulus of rigidity, E, with respect to displacement d (Takayasu, 1986).

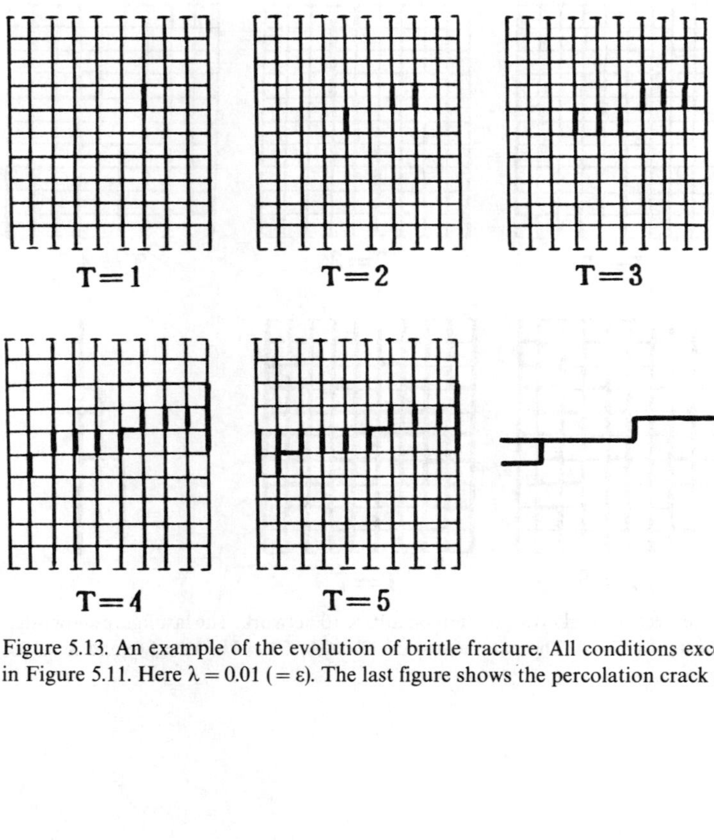

Figure 5.13. An example of the evolution of brittle fracture. All conditions except λ are identical to those in Figure 5.11. Here $\lambda = 0.01$ ($= \varepsilon$). The last figure shows the percolation crack at $T = 5$ (Takayasu, 1986).

of Figure 5.11, as obtained by Takayasu. It is seen that the breakdown pattern in Figure 5.11 grows vertically, while the fracture pattern in Figure 5.13 grows horizontally and growth stops when the fracture pattern forms a percolation cluster. It may be mathematically quite interesting that these differences originate only in parameter λ.

Fractal properties of these percolation clusters have been examined and their fractal dimension is estimated as

$$D = 1.65 \pm 0.05.$$

The average number of broken sticks $N_{(T)}$ follows a power law in this case:

$$N_{(T)} \propto T^a, \quad a = 2.4 \pm 0.2.$$

This power law may indicate that the crack growth process has fractal properties not only in its spatial patterns but also in its temporal behaviors.

While directly considering the Laplace equation of the continuum elastic material, Louis et al. (1986) have studied the fracture pattern in materials, based on a continuum elastic equation. The continuum elastic equations can be written as;

$$(\lambda + \mu)(\partial/\partial x_i)\left(\sum_j \partial u_j/\partial x_j\right) + \mu\left(\sum_j \partial/\partial x_j\right)u_i = 0, \tag{5.28}$$

where λ and μ are the Lame coefficients, $\partial/\partial x_j$ represents the partial derivative with respect to coordinate component x_j, and u_i are displacement components of the strain field. Different boundary conditions can be imposed: shear deformation, uniaxial compression, and uniform compression,.... Louis et al. have developed the discretized

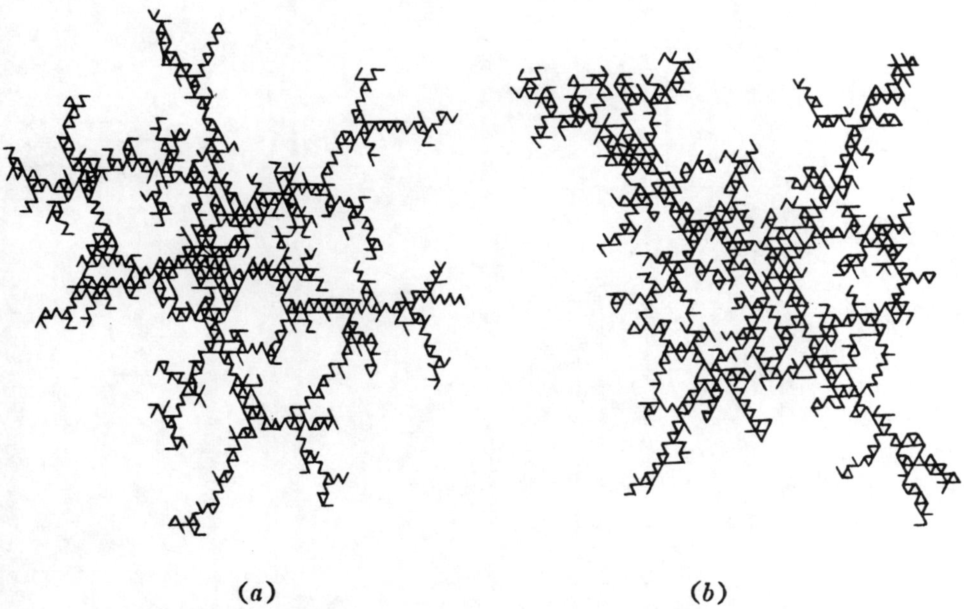

Figure 5.14. Broken bonds: (a) Expansion and (b) Shear (Louis et al., 1986).

formulation of Equation (5.28) by using a triangular lattice with springs joining the nearest neighboring lattice sites. Thus, the model can either simulate a monocrystal, in which case springs are regarded as interatomic forçes, or a polycrystal, whereby the springs are used to describe the grain boundary strength. In general, a $\sqrt{3} \times \sqrt{3}$ reconstruction can describe an isotropic system with a range of values for the ratio λ/μ, whereas anisotropy can be simulated by considering a 2×1 reconstruction. Louis et al. used the two lattice vectors $\boldsymbol{a}_1 = a(1,0)$ and $\boldsymbol{a}_2 = a(-\frac{1}{2}, \sqrt{\frac{3}{2}})$ in the triangular lattice as basis vectors, and obtained the discretization of Equation (5.28) as follows (here $\lambda = \mu$ is considered):

$$\begin{aligned} &4u_i(k,m) - u_i(k+1,m) - u_i(k-1,m) - u_i(k+1,m+1) \\ &- u_i(k-1,m-1) + 2u_j(k,m) - u_j(k+1,m+1) - u_j(k-1,m-1) \\ &= 0, \quad (i,j = 1,2), \end{aligned} \tag{5.29}$$

where $u_i(k,m)$ are displacement components along the basis vector $\boldsymbol{a}_i$ at lattice point (k,m). Lattices with sizes up to 100×100 were studied. The boundary conditions chosen were uniform expansion and shear deformation. The simulation procedure of the fracture patterns starts with a lattice whose bounds are stretched by 10% over their equilibrium length. A bond is then broken, and the lattice is allowed to relax to the new equilibrium state by using Equation (5.28). A neighboring bond of the broken bond is subsequently and randomly broken, due to the stress accumulated on it. These steps are repeated until a reasonable number of bonds is broken. Louis et al. have built up fracture patterns with up to 1200 broken bonds, which is lower than those usually achieved in DBM and DLA. Figure 5.14 shows their results. The fracture patterns turn out to be self-similar with the fractal dimension $D = 1.62 \pm 0.05$ for expansion, and $D = 1.64 \pm 0.05$ for shear.

CHAPTER 6

Multifractals

Multifractals are the generalization of self-similar fractals, which describe the local features of nonuniform fractals with self-similarity. Nonuniform fractal structures have no consistency of self-similarity in the whole, but have local self-similarity, which may be characterized by a self-similar distribution function with multi-dimensions on a geometric 'support'. Therefore, multifractal measures are related to the study of this distribution on the geometric support (or fractal objects).

A multifractal was originally introduced by Mandelbrot (1974) in the discussion of turbulence and extended by Mandelbrot (1982) to many other contexts. It has then been developed by various other authors such as Grassberger (1983), Grassberger & Procaccia (1983), Badii & Politi (1984), Aharony (1986), Pietronero et al. (1986), Blumenfeld et al. (1987), Maløy et al. (1987), and Williford (1988).

In this chapter, we first discuss a few of the basic ideas of multifractals (Pietronero et al., 1986), illustrating them with simple examples. Later, we will discuss experimental evidence for multifractal behavior, especially multifractal fracture (Williford, 1988a).

6.1 THE CANTOR SET WITH WEIGHT DISTRIBUTION

We start with a Cantor set $C_k(\lambda, m)$ where λ is the number of intervals, m is their length, and k is the order of iteration. A weight defined to each of the λ intervals of $C_1(\lambda, m)$ is given by $U_{k=1}(x)$:

$$U_{k=1}(x) = \begin{cases} W_1, & \text{if } x \in [0, m] \\ W_2, & \text{if } x \in [(1-m)/(\lambda-1), ((\lambda-1)m+1-m)/(\lambda-1)] \\ \cdots & \\ W_\lambda, & \text{if } x \in [1-m, 1] \\ 0, & \text{otherwise} \end{cases} \tag{6.1}$$

and, because of normalization,

$$\sum_{i=1}^{\lambda} W_i m = 1. \tag{6.2}$$

For the next iteration, we proceed in the same manner with the same weight distribution. For simplicity, we consider the specific case where $\lambda = 3$, $m = \frac{1}{4}$ and $(W_1, W_2, W_3) = (W, 2W, W)$ as indicated in Figure 6.1. Because of normalization, the relationship between m and W is simply $4Wm = 1$. The scaling property can then be obtained as follows

$$U_{k+1}(mx) = U_k(x)/(4m). \tag{6.3}$$

For a small x (note that the origin of the coordinates is at the edge of Figure 6.1),

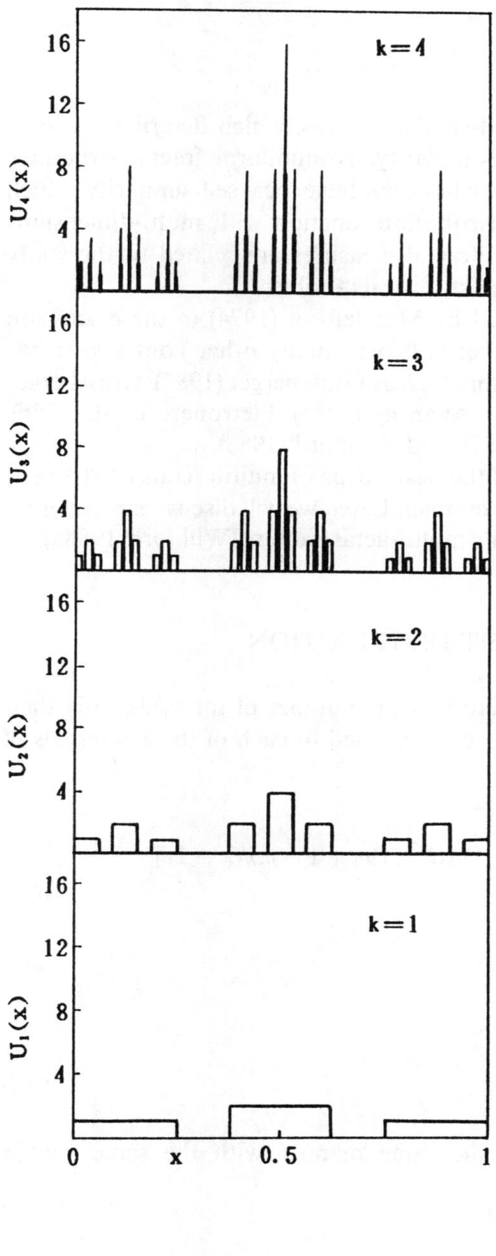

Figure 6.1. Example of a self-similar probability distribution obtained as a generalization of Cantor set construction. This distribution shows a simple example of multifractals. Around the center, $D = 0.5$, and, at the edges, $D = 1$ (Pietronero et al., 1986).

$U(x)$ can be regarded as the smooth function in the form

$$U(x) \propto x^{(D-1)}, \quad (x \to 0), \tag{6.4}$$

where $D = -\log 4/\log m = 1$ when $m = \frac{1}{4}$. These well-defined scaling properties have been derived for a particular origin. On changing the origin, one obtains different scaling properties. Let us now consider a translation $x \to x - \frac{1}{2}$, so that the origin lies in the center of the distribution. The scaling relationships then become

$$U_{k+1}(mx) = U_k(x)/(2m), \tag{6.5}$$

with $D = -\log 2/\log m = 0.5$ (for $m = \frac{1}{4}$), which is a factor of two smaller than the previous case. This origin dependence of the fractal dimension is due to the introduction of different weights and shows that a single fractal dimension is not enough to describe self-similar distributions like the one considered above.

6.2 THE CONTINUOUS SPECTRUM OF FRACTAL DIMENSIONS

Let us begin by considering the simple example of a probability distribution given by

$$P_r(x) = a_0 x^{(\alpha_0 - 1)}, \tag{6.6}$$

and defined on the interval $x \in [0, 1]$, as shown in Figure 6.2. In order to be integrable for Equation (6.6), we require that $\alpha_0 > 0$. For $0 < \alpha_0 < 1$, we have a singularity at $x \to 0$. For $\alpha_0 > 1$, $P_r(x) \to 0$ when $x \to 0$, and for $\alpha_0 = 1$, the function is regular and analytic in the entire interval. Thus, we say that the function $P_r(x)$ has a singularity of type α_0 at the point $x = 0$; also generalizing this concept to the case $\alpha_0 > 1$. After dividing the unit interval into L^{-1} segments, each of size L, as illustrated in Figure 6.2, the integration of $P_r(x)$ over each segment is defined by

$$P_{ri}(L) = \int_{(i-1)L}^{iL} P_r(x)\,dx, \tag{6.7}$$

where $i = 1, 2, \ldots,$ and L^{-1} is an index characterizing a particular segment. The first

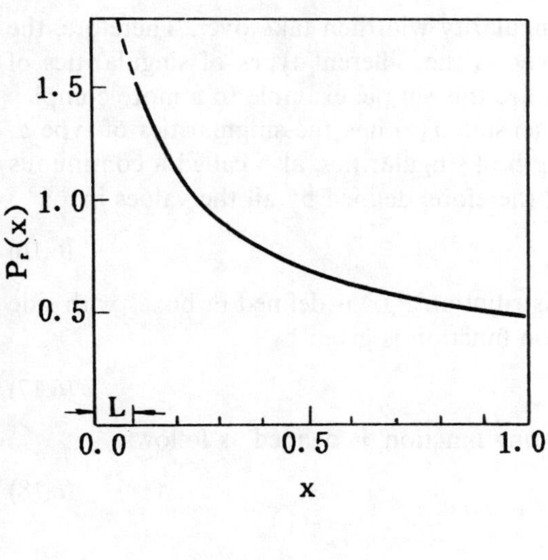

Figure 6.2. Schematic picture of a probability distribution $P_r(x)$ on the interval [0,1] with an integrable singularity at $x = 0$ (Pietronero et al., 1986).

segment contains the singularity and gives

$$P_{r1}(L) = P_{rs}(L) = \int_0^L P_r(x)\,dx = L^{\alpha}. \tag{6.8}$$

For the remaining $(L^{-1} - 1)$ segments, we have

$$P_{ri}(L) \approx L \quad (i = 2, 3, \ldots, L^{-1}). \tag{6.9}$$

We then construct the function as follows:

$$X(q, L) = \sum_{i=1}^{L^{-1}} [P_{ri}(L)]^q. \tag{6.10}$$

This function is usually called the partition function in the literature, but one could call it the 'hunting function' because it acts like a hunter of singularities. It is easy to see that in the limit of small L, this function behaves as a power of L,

$$X(q, L) \approx L^{\Gamma(q)}. \tag{6.11}$$

Visually, the contribution of the segment that contains the singularity gives

$$X_s(q, L) = L^{q\alpha_0}, \tag{6.12}$$

while the normal contribution of all the other $L^{-1} - 1 \approx L^{-1}$ segments is

$$X_N(q, L) \approx L^{(q-1)}. \tag{6.13}$$

The smallest of these two exponents for small L will provide the main contribution to Equation (6.10) and, therefore,

$$\Gamma(q) = \min[(q - 1), q\alpha_0]. \tag{6.14}$$

This result can be interpreted by saying that the distribution $P_r(x)$ consists of two types of singularities: one with $\alpha = \alpha_0$ defined on a fractal set with dimension $f = 0$ (the point $x = 0$), and one of type $\alpha = 1$ defined on a set of fractal dimensions $f = 1$ (the whole interval). By varying q from $-\infty$ to $+\infty$ in Equation (6.10), one of these two contributions dominates at the beginning; however, when the value of q is defined by

$$q - 1 = q\alpha_0, \tag{6.15}$$

the contribution of the other type of singularity will then take over. Therefore, the variation of q in Equation (6.10) behaves as the different types of singularities of the distribution. It is now easy to generalize this simple example to a more complex case. Assume that a set with fractal dimension $f(\alpha)$ has the singularities of type α. Fractal dimension $f(\alpha)$ is here the strength of singularities, also called a continuous spectrum of singularities. The system is, therefore, defined by all the values in

$$\{\alpha, f(\alpha)\}. \tag{6.16}$$

If we divide the volume on which the distribution $P_r(r)$ is defined in boxes with side length L, for the i-th box, the distribution function is given by

$$P_{ri}(L) = \int_{i\text{-th box}} P_r(r)\,dr. \tag{6.17}$$

Similarly, from Equation (6.10), the 'hunting function' is defined as follows:

$$X(q, L) = \sum_{[\text{all boxes}]} [P_{ri}(L)]^q \approx L^{\Gamma(q)}, \tag{6.18}$$

and for small L it behaves as a power of L with exponent $\Gamma(q)$. If the box i contains a singularity of type α, we have

$$[P_{ri}(L)]^q \approx L^{q\alpha}. \tag{6.19}$$

Visually, the singularities of type α are actually defined on a set with fractal dimension $f(\alpha)$. Therefore, the number of boxes that contain this type of singularity should satisfy the fractal relationship defined by Equation (2.11) (i.e., $N_{(L)} \propto L^{-f(\alpha)}$). The density of singularities of type α for boxes of size L, $n(\alpha, L)$, is defined by

$$n(\alpha, L)\mathrm{d}\alpha = \rho(\alpha)L^{-f(\alpha)}\mathrm{d}\alpha, \tag{6.20}$$

where $\rho(\alpha)$ is a regular function. Thus, Equation (6.18) can be rewritten as

$$X(q) \approx N_{\alpha(L)}[P_{ri}(L)]^q = \int \rho(\alpha)L^{[q\alpha - f(\alpha)]}\mathrm{d}\alpha. \tag{6.21}$$

As L is very small, this integral is dominated by a minimum number of the possible exponents. This means that for a given q, there will be a single pair of values $\alpha(q)$ and $f[\alpha(q)] = f(q)$ that gives the main contribution. The saddle condition is

$$(\mathrm{d}/\mathrm{d}\alpha)[q\alpha - f(\alpha)]|_{\alpha=\alpha(q)} = 0, \tag{6.22}$$

implying that

$$\mathrm{d}f/\mathrm{d}\alpha = q \tag{6.23}$$

Finally, we have

$$\Gamma(q) = q\alpha(q) - f(q) \tag{6.24}$$

and

$$\mathrm{d}\Gamma(q)/\mathrm{d}q = \alpha(q). \tag{6.25}$$

Return to the example of the self-similar distribution defined in Figure 6.1 and analyze it as a multifractal. Since the iteration procedure is strictly self-similar, we can consider a single iteration. This splits a constant distribution over a given interval into three peaks with probabilities $P_{r1} = \frac{1}{4}$, $P_{r2} = \frac{1}{2}$, and $P_{r3} = \frac{1}{4}$, each occupying an interval of length $\frac{1}{4}$ with respect to the initial segment. From Equation (6.18), we have:

$$\sum_{i=1}^{3} [P_{ri}]^q = 2(\tfrac{1}{4})^q + (\tfrac{1}{2})^q = [\tfrac{1}{4}]^{\Gamma(q)} \tag{6.26}$$

Clearly, $\Gamma(q)$ is directly determined from Equation (6.26). From Equations (6.24) and (6.25), $(\alpha, f(\alpha))$ spectrum can be obtained (Pietronero et al., 1986) as illustrated in Figure 6.3. This makes it clear that a continuum spectrum of singularities, α, and corresponding fractal dimensions, $f(\alpha)$, is necessary to characterize even the very simple self-similar distribution of Figure 6.1.

Pietronero et al. have also studied the multifractal spectrum $(\alpha, f(\alpha))$ of the DBM as shown in Figure 6.4. In Equation (5.3), for $m = 0$ the spectrum essentially consists of a single point $f = \alpha = 1$. For $m \neq 0$, the development of a continuum spectrum $(\alpha, f(\alpha))$ is broader for larger values of m. It should be noted that the results given in Figure 6.4 correspond to clusters of relatively small size.

Amitrano et al. (1986) used an electrostatic formation of the DLA problem and numerically solved the resulting equations. The $f(\alpha)$ curve they obtained is reproduced

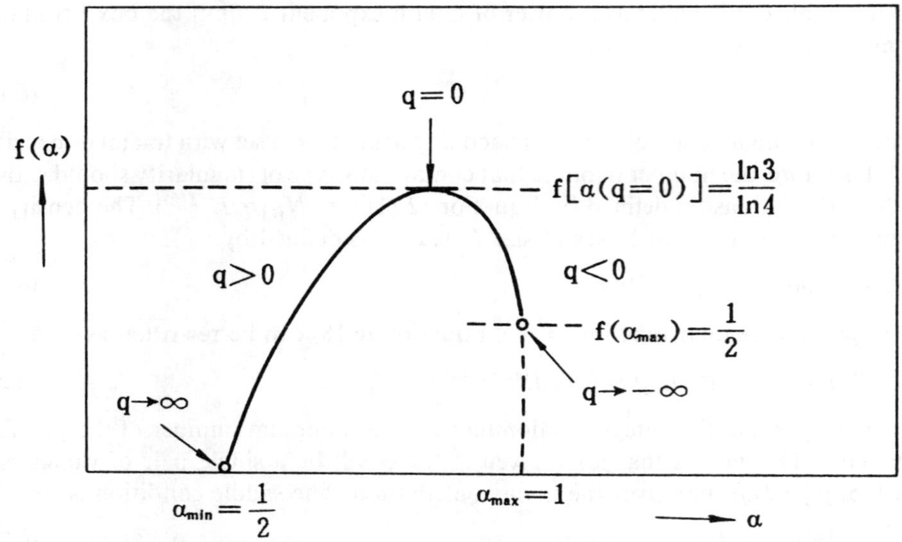

Figure 6.3. Multifractal spectrum (f, α) obtained by the analysis of the self-similar distribution shown in Figure 6.1 (Pietronero et al., 1986).

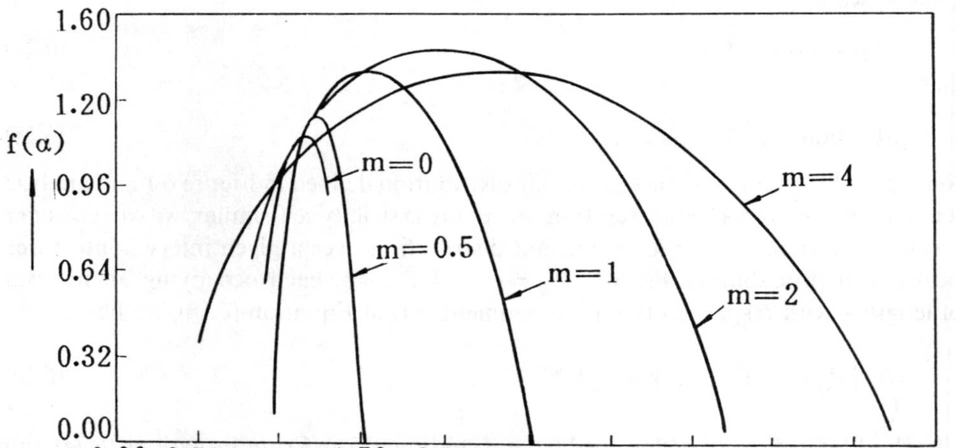

Figure 6.4. Spectrum of singularities (f, α) obtained by a multifractal analysis of the growth probability for the dielectric breakdown model for different values of the parameter m (Pietronero et al., 1986).

in Figure 6.5a. Note that $f(\alpha)$ is the fractal dimension of a subset of the sites, and negative values of $f(\alpha)$ in Figure 6.5a are artifacts that result from averaging many independent simulations for probabilities P_{ri}. Nittmann et al. (1987) have calculated the fractal dimension spectrum $f(\alpha)$ for viscous fingering structure, as shown in Figure 6.5b.

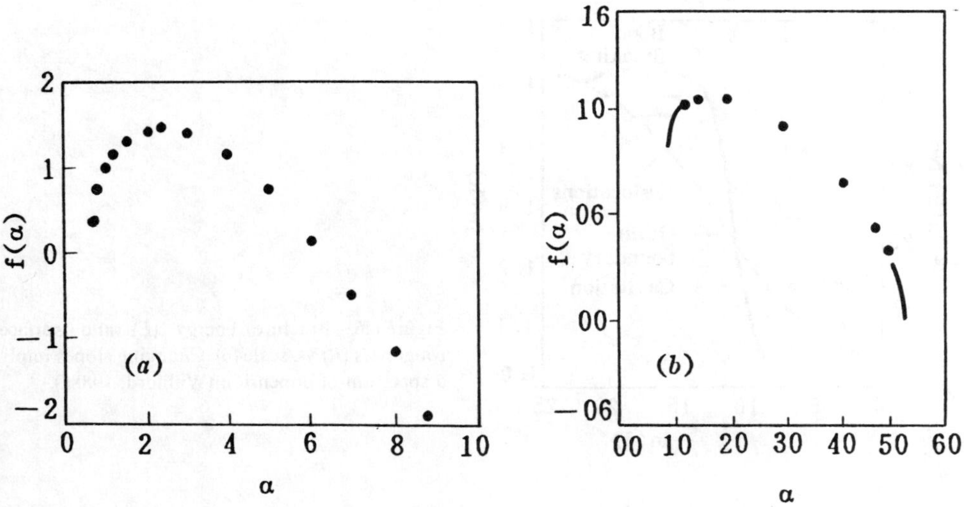

Figure 6.5. (a) The $f(\alpha)$ curve of DLA clusters. The largest cluster used in the estimate contained 150 particles (Amitrano et al., 1986); (b) The $f(\alpha)$ curve for a harmonic measure calculated for a viscous fingering structure observed in a Hele-Shaw cell (Nittmann et al., 1987).

6.3 MULTIFRACTAL FRACTURE

In Chapter 3, we have discussed the fractal characters of fracture surfaces. Only a single dimension has been used to statistically characterize the fracture surfaces. In fact, since the slope of correlation between $\log(L(\delta))$ and $\log\delta$ changes with the scale δ, a spectrum of fractal dimensions is needed to fully describe fracture surfaces at all scales. The true fracture surface area increases with increasing $L(\delta)$ and, thus, with increasing D. Letting R be the roughness of fracture surfaces, E (its unit is Nm) the total energy of fracture, and r (its unit is N/m) the surface energy, Williford (1988a) has studied the scaling similarities between E and scale δ, R vs. δ (Figure 6.6, and see Section 14.2), and the fractal dimension spectrum $f(\alpha)$ of fracture surfaces. He collected all the data of positive and negative correlation between E and D (or K_{1c} and D, G_f and D discussed in Chapter 3) and plotted an analogy to the dimension spectrum as indicated in Figure 6.7. The horizontal axis in this figure is the apparent surface energy $r = E/\delta^2$, where δ^2 is the projected surface area. A multifractal diagram for fracture is schematically shown in Figure 6.8, where the vertical axis $f(\alpha)$ describes the corresponding distribution of the measure in terms of α, which is a 'process dimension' because of its relationship to the process measure $E \propto \delta^{\alpha}$. It is easy to see that the dashed curve in Figure 6.7 shows the emergence of an unimodal pattern quite similar to $f(\alpha)$ vs. α in the multifractal diagram given in Figure 6.8. On the vertical axis, D-2 in Figure 6.7 corresponds to f in Figure 6.8. On the horizontal axis, $\ln r$ in Figure 6.7 corresponds to $(\alpha - 2)\ln\delta$ in Figure 6.8. Williford has explained in detail this comparison as follows. On the left side of Figure 6.8, $f(\alpha)$ for very brittle ceramics is low because their microcracks are flat and essentially two-dimensional (i.e., a plane of missing atoms). An uniaxial tensile bond-breaking scenario for brittle

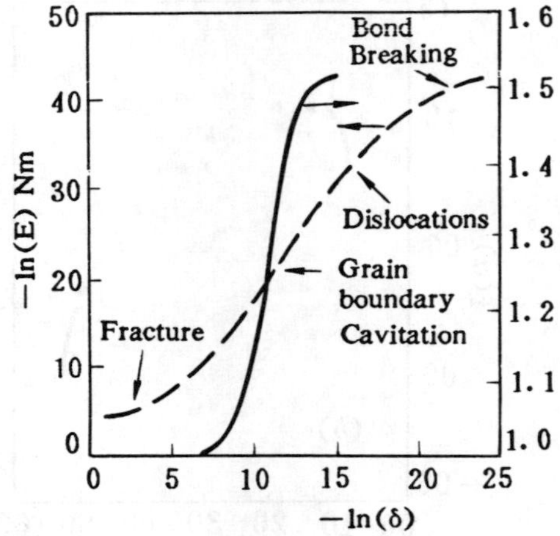

Figure 6.6. Fracture energy (E) and surface roughness (R) vs. scale (δ). Changing slopes imply a spectrum of dimension (Williford, 1988a).

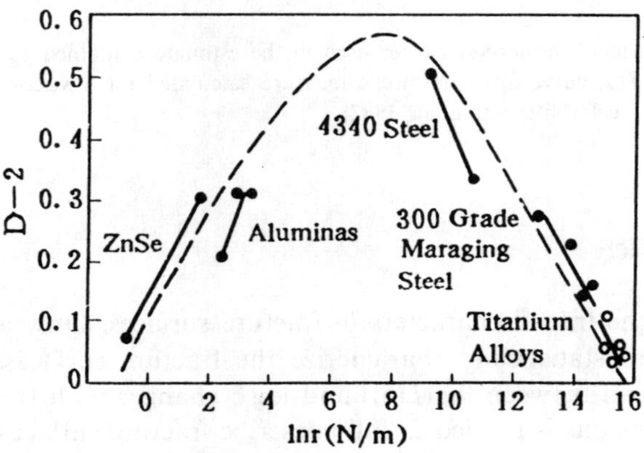

Figure 6.7. Surface dimension (D_s) vs. surface energy (r): emergence of a multifractal pattern from data (Williford, 1988a).

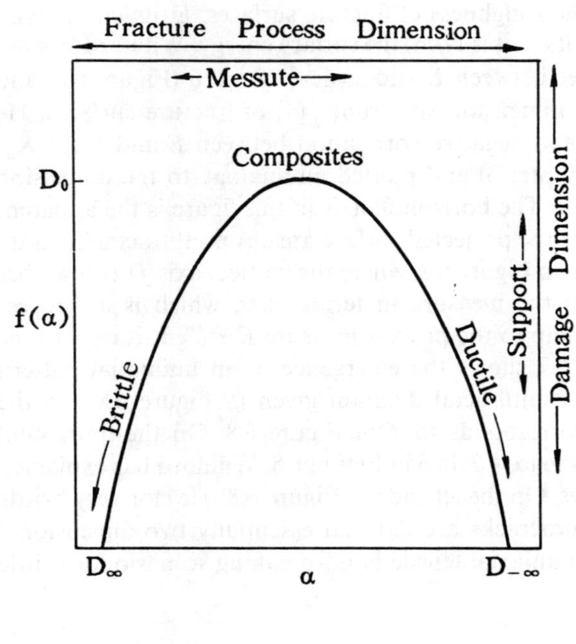

Figure 6.8. The multifractal diagram for fracture: a structure process relationship (Williford, 1988a).

fracture also leads to small $f(\alpha)$ values. For brittle materials, f (or D) increases as α (or E) increases. Thus the magnitudes and trends in $f(\alpha)$ for brittle materials seem reasonable. On the right side of Figure 6.8, f for very ductile materials is also low when damage is dominated by a tangle of dislocation cores. Each core is essentially a line with dimension one, and a tangle of cores also has a low dimension; probably less than two. A distribution of voids in a ductile material would also have low dimension, but slightly larger than a distribution of microcracks. For ductile materials, $f(\alpha)$ decreases with increasing α (or E), because the more ductile the material, the smoother the created surface. A limiting case is the large E given in the work to ideally shear (crack) a very ductile material into two halves by passing a large number of dislocations through the same plane, thus producing an atomically smooth 'fracture' surface with very low R and $D=2$. Therefore, magnitudes and trends in $f(\alpha)$ for ductile materials also seem reasonable. Heterogeneous materials such as fibrous composites which exhibit mixed fracture behavior, are expected to occupy the central position in Figure 6.8. On the horizontal axis, E for composites made of brittle fibers in a ductile matrix can fall between the E for the constituent materials, and α would be expected to behave similarly. A combination of two brittle materials in a ceramic composite can have a higher E than either constituent, whereas a metallic alloy can have an E less than its ductile constituents. On the vertical axis of Figure 6.8, an analogy between brittle bond-breaking, ductile ligament rupture, and fiber pullout mechanisms places fibrous composites at or near the top of the unimodal curve. Compared to other materials, these composites can sustain very large crack openings and exhibit very rough surfaces (high D), so their f values should be maximized near unity. These higher dimensions may occur because composite damage appears to be dominated by the multiple, large-scale, interactive mechanisms that are sometimes called 'nonlocal effects' in continuum mechanics. These nonlocal effects distribute the damage over larger distances than do the dominantly single mechanisms in more homogeneous materials.

From the above arguments, it can be concluded that the central region of the unimodal curve in Figure 6.8 implicitly contains all interactive effects, whereas the ends (α-axis intercepts) of the curve denote only ideal brittle (tensile) or ductile (shear) mechanisms. Mixed mechanisms are actually combinations of these two ideal mechanisms, as reflected in the character of the fracture surface. Examples are cleavage facets (ref. microfracture in Chapter 10) distributed over a generally ductile fracture surface. Such a set of cleavage facets is one of many subfractals in the overall damage fractal, each with its own $(\alpha, f(\alpha))$ or (D, E) values. Because each of these subfractals corresponds to some particular mixture of tensile and shear mechanisms, these mechanisms provide a partitioning of the fracture process.

The above brief description of the multifractal concept given by Williford, based on the analyses of physical mechanisms of fracture processes in different type of materials, provides a new understanding of the correlation between the fracture toughness (or energy) and fractal dimension f (or D). In order to fully understand the correlation (ordering) between the energetics of various fracture mechanisms and the complicated geometries of material damage, Williford has given a quantitative description by a multifractal fracture model based on the brittle/ductile partition of fracture. The construction of this model is illustrated in Figure 6.9. Dividing a line of unit length into three segments at random, the middle segment is removed; two segments of lengths l_1 and l_2 remain. The same procedure is repeated for each of these remaining segments, and for the subsequently smaller segments. The segments correspond to brittle (l_1), ductile (l_2), and mixed ($l_1 l_2$) regions of the fracture surface and, thus, provide the support for a multifractal model of fracture. Each l_{ij} corresponds to a probability P_{rij} that fracture will occur on it, i.e., P_{r1}, P_{r2}, P_{r12} for l_1, l_2, and $l_1 l_2$, respectively. The mathematical expression for this model is given by Halsey in 1986 as follows:

$$\ln(n/m)\ln(l_1/l_2) - \ln(n/m - 1)\ln(l_1) = q[\ln(P_{r1})\ln(l_2) - \ln(P_{r2})\ln(l_1)], \tag{6.27}$$

$$\alpha = [\ln(P_{r1}) + (n/m - 1)\ln(P_{r2})]/[\ln(l_1) + (n/m - 1)\ln(l_2)], \tag{6.28}$$

$$f = [(n/m - 1)\ln(n/m - 1) - (n/m)\ln(n/m)]/[\ln(l_1) + (n/m - 1)\ln(l_2)], \tag{6.29}$$

and

$$(q - 1)D_q = q\alpha(q) - f(q), \tag{6.30}$$

where n and m are the numbers of each type of line segments, q is an index, and D_q is the generalization dimension (ref. Section 2.7). For given l_1, l_2, P_{r1}, P_{r2}, and q, f vs. α diagrams can be determined from Equations (6.27)-(6.29). From Equation (6.30), we can find that $f = D_0$ at $q = 0$, the dimension of the overall damage fractal; at $q = \infty$ the process dimension is $\alpha = \ln(P_{r1})/\ln l_1$, and at $q = -\infty$, $\alpha = \ln P_{r2}/\ln l_2$. These results define the intercepts D_∞ and $D_{-\infty}$ on the horizontal axis of Figure 6.8 for 'ideal brittle' and 'ideal ductile' mechanisms, respectively. They physically correspond to tensile rupture of single bonds (brittle) and bond breaking in shear by dislocation motion (ductile).

For Cantor-type models of surfaces, $D = 2 + f(q = 1)$ (Liu et al., in 1987). As an example of how to calculate $f(q = 1)$, Williford analyzed the experimental results of $D = 2.51$ of the fractal surface for AISI 4340 tempered at 200°C obtained by Underwood & Banerji (1986), where fracture behavior was dominated by ductile fracture along martensite lath boundaries. However, up to 70% quasicleavage has also been obtained in this material, and such brittle behavior is herein assumed to originate from fracture across the martensite packets of different orientations than for the above ductile mechanism. Micrographs have shown about 8/3 martensite packets of roughly equal size across a typical prior austenite grain, regardless of orientation. When the grain diameter is taken to represent unit l_{ij} in Figure 6.9, then values of l_{ij} are, thus, $l_1 = l_2 = 0.375$. Note that these l_{ij} serve as scale factors for the brittle and ductile packets, relative to the grain rather than ideal substructure sizes,

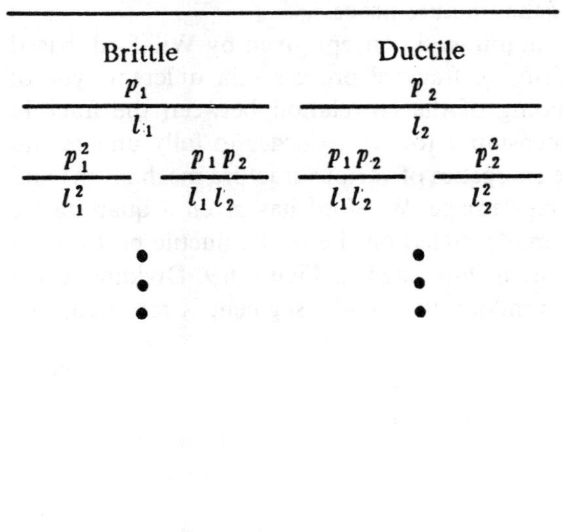

Figure 6.9. A two-scale Cantor model for partitioned fracture processes (Williford, 1988a).

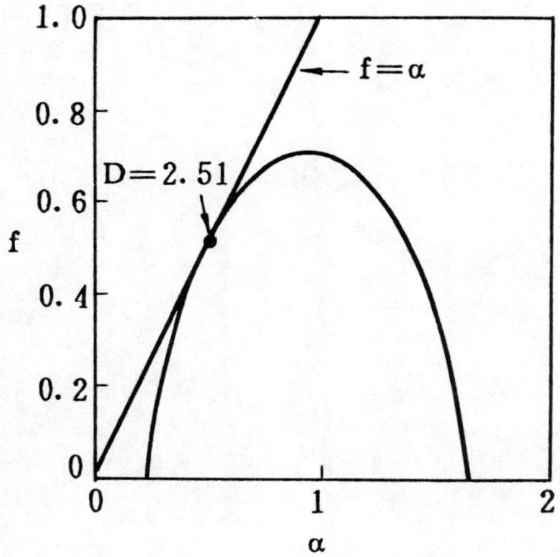

Figure 6.10. Results of model calculations for AISI 4340 (Williford, 1988a).

because here the grain diameter is in the low cutoff of the substructure sizes. Probabilities have been estimated from the hardness of martensite ($H_1 = 755 \pm 0.34$ HV) and ferrite ($H_2 = 135 \pm 0.48$ HV) on the basis that the harder material would have a lower E and a higher probability of fracture: $P_{r1} = H_1/(H_1 + H_2) = 0.8$ (martensite), and $P_{r2} = H_2/(H_1 + H_2) = 0.2$ (ferrite). The results of calculations from Equations (6.27)-(6.29) are indicated in Figure 6.10, herein the calculated $f(q = 1) = 0.51$ is in agreement with the measured $D - 2 = 0.51$.

CHAPTER 7

Self-inverse fractals*

Fractal geometry begins by dealing with 'linear fractal', i.e., the fractal with self-similarity, but the concept of fractals is not dependent on being self-similarity. Some invariant sets under transformation group with respect to the geometric inversion also behave the character of fractal, and are called self-inverse fractals. The first self-inverse fractal was introduced in the 1880s by Poincare and Klein, while self-inverse fractals have currently become a special topic of the theory of autographic functions.

In this chapter, the concept of fractal geometry derived from geometric inversion (Mandelbrot, 1982) and a celebrated self-inverse fractal are introduced. We constructed a ramified self-inverse fractal with symmetry and a self-inverse fractal dust set (Zhang & Xie, 1990) and extended the concept of fractal osculation and proposed a new notion – a fractal envelope. Finally, two examples of self-inverse fractals (soap and eggs) are discussed.

7.1 STANDARD GEOMETRIC INVERSION

Given a circle C with an origin O and radius R, an inversion with respect to C transforms point P to point P', such that P and P' lie on the same half line from O (the points P and P' differ from O), and the length $|OP|$ and $|OP'|$ satisfy $|OP| \times |OP'| = R^2$.

Inversion is a conformal, and periodic mapping with period 2 (the straight lines being regarded as circles of radius ∞). It has the following properties:

Theorem 7.1: Circles not containing O invert into circles.

Theorem 7.2: Mutually orthogonal (or tangent) circles invert into mutually orthogonal (or tangent) circles.

Theorem 7.3: The points P and P' are inverse points of each other with respect to C (the point P differ from P'); thus, circles passing through P and P' invert into themselves, and orthogonals invert to C.

Theorem 7.4: C is a circle with origin O and radius R, the points A and B (A and B differ from O) invert into A' and B' with respect to C, thus the lengths $|AB|$ and $|A'B'|$ satisfy $|A'B'| = R^2|AB|/(|OA|\cdot|OB|)$.

7.2 SELF-INVERSE SETS, SELF-INVERSION FRACTALS

7.2.1 *Self-inverse sets*

As linear fractals, we begin with a 'generator' which in the present case is made up of any number M of circles. Subsequently, the transformation group consists of a succession of inversions with respect to these circular forms, and we call it G. A self-inverse set is an invariant set under the operation of the group G.

For any nonempty set S (called a seed), the union of images under the transformation of S by all the operations of G is called the clan of S. The clan of S is a self-inverse set of group G. Given a transformation group G, if the clan of every domain S covers the entire plane, then the self-inverse set is the entire plane, and such groups are called chaotic. Nonchaotic groups were proposed by Poincare. If G is a nonchaotic group, the smallest one in the self-inverse sets of G is called the limit set, and is denoted by L. It is also the set of limit points of the transformation of any initial point under the operation of G. The limit set may reduce to a point or a circle, etc. but, in general, it is a fragmented and/or irregular fractal set.

7.2.2 *Apollonian nets and gaskets*

A set L is to be called Apollonian if it is made of an infinite number of circles plus their limit points. We now construct an Apollonian net, then show that it is self-inverse.

The three given circles are mutually tangent. There is no loss of generality in assuming that two of the given circles are exterior to each other but contained within the third (Figure 7.1). These three circles define two circular triangles with angles of 0°. The largest circle is inscribed in each circular triangle, and such a circle is called an Apollonian circle. The Apollonian circles are tangent to the circles that constitute circular triangles. Figure 7.1 shows two Apollonian circles of these circular triangles. The construction in Figure 7.1 concludes with five circles which, together, define six

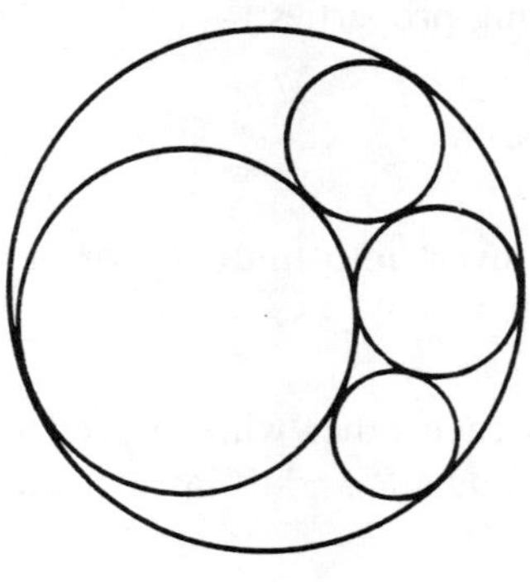

Figure 7.1. Apollonian circles.

circular triangles. By repeating the same procedure infinitely, one can obtain Apollonian packing. By adding limit points to the resulting infinite collection of circles, one obtains a set called an Apollonian net. A portion of an Apollonian gasket within a circular triangle is shown in Figure 7.2, and the boundaries of these black discs, plus their limit points, is an Apollonian net within the circular triangle (Mandelbrot, 1982).

We will now show that the Apollonian net is self-inverse. Let us return to the starting point of the construction of an Apollonian net: three circles tangent to each other, add any one of the corresponding Apollonian circles, and the resulting four circles are called Γ circles (Figure 7.3). There are four combinations of the 3 by 3 circles (called triplets), and a circle responds to each. The circle contains three points whose circles are mutually tangent in the triplet and orthogonal to each circle in the

Figure 7.2. Apollonian net.

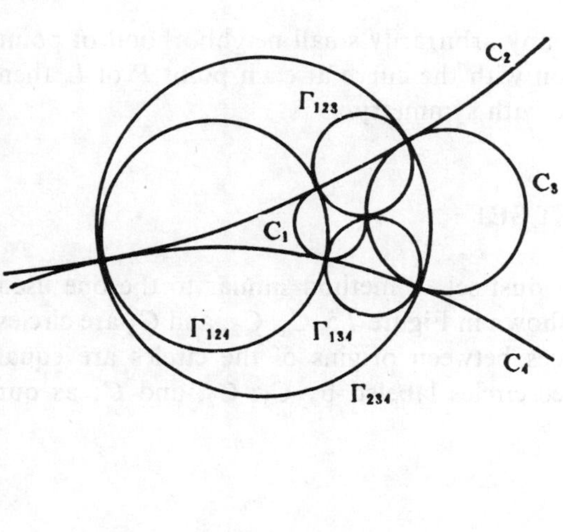

Figure 7.3. The Γ_{ijk} circles.

triplet. Four circles are defined by four triplets. We label them as $C1$, $C2$, $C3$, and $C4$ (Figure 7.3); and take the four circles as our generator. Every circle of the Γ circles is orthogonal only to the three circles in the generator, and the Γ circles orthogonal to C_i, C_j, and C_k will be labeled as Γ_{ijk} (as shown in Figure 7.3). Inversion with respect to C_i, C_j, and C_k transforms Γ_{ijk} to itself. With respect to the other generator transforms Γ_{ijk}, the Apollonian circle of the circular triangle is constructed by circles in Γ circles except Γ_{ijk}. In this way, we get four Apollonian circles and an inversion with respect to C_1, C_2, C_3, and C_4. We then transform them to 12 Apollonian circles within the circular triangles that are constructed by Γ circles and the four Apollonian circles. By infinite further repeating the same procedure, we assume that the group G is generated by transformations made of a succession of inversions with respect to the C_1, C_2, C_3, and C_4 forms. The self-inverse set, made up of the clan of operations of G plus their limit points, then becomes the Apollonian net constructed on the Γ circles. Therefore, the Apollonian nets are self-inverse.

The circular triangles of Apollonian packing are not similar to each other, hence the Apollonian cascade is not self-similar. The derivation or evaluation of the fractal dimension D is surprisingly difficult. So far, the best evaluation obtained by Boyd is that $1 < D < 1.3058$. Since the topological dimension $D_T = 1$, the Apollonian nets are self-inverse fractal curves.

7.3 RAMIFIED SELF-INVERSE FRACTALS WITH SYMMETRY

In this section, a ramified self-inverse fractal curve with symmetry will be constructed. We take 18 circles as our generator. Figure 7.4 shows them having thin curves. In the first stage, let us replace every circle C_i by the inverses in C_i of the circles other than C_i, thus creating 306 circles. Figure 7.4 shows them as boundaries of these black discs. In the second stage, replace the 306 circles by 5202 new circles (first appearing at this stage) that are inverted with respect to the 18 circles of the generator. By infinite repetition of the same procedure, the final outcome is a limit set denoted by L. It is self-inverse.

It is clear that L is also symmetric. In addition, according to the construction of the limit set L, we suppose the fractal dimension D of L is less than two, but not less than the fractal dimension of the Apollonian nets. That is to say,

$$1.3058 \leqslant D < 2.$$

On the other hand, if the boundary of any arbitrarily small neighborhood of point P has more than two points in common with the curve at each point P of L, then L is a ramified self-inverse fractal curve with symmetry.

7.4 SELF-INVERSE FRACTAL DUST SET

We now construct a self-inverse fractal dust set. A method similar to the one used in the previous section is used here. As shown in Figure 7.5, C_1, C_2, and C_3 are circles with the same radius, and the distances between origins of the circles are equal and nonoverlapping. We take the three circles labeled by C_1, C_2, and C_3 as our

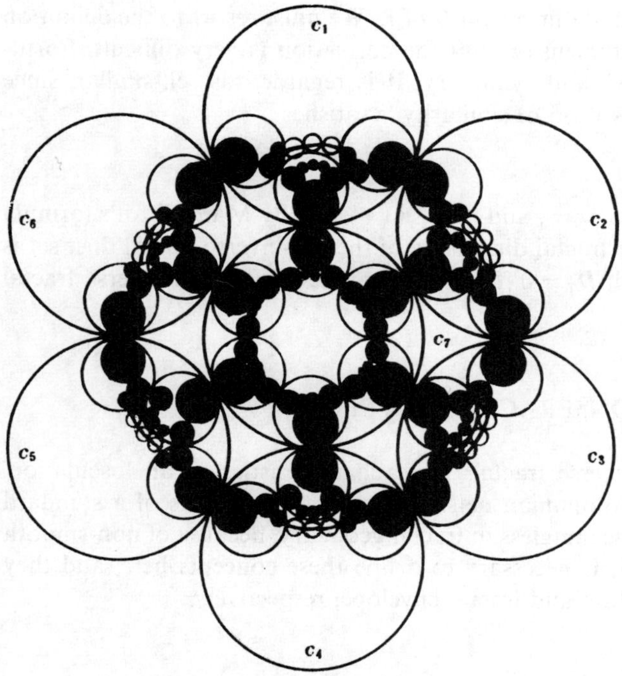

Figure 7.4. A ramified self-inverse fractal set.

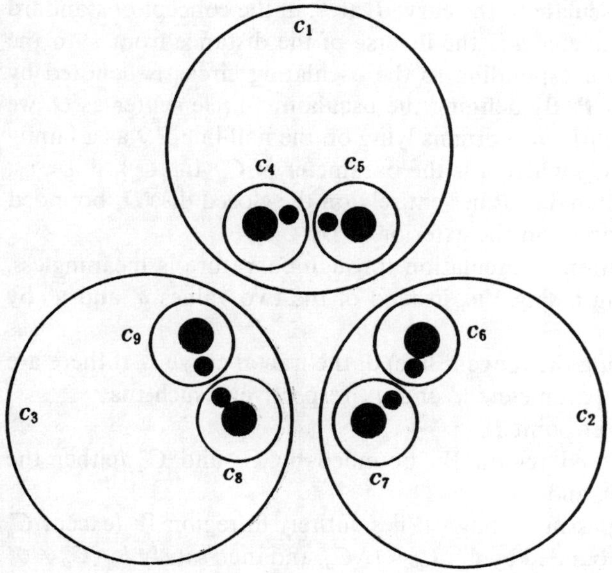

Figure 7.5. A self-inverse fractal dust set.

generator. In the first stage, we replace C_i by the inverses in C_i of C_j and C_k, thus creating six circles defined of $C_4, C_5, \ldots, C_9$. In the second stage, we replace the 6 circles by the 12 new circles ($C_{10}, C_{11}, \ldots, C_{21}$) that are inverted with respect to the 3 circles of the generator. By infinitely repeating the same procedure, the result is a set of limit points denoted by L, and it is self-inverse.

Let us now evaluate the fractal dimension D of L. We must resort to the definition of the Hausdorff-Besicovitch dimension, but the derivation is very difficult. Fortunately, L has good 'similarity', and symmetry. It is regarded as self-similar, since $N=2$ and by Theorem 7.4, its ratio of similarity, r satisfies

$$\tfrac{1}{4} < r_i < \tfrac{2}{5}.$$

For the limit case we find ratio $r_1 = \frac{1}{4}$ and $r_2 = 2/(3+\sqrt{5})$. By Mandelbrot's formula $\sum r_i^D = 1$, we conclude that the fractal dimension of the self-inverse fractal dust set is $D = 0.597$. On the other hand, $D_T = 0$, thus, the limit set L is a self-inverse fractal dust set.

7.5 FRACTAL OSCULATION, FRACTAL ENVELOPE

It is noted that many self-inverse fractals have characteristics of an 'osculation' and/or 'envelope' similar to Apollonian nets. However, the concepts of a standard osculation and envelope are meaningless in fractal geometry, because of non-smooth of the fractal curves. Hence, it is necessary to define these concepts here, and they will be called a fractal osculation and fractal envelope, respectively.

7.5.1 *Osculation, fractal osculation*

In order to describe the circle osculate to the curve L at P, in the concept of standard osculation, the convenient parameter u is the inverse of the distance from P to the circle center. This parameter, corresponding to the osculating circle, is denoted by u_0, and is the curvature of L at P. By defining the osculating circle center as O, we accept all circles passing through P with origins lying on the half line PO as a family of circles, and denote them by C_u, where u is the parameter of C_u. In $\{C_u\}$, if $u < u_0$, a small portion of the curve centered at P lies entirely on the closed disc D_u bounded by C_u, while if $u > u_0$ it lies entirely on the exterior of $D_u \backslash C_u$.

For fractal curves, the definition of osculation through curvature is meaningless. We now define osculation using a single u_0 instead of the two values u' and u'' by the following definition:

For the regular point P (having the tangent line) of the fractal curve L, if there are two circles C'_u and C''_u with the parameters u' and u'', respectively, such that:

(1) C'_u and C''_u are tangent L at point P;

(2) L lies entirely on the closed region W, bounded by C'_u and C''_u (either the bounded or unbounded region); and

(3) For any circle, when C_u passing though P lies entirely in region W (except C'_u and C''_u), let D_u be a closed disc bounded by C_u, $L_u = L \backslash C_u$, and then satisfy $L_u \backslash C_u \neq \varnothing$ and $L \backslash L_u = \varnothing$; C'_u and C''_u are both called (global) fractal osculating circles of L.

The Apollonian circles are fractal osculating circles of the Apollonian nets.

It is necessary to point out that one of the osculating circles may degenerate to a point or a straight line (the point being regarded as a degenerate circle, the straight line being regarded as a circle of radius ∞) the above definition of fractal osculation is global, and fractal osculation may be locally redefined.

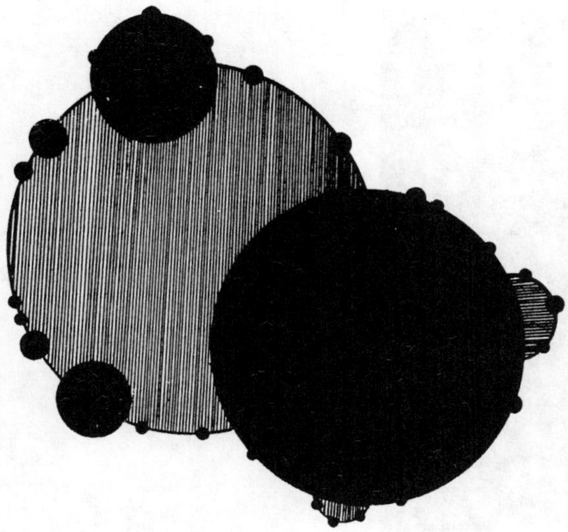

Figure 7.6. Fractal envelope.

7.5.2 *Envelope, fractal envelope*

If every point on curve *L* is tangent to a curve of a family of curves, and each of these family of curves is tangent to *L*, then *L* is called the envelope of the family of curves.

If curve *L* is the envelope of a family of curves, then *L* is a smooth curve. It is harsh terms that the fractal curves must be smooth. Replacing smooth by a.e. (almost everywhere) smooth, we define the concept of fractal envelopes as follows:

If a fractal curve *L* meets the following criteria:

(1) the points on L are regular except finite or countable;

(2) for any regular point *p* of *L*, there are curves of the family of curves tangent to *L* at *p*, and

(3) each of the family of curves is tangent to *L*; then *L* is called the fractal envelope of the family of curves. The outline of the shadow (black and gray) in Figure 7.6 is a fractal curve *L*, and *L* is the envelope of the family of curves that is composed of the circles of the shadow.

7.6 THE APOLLONIAN MODEL OF LIQUID CRYSTALS

In this section two examples of self-inverse fractals (eggs and soap) are given. We will first use the Apollonian model of eggs because they are a practical example of a fractal playing a role in the description of a category of 'liquid crystals.'

Using an egg custard for our illustration, we first of all place the eggs in a bowl, and slightly hot water or milk, and then beat the eggs with a beater. After beating, we discover that the surface of the 'eggs' forms a foam layer (Photo 7.1 shows a picture of this foam layer on negative film). This is our Appolonian model. The surface of 'eggs' is covered by large and small bubbles, and the shape is relatively stable. It is only when this thick soupy custard is heated that these bubbles are released.

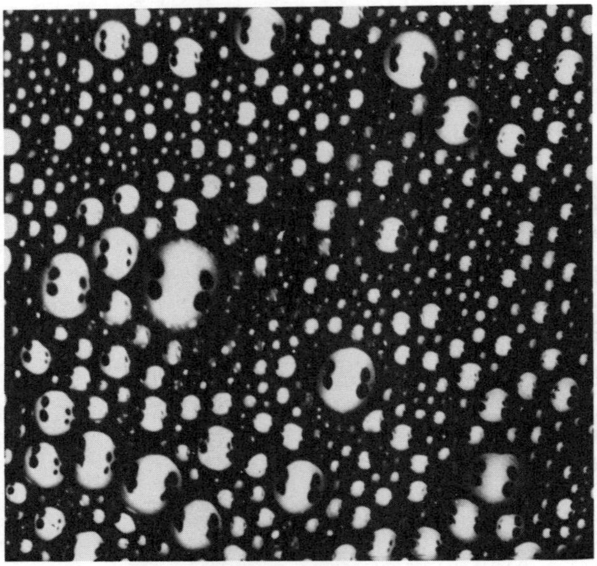

Photo 7.1. Foam layer on negative film.

Since liquid crystals are relatively unknown, we describe them by paraphrasing Bragg in 1934. These beautiful and mysterious substances are liquid in their mobility but crystalline in their optical behavior. Their molecules are relatively complicated structures, being lengthy and chain-like.

Some liquid crystal phases constitute a model of a soap-like organic system. A smectic liquid crystal is made of molecules arranged side by side, like corn in a field. The thickness of the layer is the molecules length. The resulting layers are very flexible and strong, tending to straighten out when being bent and then released.

There are rich substances in which self-inverse fractals are regarded as the invariant set under the operation of the inversion group. However, self-inverse fractals are not wedded to geometric inversion. It is also a self-inverse fractal when the invariant set under the operation of a transformation group, particularly, a conforming transformation group. For example, the transformation groups generated by Mobius transformation or compound Mobius transformation with conjugate transformations, in which a large amount of self-inverse fractals may be constructed. For the application of self-inverse fractals, we have confidence in fractal simulations of macromolecule chains and the ductile damage evolution of material.

CHAPTER 8

Fuzzy fractals*

Natural phenomena may be roughly divided into two categories: one is determinacy, and the other is indeterminacy. Indeterminacy can also be divided into randomness and fuzziness. Today one can investigate some irregular phenomena with determinacy and randomness by fractal geometry. However, so far, the fractal geometric theory and its applications have only been confined to these two natural phenomena. In order to describe indeterminate fuzzy phenomena by the fractal geometric theory, it is very important to define new fractal concepts and construct new fractal sets. In this chapter, we introduce our recent work in this respect (Zhang & Xie, 1989).

8.1 DIMENSIONS OF FUZZY SETS

Letting universe U be a metric space and the set constructed by subsets of U be denoted by $F(U)$, then a fuzzy set $A \in F(U)$ is different from the common sets. The relationship between the elements of U and A is characterized by the membership function $\mu_\lambda(u)$.

Definition 8.1: Let $A \in F(U)$ and its support, supp A, be bounded. For $0 < \lambda \leq 1$, when $N_\lambda(R)$ denotes the smallest number of balls of radius R that cover the λ-cutset A_λ of A, then the lower limit is defined by

$$\operatorname*{limit\,inf}_{R \to 0}(\ln N_\lambda(R)/\ln(1/R)). \tag{8.1}$$

Equation (8.1) is also called the definition of the lower entropy dimension of the λ-cutset A_λ of A and is denoted by $\underline{D}(\lambda)$. Similarly, the upper limit,

$$\operatorname*{limit\,sup}_{R \to 0}(\ln N_\lambda(R)/\ln(1/R)), \tag{8.2}$$

is called the upper entropy dimension of the λ-cutset A_λ of A, and is denoted by $\bar{D}(\lambda)$.

When $\underline{D}(\lambda) = \bar{D}(\lambda)$, let $D(\lambda) = \underline{D}(\lambda) = \bar{D}(\lambda)$ and $D(\lambda)$ be called the entropy dimension of the λ-cutset A_λ of A.

Definition 8.2: Let $M(\lambda) = \{\ln N_\lambda(R)/\ln(1/R) \mid R \in (0, +\infty)\}$, and $\hat{D}(\lambda)$ denote the set of all accumulation points of the set $M(\lambda)$. $\hat{D}(\lambda)$ is called the generalized entropy dimension of the λ-cutset A_λ of A.

Definition 8.3: Let $A \in F(U)$ and its support, supp A, be bounded. Then $N_{\text{supp}\,A}(R)$ denotes the smallest number of balls of radius R that covers supp A of A, its lower limit,

$$\underset{R\to 0}{\text{limit inf}}(\ln N_{\text{supp}\,A}(R)/\ln(1/R))$$

is called the lower entropy dimension of supp A and is denoted by $\underline{D}$. Similarly, the upper entropy dimension $\bar{D}$ of supp A is defined as follows:

$$\bar{D} = \underset{R\to 0}{\text{limit sup}}(\ln N_{\text{supp}\,A}(R)/\ln(1/R)).$$

When $\underline{D} = \bar{D}$, let $D = \underline{D} = \bar{D}$, then D is called the entropy dimension of supp A.

Definition 8.4: Let $M = \{\ln N_{\text{supp}\,A}(R)/\ln(1/R) \mid R \in (0, \infty)\}$, and $\hat{D}$ denote the set of all accumulation points of the set M, and $\hat{D}$ be called the generalized entropy dimension of supp A. Similarly, we can also define the lower entropy dimension $\underline{D}_{\partial A}$, the supper entropy dimension $\bar{D}_{\partial A}$, the entropy dimension $D_{\partial A}$ and the generalized entropy dimension $\hat{D}_{\partial A}$ of the boundary set ∂A of A.

Definition 8.5: Let $A \in F(U)$ and supp A be bounded. We call A a fuzzy fractal set if $\lambda \in (0, 1]$ is such that the supper entropy dimension $\bar{D}(\lambda)$ is greater than the topological dimension $D_T(\lambda)$ of A_λ or the supper entropy dimension $\bar{D}$ is greater than the topological dimension D_T of supp A.

Definition 8.6: Let $\{S(\lambda) \mid \lambda \in \Lambda\}$ be a family of sets. If $\lambda \in \Lambda$ exists (here $\Lambda = (0, 1]$) such that the supper entropy dimension

$$\bar{D}(\lambda) = \underset{\lambda\to 0}{\text{limit sup}}(\ln N_\lambda(R)/\ln(1/R))$$

of set $S(\lambda)$ is greater than its topological dimension $D_T(\lambda)$, then $\{S(\lambda) \mid \lambda \in \Lambda\}$ is called a dynamic fractal set. A fuzzy fractal set is a dynamic fractal set (for details, see the paper of Zhang & Xie (1989)).

8.2 A FUZZY FRACTAL SET, ITS DIMENSION OF PROJECTION AND SECTION

In linear fractals (i.e. self-similarity fractals), the triadic Cantor set is a typical example (Chapter 2). We can construct a Cantor set as follows: let $[0, 1]$ be the initiator and its generator be as shown in Figure 8.1. Here $N = 2$, but we choose r to satisfy $r \in [0, \frac{1}{2}]$. The ultimate construction of this Cantor set is denoted by C_r.

Let universe $U = V = [0, 1]$ and $B_r = \{(x, 2r) \mid x \in C_r,\ r \in [0, \frac{1}{2}]\}$, $A \in F(U \times V)$. Note that A is a fuzzy fractal set and its membership function becomes

$$\mu_A(u, v) = \begin{cases} 1 - 2r & \text{if } (u, v) \in B_r \\ 0 & \text{otherwise.} \end{cases}$$

Then

$$A_\lambda = \bigcup_{r \in [0, (1-\lambda)/2]} B_r, \quad 0 < \lambda \leq 1,$$

N=2; $r \in [0,1/2]$

Figure 8.1. A fuzzy fractal set based on the Cantor set.

hence

$$\underline{D}(\lambda) = \bar{D}(\lambda) = 1 + \ln 2/\ln(2/(1-\lambda)).$$

Support supp A of A has $D = 2$ and the boundary set ∂A of A has the entropy dimension $D_{\partial A} = 2$.

Let us now discuss the dimension of projection and section of the fuzzy fractal set.

For horizontal projection A_v of A, since

$$\mu_{A_v}(v) = 1 - v,$$

then the entropy dimension $D(\lambda)$ of λ-cutset $(A_v)_\lambda$ of A_v becomes

$$D(\lambda) = \begin{cases} 1, & \text{if } 0 < \lambda < 1; \\ 0, & \text{if } \lambda = 1. \end{cases}$$

Obviously, the support supp A_v of A_v has the entropy dimension $D = 1$ and the boundary set ∂A_v of A_v has the entropy dimension $D_{\partial A_v} = 1$.

For the horizontal section $A|_v$:

$$\mu_{A|_v}(u) = \begin{cases} 1 - v, & \text{if } u \in C_{r/2}; \\ 0, & \text{otherwise}; \end{cases}$$

then

$$(A|_v)_\lambda = \begin{cases} B_{v/2}, & \text{if } 0 < \lambda \leq v; \\ 0, & \text{if } v < \lambda \leq 1; \end{cases}$$

hence, the entropy dimension $D(\lambda)$ of $(A|_v)_\lambda$ is defined by

$$D(\lambda) = \begin{cases} \ln 2/\ln(2/v), & \text{if } 0 < \lambda \leq v; \\ -1, & v < \lambda \leq 1; \end{cases}$$

where the dimension of empty set $\varnothing$ equals -1 which is appointment in the axiom of dimensions.

We can obtain the entropy dimension D of supp $A|_v$:

$$D = \begin{cases} \ln 2/\ln(2/v), & \text{if } 0 < \lambda \leq 0; \\ 0, & \text{if } v = 0; \\ -1, & \text{if } v = 1; \end{cases}$$

and the entropy dimension $D_{\partial A|_v}$ of the boundary set $\partial A|_v$ of $A|_v$

$$D_{\partial A|_v} = \begin{cases} \ln 2/\ln(2/v), & \text{if } 0 < \lambda \leq 1; \\ -1, & \text{if } v = 0 \text{ or } v = 1. \end{cases}$$

For the vertical projection A_U of A, since

$$\mu_{A_U}(u) = \begin{cases} 1 - 2u, & \text{if } u \in [0, \frac{1}{2}], \\ 2u - 1, & \text{if } u \in [\frac{1}{2}, 1]; \end{cases}$$

then $(A_U)_\lambda = [0, (1-\lambda)/2] \cup [(1+\lambda)/2, 1]$. Hence, the entropy dimension $D(\lambda)$ of $(A_U)_\lambda$ becomes:

$$D(\lambda) = \begin{cases} 1, & \text{if } 0 < \lambda < 1, \\ 0, & \text{if } \lambda = 1. \end{cases}$$

It is clear that the support supp A_U of A_U has $D = 1$ and the boundary set ∂A_U of A_U has the entropy dimension $D_{\partial A_U} = 1$.

Here, we only give the mathematical description of fuzzy fractal sets. Its applications to the nature should be very interesting to further investigate and explore.

CHAPTER 9

The theory of large deformation and its application

Strain and rotation are basic characteristic quantities which determine the form of motion of a deformable body. The concept of local rotation at a point in a deforming body during its motion is quite different from that of intrinsic rigid body motion. The latter is a special case of the former.

We should first note that the statement about the existence of an infinite number of definitions of a finite strain is a misleading idea. The reasons are:

(1) strain components must be tensor components, they cannot be arbitrarily defined;

(2) they must be zero when the displacement function is a rigid motion;

(3) when a strain energy function is formed, it should have physical meaning; and

(4) the geometric meaning of the definition fits physical measurements.

Until now, there remain only three definitions of finite strain which are rigorous in mathematical character:

(a) finite strain defined by a metric tensor, the so-called Green's strain;

(b) defined by the polar decomposition theorem; and

(c) defined by the S-R decomposition theorem (a new theory).

As for Green's strain, the finite strain and finite rotation are defined separately, it is not compatible in a mathematical sense. Moreover, Green's strain is not suitable for engineering measures; its defects are now widely known. The polar decomposition theorem losses its utility for non-uniqueness of decomposition owing to the non-commutative property of matrix products (Chen, 1986).

The S-R decomposition theorem (Chen, 1979, 1986) which is compatible in strain and rotation is the main discussion in this chapter. Finally, we will introduce the application of the S-R theorem to large deformation analyses in mining and geological problems.

9.1 NONLINEAR GEOMETRICAL FIELD THEORY OF MECHANICS

One of the distinguished works on the nonlinear field theory of mechanics is due to Truesdell and Noll. Another work, Continuum Physics, edited by Eringen is also worth mentioning.

The field theory of modern mechanics contains two main parts: the first part is the geometrical field theory about the continuum; the second is the material field theory about the changes of physical properties of a continuum (see Chapter 10). Here, we restrict ourselves only to the first part.

Classical Euclidean geometry studies the transformation of space figures and their invariant properties. The geometrical field theory of modern mechanics studies the characteristics of space-time motion transformation of a continuum body and its invariant properties. Applications of the geometric field theory are not only in mechanics, but also in electromagnetics and general relativity. Here we deal only with mechanics.

Let B_0 indicate a bounded continuum point set in Euclidean space R^3. Suppose two points $x_1, x_2 \in B_0$, $x_1 \neq x_2$, have the property:

$$\{X: |x_1 - x_2| > 0\}.$$

We then call B_0 a measurable point set.

Suppose a point set B_0 at time t_0 transforms to another measurable point set B at time t. We call this a transformation (denoted by f)

$$f: B_0 \to B;$$

f is called a continuous transformation if there exists a region V in B which has one-to-one correspondence with the points in the neighborhood of a point in the region U of B_0, $f(u) \subset V$. If the transformation is at the same time a continuous function of time t, then the transformation is by essence also continuous in the time process.

If the distance between any two points remains unchanged in the case of transformation

$$\{X: |x_1 - x_2| = \text{const.}\},$$

then the transformation is rigid. The motion of a rigid body is seen to be a combination of transformation with a chosen point P, as a whole, and a relative motion around P. When deformation occurs in a body, the relative strain motion of the internal particles, coupling with relative rotation, makes the picture more complex than does a rigid motion only. The rotation angle and the direction of rotation axis are no longer invariant with respect to location in the body, and become a distribution function.

Separating rotation from strain in a deforming body has been a mathematical problem since the middle of the last century. In 1845, Stokes suggested a principle which stated: rotation may be separated in a field of fluid motion without affecting the pressure distribution. He stated:

"Hence I shall assume the following principle: '... the fluid in its neighborhood in a state of relative equilibrium depends only on the relative motion of the fluid immediately about P, and the relative motion due to any motion of rotation may be eliminated without affecting the differences of the pressures above mentioned'."

Stokes defined angular velocity at a point in a viscous fluid by

$$w' = (\mathrm{d}w/\mathrm{d}y - \mathrm{d}v/\mathrm{d}z)/2, \quad w'' = \ldots$$

and separated w from the velocity field as

$$U = (\mathrm{d}u/\mathrm{d}x)x' + (\tfrac{1}{2})(\mathrm{d}u/\mathrm{d}y + \mathrm{d}v/\mathrm{d}x)y' + (\tfrac{1}{2})(\mathrm{d}u/\mathrm{d}y - \mathrm{d}v/\mathrm{d}x)y' + \cdots$$

In fact, his method of separation is by means of symmetry and skewsymmetry decomposition of a tensor. By the usual notation now in use,

$$v_{i,j}=(v_{i,j}+v_{j,i})/2+(v_{i,j}-v_{j,i})/2, \tag{9.1}$$

where $v_{i,j}$ denotes the derivative of the velocity component v_i with respect to the space coordinate x_j.

It was also indicated by Stokes that the same method can be generalized to the displacement field of an elastic body, but he pointed out that the result is correct only when the displacement is sufficiently small. Thus, terms higher than the second degree will be neglected. Letting $u_{i,j}$ be the displacement gradient, the infinitesimal strain $\varepsilon_{i,j}$ and rotation $w_{i,j}$ are defined by

$$u_{i,j}=\underbrace{(u_{i,j}+u_{j,i})/2}_{\varepsilon_{i,j}}+\underbrace{(u_{i,j}-u_{j,i})/2}_{w_{i,j}} \qquad (i,j=1,2,3). \tag{9.2}$$

Actually, the finite rotation tensor is orthogonal in character. Only when rotation is quite small can it become skewsymmetry in appearance. So, in the case of rigid rotation infinitesimal strain components will not all be equal to zero, errors may appear in computation.

To modify Stokes' method of decomposition, two aspects should be considered:

(1) For describing the displacement field, the description method of double coordinate systems is used in conforming with any degree of large deformation and rotation of a deforming body. A fixed system X^i (inertia system) is chosen in R^3 space. At the same time, an alternate reference frame, called a co-moving coordinate system, embedded in a deforming body is set up as shown in Figures 9.1 and 9.2.

This system of describing deformation of a solid was first used by Synge & Chien (1941) in the theory of finite deformation of a plate and shell;

(2) Changing the manner of decomposition by the principle of symmetry and orthogonality, instead of symmetry and skewsymmetry.

The description method of the co-moving coordinate system for a deformable body is a natural extension of the moving coordinate system used in rigid body dynamics. In rigid body dynamics, only one moving system is used, but for a deforming body, there is an infinite number of local systems with a triad of basic vectors located at every point in the body. Local frames are deformed as the body changes its configuration.

Let $\mathring{\boldsymbol{g}}_i$ be the local basic vector of the co-moving system at time t_0. It changes to $\boldsymbol{g}_i$ at time t. The transformation $f:\mathring{\boldsymbol{g}}_i\rightarrow\boldsymbol{g}_i$ is realized by the transformation matrix $\boldsymbol{F}$.

$$\boldsymbol{g}_i=F^j_i\mathring{\boldsymbol{g}}_j, \quad F^j_i=\delta^j_i+u^j|_i, \tag{9.3}$$

$$\mathring{g}_{ij}=\mathring{\boldsymbol{g}}_i\cdot\mathring{\boldsymbol{g}}_j, \quad g_{ij}=\boldsymbol{g}_i\cdot\boldsymbol{g}_j; \tag{9.4}$$

where $u^j|_i$ denotes the covariant derivative of the displacement component u^j in the local system $\mathring{\boldsymbol{g}}_i$. In general, the co-moving system is curvilinear due to deformation.

The S-R theorem proves: (Chen, 1979, 1983) 'Suppose in R^3 space, a physically possible displacement function $u(x^i, t)$ is given for a deformable continuous point set. Let u be a single valued, continuous differentiable function, then the transformation induced by the motion of point set can be decomposed into the representations of a symmetrical transformation and orthogonal transformation.'

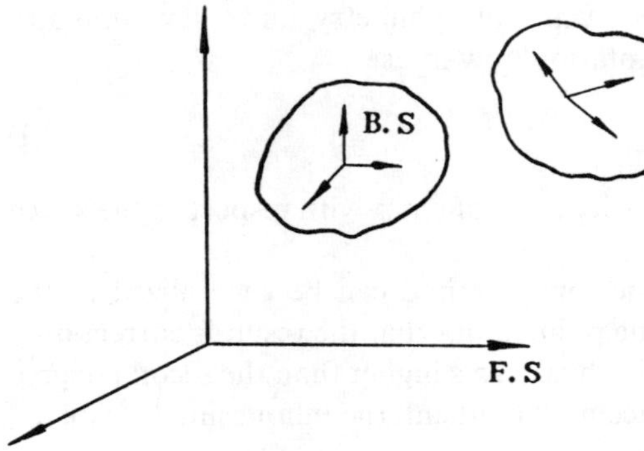

Figure 9.1. The description of motion.

	Description of motion	
Fluid Mechanics	Specification of flow field	Eulerian type $Y = F(x, y, z, t)$ Lagrangian type $Y = F(a, b, c, t)$
Rigid body mechanics	Specification of displacement field	Euler's method (1749) Two frames of reference Fixed system — Body system (embedded) 6 degrees of freedom
Deformable body mechanics	Specification of displacement	Generalized Euler's method (Hencky, Synge & Chien, Brillouin etc.) Two frames of reference Fixed system — Body system (embedded) *Variable metric* Local frame, 12 degrees of freedom

According to this theorem, $\boldsymbol{F}$ can be decomposed as

$$\begin{array}{l} F^i_j = S^i_j + R^i_j \quad (i, j = 1, 2, 3) \\ \quad \text{strain} + \text{rotation} \\ \quad\quad \downarrow \qquad\quad \downarrow \\ \text{symmetric} \quad \text{orthogonal.} \end{array} \tag{9.5}$$

Respectively S^i_j and R^i_j are defined as components of the strain tensor and rotation tensor. It is evident that the shortcoming of assuming the order of precedent of rotation

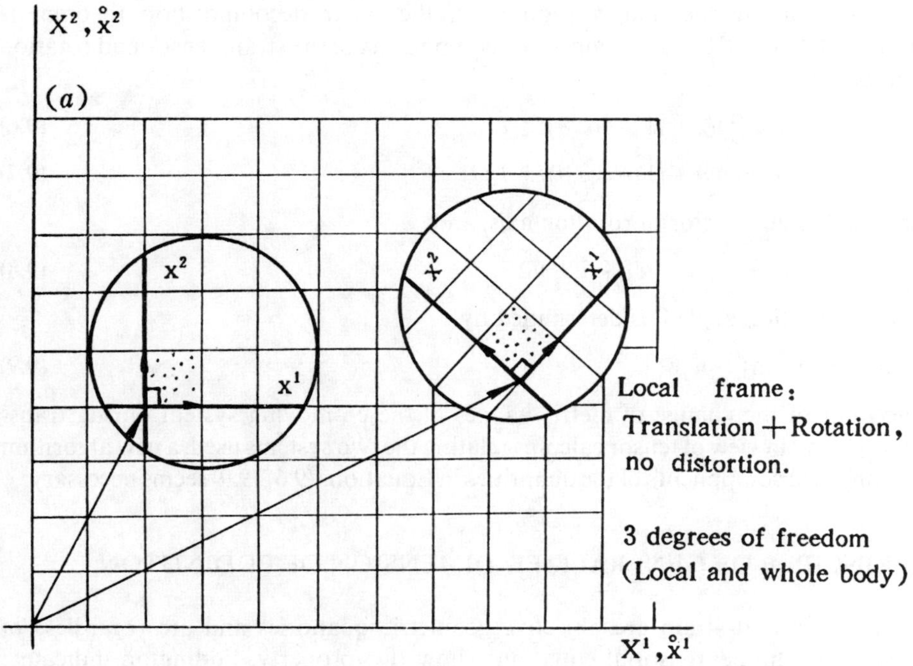

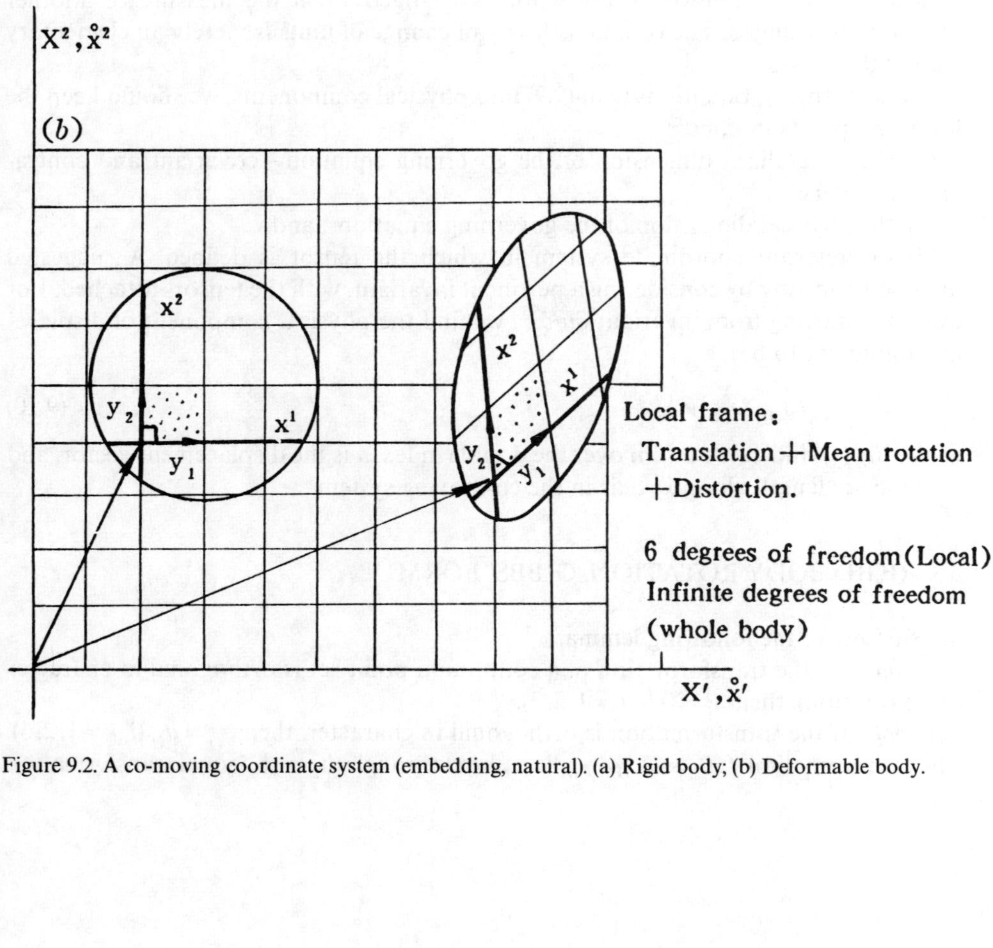

Figure 9.2. A co-moving coordinate system (embedding, natural). (a) Rigid body; (b) Deformable body.

or deformation by the matrix product in the polar decomposition theorem is eliminated. Chen (1979) has obtained the components of the strain tensor and rotation tensor as follows:

$$S^i_j = (u^i|_j + u^i|^T_j)/2 - (1 - \cos\vartheta)L^i_k \cdot L^k_j, \tag{9.6}$$

$$R^i_j = \delta^i_j + L^i_j \cdot \sin\vartheta + (1 - \cos\vartheta)L^i_k \cdot L^k_j, \tag{9.7}$$

where L is the unit vector of rotation axis, and

$$L^i_j = w^i_j/\sin\vartheta, \quad w^i_j = (u^i|_j - u^i|^T_j)/2. \tag{9.8}$$

The mean rotation angle ϑ is determined by,

$$\vartheta = \pm \arcsin[-w^i_j w^j_i]^{1/2}. \tag{9.9}$$

On account of the change of metric gauges of the co-moving system during transformation, and in view of tensor calculus relating the two systems used, a new algorithm for the physical components of the quantities in Equations (9.6)-(9.9) seems necessary.

9.2 PRINCIPLE OF GENERALIZED DIMENSION OF EDDINGTON

Local rotation and strain are objective geometric quantities and are regardless of their reference frame; tensorial equations show this property. Eddington indicated: '... Covariant and contravariant are a kind of generalized dimension, showing how the measure of one condition of the world is changed when the measure of another condition is changed. The ordinary theory of change of units is merely an elementary case of this.'

While changing Equations (9.6)-(9.9) into physical components, we should keep the following points in mind:

(1) the generalized dimension of the governing equation – covariant and contravariant orders;

(2) the physical dimension of the governing equations, and

(3) the relevant coordinate system in which the tensor is defined. A suggested method is starting by considering a pertinent invariant, with the tensors attached. For example, starting from invariant $\partial u/\partial s^1$, we find the physical component of displacement gradient to be

$$\hat{u}^j|_i = (\sqrt{g_{(jj)}/g_{(ii)}})u^j|_i, \tag{9.10}$$

where (ii), (jj) indicate no sum over the double index, u is the displacement vector, and s^1 is the arc length along x^1 axis in the co-moving system.

9.3 RIGID BODY ROTATION, GIBBS' FORMULA

We first prove the following lemma.

Lemma: If the transformation of a continuum point set is orthogonal in character (pure rotation), then $S^j_i = 0$ $(i, j = 1, 2, 3)$.

Proof: If the transformation is orthogonal in character, then $\mathring{g}_{ij} = g_{ij}$ $(i, j = 1, 2, 3)$. The transformation $f: B_0 \to B, \mathring{g}_i \to g_i$ is realized through the orthogonal transformation

Table 9.1. Gibbs' formula and formulae by S-R theorem.

Gibbs' formula (Wilson in 1901)	Formulae by S-R theorem (Chen, 1988)	
(1) Rotation transformation $\boldsymbol{r}' = \boldsymbol{\phi}\cdot\boldsymbol{r}$	$X^i = R^i_j x^j$	(a_1)
$[\phi] = \begin{bmatrix} a_{11} & a_{12} & a_{13} \\ a_{21} & a_{22} & a_{23} \\ a_{31} & a_{32} & a_{33} \end{bmatrix}$	$[R^i_j] = \begin{bmatrix} R^1_1 & R^1_2 & R^1_3 \\ R^2_1 & R^2_2 & R^2_3 \\ R^3_1 & R^3_2 & R^3_3 \end{bmatrix}$	(a_2)
$\phi = \boldsymbol{aa} + \cos q(\boldsymbol{I} - \boldsymbol{aa}) + \sin q\, \boldsymbol{I} \times \boldsymbol{a}$	$R^i_j = \delta^i_j + (1 - \cos \vartheta)L^i_k \cdot L^k_j + L^i_j \sin \vartheta$	(a_3)
(2) Unit vector along rotation axis $\boldsymbol{a} = a_1\boldsymbol{i} + a_2\boldsymbol{j} + a_3\boldsymbol{k}$	$L = L_1\boldsymbol{i} + L_2\boldsymbol{j} + L_3\boldsymbol{k}$	(a_4)
$[\boldsymbol{I} \times \boldsymbol{a}] = \begin{bmatrix} 0 & -a_3 & a_2 \\ a_3 & 0 & a_1 \\ -a_2 & a_1 & 0 \end{bmatrix}$	$[L^i_j] = \begin{bmatrix} L^1_1 & L^1_2 & L^1_3 \\ L^2_1 & L^2_2 & L^2_3 \\ L^3_1 & L^3_2 & L^3_3 \end{bmatrix} = \begin{bmatrix} 0 & -L_3 & L_2 \\ L_3 & 0 & -L_1 \\ -L_2 & L_1 & 0 \end{bmatrix}$	(a_5)
(3) Rotation angle q, ϑ $\sin q = \sqrt{(\phi_x \cdot \phi_x)}/2 = [(a_{23} - a_{32})^2 + (a_{31} - a_{13})^2 + (a_{12} - a_{21})^2]^{1/2}/2$	$\sin \vartheta = \sqrt{(-w^i_j w^j_i)} = [(R^1_2 - R^2_1)^2 + (R^2_3 - R^3_2)^2 + (R^3_1 - R^1_3)^2]^{1/2}/2$	(a_6)

coefficient matrix R,

$$g_i = R^j_i \mathring{g}_j. \tag{9.11}$$

Under this condition,

$$\delta^j_i + u^j|_i = R^j_i, \quad u^j|_i = R^j_i - \delta^j_i = L^j_i \cdot \sin \vartheta + (1 - \cos \vartheta)L^j_k \cdot L^k_j,$$

and we then have

$$S^i_j = (\tfrac{1}{2})(R^i_j + R^{iT}_j) - (1 - \cos \vartheta)L^i_k \cdot L^k_j = 0, \quad (i, j, k = 1, 2, 3).$$

The skewsymmetry property of L^j_i is used in the above proof. We see that under the condition of orthogonal transformation, the formula for rigid body rotation by Gibbs can be derived from the S-R decomposition theorem. A comparison is made in Table 9.1.

The above deductions ascertain that, when $S = 0$, the transformation reduces to rigid body rotation. If transformation is physically possible and $|\sin \vartheta| \leq 1$, then the rotation angle is real for any case.

9.4 LOCAL ROTATION, THE CONCEPT OF MEAN ROTATION ANGLE

In a general case when deformation occurs in a body, the line segment passing through a point in the body will have relative rotation. In reality, the concept of rotation is more complex in a deforming body than in a rigid body, because the angle between line segments will change (compare Figure 9.3(a) and (b)).

Let us examine displacement field $\boldsymbol{u}$ inside an elemental volume $\Delta\Omega$. Suppose $\Delta\Omega$ is a sphere of infinitesimal radius ε with its center at point P. All the diametral vectors passing through P have intersecting points on the spherical surface. The trajectories traced by the end points of these diametral vectors on the spherical surface determine the displacement vectors of all the line segments passing through P.

Applying the theorem of divergence, we find the curl of $\boldsymbol{u}$ at point P,

$$(\nabla \times \boldsymbol{u})_P = \lim_{\Delta\Omega \to 0} (1/\Delta\Omega) \iint_{\Delta s} (\boldsymbol{n} \times \boldsymbol{u})\,\mathrm{d}S, \tag{9.12}$$

where $\boldsymbol{n}$ is unit outward normal vector on surface Δs of $\Delta\Omega$. The following theorem is then established.

Theorem: The average value of curl of a displacement field in an infinitesimal neighborhood of a point in a deforming body is

$$(\nabla \times \boldsymbol{u})_P / 2 = \boldsymbol{L} \sin \vartheta. \tag{9.13}$$

Proof: Suppose the state of rotation at a point can be represented by a mean rotation angle ϑ, and let $\boldsymbol{L}$ denote the unit vector along the axis of mean rotation. As the local reference system may be arbitrarily chosen, we select a rectilinear orthogonal frame with its z axis in the direction of $\mathbf{L}$ (Figure 9.4). The displacement components of a point on the spherical surface can then be expressed as

$$u_x = -2\varepsilon \cos\phi \sin(\vartheta + \alpha/2) \sin(\alpha/2),$$

$$u_y = 2\varepsilon \cos\phi \cos(\vartheta + \alpha/2) \sin(\alpha/2),$$

and

$$u_z = 0. \tag{9.14}$$

Applying Equation (9.12), we have

$$(\nabla \times \boldsymbol{u})_P = \lim_{\Delta\Omega \to 0} (1/\Delta\Omega) \iint_{\Delta s} (\boldsymbol{n} \times \boldsymbol{u})\,\mathrm{d}S = \boldsymbol{L}(2 \sin \alpha),$$

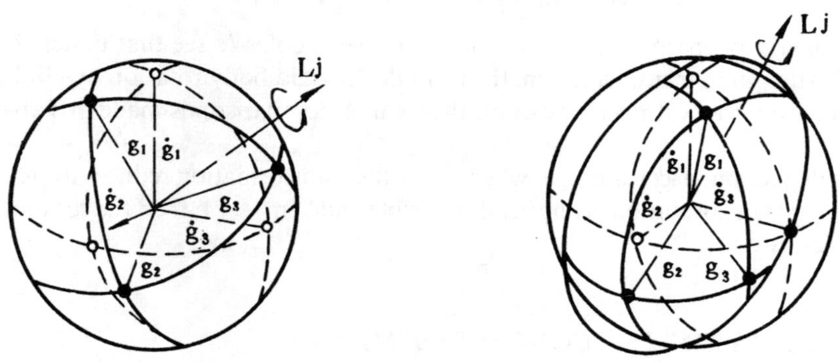

(*a*) ($\dot{g}_i = g_i$, 6 degrees of freedom)

(*b*) ($\dot{g}_{ij} = g_{ij}$, 12 degrees of freedoms at a point)

Figure 9.3. The displacement field $\boldsymbol{u}$ inside an element volume $\Delta\Omega$. (a) Rigid body; (b) Deformable body.

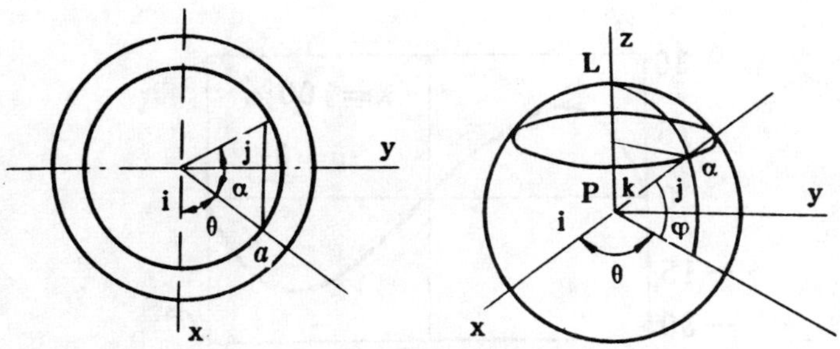

Figure 9.4. The state of rotation at a point.

or

$$(\nabla \times \boldsymbol{u})_P/2 = \boldsymbol{L} \sin \alpha, \quad |\nabla \times \boldsymbol{u}|_P/2 = |\sin \alpha|. \tag{9.15}$$

On the other hand, Equation (9.9) gives

$$|\sin \vartheta| = [(u^1|_2 - u^2|_1)^2 + (u^2|_3 - u^3|_2)^2 + (u^3|_1 - u^1|_3)^2]^{1/2}/2 = |\nabla \times \boldsymbol{u}|_P/2. \tag{9.16}$$

Therefore, we obtain $\sin \vartheta = \sin \alpha$. Thus, Equation (9.13) is proved. This result leads us to the conclusion that ϑ is a mean rotation angle at a point in a deforming body. Because $|\sin \vartheta| \leq 1$, therefore,

$$|\sin \vartheta| = |\nabla \times \boldsymbol{u}|_P/2 \leq 1. \tag{9.17}$$

No matter how large the rotation and strain, the transformation function is to this point continuous and differentiable and a definite real value of ϑ exists.

Moreover, the foregoing theorem acts as a factor in determining whether the transformation function is a possible one. If $|\sin \vartheta| > 1$, then the transformation is physically impossible.

9.5 DISTRIBUTION FUNCTION OF ROTATION OF LINE SEGMENTS

When a body is defined, the rotation of line segments passing through a point in general differs in magnitude and direction for different line orientations and forms a distribution function. Some examples will be shown below. Let x^i denote co-moving coordinates and x^i be the coordinate of the point referring to the fixed system after deformation.

The transformation function is

$$x^i = X^i(x^1, x^2, x^3; t).$$

9.5.1 *Uniaxial extension*

The transformation function of a pure tension (Figure 9.5) is supposed to be

$$X^1 = (1 + \lambda)x^1, \quad X^2 = x^2, \quad X^3 = x^3.$$

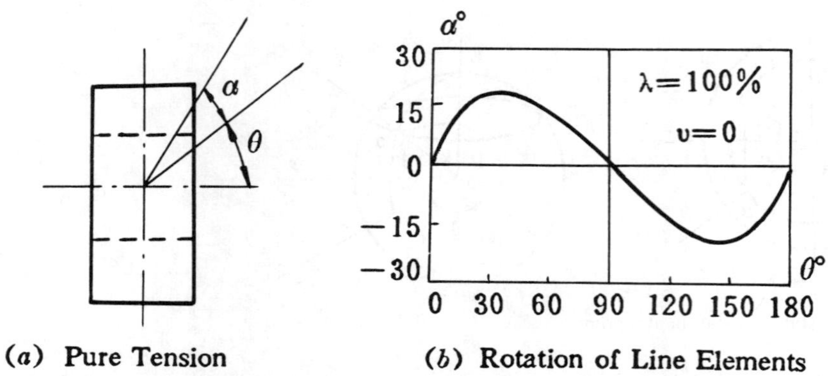

Figure 9.5. (a) Pure tension; (b) Rotation of line elements.

Determining the physical components of the strain and mean rotation angle by Equations (9.6)-(9.9), we have

$$[\hat{S}^i_j] = \begin{bmatrix} \lambda/(1+\lambda) & 0 & 0 \\ 0 & 0 & 0 \\ 0 & 0 & 0 \end{bmatrix} \quad (\vartheta = 0).$$

The rotation angle α of a line segment with initial inclination to the horizontal axis is determined by the formula:

$$\sin(\alpha) = (1+\lambda)/(\lambda+2)\cdot \mathrm{d}[\sqrt{(1-k^2\sin^2 u)}]/\mathrm{d}\theta. \tag{9.18}$$

where

$$k^2 = [(1+\lambda)^2 - 1]/(1+\lambda)^2, \quad \sin u = \sin(\pi/2 - \vartheta).$$

The average value of $\sin(\alpha)$ is equal to zero ($\vartheta = 0$), and

$$(1/\pi)\int_0^\pi \sin\alpha(\theta)\,\mathrm{d}\theta = 0.$$

By this simple example, we see that even in the case of uniaxial extension, the rotation of line segments will not be zero, but that the mean rotation angle ϑ is equal to zero. We should pay attention to the fact that in many cases the so-called average value is not the simple arithmetic mean.

9.5.2 *Pure shear*

The transformation function of a pure shear (Figure 9.6) is supposed to be

$$X^1 = x^1 + \tan\gamma x^2, \quad X^2 = \tan\gamma x^1 + x^2, \quad X^3 = x^3,$$

where γ is the shear angle. By Equations (9.6)-(9.9), we find the physical components of the strain tensor and mean rotation angle,

$$[\hat{S}^i_i] = \begin{bmatrix} 0 & \sin\gamma & 0 \\ \sin\gamma & 0 & 0 \\ 0 & 0 & 0 \end{bmatrix} \quad (\vartheta = 0).$$

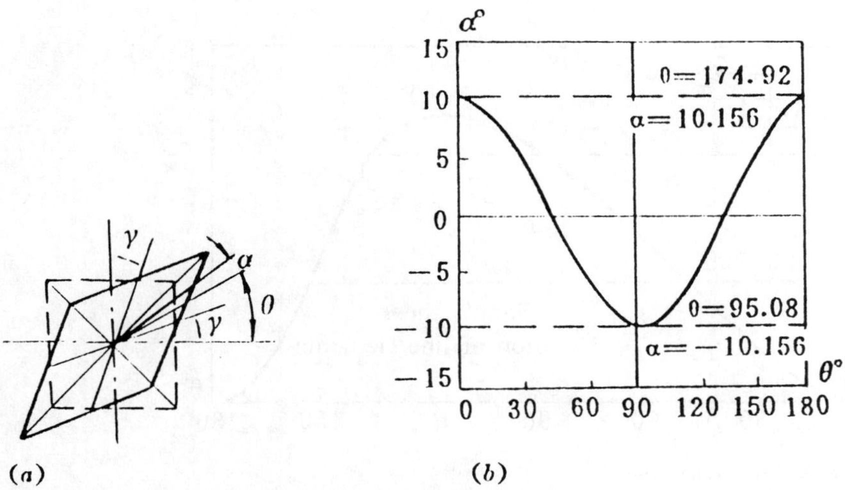

Figure 9.6. (a) Pure shear; (b) Rotation of line elements.

The relation $\alpha = \alpha(\theta)$ is determined by

$$\sin\alpha = \tan\gamma\cos 2\theta/\sqrt{(\sec^2\gamma + 2\tan\gamma\sin 2\theta)}, \tag{9.19}$$

and the mean value of $\sin\alpha$ is

$$(1/\pi)\int_0^\pi \sin\alpha(\theta)\,d\theta = 0,$$

corresponding to $\vartheta = 0$.

9.5.3 *Simple shear* (*plane sliding*)

The transformation function is supposed to be

$$X^1 = x^1 + \tan\gamma x^2, \quad X^2 = x^2, \quad X^3 = x^3.$$

From Equations (9.6)-(9.9), we find the physical components of the strain tensor and rotation angle:

$$[\hat{S}^i_j] = \begin{bmatrix} 1-\sqrt{1-(\sin^2\gamma)/4} & (\sin\gamma)/2 & 0 \\ (\sin\gamma)/2 & 1-\sqrt{1-(\sin^2\gamma)/4} & 0 \\ 0 & 0 & 0 \end{bmatrix}$$

and

$$\vartheta = -\arcsin(0.5\sin\gamma).$$

Components S^1_1 and S^2_2 appear as a second order effect in nonlinear mechanics and have been confirmed by many experiments.

The relation $\alpha = \alpha(\theta)$ is determined by

$$\sin\alpha = -\sin\gamma\sin^2\theta/\sqrt{\sin^2\theta\cos^2\gamma + \cos^2(\theta-\gamma)}. \tag{9.20}$$

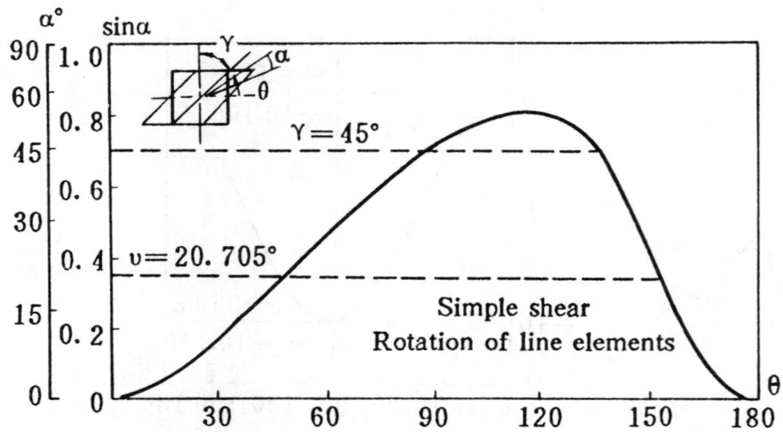

Figure 9.7. Simple shear (plane sliding).

When $\gamma = 45°$,

$$\vartheta = \arcsin[(\sin\gamma)/2] = 20.705°,$$

and the result is shown in Figure 9.7. It is interesting to note that there appears to be a region where $\alpha > 45°$.

9.6 LOCAL ANGULAR VELOCITY, GENERALIZED EULER'S KINEMATIC FORMULA FOR DEFORMING BODY

In the process of deformation, a velocity field exists in a deforming body. Let $\dot{S}$ be the strain rate and $L\dot{\vartheta}$ be the local angular velocity. The velocity gradient can then be decomposed as

$$v^j\|_i = L_i^j\dot{\vartheta} + \dot{S}_i^j, \tag{9.21}$$

with $v^j\|_i$ being the covariant derivative of velocity component v^j with respect to the co-moving coordinate x^i in the real time system. The proof of Equation (9.21) is given by Chen (1983). The strain rate and local angular velocity can be expressed as,

$$\dot{S}_i^j = (v^j\|_i + v^j\|_i^T)/2, \quad L_i^j\dot{\vartheta} = (v^j\|_i - v^j\|_i^T)/2. \tag{9.22}$$

Euler's kinematic formula for rigid body motion,

$$\boldsymbol{v} = \boldsymbol{v}_0 + \boldsymbol{w} \times \boldsymbol{r}, \tag{9.23}$$

can be derived from Equation (9.22) with $\boldsymbol{r}$ being the position vector; $\boldsymbol{v}$ the velocity, $\boldsymbol{w}$ the angular velocity of a point in a rigid body and $\boldsymbol{v}_0$ the velocity of a reference point.

In the case of a rigid body, $\boldsymbol{S} = \boldsymbol{0}$, Equation (9.21) reduces to the following differential Equation

$$v^j\|_i = L_i^j w. \tag{9.24}$$

Because deformation does not occur, the covariant derivative may be substituted by the common derivative.

Hence, by integrating the following differential Equation:

$$\partial v_j/\partial x^i = L_{ij}w, \tag{9.25}$$

we obtain Equation (9.23), a special case of Equation (9.21). We thus call Equation (9.21) the generalized Euler's kinematic formula (Chen, 1988).

9.7 NONLINEAR ANALYSIS FOR LARGE DEFORMATION OF UNDERGROUND OPENINGS IN ROCK

In underground mining, excavations and strata movements give rise to large deformations of rock mass. Tectonic movements such as folding, faulting, jointing, and cleavage, by which earth and rock mass are structurally disturbed, give rise to mass displacement deformations. Complications of deformation geometry and the difference in engineering properties of the earth's material (rock) usually cause its mathematical solutions to possess highly nonlinear characters. Under such circumstances, utilizing the classical small deformaton theory to make mathematical analyses most likely leads to great errors, sometimes even no result can be obtained. On the contrary, the finite deformation theory suitable for solving both small and large deformation problems will be possible to prevent unnecessary error from occurring in the theoretical arithmetics. This section attempts to solve problems of excavation in rock by the two-dimensional nonlinear program NEPR based on the S-R theory of finite deformation, i.e., utilizing a continuous medium model in conjunction with some special methods to solve the problems suitable to use F.E.M. on the co-moving coordinate system (Xie, 1984, 1985).

9.7.1 *Formula of nonlinear finite element method for large deformation*

The co-moving coordinate x^i, which is embedded on a body, varies with deformation of the body. The fixed coordinate X^i is used to describe the position of particles of the body in Euclidean space,

$$X^i = x^i + u^i. \tag{9.26}$$

The metric tensor of the co-moving coordinate frame is

$$g_{ij} = F^i_k F^k_j, \tag{9.27}$$

where $F^i_j = \partial X^i/\partial x^j$ is the deformation gradient.

When we use the n-node isoparametric quadrilateral element to discretize the entire region Ω, then there are relationships between the coordinate systems in the element as follows:

$$x^i = x^i(\xi^j),$$

and

$$X^i = X^i(\xi^j); \tag{9.28}$$

where ξ^j are local coordinates of the element. Thus, Equation (9.27) can be rewritten as

$$g_{ij}=(\partial X^m/\partial x^i)(\partial X^n/\partial x^j)\delta_{nm}. \tag{9.29}$$

Letting $\boldsymbol{J}$ and $\boldsymbol{J}_c$ be the Jacobi matrices in the fixed coordinate X^i and co-moving coordinate x^i, respectively, then

$$\boldsymbol{J}=\begin{bmatrix}\partial X^1/\partial\xi^1 & \partial X^1/\partial\xi^2\\ \partial X^2/\partial\xi^1 & \partial X^2/\partial\xi^2\end{bmatrix},\quad \boldsymbol{J}_c=\begin{bmatrix}\partial x^1/\partial\xi^1 & \partial x^1/\partial\xi^2\\ \partial x^2/\partial\xi^1 & \partial x^2/\partial\xi^2\end{bmatrix}. \tag{9.30}$$

The matrix form of the metric tensor can be derived from $\boldsymbol{J}$ and $\boldsymbol{J}_c$:

$$\boldsymbol{g}=\begin{bmatrix}g_{11} & g_{12}\\ g_{21} & g_{22}\end{bmatrix}=\boldsymbol{F}^T\boldsymbol{F}=(\boldsymbol{J}_c^{-1})^T\boldsymbol{J}^T\boldsymbol{J}\boldsymbol{J}_c^{-1}. \tag{9.31}$$

We now consider the derivative of velocity along the co-moving coordinate curves x^j:

$$\partial \boldsymbol{V}/\partial x^j=V^i\|_j\boldsymbol{g}_i=V^i\|_jF^k_i\mathring{\boldsymbol{g}}_k. \tag{9.32}$$

On the other hand, when $\boldsymbol{V}$ is denoted by its components on a fixed coordinate system, i.e. $\boldsymbol{V}=\tilde{V}^k\mathring{\boldsymbol{g}}_k$, then $\partial\boldsymbol{V}/\partial x^j$ is also written by

$$\partial\boldsymbol{V}/\partial x^j=(\partial\tilde{V}^k/\partial x^j)\mathring{\boldsymbol{g}}_k. \tag{9.33}$$

Comparing Equation (9.32) with Equation (9.33), we have

$$V^i\|_j=(F^i_k)^{-1}(\partial\tilde{V}^k/\partial x^j). \tag{9.34}$$

The physical components of $\tilde{V}^i\|_j$ are defined as

$$\tilde{V}^i\|_j=\sqrt{g_{(ii)}/g_{(jj)}}(F^i_k)^{-1}(\partial\tilde{V}^k/\partial x^j), \tag{9.35}$$

and the displacements and velocity in the element can be expressed in terms of shape function $N_k(\xi^i)$,

$$\Delta u^i=\sum_{i=1}^{n}N_k(\xi^i)(\Delta_k u^i),$$

and

$$\tilde{V}^i=\sum_{i=1}^{n}N_k(\xi^i)({}_k\tilde{V}^i). \tag{9.36}$$

Nonlinear strain rate $\dot{s}^i_j$ can be written by matrix form

$$\dot{\boldsymbol{S}}=[\dot{s}^1_1,\dot{s}^2_2,2\dot{s}^1_2]^{\mathrm{T}}=\boldsymbol{B}_N({}_k\tilde{\boldsymbol{V}}), \tag{9.37}$$

where

$${}_k\tilde{\boldsymbol{V}}=[{}_1V^1,{}_1V^2,{}_2V^1,{}_2V^2,\ldots,{}_nV^1,{}_nV^2]^{\mathrm{T}} \tag{9.38}$$

is the nodal velocity matrix of the element on fixed coordinates, and

$$\boldsymbol{B}_N=[\boldsymbol{B}_1,\boldsymbol{B}_2,\ldots,\boldsymbol{B}_i,\ldots,\boldsymbol{B}_n] \tag{9.39}$$

is the nonlinear strain rate matrix.

From Equations (9.35) and (9.36), we obtain

$$\boldsymbol{B}_i=\boldsymbol{Q}\tilde{\boldsymbol{N}}_i; \tag{9.40}$$

$$\boldsymbol{Q}=\begin{bmatrix} (F_1^1)^{-1} & 0 & 0 & (F_2^1)^{-1} \\ 0 & (F_2^2)^{-1} & (F_1^2)^{-1} & 0 \\ \sqrt{\frac{g_{22}}{g_{11}}}(F_1^2)^{-1} & \sqrt{\frac{g_{11}}{g_{22}}}(F_2^1)^{-1} & \sqrt{\frac{g_{11}}{g_{22}}}(F_1^1)^{-1} & \sqrt{\frac{g_{22}}{g_{11}}}(F_2^2)^{-1} \end{bmatrix}, \tag{9.41}$$

$$\tilde{\boldsymbol{N}}_i=\begin{bmatrix} \partial N_i/\partial x^1 & 0 \\ 0 & \partial N_i/\partial x^2 \\ \partial N_i/\partial x^2 & 0 \\ 0 & \partial N_i \partial x^1 \end{bmatrix} \tag{9.42}$$

and

$$\boldsymbol{B}_i=\begin{bmatrix} (F_1^1)^{-1}\partial N_i/\partial x^1 & (F_2^1)^{-1}\partial N_i/\partial x^1 \\ (F_1^2)^{-1}\partial N_i/\partial x^2 & (F_2^2)^{-1}\partial N_i/\partial x^2 \\ \sqrt{\frac{g_{22}}{g_{11}}}(F_1^2)^{-1}\frac{\partial N_i}{\partial x^1}+\sqrt{\frac{g_{11}}{g_{22}}}(F_1^1)^{-1}\frac{\partial N_i}{\partial x^2} & \sqrt{\frac{g_{22}}{g_{11}}}(F_2^1)^{-1}\frac{\partial N_i}{\partial x^2}+\sqrt{\frac{g_{11}}{g_{22}}}(F_2^2)^{-1}\frac{\partial N_i}{\partial x^1} \end{bmatrix} \tag{9.43}$$

Consider the elasto-plastic constitutive equation of a simple parameter defined by Chen (1988) in the form

$$\dot{\boldsymbol{\sigma}}=2G(1-\beta)\{\dot{\boldsymbol{S}}+[3K_{(\beta)}/(2G(1-\beta))-1]/3\cdot\dot{S}_k^k\boldsymbol{I}\}, \tag{9.44}$$

where β is a parameter of plastic deformation. In the elastic region, $\beta=0$. In the plastic region, $|\beta|>0$: $\beta>0$ for plasticity with hardening, and $\beta<0$ for plasticity with softening. G is shear modulus and $K_{(\beta)}$ is the bulk modulus, which is a function of β. If the stress rate matrix associated with the strain rate matrix is defined as

$$\dot{\boldsymbol{\sigma}}=[\dot{\sigma}_1^1,\dot{\sigma}_2^2,\dot{\sigma}_2^1]^T, \tag{9.45}$$

then Equation (9.44) is expressed in the matrix form

$$\dot{\boldsymbol{\sigma}}=\boldsymbol{E}\dot{\boldsymbol{S}}=\boldsymbol{E}\boldsymbol{B}_N({}_k\tilde{\boldsymbol{V}}), \tag{9.46}$$

where $\boldsymbol{E}$ denotes the elasto-plastic modulus matrix (or tensor) given by

$$\boldsymbol{E}=\begin{bmatrix} E_{11} & E_{12} & 0 \\ E_{21} & E_{22} & 0 \\ 0 & 0 & E_{33} \end{bmatrix}. \tag{9.47}$$

For the plane stress case: we have

$$E_{11}=E_{22}=K_{(\beta)}+\alpha_2/3+\alpha_1,$$
$$E_{12}=E_{21}=K_{(\beta)}-2\alpha_2/3+\alpha_1,$$
$$E_{33}=\alpha_2,$$

and

$$\alpha_2=G(1-\beta),\quad \alpha_1=[2G(1-\beta)/3-K_{(\beta)}]/[K_{(\beta)}+4G(1-\beta)/3].$$

For plane strain case: we have

$$E_{11} = E_{22} = K_{(\beta)} + 4G(1-\beta)/3,$$
$$E_{12} = E_{21} = K_{(\beta)} - 2G(1-\beta)/3,$$

and

$$E_{33} = G(1-\beta).$$

Consider the virtual power equation which is written with respect to the instantaneous co-moving coordinates x^i:

$$\int_\Omega \dot{\sigma}^i_j \delta(\dot{S}^j_i)\,d\Omega - \int_{\partial\Omega} \dot{P}_j \delta V^j dS - \int_\Omega (\rho \dot{f}_j)\delta V^j d\Omega = 0, \tag{9.48}$$

where $\dot{P}_i$ and $\rho\dot{f}_i$ are rates of the surface and volume forces acting on the body, respectively. Its increment form can be written as

$$\int_{(t+\Delta t)} \Delta\sigma^i_j \delta(\dot{S}^j_i)\,d\Omega - \int_{(t+\Delta t)} \Delta P_i \delta V^i dS - \int_{(t+\Delta t)} \Delta(\rho f_i)\delta(V^i)\,d\Omega = 0. \tag{9.49}$$

Substituting Equations (9.37) and (9.46) into (9.49), we can establish the finite element equations for elasto-plasticity analyses of large deformations as follows:

$$\boldsymbol{K}_k \Delta \boldsymbol{u} = {}_k\Delta \boldsymbol{R},$$
$$\boldsymbol{K} = \int_\Omega \boldsymbol{B}_N^T \boldsymbol{E} \boldsymbol{B}_N\, d\Omega, \tag{9.50}$$

and

$${}_k\Delta \boldsymbol{R} = \int_{\partial\Omega} \boldsymbol{N}^T \Delta \boldsymbol{P}_k\, ds + \int_\Omega \boldsymbol{N}^T (\Delta\rho \boldsymbol{f}_k)\, d\Omega,$$

where $\boldsymbol{K}$ is the nonlinear stiffness matrix of the element, ${}_i\Delta \boldsymbol{u}$ represents the incremental nodal displacement vector. ${}_i\Delta \boldsymbol{R}$ denotes the increment of element equivalent nodal force vector, and $\boldsymbol{N}$ is a shape function matrix. $\Delta \boldsymbol{P}_i$ and $(\Delta\rho \boldsymbol{f}_i)$ are components of the surface force and body force vectors on co-moving coordinates, respectively.

9.7.2 *Function and features of program NEPR*

The program NEPR has taken constitutive equations and the geometry of large deformation into consideration, and it can be used to exactly simulate the induced stress and displacement field to analyze mining engineering problems either in longwall

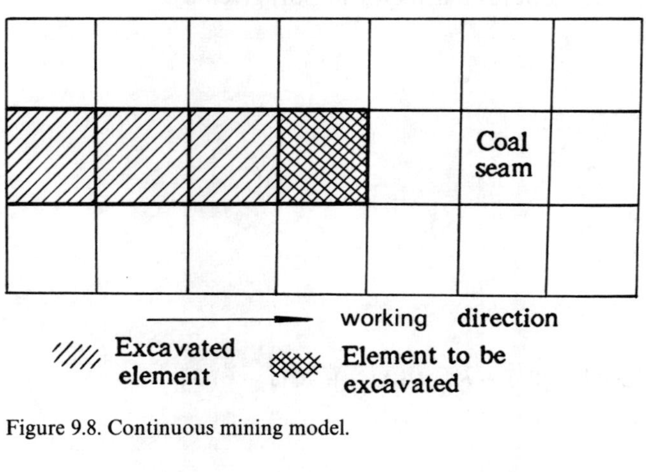

Figure 9.8. Continuous mining model.

faces or roadways and also in tectonics and so forth. As excavation is a continuously progressive work (as shown in Figure 9.8), this feature has been considered in the program.

9.7.3 *Numerical examples*

9.7.3.1 *Linear and nonlinear analyses of a tunnel*

By variational functional theory, a catenary under a gravity force takes such a curve that its potential energy is the smallest in all curves. Euler supposed that an arch shaped like a reverse catenary would give the smallest subsidence in arches, but this has no mathematical proof. A discussion of the shape of an arch is also an interesting problem

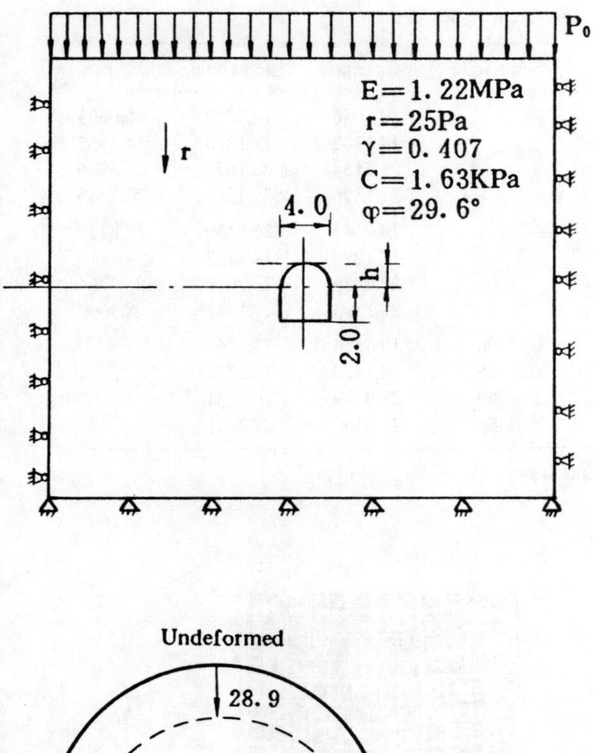

Figure 9.9. Computing model of a tunnel.

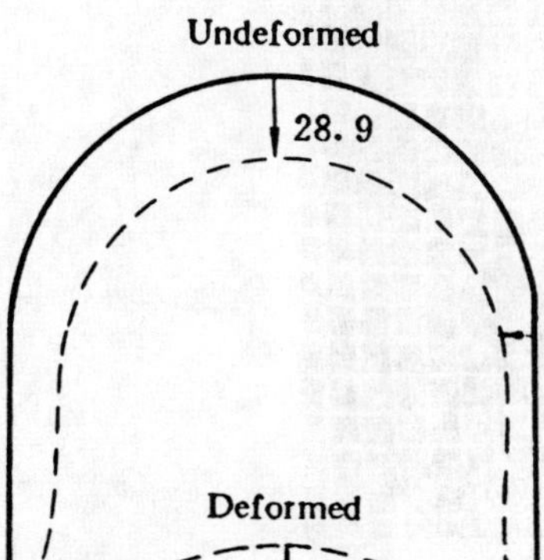

Figure 9.10. Configuration change of a tunnel.

of rock mechanics. Therefore, we make numerical examinations of different arch shapes. A model is shown in Figure 9.9, its material is in perfect plasticity. The yield criteria of Mohr-Coulomb is used. Eighteen numerical examples for different arch shapes and arch heights are examined. Results are listed in Table 9.2. The configuration change of the tunnel is given in Figure 9.10.

From Table 9.2 the following conclusions are obtained:

(a) Parabolic arch is better than the others under body force. It is in agreement with the pressure curve theory of structural mechanics;

(b) Through numerical analysis, it is clear that the catenary arch hypothesis of Euler is invalid in underground engineering, but the catenary arch is better than semicircle and ellipse archs when only deformation is considered; and

Table 9.2. Numerical analysis of different arch shapes of tunnel.

Values			Arch shape			
			Semicircle	Catenary	Parabola	Ellipse
$h = 1.50$ m	$P_0 = 5$ kN	M.S.		14.5056	14.4402	15.0603
		M.F.		14.1102	14.1050	14.1655
	$P_0 = 10$ kN	M.S.		28.3154	28.1872	29.3906
		M.F.		27.1396	27.1281	27.2546
$h = 1.80$ m	$P_0 = 5$ kN	M.S.		14.0547	13.9150	14.8139
		M.F.		14.1290	14.1193	14.2118
	$P_0 = 10$ kN	M.S.		27.4836	27.2079	28.9764
		M.F.		27.1893	27.1681	28.3600
$h = 2.00$ m	$P_0 = 5$ kN	M.S.	14.7300	13.8316	13.6261	
		M.F.	14.2508	14.1472	14.1321	
	$P_0 = 10$ kN	M.S.	28.8769	27.0979	26.6881	
		M.F.	27.4483	27.2367	27.2041	

Note: M.S. – Max. subsidence; M.F. – Max. heave.

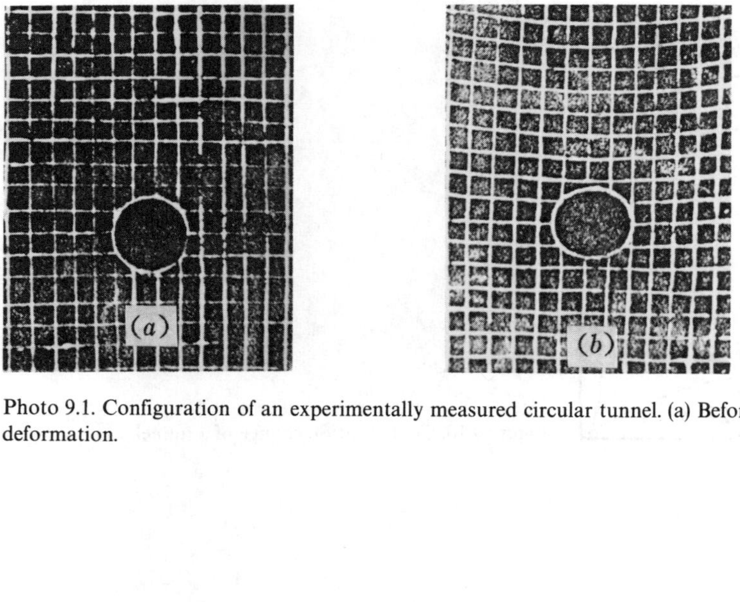

Photo 9.1. Configuration of an experimentally measured circular tunnel. (a) Before deformation; (b) After deformation.

(c) Change of subsidence and floor heave caused by different arch shapes is not severe.

To examine large deformations of a tunnel, a hollow circular tunnel with a boundary of finite width under gravity was used as the model. In our experiments gelatine material was used, and the configuration was measured by the network method (Xie, 1984) as shown in Photo 9.1. The numerical example is then analyzed by using the NEPR and ADINA program.

In this example, both numerical results are almost the same as shown in Figures 9.11 and 9.12 because of the moderate deformation and small rotation. However in the case of large deformation and large rotation, the NEPR program has greater accuracy than that of the ADINA program (Xie, 1985).

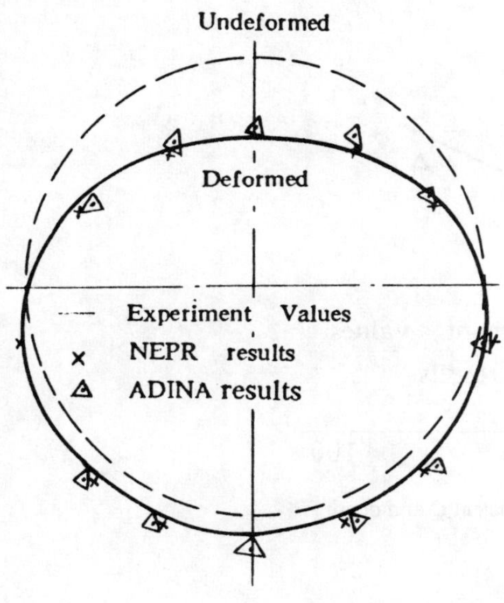

Figure 9.11. Configuration of a circular tunnel.

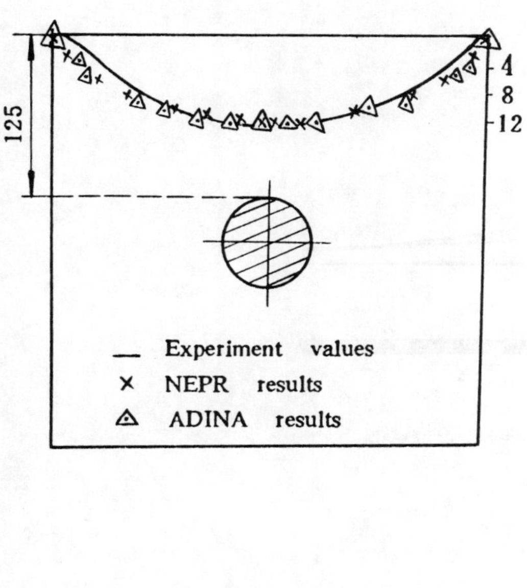

Figure 9.12. Subsidence induced by the underlying tunnel.

9.7.3.2 *Modelling for subsidence and its numerical analysis*

First, modelling for subsidence analysis is examined with analogue materials (Hao, 1985), which are prepared with six different compounds. Mechanical properties of the materials are measured in the laboratory (Xie, 1984). The numerical example, which was designed in four different mining states is then analyzed by the NEPR. Therefore, four groups of solutions of different working lengths are obtained. The relationship between the subsidence coefficient Q and depth Z is given in Figure 9.13. Stress distribution along the central line of the coal seam is shown in Figure 9.14. Stress concentration phenomenon has apparently been observed, which is congruous with the results of the three-dimensional program (Lshi in 1983).

The fracture region of the experimental model as predicted by the NEPR program

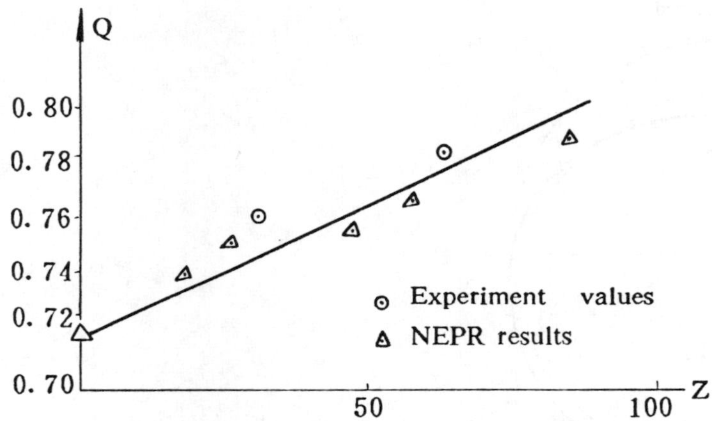

Figure 9.13. The relationship between subsidence coefficient Q and depth Z.

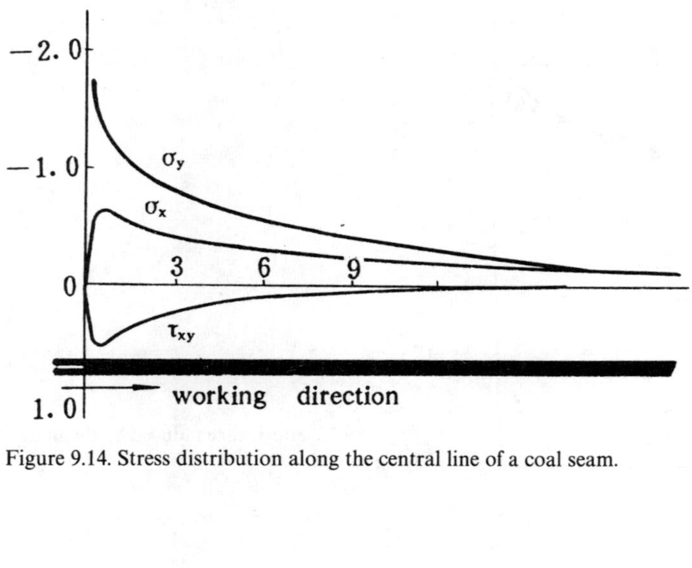

Figure 9.14. Stress distribution along the central line of a coal seam.

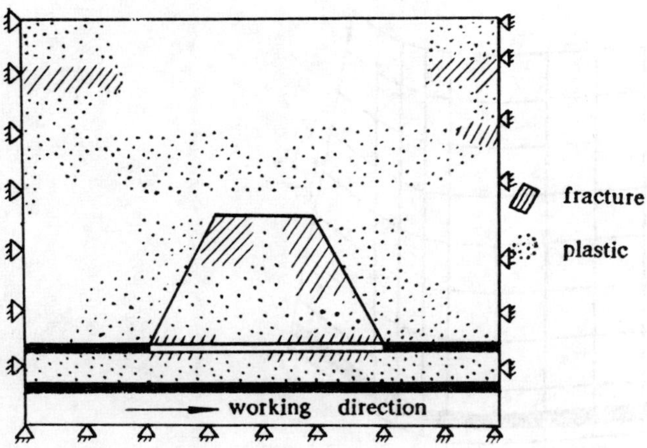

Figure 9.15. Fracture and plastic regions obtained by the NEPR program.

is shown in Figure 9.15. A fracture arch occurs above the gob area, which has the same tendency to occur when performing a simulated test of analogous material.

9.7.3.3 *Modelling thickness change of a coal seam*
A coal seam is a deformable soft bed. Under a certain tectonic stress, it would experience elasto-plastic large deformation and solid flowing. Coal components would

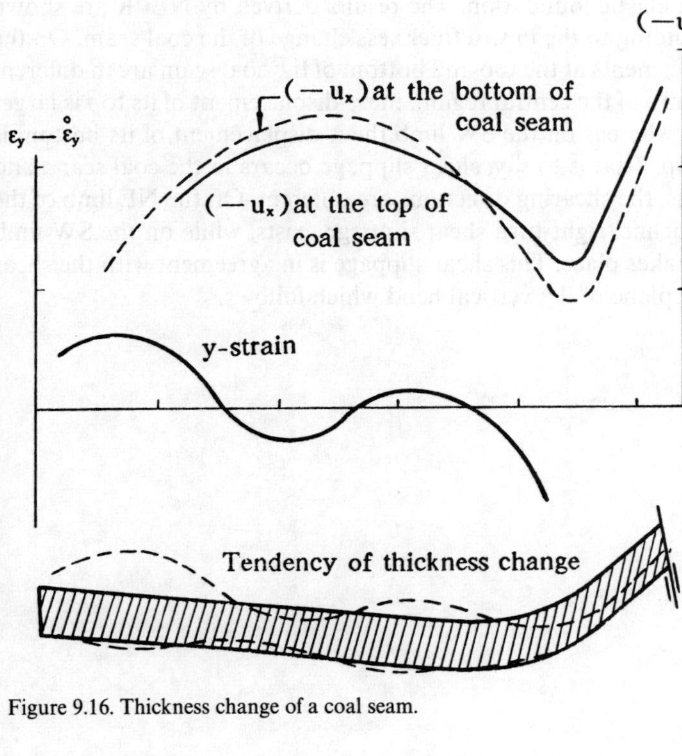

Figure 9.16. Thickness change of a coal seam.

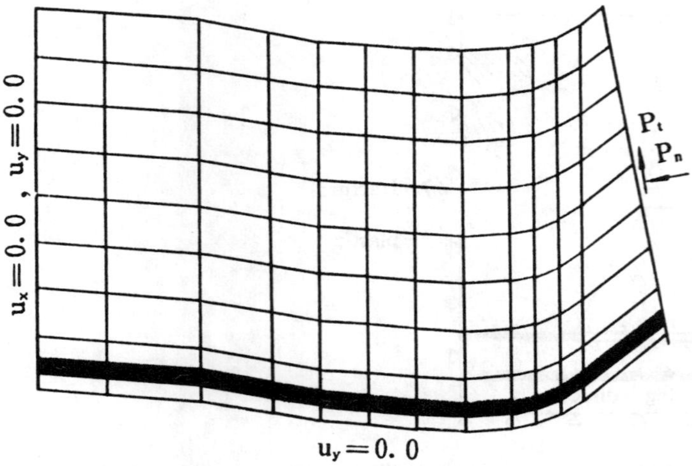

Figure 9.17. Model and mesh.

move from the high to low stress zone, so that the coal seam would be thickened in the low stress zone, and thinned in the high stress zone. This phenomenon is examined in the Shang Zhuanf region of China (Xu, 1985). Based on in situ data, a computing model is designed as shown in Figure 9.17. The coal seam is five meters thick. The overlying and underlying strata of the coal seam are 500 meters of sandstone and about 9 meters of arenaceous mudstone, respectively. On the left of the model is the fixed boundary. The right portion of the model is loaded by P_n and P_t. The bottom of the model is regarded as an elastic foundation. The results derived by NEPR are shown in Figure 9.16, corresponding to the in situ thickness change of the coal seam. On the other hand, the x-displacements at the top and bottom of the coal seam are in different directions. On the NE limb of the central region, the x-displacement of its top is larger than that of its bottom, whereas on the SW limb the x-displacement of its bottom is larger than that of its top. That is to say, shear slippage occurs in the coal seam, and in different tectonic areas, the shearing directions are different. On the NE limb of the central region of the syncline, right-turn shear slippage exists, while on the SW limb left-turn shear slippage takes place. This shear slippage is in agreement with the shear slippage on the bedding plane of the vertical bend which follows.

CHAPTER 10

Damage mechanics of rock

The micro-structural change which causes material strength deterioration is usually termed damage. Damage is generally induced by the nucleation of micro-voids (micro-cracks and cavities) with their growth and coalescence giving rise to the initiation of macroscopic cracks in the material. The ultimate state of a damage process corresponds to macrocrack initiation or the failure of the material.

Rock is a natural damage material and its mechanical properties are strongly dependent on micro-pores and micro-cracks existing in it. The problem of the description of inelastic behavior of rock is essential for the solution of numerous boundary-value problems in mining or nuclear engineering. It is evident that classical elastic-plastic analysis does not provide realistic information on the stress distribution within elements of the material or on the assessment of the safety factor against failure. On the other hand, the limit analysis approach, based on a perfect plasticity model with unlimited ductility, may overestimate the real load carrying capacity and does not provide any information on magnitudes of strains and displacements occurring before failure. Thus, there is a need for new ideas, new methods, and new material models which would sufficiently and accurately simulate the stress-strain behavior and describe progressive fracturing during deformation. It is the purpose of this chapter to introduce a new theory in rock mechanics–damage mechanics of rock.

10.1 DAMAGE MECHANICS OF BRITTLE MATERIALS

Damage denotes the process of gradual material deterioration induced by nucleation and growth of the multitude of micro-defects (such as micro-cracks and voids) (Li, 1986). Material defects can be roughly classified with respect to their geometry into point defects (vacancies, and interstitial and impurity atoms), line defects (dislocations), plane defects (slip planes and cracks), and volume defects (pores and cavities). Therefore, we can establish different analytical models of damage (micro, meso and macro) in terms of defect scales, as shown in Table 10.1. It is seen that damage mechanics is used mainly to study the effect of material deterioration induced by micro-defects before macro-cracks occur in the materials, whereas fracture mechanics deals only with the influence of macro-cracks. Microdamage mechanics is concerned with the damage analyses of materials at the atomic or molecular scale, commonly using the statistical method. Mesodamage mechanics describes the propagation of cavities or pores and its effects on material properties at the grain scale. However, it is difficult to apply both of them to quantitatively predict the material damage degree such as the life-time of final fracture, because of the effect of size difference between the micro- or mesoscopic quantity and the macroscopic one. For continuum damage mechanics (or macroscopic damage mechanics) it is expected that the mechanics analysis of a material or an engineering structure not only considers damage effects on the material behavior, but also still lies in the framework of continuum mechanics. As an illustration, now let us consider a body B which contains distributed micro-cracks as shown in Figure 10.1 and consider an arbitrary point x in it. We assume that a volume element V exists around x which is large enough to contain sufficient number of micro-cracks and micro-cavities, but is also sufficiently small so that the states of stress, strain, and microcrack distribution are sufficiently uniform. The notion of continuum damage mechanics hypothesizes that the effects of these micro-cracks and micro-cavities at any arbitrary point x can be characterized by an appropriate mechanics variable $w(x)$, and we call it the damage field or damage variable which is the same as the notion of stress, strain and temperature field. It may be a scalar, a vector, or a tensor, depending on the case of the spacial distribution of the damage field. Since the damage variable is a macroscopic variable which mechanically describes the state of material deterioration, it is exactly an internal state variable in

Table 10.1. Analytical models and defect scales.

Scale	Defects	Material elements	Analytical model
Micro	Vacancies and dislocations	Atoms, chains of molecules	Micro damage mechanics or continuum damage mechanics
Meso	Microcracks, pores, slips	Ensemble of grains	Meso damage mechanics or continuum damage mechanics
Macro	Marcocracks, faults	Test specimen	Fracture mechanics

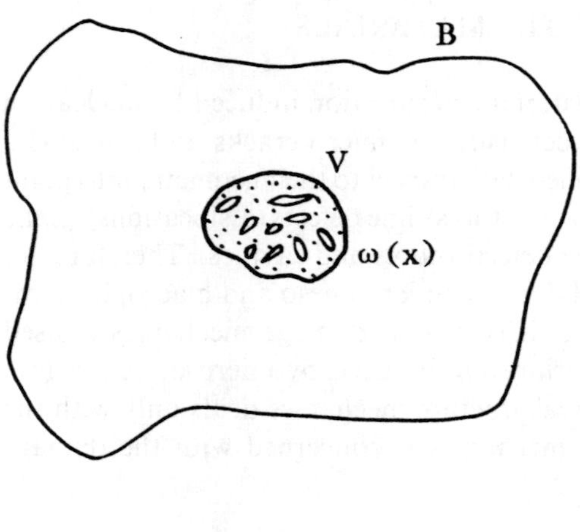

Figure 10.1. The element volume of damage materials.

thermodynamical theories of constitutive equations. According to these assumptions, continuum damage mechanics (CDM) can be established in the framework of continuum mechanics by the following procedure:

(1) Choose appropriate mechanical variables, i.e., damage variables, to describe the mechanics effects of the micro-cracks and micro-cavities distributed in materials;

(2) Establish the evolution equations of these damage variables in terms of experiments or certain theories;

(3) Formulate the constitutive equations containing these damage variables; and finally;

(4) Solve the boundary-value problems constituted by these equations.

The notion of continuum damage mechanics was first proposed by Kachanov (1958) when he studied creep rupture time. If we assume that the effective sectional area reduction due to micro-voids and micro-cracks, the corresponding microstress concentrations in the vicinity of discontinuities, and the interactions between closed defects is the principal mechanism of material deterioration or damage under uniaxial stress along a straight bar, then the state of deterioration may be represented by a state variable ψ, known as the continuity variable and it may be physically interpreted as the ratio of the effective sectional area A_e of the bar to the current sectional area A:

$$\psi = A_e/A \text{ or } A_e = \psi A. \tag{10.1}$$

If σ is the Cauchy stress, the force acted on the section is

$$P = \sigma A = \sigma_e A_e. \tag{10.2}$$

Equation (10.2) defines the effective stress σ_e. Thus, we have

$$\sigma_e = \sigma/\psi. \tag{10.3}$$

Equation (10.3) implies that a stress is magnified by the effective area reduction as indicated in Figure 10.2, where $\psi = 1$ and $\psi = 0$ respectively specify the initial undamaged state and the final local failure state. Actually the failure is characterized by $\psi = \psi_c$ with $0.2 \leq \psi_c \leq 0.7$ for rocks (ψ_c is a critical value).

Rabotnov (1969) introduced an alternative variable,

$$w = 1 - \psi, \tag{10.4}$$

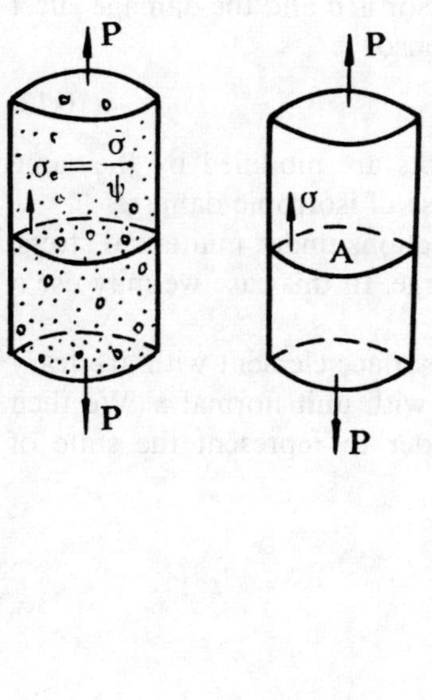

Figure 10.2. Damage of a bar under tension.

when he studied the constitutive equations for creep. The so-called damage variable is w, and $w = 0$ and $w = 1$ respectively represent the undamaged and failure states. Actually, $w = w_c = 1 - \psi_c$ at local failure. From Equations (10.1) and (10.4), we have

$$w = (A - A_e)/A. \tag{10.5}$$

We also use the damage effect variable,

$$\Phi = \psi^{-1} = 1/(1 - w), \tag{10.6}$$

and thus,

$$\sigma_e = \sigma/(1 - w) = \sigma\Phi. \tag{10.7}$$

Broberg (1975) defined the damage variable w as

$$w = \ln(A/A_e); \quad (\text{or } dw = -dA_e/A_e); \tag{10.8}$$

If the area A effectively reduces first to A' and then to A'', the damage in these two steps are

$$w_1 = \ln(A/A'), \quad w_2 = \ln(A'/A''),$$

and the total damage is

$$w = w_1 + w_2.$$

Thus, the damage values are additive according to Broberg's definition. Using Equations (10.2) and (10.8), we have

$$\sigma_e = \sigma \exp(w). \tag{10.9}$$

For an incompressible material, the strain of a straight bar is

$$\varepsilon = \ln(l/l_0) = \ln(A_0/A), \tag{10.10}$$

where l_0 and A_0 are the unstressed length and sectional area of the bar, respectively, and l is its current length. From Equation (10.10), the normal stress is

$$\sigma/\sigma_0 = A_0/A = \exp(\varepsilon). \tag{10.11}$$

Combining Equations (10.9) and (10.11), we obtain

$$\sigma_e = \sigma_0 \exp(\varepsilon + w). \tag{10.12}$$

In the case of multiaxial stress, if the Cauchy stress tensor is $\boldsymbol{\sigma}$ and the damage effect is the same in all directions, then the effective stress tensor is

$$\boldsymbol{\sigma}_e = \boldsymbol{\sigma}/\psi = \boldsymbol{\sigma}/(1 - w) = \boldsymbol{\sigma}\Phi. \tag{10.13}$$

Equation (10.13) implies that all the stress components are modified by the same value to give the effective stress components in the case of isotropic damage.

Usually the damage effect is not the same in all directions under multiaxial stress, especially when the micro-cavity volume fraction is large. In this case we may use a tensor to measure anisotropic damage.

Under a multiaxial stress state, we must consider a surface element with arbitrary orientation. Let $\boldsymbol{A} = A\boldsymbol{n}$ be the vector of the element with unit normal $\boldsymbol{n}$. We then use a symmetric continuity tensor of the second order to represent the state of deterioration and change Equation (10.1) into a tensorial equation,

$$\boldsymbol{A}_e = A_e\boldsymbol{n} = \boldsymbol{\psi}\boldsymbol{A} = \boldsymbol{\psi}A\boldsymbol{n}, \tag{10.14}$$

where $\boldsymbol{A}_e$ is the effective area vector of a fictition surface element with unit normal $\boldsymbol{n}$. Since

$$\boldsymbol{\psi} = \psi_1\boldsymbol{e}_1 * \boldsymbol{e}_1 + \psi_2\boldsymbol{e}_2 * \boldsymbol{e}_2 + \psi_3\boldsymbol{e}_3 * \boldsymbol{e}_3, \tag{10.15}$$

ψ_i and $\boldsymbol{e}_i$ $(i = 1, 2, 3)$ are the principal values and the unit base vectors along principal directions of the symmetric tensor. We see from Equations (10.14) and (10.15) that

$$\boldsymbol{\psi}\cdot\boldsymbol{e}_i = \psi_i\boldsymbol{e}_i \quad (i = 1, 2, 3), \tag{10.16}$$

which implies that a unit area normal to a principal direction $\boldsymbol{e}_i$ of $\boldsymbol{\psi}$ is reduced to an area of magnitude equal to the corresponding principal value ψ_i without change of orientation. Moreover, we have $0 \le \psi_i \le 1$. But in general, both the orientation and the area of the surface element are changed through the linear transformation (Equation (10.14)). Define the symmetric damage tensor $\boldsymbol{w}$ as

$$\boldsymbol{w} = \boldsymbol{I} - \boldsymbol{\psi} = w_1\boldsymbol{e}_1 * \boldsymbol{e}_1 + w_2\boldsymbol{e}_2 * \boldsymbol{e}_2 + w_3\boldsymbol{e}_3 * \boldsymbol{e}_3, \tag{10.17}$$

where $w_i = 1 - \psi_i$, $0 \le w_i \le 1$ $(i = 1, 2, 3)$, and $\boldsymbol{I}$ is the unit tensor. We then have

$$\boldsymbol{A}_e = A_e\boldsymbol{n} = (\boldsymbol{I} - \boldsymbol{w})\boldsymbol{A} = (\boldsymbol{I} - \boldsymbol{w})A\boldsymbol{n}. \tag{10.18}$$

If $\boldsymbol{\sigma}$ is the Cauchy stress tensor, then using Cauchy's formula and Equation (10.18), the force acted on the surface element with area vector $\boldsymbol{A}$ or on the fictition surface element with area vector $\boldsymbol{A}_e$ is

$$\boldsymbol{P} = \boldsymbol{\sigma}\boldsymbol{A} = \boldsymbol{\sigma}_e\boldsymbol{A}_e = \boldsymbol{\sigma}_e\boldsymbol{\psi}\boldsymbol{A} = \boldsymbol{\sigma}_e(\boldsymbol{I} - \boldsymbol{w})\boldsymbol{A}. \tag{10.19}$$

Equation (10.19) defines the effective stress tensor $\boldsymbol{\sigma}_e$. Since this equation is correct for an arbitrary area vector $\boldsymbol{A}$, we have

$$\boldsymbol{\sigma} = \boldsymbol{\sigma}_e\boldsymbol{\psi} = \boldsymbol{\sigma}_e(\boldsymbol{I} - \boldsymbol{w}), \tag{10.20}$$

or

$$\boldsymbol{\sigma}_e = \boldsymbol{\sigma}\boldsymbol{\Phi} = \boldsymbol{\sigma}\boldsymbol{\psi}^{-1} = \boldsymbol{\sigma}(\boldsymbol{I} - \boldsymbol{w})^{-1}, \tag{10.21}$$

where

$$\begin{aligned} \boldsymbol{\Phi} &= \boldsymbol{\psi}^{-1} = (\boldsymbol{I} - \boldsymbol{w})^{-1} = \Phi_1\boldsymbol{e}_1 * \boldsymbol{e}_1 + \Phi_2\boldsymbol{e}_2 * \boldsymbol{e}_2 + \Phi_3\boldsymbol{e}_3 * \boldsymbol{e}_3, \\ \Phi_i &= 1/\psi_i = 1/(1 - w_i), \quad (i = 1, 2, 3) \end{aligned} \tag{10.22}$$

and the operator $*$ is called the tensor product, and defined as follows:

$$\boldsymbol{n}_1 * \boldsymbol{n}_1 = \begin{bmatrix} 1 & 0 & 0 \\ 0 & 0 & 0 \\ 0 & 0 & 0 \end{bmatrix}; \quad \boldsymbol{n}_1 * \boldsymbol{n}_2 = \begin{bmatrix} 0 & 1 & 0 \\ 0 & 0 & 0 \\ 0 & 0 & 0 \end{bmatrix}.$$

The symmetric tensor $\boldsymbol{\Phi}$ is called the damage effect tensor. In general, the effective stress tensor is asymmetric. This fact is a disadvantage, it is awkward to use an asymmetric tensor $\boldsymbol{\sigma}_e$ in constitutive equations with a symmetric strain or strain rate tensor. We may symmetrize $\boldsymbol{\sigma}_e$ by using the symmetric part of $\boldsymbol{\Phi}\boldsymbol{\sigma}$,

$$\boldsymbol{\sigma}_e = (\boldsymbol{\sigma}\boldsymbol{\Phi} + \boldsymbol{\Phi}\boldsymbol{\sigma})/2. \tag{10.23}$$

Sometimes the influence of a stress state inside a damage material depends not only upon the net area reduction caused by micro-defect formation but also upon the three-dimensional arrangement or configuration of micro-defects. We should then use a higher order tensor, say the fourth order tensor specified by the damage state, to do the linear transformation for an effective stress tensor $\boldsymbol{\sigma}_e$,

$$\boldsymbol{\sigma}_e = \boldsymbol{M}:\boldsymbol{\sigma}, \tag{10.24}$$

where $\boldsymbol{M}$ is the fourth order tensor function with the tensor $\boldsymbol{w}$ or $\boldsymbol{\Phi}$ as an argument. If the tensor $\boldsymbol{\sigma}_e$ is asymmetric, we may symmetrize it in the usual way,

$$\boldsymbol{\sigma}_e = [(\boldsymbol{M}:\boldsymbol{\sigma}) + (\boldsymbol{M}:\boldsymbol{\sigma})^{\mathrm{T}}]/2. \tag{10.25}$$

According to the representation theory, the most general expression of $\boldsymbol{M}(\Phi)$ is given by a linear combination of fourth order basic tensors formed by Kronnecker delta δ_{ij} and ϕ_{ij},

$$\begin{aligned} M_{ijkl} = {} & \lambda\delta_{ij}\delta_{kl} + \mu(\delta_{ik}\delta_{jl} + \delta_{il}\delta_{jk}) + \nu\delta_{ij}\phi_{kl} + \pi\delta_{kl}\phi_{ij} \\ & + \rho(\delta_{ik}\phi_{jl} + \delta_{il}\phi_{jk} + \delta_{jk}\phi_{il} + \delta_{jl}\phi_{ik}) + A\phi_{ij}\phi_{kl} + B\delta_{ij}x_{kl} \\ & + C\delta_{kl}x_{ij} + D(\delta_{ik}x_{jl} + \delta_{il}x_{jk} + \delta_{jk}x_{il} + \delta_{jl}x_{ik}) \\ & + G\phi_{ij}x_{kl} + H\phi_{ik}x_{jl} + Kx_{ij}x_{kl}; \\ x_{ij} = {} & \phi_{im}\phi_{mj}. \end{aligned} \tag{10.26}$$

where λ, μ, ν, π, and ρ are constants, and A, B, C, D, E, F, G, H, and K are polynomials of the scalar invariants of ϕ_{ij}. If we let $\rho = \frac{1}{4}$ and neglect the remaining terms in Equation (10.26), this equation is reduced to a simpler form:

$$M_{ijkl} = (\delta_{ik}\phi_{jl} + \delta_{il}\phi_{jk} + \delta_{jk}\phi_{il} + \delta_{jl}\phi_{ik})/4. \tag{10.27}$$

In this case, Equation (10.25) takes the form of Equation (10.23).

Leckie & Onat (1980) discussed the tensorial nature of damage variables. They assumed that the internal state and orientation of a deformed and damage material element can be described by a number of parameters known as internal state variables,

$$\boldsymbol{\alpha} = \{\boldsymbol{\sigma}, \boldsymbol{q}, \boldsymbol{r}\}, \tag{10.28}$$

where $\boldsymbol{q}$ represents the parameters that reveal the distribution of dislocations, $\boldsymbol{r}$ represents the parameters that measure the size and distribution of micro-cracks and the number and distribution of sites for micro-crack growth. These parameters reveal not only the magnitude of the quantities but also the orientations. Suppose that $\dot{\boldsymbol{q}}$ depends strongly on stress $\boldsymbol{\sigma}$ and $\boldsymbol{q}$, but weakly on $\boldsymbol{r}$. Under constant stress, $\boldsymbol{q}$ attains its saturation value within a short time in comparison to the time for rupture. Thus, in the presence of slowly varying stress,

$$\boldsymbol{q} = F(\boldsymbol{\sigma}). \tag{10.29}$$

Therefore, we get

$$\dot{\boldsymbol{\sigma}} = f(\boldsymbol{\sigma}, \boldsymbol{r}) + \boldsymbol{E}:\dot{\boldsymbol{\varepsilon}}$$

and

$$\boldsymbol{r} = g(\boldsymbol{\sigma}, \boldsymbol{r}), \tag{10.30}$$

where $\boldsymbol{E}$ is the elasticity tensor.

Besides, we may have a damage criterion in the form of an inequality,

$$R(\boldsymbol{\sigma}, \boldsymbol{r}) \leq 0. \tag{10.31}$$

Now, we further divide $\boldsymbol{r}$ into two groups $\boldsymbol{V}$, and $\boldsymbol{N}$,

$$\boldsymbol{r} = \{\boldsymbol{V}, \boldsymbol{N}\}, \tag{10.32}$$

where $\boldsymbol{V}$ represents the parameters that measure the size and distribution of micro-cracks, $\boldsymbol{N}$ represents those that measure the number and distribution of sites for micro-crack growth.

Consider a sphere S of unit radius, and let $\boldsymbol{n}$ be the position vector of a point on this sphere. An area element $\mathrm{d}A$ on the sphere about the point $\boldsymbol{n}$ represents a bundle of direction about $\boldsymbol{n}$. The total volume of micro-cracks orthogonal to this bundle of directions will be given by,

$$V(\boldsymbol{n})\mathrm{d}A,$$

where V is the density of distribution of micro-crack volume which depends on $\boldsymbol{n}$. Obviously,

$$V(\boldsymbol{n}) = V(-\boldsymbol{n}). \tag{10.33}$$

We may try to represent $\boldsymbol{V}$ in terms of the moments of $V(n)$,

$$V_0 = \int_s V(\boldsymbol{n})\,\mathrm{d}A, \quad V_i = \int_s V(\boldsymbol{n})n_i\,\mathrm{d}A, \quad \text{and} \quad V_{ij} = \int_s V(\boldsymbol{n})n_i n_j \mathrm{d}A, \ldots \tag{10.34}$$

where n_i $(i = 1, 2, 3)$ are the components of $\boldsymbol{n}$. V_0, V_i, and $V_{ij}, \ldots$, are evidently a scalar and the components of a vector, a second order tensor, and higher order tensors, respectively. We find from Equation (10.33) that the odd order tensors in Equation (10.34) vanish,

$$V_i = 0, \quad V_{ijk} = 0, \ldots, \tag{10.35}$$

and that the even order tensors are completely symmetric:

$$V_{ij} = V_{ji}, \quad V_{ijkl} = V_{jikl} = V_{jkil} = \ldots, \ldots. \tag{10.36}$$

Note that V_0 measures the total volume of micro-cracks per unit volume. Moreover, from the Equation (10.34), we have

$$V_{kk} = V_0, \quad V_{ijkk} = V_{ij}, \ldots, \tag{10.37}$$

if we use Einstein's summation convention. Thus,

$$\boldsymbol{V} = \{V_0, V_{ij}, V_{ijkl}, \ldots\} \tag{10.38}$$

is a sequence of irreducible even order tensors.

Similarly, we have

$$\boldsymbol{N} = \{N_0, N_{ij}, N_{ijkl}, \ldots\}, \tag{10.39}$$

and, hence, arrive at the conclusion that the sequence of physically significant tensors

$$\boldsymbol{r} = \{V_0, V_{ij}, V_{ijkl}, \ldots; N_0, N_{ij}, N_{ijkl}, \ldots\}, \tag{10.40}$$

describes the damage events well enough.

Murakami & Ohno (1981) developed a continuum theory of creep and creep damage of polycrystalline metals. Let us consider a characteristic volume element V in the material as illustrated in Figure 10.3 and denote the area of a grain boundary element occupied by the k-th cavity and the unit normal vector to it by $dS_g^{(k)}$ and $V^{(k)}$ $(k = 1, 2, \ldots, N)$, respectively. The state of material damage may then be described by a second order symmetric tensor,

$$\mathbf{\Omega} = [3/S_g(V)] \sum \int_V [V^{(k)} * V^{(k)}] dS_g^{(k)}, \tag{10.41}$$

where $S_g(V)$ is the total area of the grain boundary in V.

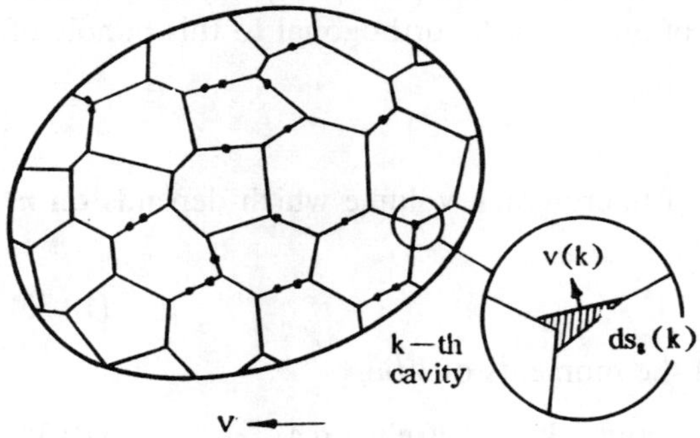

Figure 10.3. Volume element with grain boundary cavities (Murakami & Ohno, 1981).

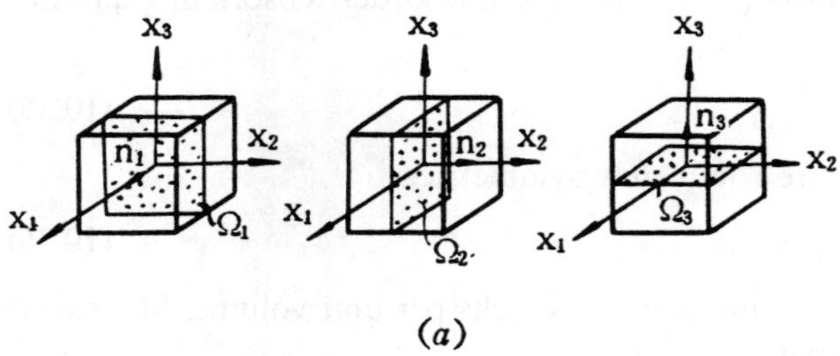

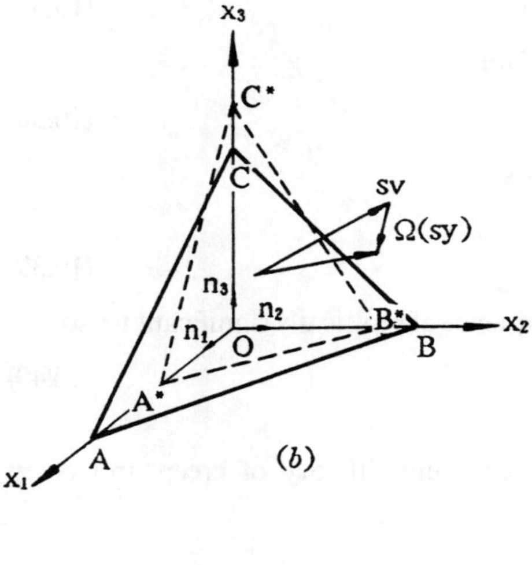

Figure 10.4. Local interpretation of the damage tensor $\mathbf{\Omega}$. (a) Net area reduction on the principal plane of the damage tensor Ω; (b) Net area reduction on arbitrary planes (Murakami & Ohno, 1981).

A second order symmetric tensor has three real principal values and corresponding principal directions. Let Ω_j and n_j respectively denote the principal values and the principal directions of the tensor $\mathbf{\Omega}$. Equation (10.41) can then be rewritten by,

$$\mathbf{\Omega} = \sum_{j=1}^{3} \Omega_j n_j * n_j. \tag{10.42}$$

It is seen from Equation (10.42) that the effect of material damage can be characterized by the cavity area density Ω_j. The principal directions correspond to the damage state, as indicated in Figure 10.4a, $\Omega_j = 0$ corresponds to the undamaged state, and $\Omega_j = 1$ corresponds to the state of creep rupture occurring on the j-th plane. Thus, the tensor $\mathbf{\Omega}$ represents the three-dimensional area density of the grain boundary cavities due to damage.

Figure 10.4b shows that the area vector SV of an arbitrary surface element ABC is transformed into the corresponding net area vector $S^* V^*$ (surface element A*B*C*) because of the damage

$$S^* V^* = (I - \Omega) S V, \tag{10.43}$$

where $\Delta OB^*C^* = (1 - \Omega)\Delta OBC$, $\Delta OC^*A^* = (1 - \Omega_2)\Delta OCA$, and $\Delta OA^*B^* = (1 - \Omega_3)\Delta OAB$.

From the definition of the effect stress tensor, we have

$$\boldsymbol{\sigma}_e = \boldsymbol{\sigma}:(\boldsymbol{I} - \boldsymbol{\Omega})^{-1} = \boldsymbol{\sigma}:\boldsymbol{\Phi}, \tag{10.44}$$

and

$$\boldsymbol{\Phi} = (\boldsymbol{I} - \boldsymbol{\Omega})^{-1}. \tag{10.45}$$

Note that symbol (:) indicates the product of tensors and/or vectors. The effective stress tensor defined by Equation (10.44) is asymmetric. It is to symmetrized in the following way

$$\boldsymbol{\sigma}_e^* = (\boldsymbol{\sigma}:\boldsymbol{\Phi} + \boldsymbol{\Phi}:\boldsymbol{\sigma})/2. \tag{10.46}$$

The evolution equation of damage may then be expressed as follows:

$$\boldsymbol{\Omega} = \boldsymbol{H}(\boldsymbol{\sigma}_e^*, \boldsymbol{\Phi}, K, T), \tag{10.47}$$

where K and T are a work-hardening parameter of the material matrix and temperature, respectively.

10.1.1 *Damage and elasticity*

Sometimes, the free energy of a material may be decomposed into two parts,

$$\phi(\varepsilon^e, T, \boldsymbol{w}, \boldsymbol{V}) = \phi_e(\varepsilon^e, T, \boldsymbol{w}) + \phi_p(T, \boldsymbol{V}), \tag{10.48}$$

where the damage tensor $\boldsymbol{w}$ and $\boldsymbol{V}$ are internal variables and ϕ_e is the elastic energy. Then the complementary energy,

$$\psi(\boldsymbol{\sigma}, T, \boldsymbol{w}, \boldsymbol{V}) = \psi_e(\boldsymbol{\sigma}, T, \boldsymbol{w}) + \psi_p(T, \boldsymbol{V}), \tag{10.49}$$

where the complementary elastic energy

$$\psi_e(\boldsymbol{\sigma}, T, \boldsymbol{w}) = (1/\rho)\boldsymbol{\sigma}:\boldsymbol{\varepsilon}^e - \phi_e(\varepsilon^e, T, \boldsymbol{w}). \tag{10.50}$$

The normality relationships of thermodynamics give the elastic constitutive relationships:

$$\boldsymbol{\sigma} = \rho\partial\phi_e/\partial\boldsymbol{\varepsilon}^e, \tag{10.51}$$

and

$$\boldsymbol{\varepsilon}^e = \rho\partial\psi_e/\partial\boldsymbol{\sigma}. \tag{10.52}$$

For an undamaged material under isothermal conditions, the linear elastic energy is

$$\rho\phi_e = [\boldsymbol{\varepsilon}^e:\boldsymbol{E}:\boldsymbol{\varepsilon}^e]/2, \tag{10.53}$$

where $\boldsymbol{E}$ is the fourth order tensor of elasticity. It should be symmetric. Substituting Equation (10.53) into Equation (10.51), we have

$$\boldsymbol{\sigma} = \boldsymbol{E}:\boldsymbol{\varepsilon}^e. \tag{10.54}$$

For an isotropic material, the component of elasticity tensor $\boldsymbol{E}$ is

$$E_{ijkl} = (E/(1+\nu))[(\nu/(1-2\nu))\delta_{ij}\delta_{kl} + \delta_{ik}\delta_{jl}], \tag{10.55}$$

where E and ν are Young's modulus and Poisson's ratio of the material. Substituting Equation (10.55) into Equations (10.53) and (10.54), we get

$$\rho\phi_e = [E/(2(1+\nu))\{(\nu/(1-2\nu))(\mathrm{tr}\,\boldsymbol{\varepsilon}^e)^2 + (\boldsymbol{\varepsilon}^e)^2\}, \tag{10.56}$$

and

$$\boldsymbol{\sigma} = [E/(1+\nu)][(\nu/(1-\nu))\boldsymbol{I}\,\mathrm{tr}\,\boldsymbol{\varepsilon}^e + \boldsymbol{\varepsilon}^e]. \tag{10.57}$$

Where $\mathrm{tr}(\varepsilon^e)$ means the trace of $\boldsymbol{\varepsilon}^e$. The complementary linear elastic energy is

$$\rho\psi_e = \boldsymbol{\sigma}:\boldsymbol{E}^{-1}:\boldsymbol{\sigma}/2. \tag{10.58}$$

The compliance tensor $\boldsymbol{E}^{-1}$ is also symmetric. Substituting Equation (10.58) into Equation (10.52), we obtain

$$\boldsymbol{\varepsilon}^e = \boldsymbol{E}^{-1}:\boldsymbol{\sigma}. \tag{10.59}$$

For an isotropic material, the component of $\boldsymbol{E}^{-1}$ is

$$E^{-1}_{ijkl} = [(1+\nu)/E]\delta_{ik}\delta_{jl} - (\nu/E)\delta_{ij}\delta_{kl}. \tag{10.60}$$

Substituting Equation (10.60) into Equations (10.59) and (10.58), we get

$$\rho\psi_e = [(1+\nu)/E]\boldsymbol{\sigma}^2 - [\nu/(2E)](\mathrm{tr}\,\boldsymbol{\sigma})^2, \tag{10.61}$$

and

$$\boldsymbol{\varepsilon}^e = [(1+\nu)/E]\boldsymbol{\sigma} - (\nu/E)\boldsymbol{I}\,\mathrm{tr}\,\boldsymbol{\sigma}. \tag{10.62}$$

The influence of damage on the linear elastic stress-strain relationship may be expressed by the concept of the effective elasticity tensor $\tilde{\boldsymbol{E}}_{(w)}$ (i.e., the material deterioration caused by damage behaves as its macroscopic expression in the elastic softening – the elastic rigidity moduli decrease), we have

$$\rho\phi_e(\boldsymbol{\varepsilon}^e, \boldsymbol{w}) = (\tfrac{1}{2})\boldsymbol{\varepsilon}^e:\tilde{\boldsymbol{E}}_{(w)}:\boldsymbol{\varepsilon}^e, \tag{10.63}$$

and

$$\rho\psi_e(\boldsymbol{\sigma}, \boldsymbol{w}) = (\tfrac{1}{2})\boldsymbol{\sigma}:\tilde{\boldsymbol{E}}^{-1}_{(w)}:\boldsymbol{\sigma}. \tag{10.64}$$

We then get the constitutive equation from the normality relationships:

$$\boldsymbol{\sigma} = \rho\partial\phi_e/\partial\boldsymbol{\varepsilon}^e = \tilde{\boldsymbol{E}}_{(w)}:\boldsymbol{\varepsilon}^e, \tag{10.65}$$

and

$$\boldsymbol{\varepsilon}^e = \rho\partial\psi_e/\partial\boldsymbol{\sigma} = \tilde{\boldsymbol{E}}^{-1}_{(w)}:\boldsymbol{\sigma}. \tag{10.66}$$

The tensors $\tilde{\boldsymbol{E}}_{(w)}$ and $\tilde{\boldsymbol{E}}^{-1}_{(w)}$ should also be symmetric.

For the constitutive equations of damaged materials, Lemaitre & Chaboche (1974) proposed a hypothesis of strain equivalence which states that the behavior of a damaged material is represented by the constitutive equation of the undamaged material in which the stress is just replaced by the effective stress. In other words, any strain constitutive equation for a damaged material ($w \neq 0$) is derived from the same potentials as for a virgin material ($w = 0$), except that all the stress variables are replaced by effective stresses, as illustrated in Figure 10.5.

If the elastic strain-stress relationship for an undamaged material is

$$\boldsymbol{\varepsilon}^e = \boldsymbol{\varepsilon}^e(\boldsymbol{\sigma}, 0), \tag{10.67}$$

then that for a damaged material is

$$\boldsymbol{\varepsilon}^e = \boldsymbol{\varepsilon}^e(\boldsymbol{\sigma}, \boldsymbol{w}) = \boldsymbol{\varepsilon}^e(\boldsymbol{\sigma}_e, 0). \tag{10.68}$$

Therefore from Equation (10.24) for a damaged material with the linear elastic strain-stress relationship, we have

$$\boldsymbol{\varepsilon}^e = \boldsymbol{E}^{-1}:\boldsymbol{\sigma}_e = \boldsymbol{E}^{-1}:\boldsymbol{M}_{(w)}:\boldsymbol{\sigma} = \tilde{\boldsymbol{E}}^{-1}_{(w)}:\boldsymbol{\sigma}.$$

The effective elasticity tensor is then

$$\tilde{\boldsymbol{E}}_{(w)} = \boldsymbol{M}^{-1}_{(w)}:\boldsymbol{E} \tag{10.69a}$$

or

$$\tilde{\boldsymbol{E}}^{-1}_{(w)} = \boldsymbol{E}^{-1}:\boldsymbol{M}_{(w)}; \tag{10.69b}$$

and we have

$$\boldsymbol{M}_{(w)} = \boldsymbol{E}:\tilde{\boldsymbol{E}}^{-1}_{(w)}. \tag{10.70}$$

We then determine the components of $\boldsymbol{M}_{(w)}$ by measuring the components of $\boldsymbol{E}$ and $\tilde{\boldsymbol{E}}_{(w)}$ by experiments.

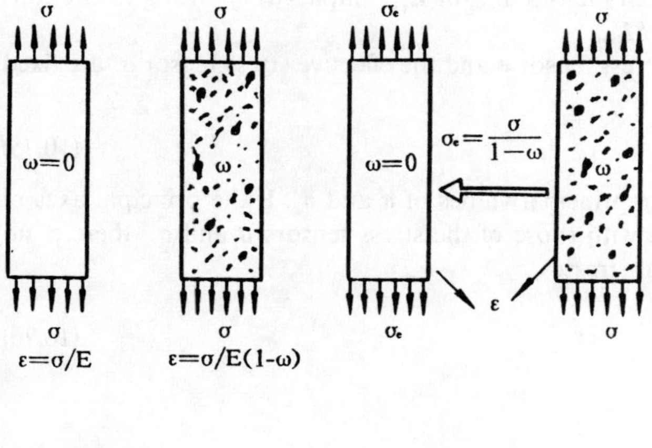

Figure 10.5. Interpretation of strain equivalence principle.

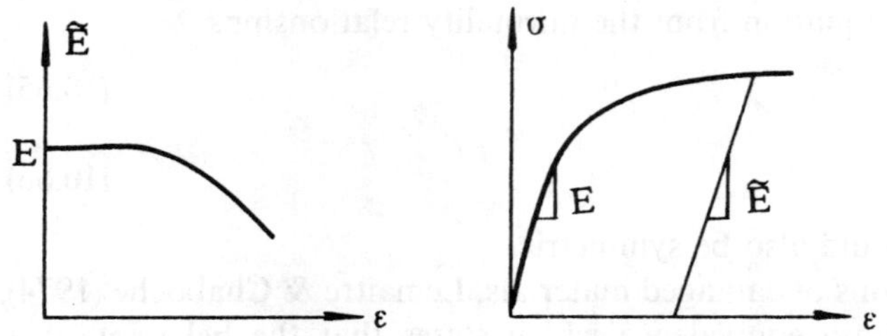

Figure 10.6. Variation of Young's Modulus (Li, 1986).

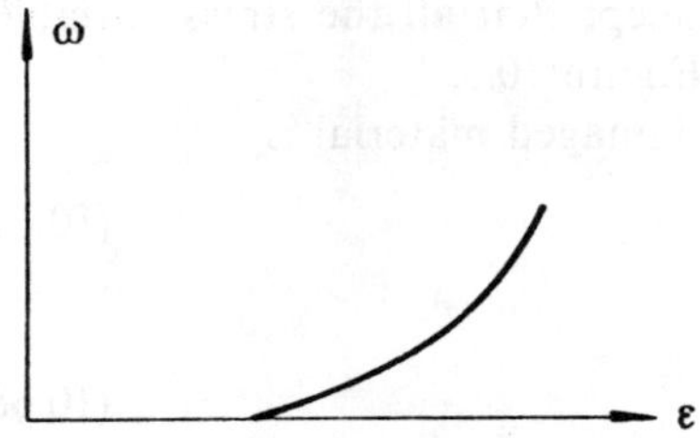

Figure 10.7. Variation of the damage variable (Li, 1986).

For example, under uniaxial stress (Figures 10.6 and 10.7), we have

$$M = 1/(1 - w) = E/\tilde{E}, \tag{10.71}$$

therefore

$$w = 1 - \tilde{E}/E. \tag{10.72}$$

Thus, we may measure E and $\tilde{E}$ during loading and unloading in a tension or compression test of a material to determine the damage variable w.

For an isotropic material, we obtain

$$\boldsymbol{\varepsilon}^e = [(1+\nu)/E]\boldsymbol{\sigma}_e - (\nu/E)\boldsymbol{I}\mathrm{tr}\boldsymbol{\sigma}_e = [(1+\nu)/E]\boldsymbol{M}{:}\boldsymbol{\sigma} - (\nu/E)\boldsymbol{I}\mathrm{tr}(\boldsymbol{M}{:}\boldsymbol{\sigma}), \tag{10.73}$$

and

$$\boldsymbol{\sigma}_e = \boldsymbol{M}{:}\boldsymbol{\sigma} = [E/(1+\nu)][(\nu/(1-2\nu))\boldsymbol{I}\mathrm{tr}\boldsymbol{\varepsilon}^e + \boldsymbol{\varepsilon}^e]. \tag{10.74}$$

Unfortunately, according to Equation (10.69) or Equation (10.70), the symmetric condition of the effective elasticity tensor $\tilde{\boldsymbol{E}}_{(w)}$ or $\tilde{\boldsymbol{E}}_{(w)}^{-1}$ implies very strong restrictions on the transformation tensor $\boldsymbol{M}_{(w)}$.

If the principal axes of the stress tensor $\boldsymbol{\sigma}$ and the effective stress tensor $\boldsymbol{\sigma}_e$ are fixed and coincide, we may write

$$\sigma_{ei} = M_{ij}\sigma_j, \tag{10.75}$$

where σ_i and σ_{ei} $(i = 1, 2, 3)$ are principal values of $\boldsymbol{\sigma}$ and $\boldsymbol{\sigma}_e$. If the principal axes of the damage tensor $\boldsymbol{w}$ coincide with those of the stress tensors $\boldsymbol{\sigma}$ and $\boldsymbol{\sigma}_e$, there is no coupling between the principal stresses,

$$M_{ij} = \begin{cases} 1/(1 - w_i), & i = j \\ 0, & i \neq j \end{cases}, \tag{10.76}$$

and we have

$$\sigma_{e1} = \sigma_1/(1-w_1); \quad \sigma_{e2} = \sigma_2/(1-w_2); \quad \sigma_{e3} = \sigma_3/(1-w_3); \tag{10.77}$$

where w_1, w_2, and w_3 are the principal damages. This is an example of uncoupled anisotropic damage.

10.1.2 *Constitutive equations of irreversible thermodynamics of damage materials*

Irreversible thermodynamics and the concept of internal state variables are indispensable in studying continuum damage mechanics.

During deformation of material, the energy dissipated by material damage, in fact, transforms into heat influx. Suppose that under constant temperature, $\mathfrak{C}$ is the heat equivalent of the dissipated energy per unit volume. The heat work equivalent of the damaged material may then be defined by $\int_\Omega \mathfrak{C}\,\mathrm{d}\Omega$. Therefore, the heat transformation of the system becomes,

$$\dot{Q} = -\int_\Omega q^i \|_i \mathrm{d}\Omega - \int_\Omega \dot{\mathfrak{C}}\,\mathrm{d}\Omega,$$

where q^i is the component of the heat influx vector supplied by adjacent materials, and $\dot{\mathfrak{C}} = -Y^i \dot{w}_i$ is the rate of the dissipated energy, in which Y^i denotes the generalized force of thermodynamics with respect to the damage variable w_i.

Now, let $\dot{\eta}$ represent the rate of entropy per unit volume on the deformed configuration of the damaged material. From the second law of thermodynamics, we then have

$$\int_\Omega \dot{\eta}\,\mathrm{d}\Omega \geq -\int_{\partial\Omega} (q^i n_i/T)\mathrm{d}\Gamma - \int_\Omega (\dot{\mathfrak{C}}/T)\mathrm{d}\Omega. \tag{10.78}$$

From Equation (10.78), the Clausius-Duham inequality of the damaged material can be obtained as follows:

$$T\dot{\eta} \geq -q^i\|_i + (q_i/T)T_{,i} - \dot{\mathfrak{C}}; \tag{10.79}$$

where T is the absolute temperature.

Letting $\dot{U}_e$ and $\dot{\Psi}_e$ represent the rate of internal energy and strain energy per unit volume, respectively, we then obtain the relationship

$$\dot{U}_e = \dot{\Psi}_e - q^i\|_i - \dot{\mathfrak{C}}, \quad \dot{\Psi}_e = \sigma^i_j \dot{S}^j_i. \tag{10.80}$$

From Equations (10.79) and (10.80), we have

$$(\dot{U}_e - T\dot{\eta}) \leq \dot{\Psi}_e - (q_i/T)T_{,i}. \tag{10.81}$$

The Helmholtz free energy $\phi_e(S^i_j, w_i, T)$ per unit volume is defined as

$$\phi_e = U_e - T\eta. \tag{10.82}$$

Its time derivative is

$$\dot{\phi}_e = \dot{U}_e - T\dot{\eta} - \dot{T}\eta = (\partial\phi_e/\partial S^i_j)\dot{S}^i_j + (\partial\phi_e/\partial T)\dot{T} + (\partial\phi_e/\partial w_i)\dot{w}_i. \tag{10.83}$$

Substituting Equation (10.83) into Equations (10.80) and (10.81) we then have

$$[\partial\phi_e/\partial S^i_j - \sigma^i_j]\dot{S}^i_j + [\partial\phi_e/\partial T + \eta]\dot{T} + [\partial\phi_e/\partial w_i - Y^i]\dot{w}_i + T\dot{\eta} + q^i\|_i = 0, \tag{10.84}$$

$$[\partial\phi_e/\partial S^i_j - \sigma^i_j]\dot{S}^i_j + [\partial\phi_e/\partial T + \eta]\dot{T} + [\partial\phi_e/\partial w_i]\dot{w}_i + (q_i/T)T_{,i} \leq 0, \tag{10.85}$$

For elastic damage materials (such as rock and concrete), we define the Helmholtz free energy as

$$\phi_e(\boldsymbol{S}, \boldsymbol{w}, T) = \int_0^s \boldsymbol{S}:\boldsymbol{M}^{-1}(\boldsymbol{w}):\boldsymbol{E}:\boldsymbol{M}^{T,-1}:\mathrm{d}\boldsymbol{S} + f(T), \tag{10.86}$$

where $\boldsymbol{M}$ is a fourth order tensor function with respect to tensor $\boldsymbol{w}$. For a two-dimensional problem of rock materials, $\boldsymbol{M}$ may be defined by (Xie, 1987a, 1990)

$$\boldsymbol{M}_{(w)} = \begin{bmatrix} 1/(1-w_1) & 0 & 0 \\ 0 & 1/(1-w_2) & 0 \\ 0 & 0 & 1 \end{bmatrix}. \tag{10.87}$$

Thus, constitutive equations of the elastic damage material can be written by

$$\sigma_j^i = \partial\phi_e/\partial S_j^i = \tilde{E}_{jl}^{ik} S_k^l; \quad Y^i = \partial\phi_e/\partial w_i, \tag{10.88}$$

and

$$\tilde{E}_{jl}^{ik} = (M_{jn}^{im})^{-1} E_{mq}^{np}((M_{pl}^{qk})^T)^{-1}; \tag{10.89}$$

and

$$(\partial\phi_e/\partial w_i)w_i < 0 \text{ or } -Y^i w_i > 0 \quad \text{(for constant temperature)}. \tag{10.90}$$

Their rate forms can be deduced as

$$\dot{\sigma}_j^i = \tilde{E}_{jl}^{ik}\dot{S}_k^l - 2(M_{jn}^{im})^{-1}(\partial M_{nl}^{mk}/\partial w_q)\dot{w}_q\sigma_k^l, \tag{10.91a}$$

and

$$\dot{Y}^k = -2(M_{jq}^{ip})^{-1}(\partial M_{pl}^{qm}/\partial w_k)\sigma_i^j\dot{S}_m^l \tag{10.91b}$$

Equation (10.90) is in the form of the second law of thermodynamics for elastic damage materials. It should be pointed out that Equations (10.88)-(10.91) are valid not only for small deformation, but also for large deformation analysis of elastic damage materials. As an example of small deformation, we choose the specific free energy ϕ_e in the form (for the case of isotropic damage and constant temperature):

$$\phi_e = (1-w)E_{ijkl}\varepsilon_{ij}\varepsilon_{kl}.$$

Stress is then defined as

$$\sigma_{ij} = \partial\phi_e/\partial\varepsilon_{ij} = (1-w)E_{ijkl}\cdot\varepsilon_{kl} = (1-w)[\lambda\varepsilon_{kk}\delta_{ij} + 2\mu\varepsilon_{ij}],$$

or

$$\mathrm{d}\sigma_{ij} = (1-w)[\lambda \mathrm{d}\varepsilon_{kk}\delta_{ij} + 2\mu \mathrm{d}\varepsilon_{ij}] - [\lambda\varepsilon_{kk}\delta_{ij} + 2\mu\varepsilon_{ij}]\mathrm{d}w, \tag{10.92}$$

and

$$Y = \partial\phi_e/\partial w = -E_{ijkl}\varepsilon_{ij}\varepsilon_{kl}/2. \tag{10.93}$$

Where λ and μ are Lame coefficients. From Equation (10.92), if we consider the state of constant stress ($\mathrm{d}\sigma_{ij} = 0$), we then have

$$\mathrm{d}\varepsilon_{ij} = \varepsilon_{ij}\mathrm{d}w/(1-w). \tag{10.94}$$

On the other hand, the increment of the density of elastic strain energy is defined by

$$\mathrm{d}W_e = \sigma_{ij}\mathrm{d}\varepsilon_{ij}, \tag{10.95}$$

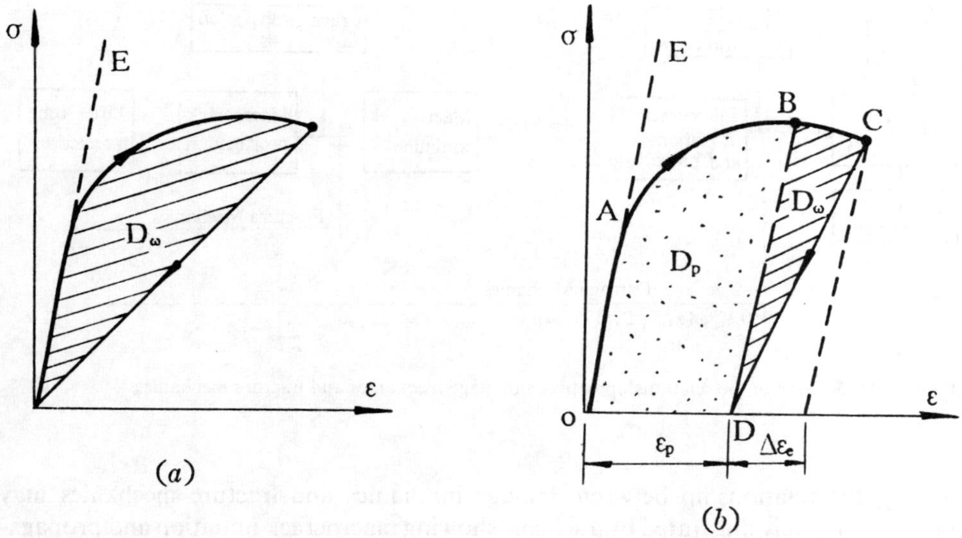

Figure 10.8. Schematic of dissipation during damaging process. (a) Elastic damage process; (b) Elastic-plastic damage process.

substituting Equation (10.94) into Equation (10.95), we have

$$\mathrm{d}W_e/\mathrm{d}w|_{\substack{\sigma=\text{const.}\\ T=\text{const.}}} = E_{ijkl}\varepsilon_{ij}\varepsilon_{kl}. \tag{10.96}$$

Comparing Equation (10.93) with Equation (10.96), we can rewrite Equation (10.93) as

$$Y = -\tfrac{1}{2}(\mathrm{d}W_e/\mathrm{d}w)|_{\substack{\sigma=\text{constant}\\ T=\text{constant}}} \tag{10.97}$$

The variable $-Y$ can then be considered as the elastic strain energy release rate associated with a unit damage growth, and $-Y$ may also be regarded as a function of the hydrostatic stress $\sigma_H = \mathrm{tr}\boldsymbol{\sigma}/3$ and the Von Mises equivalent stress $\sigma_{eq} = (\frac{2}{3}\cdot\boldsymbol{\sigma}':\boldsymbol{\sigma}')^{1/2}$, where $\boldsymbol{\sigma}'$ is the stress deviator (Lemaitre, 1984)

$$-Y = [\sigma_{eq}^2/(2E(1-w)^2)][(\tfrac{2}{3})(1+\nu) + 3(1-2\nu)(\sigma_H/\sigma_{eq})^2].$$

The relationship (Equation (10.97)) defining $-Y$ as the elastic strain energy release rate and its dissipation aspects is indicated in Figure 10.8, where D_p represents the plastic dissipation energy due to the contribution of permanent strain; $D_w = -Y\delta w$, denoting the energy released by the system during the damage process which is eventually converted into heat.

10.1.3 *Damage and fracture*

From the discussion above, it is seen that damage mechanics deals with microcrack initiation, propagation, and coalescence characterized by the damage variable w ($w = 0$, for the undamaged state; $w = w_c$ is a critical value corresponding to rupture of the element – macrocrack initiation), whereas the basic tenet of fracture mechanics is that the strength of the material is governed by the presence and propagation of macrocracks or flaws. If the macrocrack growth is the succession of points for which

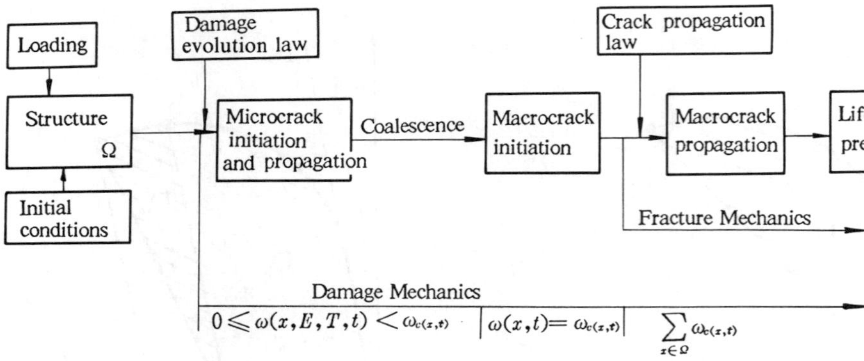

Figure 10.9. Scheme of the relationship between damage mechanics and fracture mechanics.

$w = w_c$, the relationship between damage mechanics and fracture mechanics may be approximately illustrated by a scheme showing macrocrack initiation and propagation (Figure 10.9).

Let us recall the basic concepts and principles of fracture mechanics in order to further discuss damage and fracture.

The main interest in fracture mechanics is to investigate whether macrocrack growth can be initiated under a given load at a given site and to predict the further development of the crack. Griffith (1920) firstly proposed an idea of the energy criterion for fracture, and later Orowan hypothesized that fracture occurred when sufficient energy was released (from the stress field) by growth of the crack to supply the requirements of the new fracture surfaces. The energy released comes from stored elastic or potential energy of the loading system and can, in principle, be calculated for any type of test piece. Therefore, this approach provides a measure of the energy required to extend a macrocrack over unit area, and this is termed the fracture energy (or critical strain-energy release rate) and is denoted as G_c. Irwin (1964) found that the stress field around a sharp macrocrack in a linear elastic material could be uniquely defined by parameter K (named the stress intensity factor, K has three forms defined as K_{I}, K_{II}, and K_{III} by means of the loading state), and stated that fracture occurs when the value of K exceeds some critical value, K_c. Thus, K is a stress field parameter independent of the material whereas, K_c (K_{Ic}, K_{IIc}, and K_{IIIc}) is often referred to as the fracture toughness (as mentioned in Chapter 3), and is a measure of a material property.

The energy criterion for fracture describes quasi-static crack propagation as conversion of the work done, W, by the external force and the available elastic energy, stored in the bulk of the sample, U, into surface free-energy, r_s. It may be written

$$G = \partial(W - U)/\partial a \geq r_s \partial A/\partial a, \tag{10.98}$$

where ∂A is the increase in surface area associated with an increment of crack growth ∂a. For a crack propagating in a lamina of thickness b, the criterion becomes

$$(1/b)\partial(W - U)/\partial a \geq 2r_s, \tag{10.99}$$

since $\partial A = 2b\partial a$.

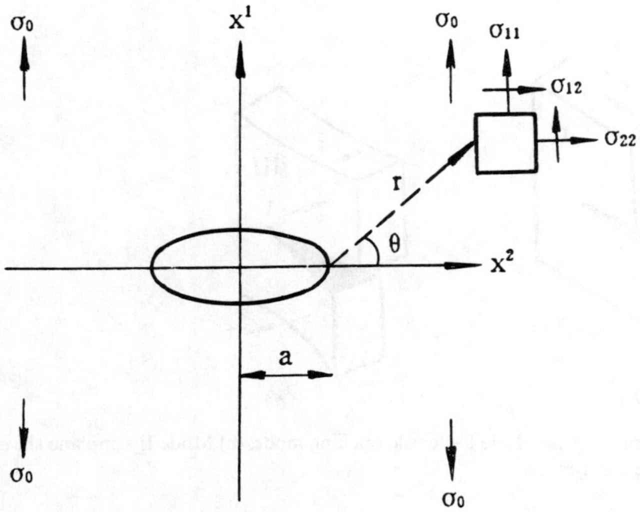

Figure 10.10. Sharp crack in a uniformly stressed, infinite lamina.

If it is assumed that energy dissipation around the crack tip occurs in a manner independent of both the test geometry and the way in which forces are applied to the specimen, then the fracture energy G_c is defined by

$$G_c = 2r_s. \tag{10.100}$$

The value of G_c includes all the energy losses incurred around the crack tip and is, therefore, the energy required to increase the crack by unit length in a specimen of unit width. For this reason, G_c is also named the critical extension force of crack. Thus, Equation (10.99) becomes

$$(1/b)\partial(W - U)/\partial a \geq G_c. \tag{10.101}$$

Some measurements on materials, however, have revealed that the energy required to cause crack growth is more than twice the surface free-energy, $2r_s$ (Kinlock & Young, 1983). One of the main reasons for this discrepancy is the fractal effect (we will discuss this concept in Chapters 13 and 14).

The concept of linear elastic fracture mechanics (LEFM) may be applied to those materials which obey Hook's law so that stress is proportional to strain. It should be pointed out that the basic LEFM analyses may also be applied to materials which exhibit inelastic deformations around the crack tip, provided that such deformations are confined to the immediate vicinity of the tip.

Figure 10.10 shows a sharp crack in a uniformly stressed infinite lamina. If the material is linear elastic, the stress solution at regions close to the crack tip can be found in any text book of fracture mechanics as follows:

$$\sigma_{ij} = K/(2\pi r)^{1/2} \cdot f_{ij}(\theta). \tag{10.102}$$

where r and θ are the polar coordinates of a point, and $2a$ is the length of the crack.

A crack in materials may be stressed in three different modes (as illustrated in Figure 10.11) called mode I, II, and III. The superposition of these three modes constitutes the general case of crack loading.

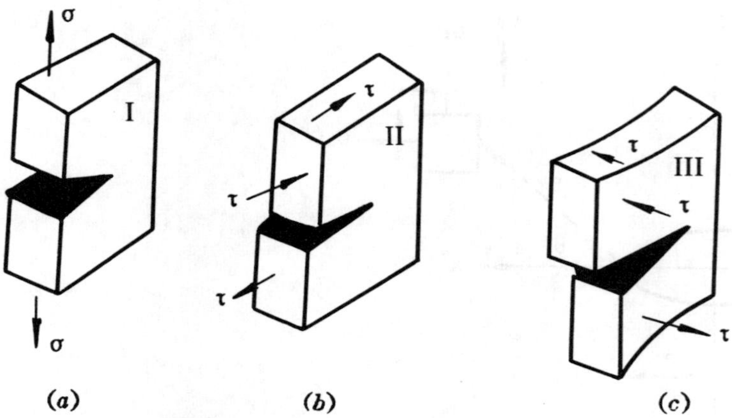

Figure 10.11. The three modes of cracking. (a) Mode I – tensile opening mode; (b) Mode II – inplane shear mode; and (c) Mode III – antiplane shear mode.

For the mode I case, the crack tip stresses can be written by

$$\begin{Bmatrix} \sigma_{11} \\ \sigma_{12} \\ \sigma_{22} \end{Bmatrix} = \frac{K_\mathrm{I}}{(2\pi r)^{1/2}} \cos(\theta/2) \begin{Bmatrix} 1+\sin(\theta/2)\sin(3\theta/2) \\ \sin(\theta/2)\cos(3\theta/2) \\ 1-\sin(\theta/2)\sin(3\theta/2) \end{Bmatrix}; \tag{10.103}$$

and

$$\sigma_{23} = \sigma_{13} = 0,$$

$$\sigma_{33} = \begin{cases} 0, & \text{(plane stress);} \\ \nu(\sigma_{11}+\sigma_{22}), & \text{(plane strain).} \end{cases}$$

From Equation (10.103), it is seen that the stress tensor $\sigma_{ij} \to \infty$ as $r \to 0$, so that it does not make a reasonable local fracture criterion, whereas K_I uniquely defines the stress field around the crack. Irwin's fracture criterion is defined by

$$K_1 \geq K_{1c}. \tag{10.104}$$

Visually, K_I is a factor which characterizes the intensity of the stress field ahead of a crack.

There is a simple relationship between G_c and K_c in the case of LEFM. For a crack under plane strain conditions, this relationship is given by

$$G_c = [(1-\nu^2)/E]K_{\mathrm{I}c}^2 + [(1-\nu)^2/E]K_{\mathrm{II}c}^2 + [(1+\nu)/E]K_{\mathrm{III}c}^2,$$

or

$$G_c = G_{\mathrm{I}c} + G_{\mathrm{II}c} + G_{\mathrm{III}c}.$$

Especially, for mode I

$$G_c = \begin{cases} K_{\mathrm{I}c}^2/E & \text{(plane stress)} \\ K_{\mathrm{I}c}^2(1-\nu^2)/E & \text{(plane strain)} \end{cases}. \tag{10.105}$$

Thus, for LEFM, a critical G_c criterion is identical to a critical $K_{\mathrm{I}c}$ criterion.

Let us now go back to the topics of damage and fracture in this section. We will now discuss the concepts proposed by Legendre & Mazars (1984). Considering an energy balance for a cracked body, subjected to generalized load Q (Figure 10.12), we have

$$\Pi = W - U = Qq - Qq/2 = Rq^2/2,$$

where q is the generalized displacement associated with Q, and $R = Q/q$ is the global stiffness of the structure. The equations of this state give:

$$G = -(\partial\Pi/\partial a)|_{q=\text{const.}} = -\tfrac{1}{2}q^2\partial R/\partial a.$$

Thus, for a constant load, an increment δa of the crack area gives the stiffness variation,

$$\delta R/\delta a = -(Q/q^2)(\delta q/\delta a),$$

and the strain energy release rate,

$$G = \tfrac{1}{2}Q\partial q/\partial a.$$

For brittle material, the propagation law is defined by

$$G < G_c \quad \text{for no propagation } (\dot{a} = 0),$$

and

$$G = G_c \quad \text{for propagation } (\dot{a} > 0).$$

Whereas the state equations of damage (Figure 10.13) give (from Equations (10.92)

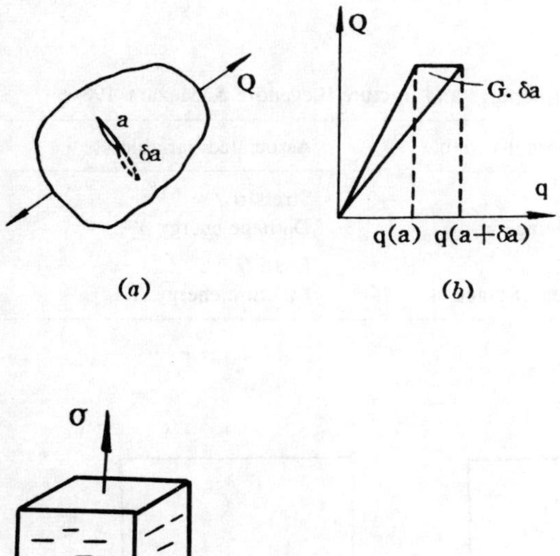

Figure 10.12. (a) Generalized loading of a cracked body; (b) Q-q correlation (Legendre & Mazars, 1984).

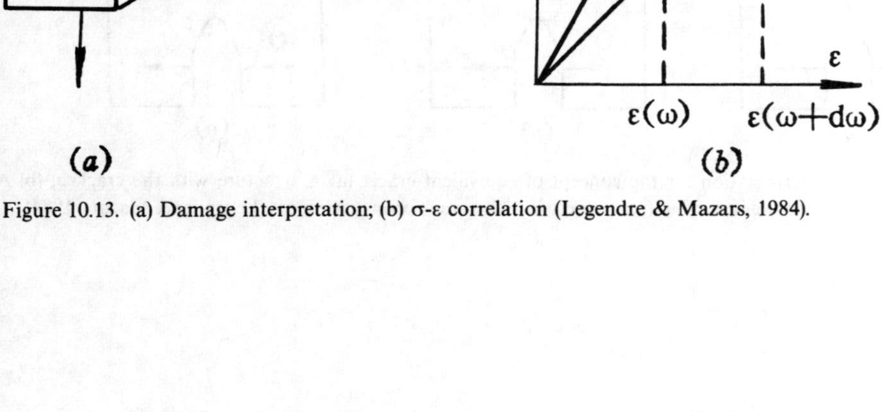

Figure 10.13. (a) Damage interpretation; (b) σ-ε correlation (Legendre & Mazars, 1984).

and (10.93)):

$$\sigma_{ij} = E_{ijkl}\varepsilon_{kl}(1-w),$$
$$-Y = \tfrac{1}{2}E_{ijkl}\varepsilon_{ij}\varepsilon_{kl} = \tfrac{1}{2}(\mathrm{d}W_e/\mathrm{d}w)|_{\sigma=\mathrm{const.}} = \tfrac{1}{2}(\mathrm{d}\Pi/\mathrm{d}w)|_{\sigma=\mathrm{const.}}.$$

Comparing the above states of both damage and fracture, their variables are listed in Table 10.2. In order to show the combination of damage mechanics with fracture mechanics, a structure with the crack a (the material is virgin everywhere else) is considered as shown in Figure 10.14a. Clearly, a loading of Q leads to a damage zone (Figure 10.14b), denoted by the local evolution of damage, δw. Thus, we can calculate the total energy dissipation as follows:

$$\mathfrak{C}_{(\delta w)} = -\int_v Y\delta w\,\mathrm{d}V = \tfrac{1}{2}\int_v E_{ijkl}\varepsilon_{ij}\varepsilon_{kl}\delta w\,\mathrm{d}V.$$

Supposing that the material is elastic and perfectly brittle, the energy dissipation caused by a load Q corresponds to an increase δa in the pre-existent crack (Figure 10.14c)

$$\mathfrak{C}_{(\delta a)} = G_c\cdot\delta a.$$

Therefore, it may be that the crack propagation δa is equivalent to damage evolution δw in volume V when

$$\mathfrak{C}_{(\delta w)} = \mathfrak{C}_{(\delta a)}.$$

Table 10.2. The classification of the variables for damage and fracture (Legendre & Mazars, 1984).

	Observable variable	Internal variable	Associated variable
Damage	deformation ε_{ij}		Stress σ_{jj}
		Damage w	Damage energy Y
Fracture	Displacement q		Load Q
		Area of crack A	Fracture energy G

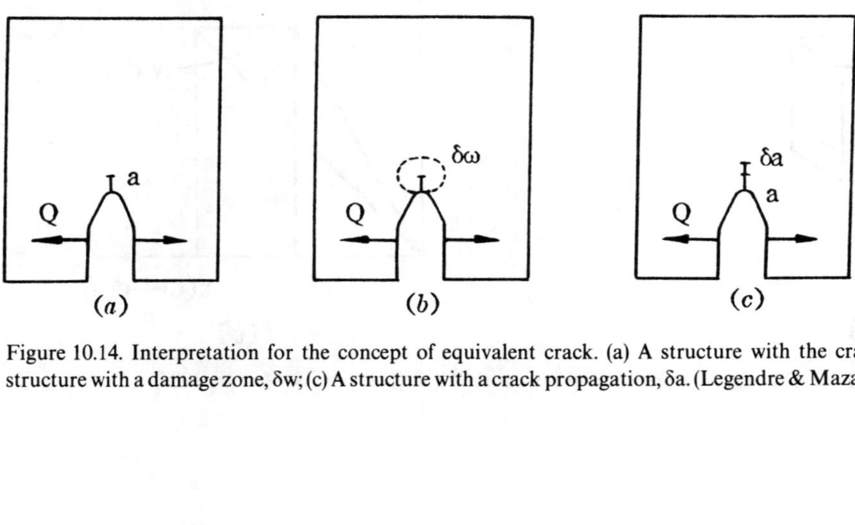

Figure 10.14. Interpretation for the concept of equivalent crack. (a) A structure with the crack a; (b) A structure with a damage zone, δw; (c) A structure with a crack propagation, δa. (Legendre & Mazars, 1984).

Thus,

$$\delta a = \tfrac{1}{2}\int_v E_{ijkl}\varepsilon_{ij}\varepsilon_{kl}\delta w \mathrm{d}V/G_c.$$

This means that the crack 'equivalent' to a damaged-cracked zone is the crack for which the energy necessary for its creation is equal to the energy dissipated in the formation of the damaged zone.

10.1.4 *A damage statistical model of perfectly brittle materials*

Krajcinovic & Silva (1982) and Krajcinovic (1989) proposed the parallel bar model to address the fundamental aspects of a link connecting the continuous damage theory and statistical strength theories.

Consider the uniaxial tension of a perfectly brittle specimen modeled by a system of parallel bars illustrated in Figure 10.15. Each bar remains completely elastic until it ruptures when the tensile force in the bar reaches its rupture strength $f_i = f_{Ri}$. If the rupture strength is bar dependent, an increment of the external load F acting on the system causes rupture of a certain fraction of bars and attendant redistribution of forces among the bars which did not rupture.

Assuming that each bar in the system has the same stiffness, (K/N), and its rupture strength is f_{Ri}/N (where N is the original number of bars), we can obtain the force-deflection relationship as follows:

$$f_i = Kx/N \quad \text{for } 0 \le Kx \le f_{Ri} \quad \text{and} \quad x \ge 0,$$

and

$$f_i = 0 \quad \text{for } Kx \ge f_{Ri} \quad \text{and} \quad x \ge 0;$$

where x is the axial displacement of the bar system. Thus the equilibrium equation is defined by

$$F = \sum_{n+1}^{N} Kx/N = Kx(1 - n/N) = Kx(1 - w), \tag{10.106}$$

where n represents the number of rupture bars, the non-negative damage variable w

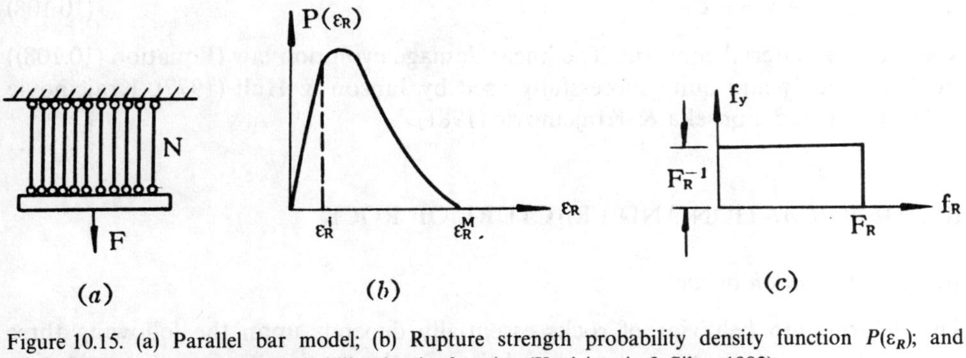

Figure 10.15. (a) Parallel bar model; (b) Rupture strength probability density function $P(\varepsilon_R)$; and (c) Uniform rupture strength probability density function (Krajcinovic & Silva, 1982).

is defined as the ratio (n/N) of the number of already ruptured bars and the original number in the bar system. If A_0 and L represent the original cross-sectional area and the length of the specimen, it is seen that $F = \sigma A_0$, $K = EA_0/L$, and $x = \varepsilon L$. Thus, Equation (10.106) becomes the standard damage constitutive equation in the one-dimensional case:

$$\sigma = E\varepsilon(1 - w),$$

where $w = A_v/A_0$, with A_v representing the surface void density.

More precisely, for a very large number of elements (bars) N, Equation (10.106) should be recast into an integral form

$$F = Kx \int_{Kx}^{\infty} P(f_R)\mathrm{d}f(R),$$

where $P(f_R)$ is the rupture strength probability density function and $P(f_R)\mathrm{d}f_R$ is the fraction of the total number of elements with rupture strengths in the range $[f_R, f_R + \mathrm{d}f_R]$. Since

$$\int_0^{F_R} P(f_R)\mathrm{d}f_R = 1,$$

the number of bars which will rupture at stress level Kx is

$$\int_0^{Kx} P(f_R)\mathrm{d}f_R.$$

Thus, the damage evolution equation is obtained as follows:

$$w(Kx) = \int_0^{Kx} P(f_R)\mathrm{d}f_R \Big/ \int_0^{F_R} P(f_R)\mathrm{d}f_R,$$

and

$$= P(Kx) = P_r(f_R \le Kx), \qquad (10.107)$$

where $P(Kx)$ is the cumulative distribution function for the given rupture strength probability density function $P(f_R)$. If the strength distribution commonly identified with the structural failure application is the Weibull distribution, then, the damage evolution equation (Equation (10.107)) can be rewritten by

$$w(Kx) = 1 - \exp[(-Kx/\theta)^m],$$

where m and θ are the shape and the scale parameter. The exponential distribution corresponds to the case $m = 1$ and leads to the damage evolution law with a linear relationship as $Kx/\theta \ll 1$:

$$w(Kx) = Kx/\theta = C\varepsilon, \qquad (10.108)$$

where C is a material constant. The linear damage evolution law (Equation (10.108)) has been widely and quite successfully used by Janson & Hult (1977), Krajcinovic (1979, 1989), and Fonseka & Krajcinovic (1981).

10.2 DEFORMATION AND FRACTURE OF ROCK

10.2.1 *Deformation of rock*

The deformation behavior of rocks essentially depends upon the following three external factors: stress, temperature, and time. In this section we focus our attention on the deformation corresponding to stresses because the deformation is mainly controlled by the stress. During the loading of a rock in a compression test, the strain value ε corresponds to each value of the stress σ. Figure 10.16 shows the $\sigma - \varepsilon$ stress-strain curve of marble obtained by using a servo-controlled testing system. It is seen that stress-strain curves for a variety of rocks can be divided into four distinct stages. The first stage (curve OA) is a nonlinear process in which natural micro-cracks in the specimen are compacted, and the stiffness of the system increases. In the second stage, strain remains small while stress increases, and if stress is removed (unloaded) the rock specimen recovers its initial shape. This is the elastic domain (curve AB). Beyond the elastic limit, straining is irreversible and the rock shows two distinct types of deformation behavior: either failure occurs at once (Figure 10.17a, which is perfectly brittle), or the strain increases without a significant increase in stress, which corresponds to inelastic strain (curve BC in Figure 10.17b). In the fourth stage (curve

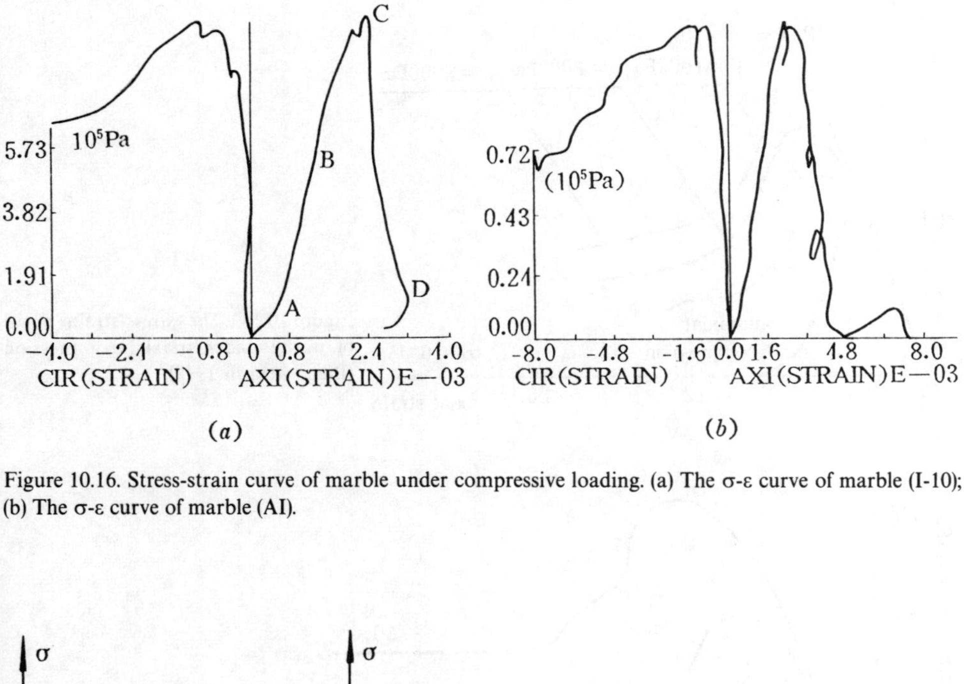

Figure 10.16. Stress-strain curve of marble under compressive loading. (a) The σ-ε curve of marble (I-10); (b) The σ-ε curve of marble (AI).

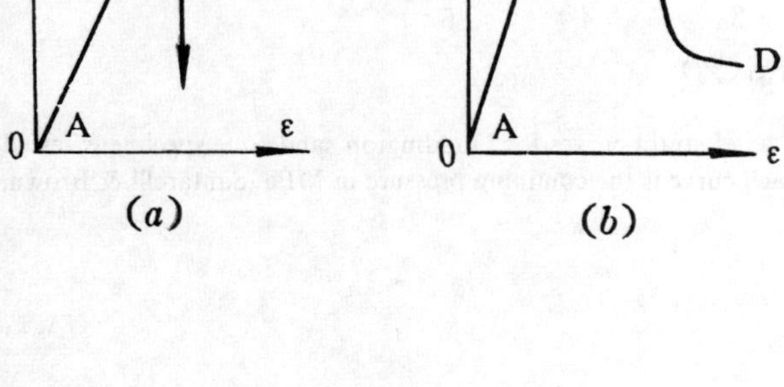

Figure 10.17. Typical stress-strain curves of rock. (a) Ideal brittle; (b) Strain-softening.

CD in Figure 10.16), after the stress reaches its maximum value (compressive strength of rock), strain increases as the stress drops in, which behaves as a strain softening process in which the initiation, unstable propagation, and coalesences of micro-cracks in the specimen lead to final failure. This is actually the process of damage and fracture.

It is necessary to mention that, in nature, a rock is under triaxial stress, and its behavior depends on the relative values of σ_1, σ_2, and σ_3, and temperature. Figures 10.18 and 10.19 show the experimental results of a rock element under triaxial compression obtained by Bieniawski in 1972 and Santarelli & Brown (1989) using a servo-controlled testing system. Having conducted many experiments on rock under triaxial compression, we can conclude the following three characteristics:

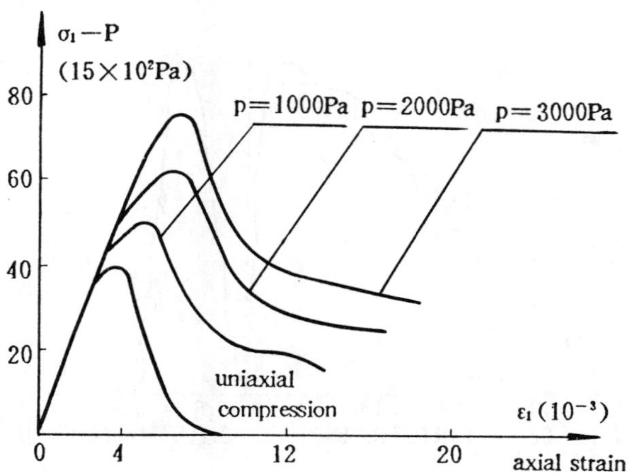

Figure 10.18. The stress-strain curve of norite under triaxial compression (Bieniawski in 1972).

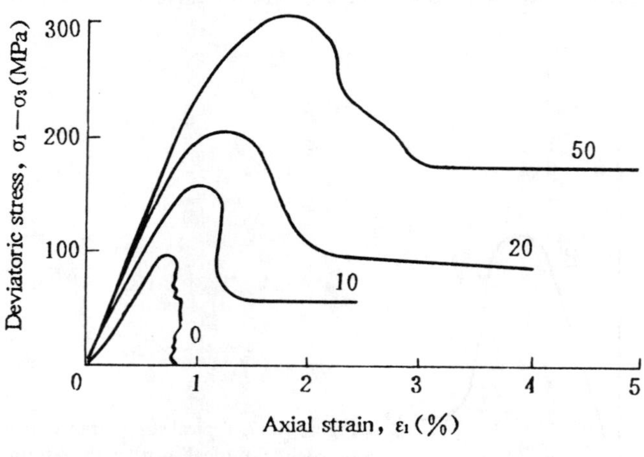

Figure 10.19. Selected deviatoric stress vs axial strain curves for Doddington sandstone specimens tested in triaxial compression. The number on each curve is the confining pressure in MPa (Santarelli & Brown, 1989).

(1) The Young's modulus E and Poisson ratio ν in the elastic region of the curve are approximately the same as those under uniaxial compression. This characteristic is so important that the elastic constants (E and ν) under a complicated stress condition can be determined from a simple uniaxial test;

(2) The phenomenon of plastic flow will occur under certain confining pressures for some rocks;

(3) The ultimate strength and ductility are directly enhanced as the confining pressure increases.

Accordingly, as $\sigma_2 = \sigma_3 = p$ increases from zero, the behavior changes successively: brittle, inelastic behavior with strain softening; ductile behavior without hardening; and ductile behavior with hardening (as shown in Figures 10.20 and 10.21).

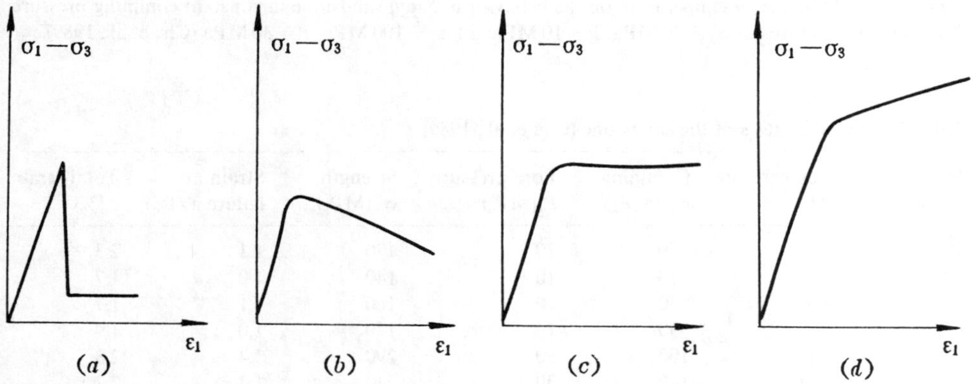

Figure 10.20. Changes from brittle to ductile with hardening as the confining pressure increases: (a) Brittle behavior; (b) Inelastic behavior with strain softening; (c) Ductile behavior without work hardening; and (d) Ductile behavior with hardening (Bles & Feuga, 1986).

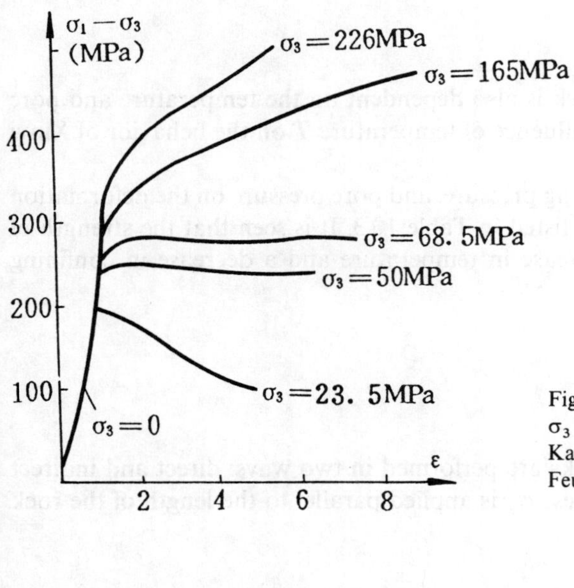

Figure 10.21. Influence of confining pressure σ_3 on the behavior of Carrara marble (Von Karman, 1911 in: Panet, 1975, also in: Bles & Feuga, 1986).

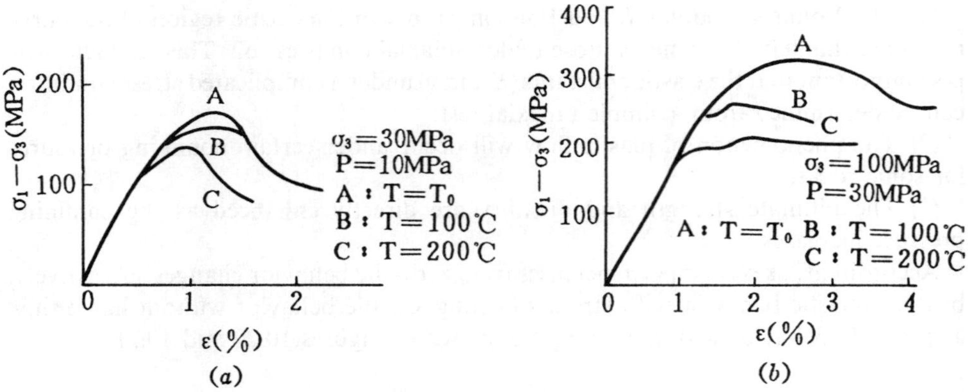

Figure 10.22. Influence of temperature on the behavior of Xinyi sandstone in constant confining pressure; T_0-room temperature. (a) $\sigma_3 = 30$ MPa, $P = 10$ MPa; (b) $\sigma_3 = 100$ MPa, $P = 30$ MPa (Che et al., 1989).

Table 10.3. Triaxial tests of the sandstone (Che et al., 1989).

No. of specimens	Temperature T(°C)	Confining σ_1 (MP$_a$)	Pore pressure P (MP$_a$)	Strength σ_c (MP$_a$)	Strain at failure ε_f (%)	Total strain ε (%)
17	T_0	30	10	170	1.1	2.3
19	200	30	10	140	1.0	1.7
20	100	30	10	160	1.1	1.7
21	T_0	30	10	170	1.0	1.9
23	100	100	30	290	1.4	2.8
27	200	100	30	230	1.3	2.2
29	100	100	30	290	1.5	2.4
40	T_0	100	30	320	2.3	4.3
42	100	100	30	260	1.7	2.6
43	200	100	30	210	1.6	3.3
50	T_0	100	10	330	3.0	4.7
51	T_0	100	50	220	1.9	3.6

The deformation behavior of rock is also dependent on the temperature and pore pressure. Figure 10.22 shows the influence of temperature T on the behavior of Xinyi sandstone (Che et al., 1989).

The effect of temperature, confining pressure, and pore pressure on the deformation and failure process of sandstone is listed in Table 10.3. It is seen that the strength of sandstone is dropped with an increase in temperature and a decrease in confining pressure.

10.2.2 *Macroscopic failure of rock*

10.2.2.1 *Tensile failure of rock*

Experiments of tensile tests on rocks are performed in two ways: direct and indirect tension. In direct tensile tests, a stress σ_1 is applied parallel to the length of the rock sample. These tests can be carried out with or without the confining pressure. Figure 10.23a gives the strain-stress curve of marble under direct tension. In indirect tension tests (termed the Brazilian test), compressive load is applied to the rock sample using two diametrically opposed jaws. In all cases, tensile fracture takes place in a plane perpendicular to the tensile stresses (see Figures 10.23b and 10.24).

In the Brazilian test (Bles & Feuga, 1986), elastic theory shows that tension occurs normal to the plane containing the two jaw faces by which the load is applied (Figure 10.24A). Fracture occurs normal to this tension plane. Tensile fracture starts at the center of the specimen, where the tensile stresses are the greatest, and then spreads to the outer ends of the plane. In reality, the phenomena are more complicated than elastic theory would suggest. It has been determined that this theory is valid only

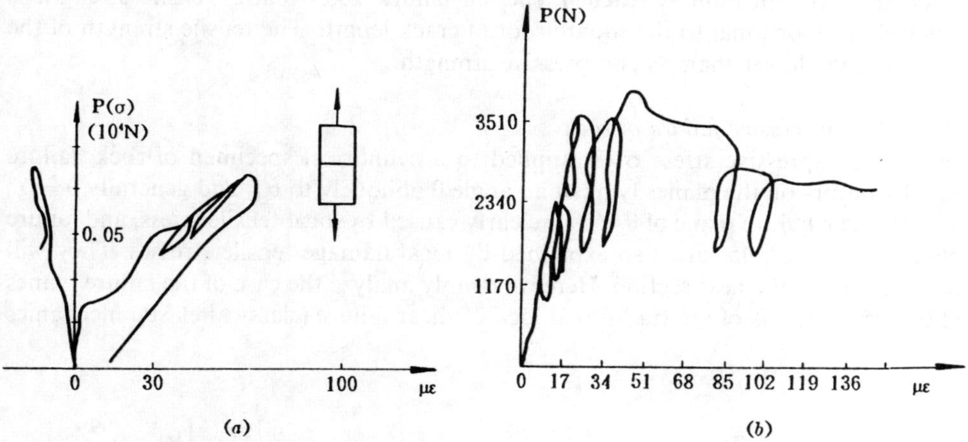

Figure 10.23. The strain-stress curve of rock in tensile tests. (a) Marble under direct tension; and (b) Coal under indirect tension.

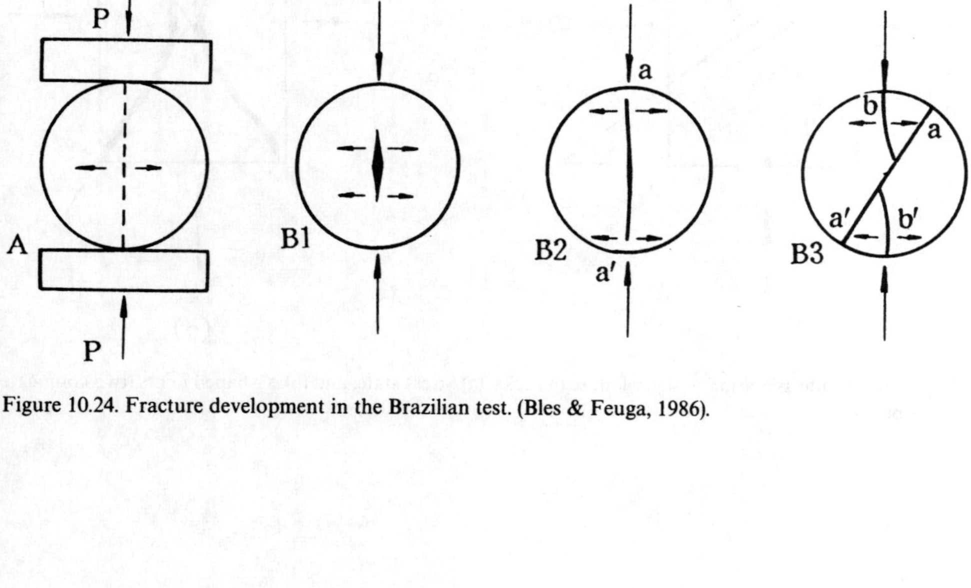

Figure 10.24. Fracture development in the Brazilian test. (Bles & Feuga, 1986).

for small stresses, well below failure values. In practice, stresses are very high at the points where the loads are applied (theoretically infinite), and fracture starts from there. After breaking a specimen, as illustrated in Figures 10.24(B1) and (B2), a new fracture in the specimen can be induced by local stress concentration. If the first fracture, *aa'* is obviously to the transverse plane. Next, the new fracture *bb'* will occur, again, in the transverse plane, after the first fracture has closed up, sometimes with a displacement of the point where the two fractures intersect, as indicated in Figure 10.24(B3). This could give the impression of a shear of *bb'* relative to *aa'*, but this is not the case, since *bb'* was formed after *aa'*.

In direct tensile testing, the fracture plane is approximately perpendicular to the tension stress plane. However, this is not always the case, and it seems likely that the actual site of fracture is influenced by micro-cracks in the specimen. Stresses are concentrated at the point of any microcrack lying perpendicular to the stress, and when the strength limit is reached, sudden failure takes place. Tensile strength is inversely proportional to the square root of crack length. The tensile strength of the rock is much lower than its compressive strength.

10.2.2.2 *Compressive failure of rock*

When a compressive stress σ_1 is applied to a cylindrical specimen of rock, failure usually occurs on the planes lying at an angle θ obliquely to σ_1, and generally $\theta = 0°$, $30°$, $45°$. The failure plane of $\theta = 0°$ is clearly caused by local tensile stress, and failure planes of $\theta = 30°$, $45°$ are also explained by local damage tensile stresses. They will be described in the next section. Here we mainly analyze the case of the failure planes of $\theta = 45°$ in terms of the traditional idea of shear failure (classical elastic mechanics

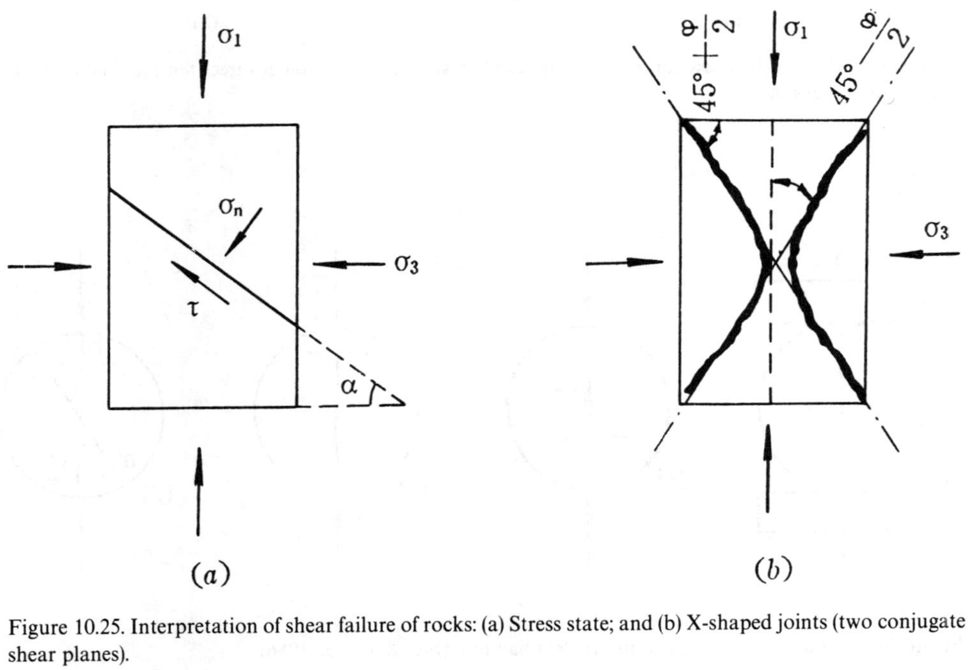

Figure 10.25. Interpretation of shear failure of rocks: (a) Stress state; and (b) X-shaped joints (two conjugate shear planes).

analysis). More commonly, we consider the biaxial stress state with a confining pressure σ_3, where the stresses on α-plane (Figure 10.25a) can be obtained as follows:

$$\sigma_n = (\sigma_1 + \sigma_3)/2 + (\sigma_1 - \sigma_3)/2 \cdot \cos 2\alpha \quad \text{and} \quad \tau = (\sigma_1 - \sigma_3)/2 \cdot \sin 2\alpha.$$

Clearly, the shear reaches a maximum, $\tau_{max} = (\sigma_1 - \sigma_3)/2$, when $\alpha = 45°$, and failure should take place at an angle 45° to the principal stresses. The ultimate stress state can be graphically represented by using the Mohr circle (Figure 10.26). Letting ϕ be the internal friction angle of the rock material, from Coulumb criterion $\tau = C + \sigma \tan \phi$, we can see that in the Mohr circle:

$$\begin{aligned} &\sin \phi = \{(\sigma_1 - \sigma_3)/2\}/\{(\sigma_1 + \sigma_3)/2 + C \operatorname{cotan} \phi\}, \\ &\sigma_1 = 2C \cos \phi/(1 - \sin \phi) + \sigma_3(1 + \sin \phi)/(1 - \sin \phi); \end{aligned} \tag{10.107}$$

where C is the cohesion, i.e., the shear strength without normal pressure. In the uniaxial stress state, $\sigma_3 = 0$, Equation (10.107) becomes

$$\sin \phi = \sigma_c/2\{\sigma_c/2 + C \operatorname{cotan} \phi\}$$

and

$$\sigma_c = 2C \cos \phi/(1 - \sin \phi). \tag{10.108}$$

Thus, the failure plane will take place at an angle $\alpha = 45° \pm \phi/2$ obliquely to principle stress σ_1. The failure plane appears as a pair of conjugate shear planes due to the symmetry, which behaves as X-shaped joints commonly mentioned in geomechanics (as shown in Figure 10.25b).

The macroscopic failure of rocks in triaxial compression is the evolution of the failure pattern with confining pressure. A commonly accepted classification has three stages, summarized by Santarelli & Brown, as shown in Figure 10.27:

(1) In uniaxial compression and at very low confining pressures, the specimen fails by axial splitting. The deviatoric stress vs. axial strain curve shows that the specimen has lost almost all cohesion at failure time;

(2) For intermediate confining pressures, the specimen fails by shear along a single plane. The deviatoric stress vs. axial strain curve shows that the specimen has lost only part of its cohesion at the end of the test;

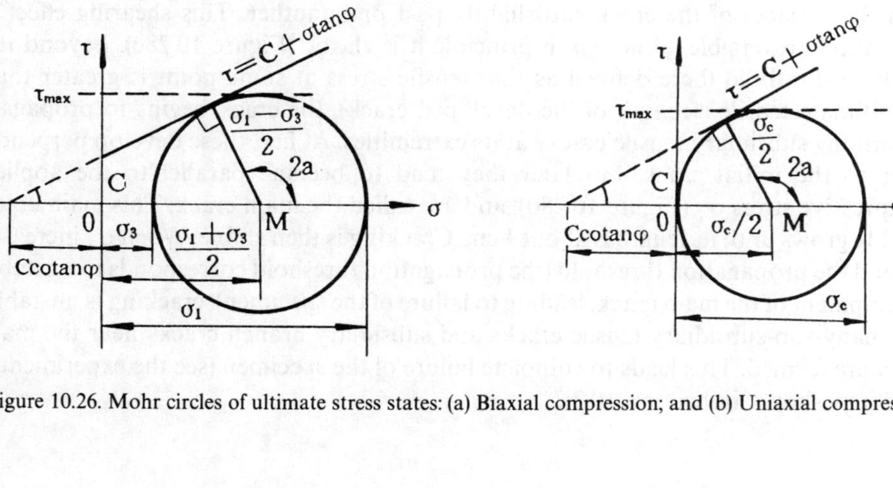

Figure 10.26. Mohr circles of ultimate stress states: (a) Biaxial compression; and (b) Uniaxial compression.

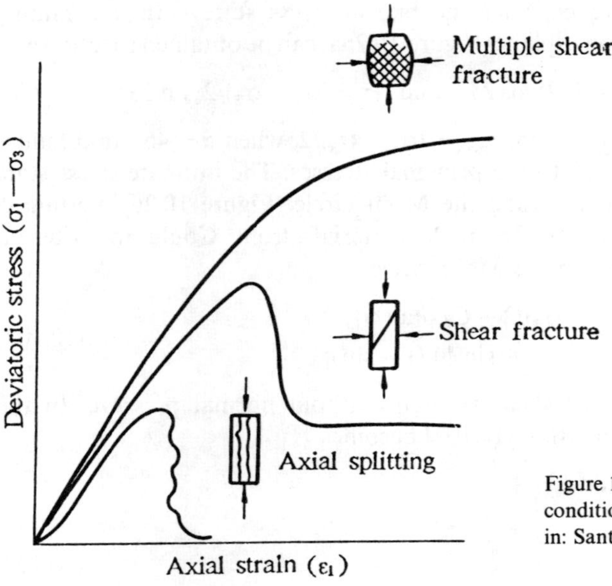

Figure 10.27. Classification of failure in triaxial conditions according to Jaeger & Cook (1979) in: Santarelli & Brown (1989).

(3) For high confining pressure, the specimen fails along the multiple shear plane. The deviatoric stress vs. axial strain curve shows that the specimen has not lost any cohesion at the end of the test.

This progression has been observed by a large number of investigators. Stages (1) and (2) are usually referred to as brittle failure while stage (3) is called ductile failure.

10.2.2.3 *Crack development under compressions*

It is worth examining how a crack develops under compressive stress conditions for geomechanics and mining engineering. First, we consider an open crack which has the ideal shape of a very flat ellipse within a specimen subjected to uniaxial compression (Figure 10.28a). Several successive stages have been summarized by Bles & Feuga (1986), as increases towards failure:

In this first stage, the crack is closed up under compressive stress (Figure 10.28a); then the surfaces of the crack slip slightly past one another. This shearing effect is not entirely reversible, although in principle it is elastic (Figure 10.28c). Beyond the cracking threshold (here defined as that tensile stress at some point is greater than the ultimate tensile strength of the developed crack), the crack begins to propagate by forming subsidiary tensile cracks at its extremities. At first these develop perpendicular to the initial crack, but later they tend to become parallel to the applied compressive stress σ_1 (Figure 10.28d), and are called the main crack. This main crack quickly grows until its length is about 1 cm. Cracking is then stable. When σ_1 increases beyond the propagation threshold (the propagation threshold corresponds to unstable development of the main crack, leading to failure of the specimen), cracking is unstable, and many sub-subsidiary tensile cracks and subsidiary branch cracks near the main crack are formed. This leads to complete failure of the specimen (see the experimental results as shown in Section 10.3).

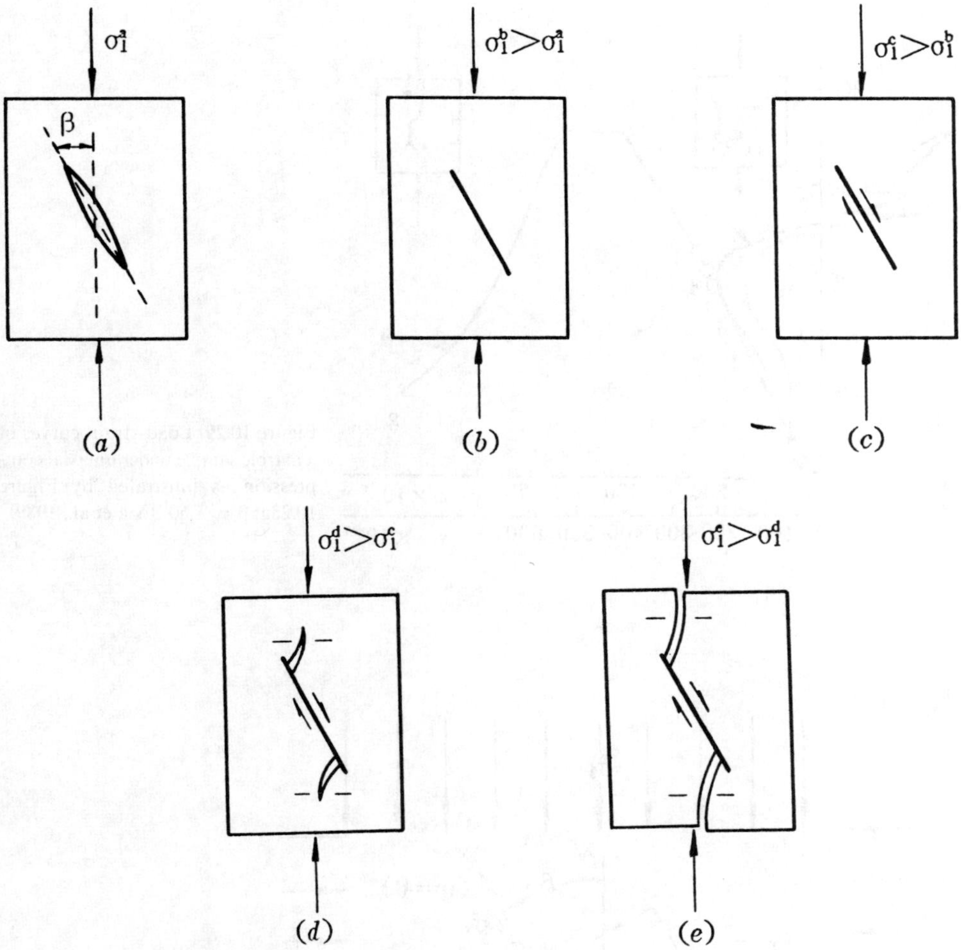

Figure 10.28. Simple compressive failure of a specimen along a pre-existing crack: (a) Initial condition; (b) Closing of the crack; (c) Sliding along the crack surfaces; (d) Crack propagation; (e) Sliding along the crack surfaces and fracture (Bles & Feuga, 1986).

This problem is also regarded as the mixed mode (I, II) fracture in fracture mechanics. Xia et al. (1985) examined the load-strain curve of marble sample under uniaxial compression by fracture tests as shown in Figure 10.29 and Table 10.4. The experimental results of fracture mechanics are in agreement with the discussions given above. From Table 10.4, it is seen that, the fracture property varies with the angle β, and the fracture toughness (K_{Ic}, K_{IIc}) enhanced as β increases, but the stress at failure is not the case.

We now consider a flat, elliptical opening crack in a two-dimensional body which is subjected to a stress system as illustrated in Figure 10.30. The stress field around an elliptical opening is related to the elliptical coordinates ξ and η, which are defined by the following equations of transformation of a rectangular system of coordinates

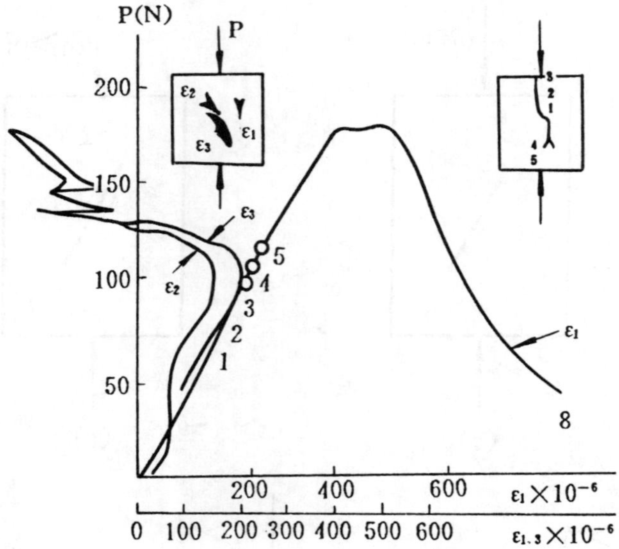

Figure 10.29. Load-strain curves of a marble sample under uniaxial compression as illustrated by Figure 10.28a; β = – 60° (Xia et al., 1989).

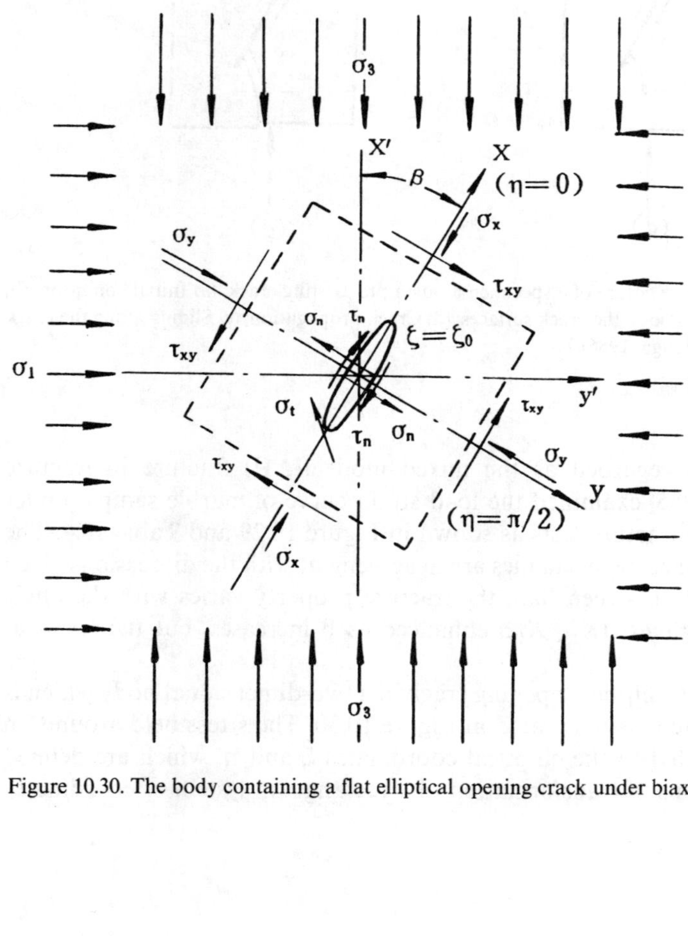

Figure 10.30. The body containing a flat elliptical opening crack under biaxial compression.

Table 10.4. Fracture properties of marble sample shown in Figure 10.28a.

β (degree)	5	10	15	20	30	40	45	50	55	60
Stress at fracture (MP_a)	310	232	208	215	285	242	275	345	345	396
K_{Ic} (J)	3.0	8.7	17.4	67.2	89.0	125.0	171.9	215.6	323.4	371.3
K_{IIc}(J)	33.6	49.6	65.0	116.4	154.3	149.0	171.9	215.6	186.6	214.4

x and y:

$$x = C \cosh \xi \cos \eta,$$

and

$$y = C \sinh \xi \sin \eta.$$

When $\xi = \xi_0 =$ constant, we have

$$x^2/(C^2 \cosh^2 \xi_0) + y^2/(C^2 \sinh^2 \xi_0) = 1.$$

The long and short semi-axis of the elliptic crack are respectively

$$a = C \cosh \xi_0, \quad \text{and } b = C \sinh \xi_0.$$

Thus, we can find that

$$y/x = \tan \theta = \tanh \xi_0 \tan \eta = (b \tan \eta)/a = m \tan \eta,$$

and

$$\tan \eta = \frac{a}{b} \tan \theta = \frac{1}{m} \tan \theta;$$

where m is the ratio of the long axis to the short axis. It is clear that η has a geometric character of the encentric angle as shown in Figure 10.31.

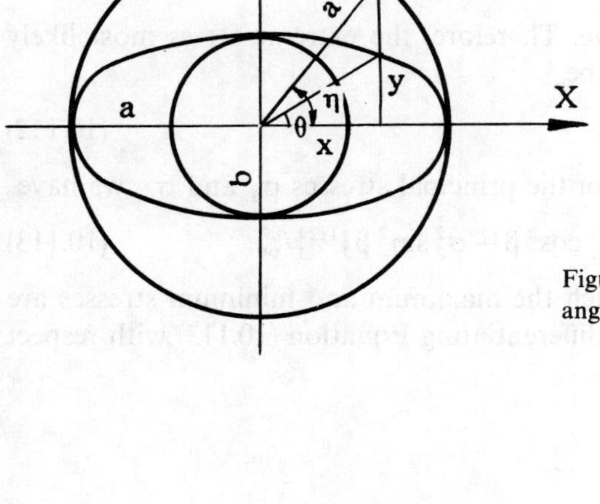

Figure 10.31. Geometric character of η (encentric angle).

In Figure 10.30, the normal stresses σ_x, σ_y and the shear stress τ_{xy} are related to the principal stresses σ_1 and σ_3 by the following equations:

$$\sigma_x = \sigma_1 \sin^2 \beta + \sigma_3 \cos^2 \beta, \quad \sigma_y = \sigma_1 \cos^2 \beta + \sigma_3 \sin^2 \beta,$$

and

$$\tau_{xy} = (\sigma_1 - \sigma_3)/2 \cdot \sin 2\beta.$$

The tangential stress σ_t around the boundary of an open elliptical crack, due to the stresses σ_x, σ_y, and τ_{xy}, can be obtained as follows:

$$\sigma_t = \{2\sigma_y \sinh 2\xi_0 + 2\tau_{xy}[1 + \sinh(2\xi_0)\operatorname{cotan} 2\beta - \exp(2\xi_0)\cos 2(\beta - \eta) \cdot \operatorname{cosec} 2\beta]\}/(\cosh 2\xi_0 - \cos 2\eta). \tag{10.109}$$

For a flat ellipse, ξ_0 is very small, thus,

$$\sinh \xi_0 \approx \xi_0, \quad \cosh \xi_0 \approx 1, \quad \tanh \xi_0 \approx \xi_0 \approx m,$$
$$\sinh 2\xi_0 = 2 \sinh \xi_0 \cosh \xi_0 \approx 2\xi_0,$$
$$\cosh 2\xi_0 = \cosh^2 \xi_0 + \sinh^2 \xi_0 \approx 1 + \xi_0^2,$$
$$\sinh 2\xi_0 = [e^{(2\xi_0)} - e^{(-2\xi_0)}]/2; \quad \cosh 2\xi_0 = [e^{(2\xi_0)} + e^{(-2\xi_0)}]/2,$$

and

$$\exp 2\xi_0 = e^{(2\xi_0)} = \sinh 2\xi_0 + \cosh 2\xi_0 \approx 2\xi_0 + 1 + \xi_0^2 = (1 + \xi_0)^2,$$

where ξ_0 is the value of the elliptical coordinate ξ on the crack boundary.

The maximum boundary stresses, both tensile and compressive, occur near the ends of the crack, i.e., when η is very small. Thus, we have

$$\sin \eta \approx \eta, \quad \cos \eta \approx 1, \quad \cos 2\eta \approx 1 - \eta^2, \quad \tan \eta = \eta.$$

According to the relationships given above, Equation (10.109) can be simplified to following form:

$$\sigma_t = 4(\xi_0 \sigma_y - \eta \tau_{xy})/(\xi_0^2 + \eta^2). \tag{10.110}$$

Differentiation of Equation (10.110) with respect to η and setting equation $\partial\sigma_t/\partial\eta$ to zero results in the coordinate of the first extreme value point as follows:

$$\eta = \xi_0/\tau_{xy} \cdot \{\sigma_y \pm [(\sigma_y)^2 + (\tau_{xy})^2]^{1/2}\}. \tag{10.111}$$

Substituting Equation (10.111) into Equation (10.110), we can determine the maximum and minimum stresses at the first extreme value as

$$\sigma_{ti} = \{\sigma_y + (\sigma_y^2 + \tau_{xy}^2)^{1/2}\}/\xi_0, \quad (i = 1, 2),$$

where tensile stress is taken as negative. Therefore, the extreme stress most likely inducing the fracture of objects should be

$$\sigma_{t1} = \{\sigma_y - (\sigma_y^2 + \tau_{xy}^2)^{1/2}\}/\xi_0. \tag{10.112}$$

Expressing Equation (10.112) in terms of the principal stresses σ_1 and σ_3, we have

$$\sigma_{t1} = \{(\sigma_1 \cos^2 \beta + \sigma_3 \sin^2 \beta) - [\sigma_1^2 \cos^2 \beta + \sigma_3^2 \sin^2 \beta]^{1/2}\}/\xi_0. \tag{10.113}$$

The critical crack orientation β_c at which the maximum and minimum stresses are induced near the crack tip is found by differentiating Equation (10.113) with respect to β and letting $\partial\sigma_{t1}/\partial\beta = 0$. This gives:

$$\beta_c = 0, \quad \beta_c = \pi/2, \quad \text{and } \cos 2\beta_c = (\sigma_1 - \sigma_3)/[2(\sigma_1 + \sigma_3)].$$

From Equation (10.113), when $\beta_c = 0$, if $\sigma_1 > 0$, $\sigma_3 > 0$, we will have the second extreme stress $\sigma_{t2} = 0$, which is obviously not a failure stress; if $\sigma_1 < 0$, $\sigma_3 < 0$, we will have

$$\sigma_{t2} = -2|\sigma_1|/\xi_0.$$

When $\beta_c = \pi/2$, similarly, if $\sigma_3 > 0$, $\sigma_1 > 0$, $\sigma_{t2} = 0$, and if $\sigma_3 < 0$, no matter if σ_1 is positive or negative, we have

$$\sigma_{t2} = -2|\sigma_3|/\xi_0.$$

When $\cos 2\beta_c = (\sigma_1 - \sigma_3)/[2(\sigma_1 + \sigma_3)]$, we will have

$$\sigma_{t2} = -(\sigma_1 - \sigma_3)^2/[4(\sigma_1 + \sigma_3)\xi_0]. \tag{10.114}$$

Meanwhile, since the absolute value $|\cos 2\beta| \leq 1$, i.e., $(\sigma_1 - \sigma_3)/[2(\sigma_1 + \sigma_3)] \leq 1$, we find the limit condition:

$$\sigma_1 + 3\sigma_3 \geq 0 \quad \text{or} \quad \sigma_3/\sigma_1 \geq -0.33.$$

This means that Equation (10.114) is only meaningful for values of $\sigma_3/\sigma_1 \geq -0.33$.

From the discussions above, we can conclude that the conditions leading to fracture or the critical conditions of fracture are: when $\beta = 0$, $\sigma_1 < 0$ and $\sigma_3 < 0$ while $|\sigma_1| = \sigma_t$ (tension strength) or $|\sigma_3| = \sigma_t$; or when $\beta = \pi/2$, $\sigma_3 < 0$ and $|\sigma_3| = \sigma_t$ no matter if σ_1 is positive or negative. In both cases, the second extreme stress leading to fracture at extreme points has the same form,

$$\sigma_{t2} = -2\sigma_t/\xi_0. \tag{10.115}$$

Equation (10.114), which is valid for $\sigma_3/\sigma_1 \geq -0.33$, is of limited practical use because the term $\sigma_{t2} \cdot \xi_0$ can not be evaluated in the laboratory. However, this term is expressed in terms of the uniaxial tensile strength of the material, according to Equation (10.115). Thus, the fracture criterion becomes

$$\begin{aligned} &(\sigma_1 - \sigma_3)^2/[8(\sigma_1 + \sigma_3)] = \sigma_t, \quad \sigma_3/\sigma_1 \geq -0.333; \\ &\quad -\sigma_3 = \sigma_t, \\ &\sigma_3/\sigma_1 < -0.333. \end{aligned} \tag{10.116}$$

This is Griffith's fracture criterion (Griffith, 1924), which can also be represented by a parabolic Mohr envelope (as indicated in Figure 10.32) defined by

$$\tau^2 = 4\sigma_t(\sigma_t + \sigma),$$

where τ is the shear stress acting along the fracture surface, and σ is the normal stress perpendicular to the fracture surface. F.A. McClintock and E. Hoek proposed modifications to Griffith's criterion in 1962 and 1965 separately. The main point of review in their modifications is that material will no longer fracture by tension, but by confined shearing, when the cracks are closed by compression. If it is assumed that the inherent cracks are initially closed, the relationship between the principal

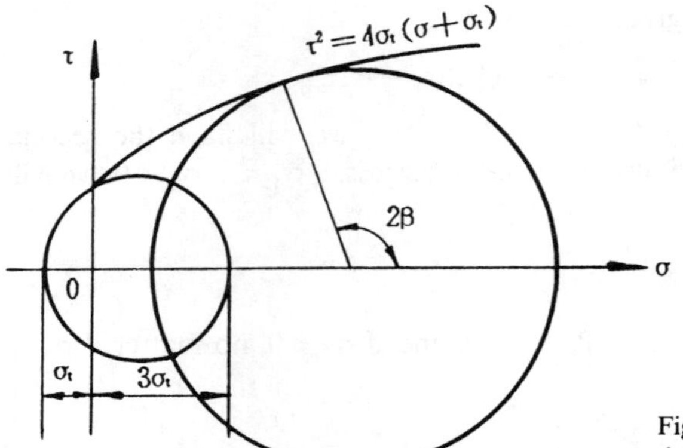

Figure 10.32. Griffith's fracture criterion on the σ-τ plane.

stresses required to initiate fracture is

$$\sigma_1 = \sigma_c + [(1+f^2)^{1/2} + f]/[(1+f^2)^{1/2} - f]. \tag{10.117}$$

Where f is the coefficient of friction between the crack faces and σ_c is the uniaxial compressive strength of the material.

In order to establish whether the original (Equation (10.116)) and modified (Equation (10.117)) Griffith fracture theory are applicable to the prediction of rock fracture behavior, Hoek & Bieniawski (1984) collected a survey of published rock fracture data and facilitated a comparison of the data in terms of a dimensionless form by dividing by the compressive strength. The plot of these dimensionless strength values, together with the original and modified Griffith fracture criterion is reproduced in Figure 10.33. It is seen from Figure 10.33 that there is a remarkable agreement between the experimental results and the fracture initiation behavior predicted by the modified Griffith theory.

It must be pointed out that both the original and modified fracture theories can predict fracture initiation from a single crack, but that they can not be applied to the fracture of the entire specimen. It has already been suggested by W.F. Brace and E.G. Bombolakis that fracture propagation from a single crack follows a more complex path than is generally assumed, and that it is the presence of favorable microcrack arrays which coalesce to form the macroscopic fracture surface (this is a damage evolution process), that makes the Griffith theory applicable to predicting fracture of rocks.

In biaxial compression, fracture initiates at a point on the crack boundary near, but not at, the crack tip and follows a curved path (as shown in Figure 10.34 obtained by Hoek & Bieniawski, 1984). Generally, fracture propagation ceases when the crack path becomes parallel to the major principal stress direction. In all cases, the applied stress was increased to at least three times the fracture initiation stress. Figure 10.34 shows the length of the stable cracks related to the ratio of the applied principal stresses (this result is similar to that for the propagation of cracks from a circular hole in a biaxial compressive stress field). After many test results in the case of biaxial compression, Hoek & Bieniawski have concluded that a single Griffith crack can not account for the failure of a specimen in a compressive field unless the ratio of applied

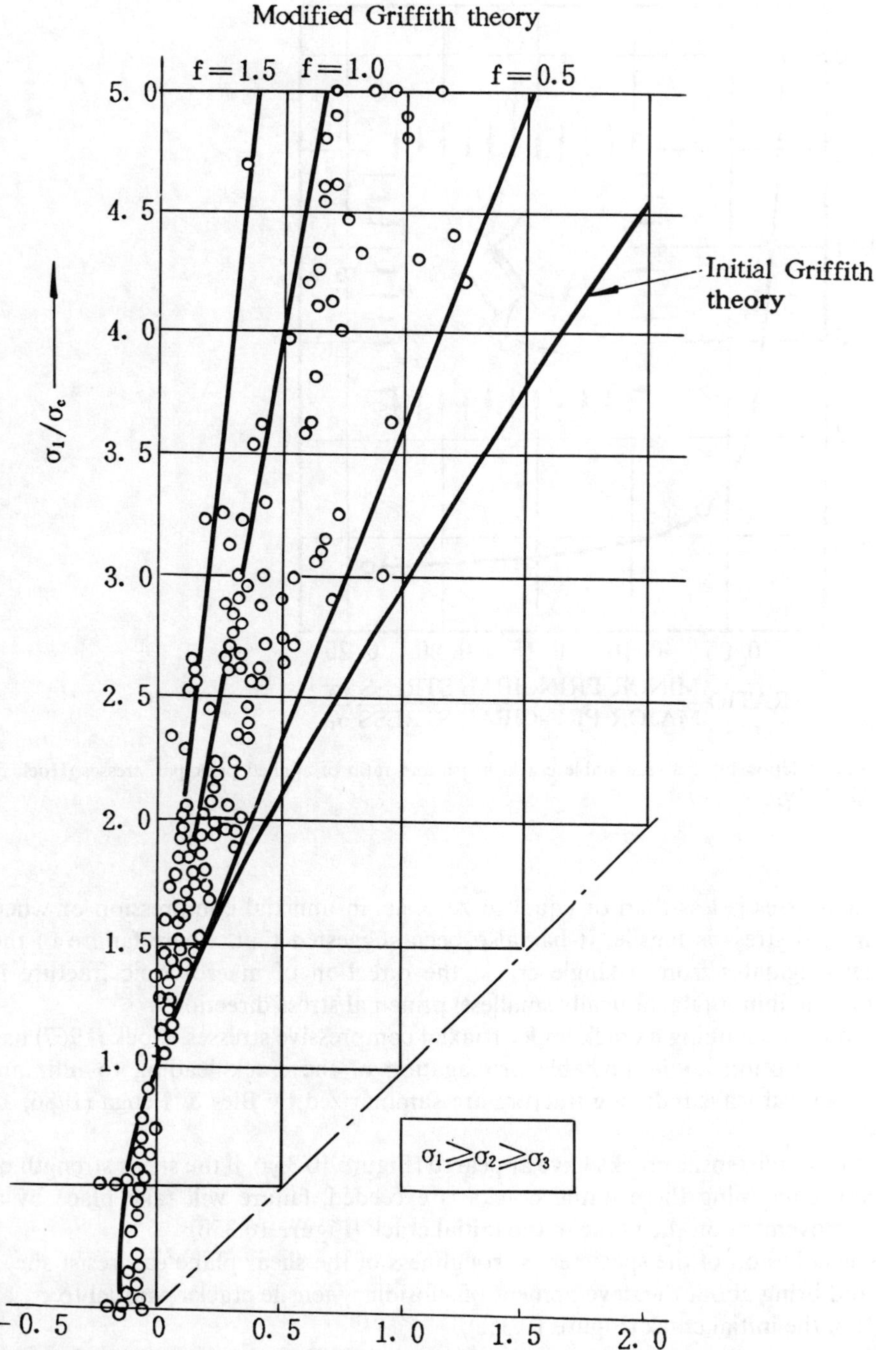

Figure 10.33. Fracture data of rock materials vs. Griffith fracture theories (Hoek & Bieniawski, 1984).

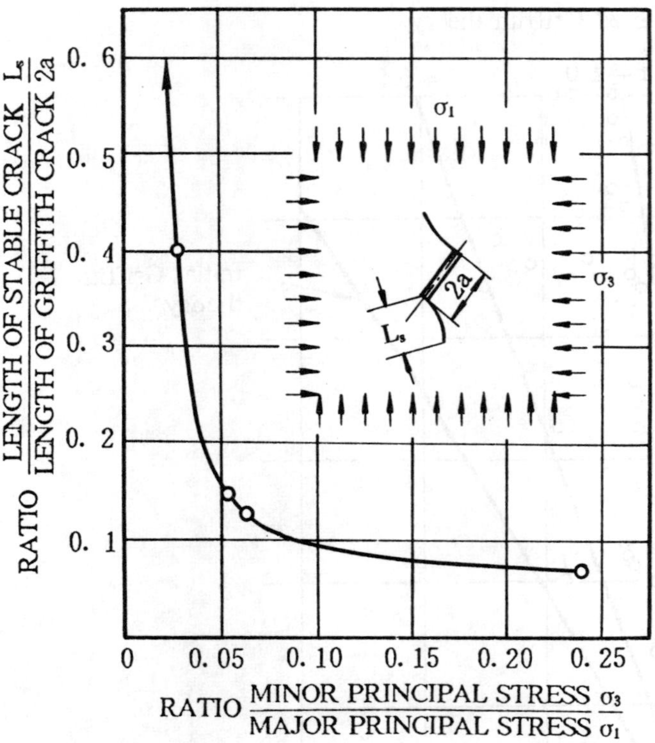

Figure 10.34. Relationship between stable crack length and ratio of applied principal stresses (Hoek & Bieniawski, 1984).

principal stresses is less than or equal to zero, i.e. in uniaxial compression or when one principal stress is tensile. It has also been suggested that, where failure of the specimen originates from a single crack, the direction of macroscopic fracture is normal to the minor (algebraically smallest) principal stress direction.

For a body containing a crack under triaxial compressive stresses, Hoek (1967) has given a description about unstable propagation of the crack leading to ultimate fracture. Several ways inducing fracture are summarized by Bles & Feuga (1986) as follows:

(1) After stable tensile cracks have appeared (Figure 10.35a), if the shear strength of the region containing these initial cracks is exceeded, failure will take place by a shearing movement on the plane of the initial crack (Figure 10.35b);

(2) The cohesion of the specimen or roughness of the shear plane can resist shear failure and bring about the development of subsidiary tensile cracks parallel to σ_1 at the ends of the initial crack (Figure 10.35c);

(3) The failure due to a combination of the two previous processes can also take place along a surface intermediate in the direction between the cracks parallel to σ_1 and the shear plane corresponding to the initial cracks;

(4) Finally, changes that take place either in the stress field or in the properties of the material during the development of cracking can stabilize the cracks (Figure 10.35d)

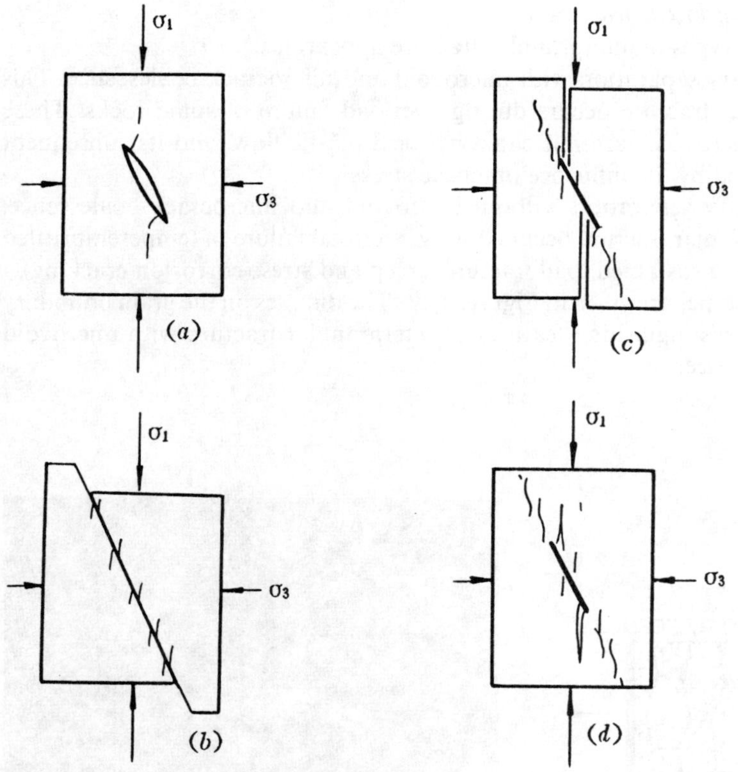

Figure 10.35. Possible failure modes under triaxial compression: (a) Initial condition (stable cracking); (b) Shear fracture takes place after stable cracks have appeared and when the shear strength of the zone containing the initial cracks has been exceeded. Conditions for the onset of shearing are: $\tau = C + \sigma \tan \phi$ (Coulomb criterion); (c) Tensile fracture occurs when cohesion in the material or interference between asperities prevents shear movement along the initial crack; and (d) Stable cracking resulting from a change in the stress conditions or in the properties of the material. For cracking to develop, σ_1 must increase (Bles & Feuga, 1986).

so that a further increase in the principal stress σ_1 is required before cracking can continue.

10.2.3 *Micromechanisms of rock failure*

Metallurgists have found fractographic examination of fracture surfaces to be a valuable tool in failure analyses, because fracture surfaces exhibit much more information of the micromechanisms of fracture. After rock fractures, we can go back to determine its fracture mechanisms by fractographic examination of the fracture surfaces of the rock, using a scanning electric micrograph.

The results of fractography of rock (Xie, 1988, Xie & Chen, 1989) show that the microscopical fractures of rock are mainly characterized by intergranular brittle fracture, transgranular brittle fracture, coalescence fracture of microcracks and microvoids, and their combinations, which manifest themselves in a tensile fracture.

10.2.3.1 *Intergranular brittle fracture*

There are two main types of intergranular fracture appearance:

(1) Grain boundary separation with microvoid and microcrack coalescence. This type of intergranular fracture occurs during overload failure of some rocks. These microcracks and microvoids are nucleated by local plastic flow, and its subsequent propagation is caused by the influence of tensile stress;

(2) Grain boundary separation without microvoid and microcrack coalescence. This type of intergranular fracture occurs during overload failure of temperembrittled rocks, and also during sustained load fracture (creep and stress corrosion cracking).

Examples of both types are given in Figure 10.36. The dimples on the grain boundary facets are the main distinguishing features of intergranular fracture with microvoid and microcrack coalescence.

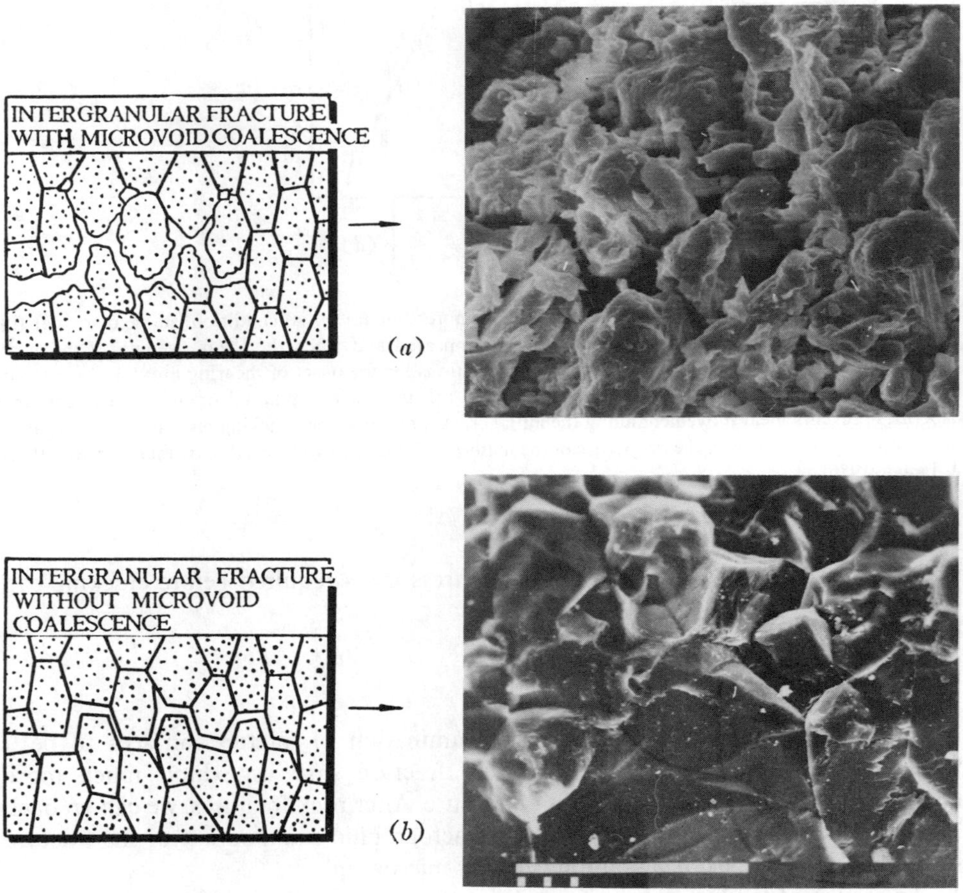

Figure 10.36. Intergranular fracture with and without microcrack and microvoid coalescence. SEM (Scanning Electron Micrograph) fractographs. (a) Mudstore under compressive load, × 770; and (b) Marble under three point bending, × 50.

10.2.3.2 *Brittle transgranular fracture* (*cleavage*)

Cleavage fracture causes new free surfaces by the rupture of atomic bonds across well defined, low index crystallographic well-defined planes. From the macroscopic point of view, it is often a brittle type of fracture without significant plastic deformation but providing microplasticity. Crack propagation occurs with high velocity in a transcrystalline manner. In polycrystalline materials, cleavage fracture is characterized by facets corresponding to the individual grains or regions of crystallographic continuity. Ideally, a cleavage fracture would have perfectly matching faces and be completely flat and featureless. However, structural materials are characteristically polycrystalline, with the grains more or less randomly oriented with respect to each other. Thus, cleavage propagation through one grain will probably have to change direction as it crosses a grain or subgrain boundary (subgrains are regions within a grain that differ slightly in crystal orientation). Such changes in direction result in the faceted fracture surface shown in Photo 10.1.

In addition, most structural rock materials contain particles, precipitates, or other imperfections that further complicate the fracture path so that truly featureless cleavage is rare, even within a single grain or subgrain. The changes of orientation between grains and subgrains and the various imperfections produce markings on the fracture surface that are characteristically associated with cleavage.

Figure 10.37 illustrates some typical features associated with cleavage. A principal feature is river patterns, which are steps between cleavages on parallel planes. River pattern always converge in the direction of local crack propagation. If the grains or subgrains are connected by a tilt boundary, meaning that the grains are misoriented about a common axis, the river patterns are continuous across the boundary. However, if adjacent grains or subgrains are axially misoriented, i.e. they are connected by a twist boundary, the river patterns do not cross the boundary but originate from it. Besides river patterns, a distinct feature of cleavage is feather markings. The apex of these

Photo 10.1. Brittle fracture of a coal specimen.

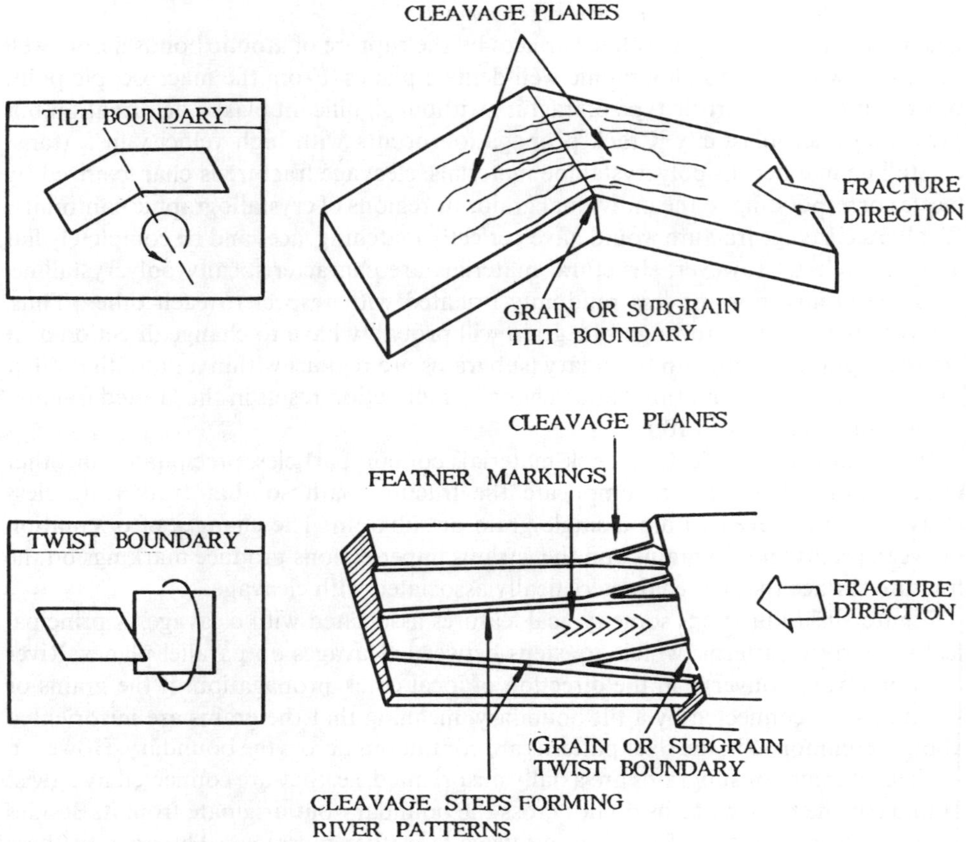

Figure 10.37. Typical features associated with cleavage (Ewalds & Wanhill, 1984).

fan-like marking points back to the fracture origin and, therefore, this feature can also be used to determine the local direction of crack propagation.

The cleavage facets of marble under different loading are shown in Photo 10.2.

10.2.3.3 *Microcrack and microvoid coalescence*

Pre-existing microcracks and microvoids in rock propagate and coalesce during loading, and also induce some new microcracks and microvoids, which ultimately lead to the formation of a continuous fracture surface (see Figure 10.38). Examples of this type of fractures are shown in Photo 10.3. The theoretical description is presented by Xie & Chen (1989).

10.2.4 *Tensile failure of local damage of rock*

Rock is different from metal materials in its mechanical properties. The tensile strength of rock is much lower than its compressive strength. Therefore, rock is regarded as an untensile material in engineering analyses. On the other hand, a rock specimen under

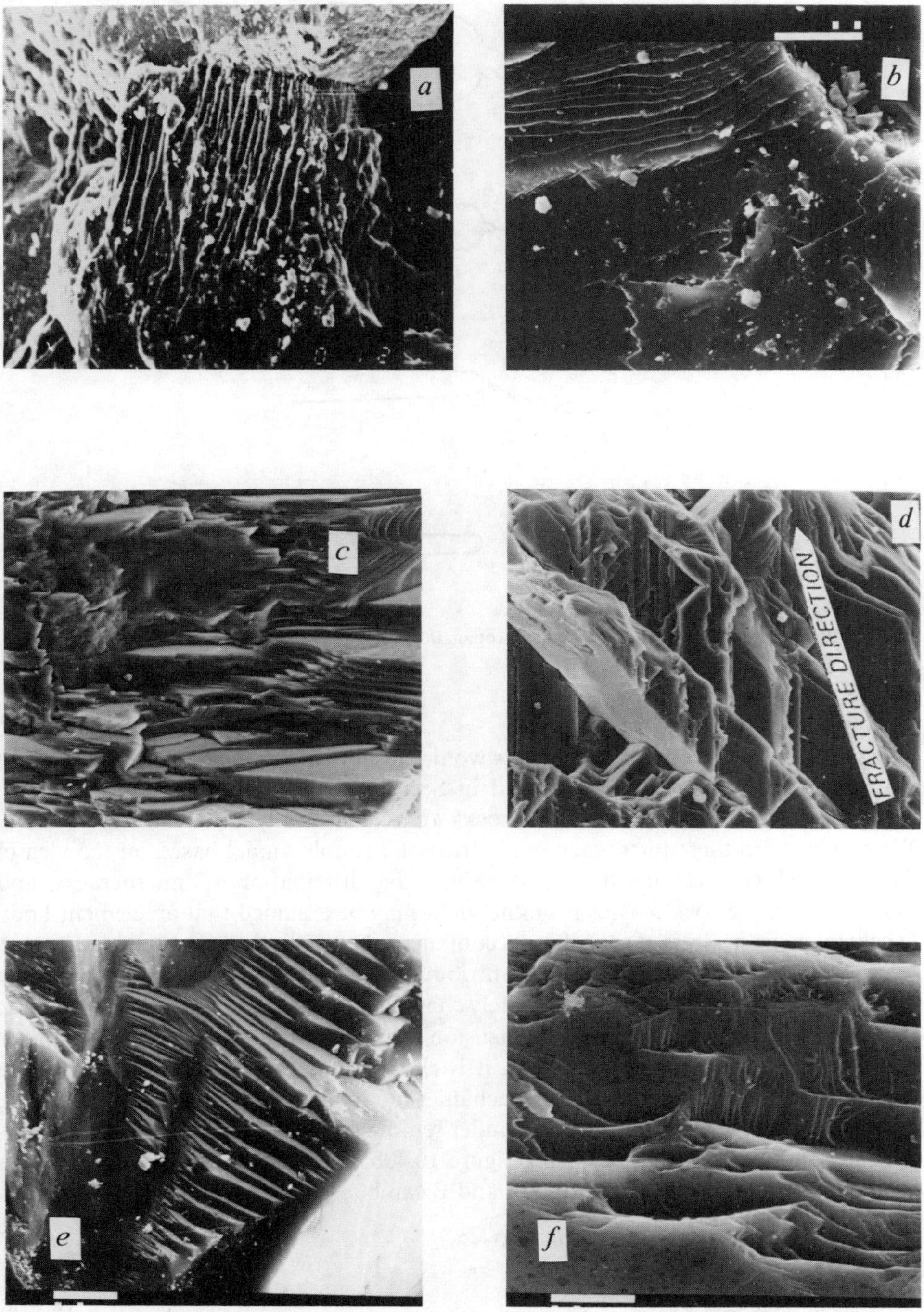

Photo 10.2. Cleavage facets of marble under different loadings (SEM fractography): (a) Explosive load, × 475, river lines; (b) Three point bending, × 2105, river lines; (c) Three point bending, × 950, irregular steps; (d) Compressive load, × 427, river lines; (e) Three point bending, × 1425, irregular steps; and (f) Shear, × 1900, river lines and steps.

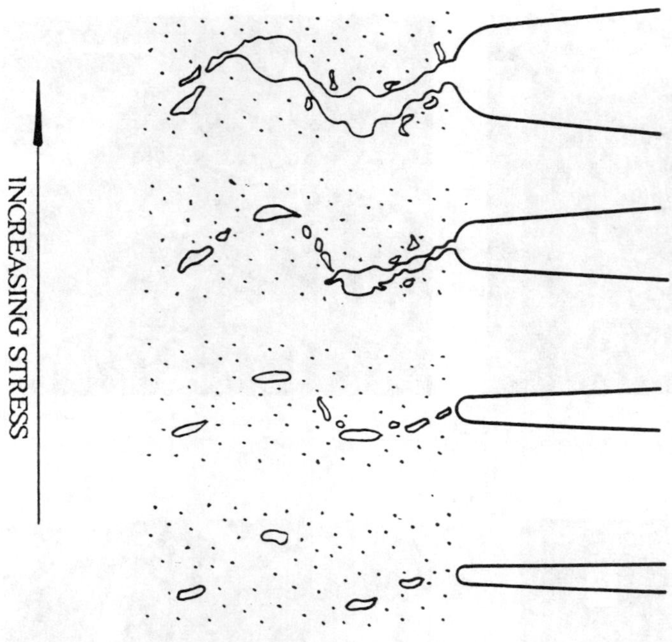

Figure 10.38. Schematic presentation of crack propagation by void coalescence.

uniaxial and triaxial compressive stress would fracture along the plane parallel to the maximum applied stress (as discussed in Section 10.2.2). The directions of brittle fracture of four basic experiments of rocks are schematically shown in Figure 10.39. We now explain these phenomenon in terms of a simple model based on the idea of damage and classical elastic theory. Since the distribution of microcracks and microvoids in the rock masses is unknown, it may be assumed that an element body (a cell) taken from stressed rock masses contains only an elliptical microcrack. In other words, the total damage effect of the contribution of microcracks and microvoids over all the element body can be approximately characterized by the elliptical microcrack as shown in Figure 10.40a. Since the dimension of the elliptical microcrack is a smaller quantity than the element body itself, it is clear that the damage stress field of the element body will be looked as the mechanics model as shown in Figure 10.40b, i.e., an elliptical crack in an infinite body under remote stresses.

Based on the mechanics model in Figure 10.40b, considering nonlinear effects, the local concentrated stresses at points A and B can be written as

$$\sigma_T^A = K_1\sigma_1 - K_2\sigma_2 + K_3 f_A(\sigma_1, \sigma_2, \alpha_i),$$

and

$$\sigma_T^B = -K_4\sigma_1 - K_5\sigma_2 + K_6 f_B(\sigma_1, \sigma_2, \alpha_i); \qquad (10.118)$$

where f_A and f_B are nonlinear functions of stresses σ_1, σ_2, and internal variables α_i. K_i $(i = 1, 2, \ldots, 6)$ is the coefficient of stress concentration, which mainly depends on the material properties of rock.

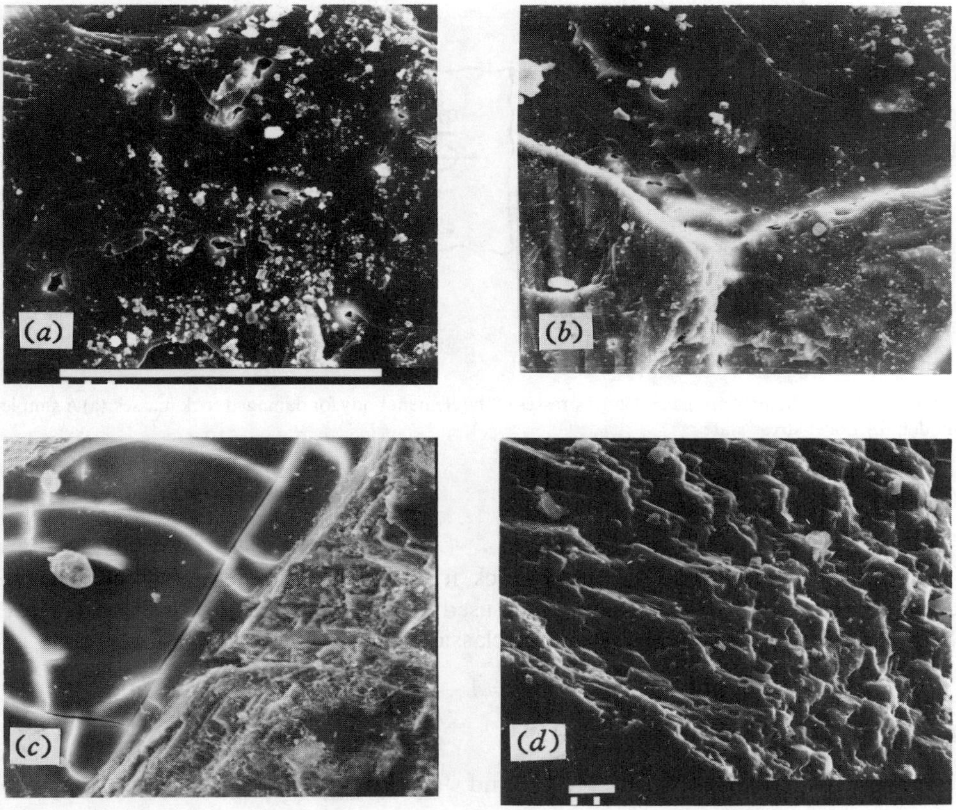

Photo 10.3. Microvoid and microcrack coalescence of marble. (a) Void nucleation and coalescence in a crystallographic plane of marble under three point bending (SEM × 665); (b) Void nucleation and coalescence in a grain boundary of marble under three point bending (SEM × 665); (c) Crack nucleation and coalescence in a crystallographic plane of marble under shear stress (SEM × 475); and (d) The dimple topography caused by microcavities coalescence in marble under three point bending (SEM × 950).

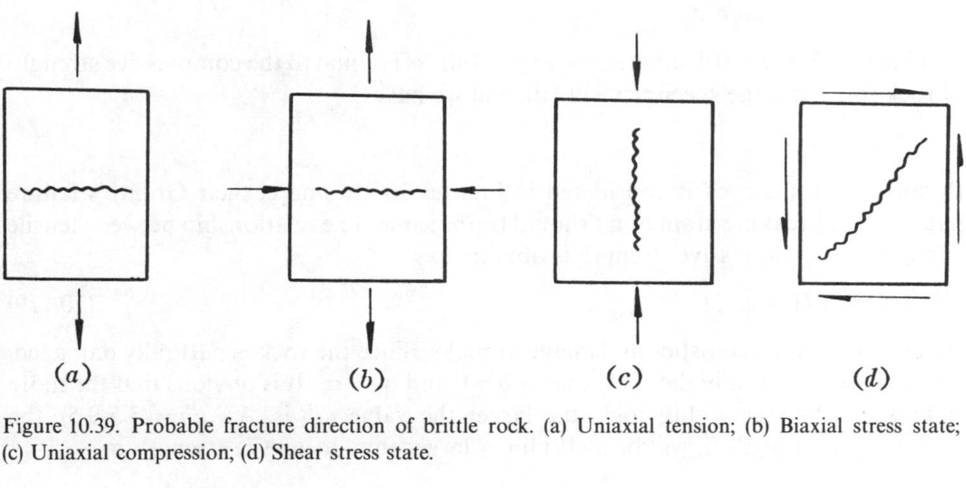

Figure 10.39. Probable fracture direction of brittle rock. (a) Uniaxial tension; (b) Biaxial stress state; (c) Uniaxial compression; (d) Shear stress state.

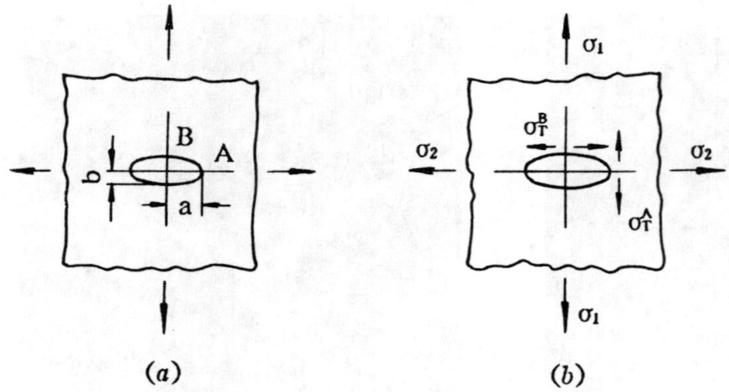

Figure 10.40. Mechanical model and local stresses of the element body for damaged rock masses. (a) A simple model; (b) Local stress state.

To clearly show the damage effect of rock, it is assumed that the rock mass is elastic, and the theory of small deformation is used. The solution of the problem (Figure 10.40b) can be directly derived from the classical elastic theory as

$$K_2 = K_4 = 1; \quad K_3 = K_6 = 0; \quad K_1 = 1 + 2a/b; \quad \text{and} \quad K_5 = 1 + 2b/a.$$

Equation (10.118) becomes

$$\sigma_T^A = \sigma|_{\max} = (1 + 2a/b)\sigma_1 - \sigma_2, \quad \text{and} \quad \sigma_T^B = \sigma|_{\max} = -\sigma_1 + (1 + 2b/a)\sigma_2.$$

Therefore, we can obtain four types of local failure stress and its probable fracture directions can be determined by the criterion of Griffith's tensile failure, as shown in Figure 10.41.

Let us now analyze the two cases shown in Figure 10.41a and c. In Figure 10.41a, when the applied stress is equal to the tensile strength of rock ($\sigma_1 = \sigma_t$), the rock specimen will fracture, and its local damage fracture stress is then

$$\sigma_{\text{Grif}} = (1 + 2a/b)\sigma_t.$$

Similarly, in Figure 10.41c, when the applied stress is equal to the compressive strength of rock ($\sigma_1 = \sigma_c$), the specimen will fail, and we have

$$\sigma_{\text{Grif}} = \sigma_c.$$

Because the same rock is considered in Figures 10.41a and c, their Griffith's failure stress on the breaking atom bond should be the same. The relationship between tensile strength and compressive strength is obtained as

$$\sigma_c/\sigma_t = (1 + 2a/b). \tag{10.119}$$

Where $(2a/b)$ is the statistics of damage in rocks. Since the rock is naturally damaged, i.e., microcracks exist in the rock, then $a/b > 0$ and $\sigma_c > \sigma_t$. It is obvious that the more serious the damage within rock, the larger the value a/b is. As $a/b = (3.5\text{-}9.5)$, the compressive strength σ_c will be (8-19) times larger than its tensile strength σ_t.

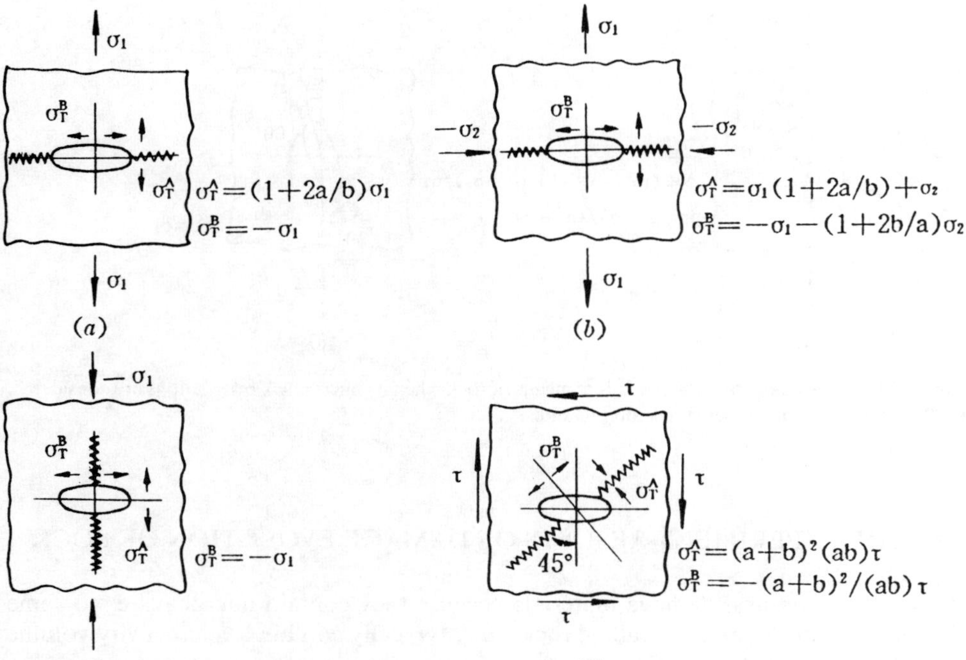

Figure 10.41. Local damage stress and probabie fracture direction (wave line) of the rock specimens shown in Figure 10.39. (a) Uniaxial tension; (b) Biaxial stress state; (c) Uniaxial compression; (d) Shear stress state.

From the above analysis it is clear that when rock is thought of as damaged material, its different failure phenomena (Figure 10.39) can be reasonably interpreted (Figure 10.41). When the orientation of the microcracks in the element body is considered, we can similarly explain other rock fracture forms (i.e., shear failure) as shown in Figure 10.25. For instance, the fracture plane in uniaxial compression is sometimes inclined to applied stress, but their failure stress is the local tensile stress of the damage (Xie, 1987a). Figure 10.42 shows these cases where the orientation of the microcracks in the element body is considered. In Figure 10.42b, suppose that the axis of the elliptical microcrack lies at an angle 15° obliquely to σ_1. From the above analyses of local damage, we can determine that the fracture may occur on the plane lying at an angle 30° obliquely to σ_1, which is in agreement with many experimental results under uniaxial compression with the fracture plane at an angle about 25°. Figure 10.42a gives the element body containing the elliptical microcrack with an inclination angle of 30° to the principal stress σ_1 under biaxial compression. Similarly, we can obtain its fracture plane inclining at an angle 60° to σ_1 in terms of the above analytical method. It may be said that, rock failure is local tensile failure caused by damage, although the rock may apparently be in compressive stress states. The Griffith fracture theory is still applicable to the local stress field of the damage in rock masses. The phenomenon for the compressive strength of rock higher than its tensile strength is caused by the damage effect of rock.

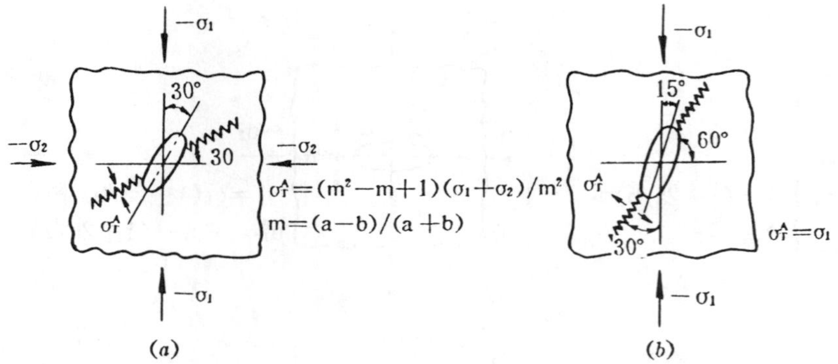

Figure 10.42. The damage stress and failure means of rock when its microcrack orientations are considered. (a) Biaxial stress state; (b) Uniaxial compression.

10.3 THE EXPERIMENT RESULTS OF DAMAGE EVOLUTION OF ROCK

All rocks are natural damage materials because they contain microcavities to some degree. Sandstones and crystalline rocks may typically contain a microcavity volume of 5-30% and < 1% respectively. Details of naturally occurring crack and pore porosities determined by Sprunt & Brace (1974) are presented in Table 10.5. Pre-existing or natural microcracks in quartzite are shown in Photo 10.4.

From the discussion in the previous sections, microvoids and microcracks (damage) have been shown to have a significant effect on rock properties. Adams & Williamson (1923) were the first to recognize the effect of microcracks on elastic properties in granite. Several possible mechanisms of microcrack growth (damage evolution) have been suggested as an explanation for dilatancy and inelasticity (see Figure 10.28, and also the paper of Kachanov, 1980):

(1) Reopening of microcracks previously formed but healed during the earlier history of rock;

(2) Microcracking resulting from local stress concentrations at inhomogeneities;

Table 10.5. Naturally occurring crack and pore porosities in crystalline rocks (Sprunt & Brace, 1974).

Rock	Porosity (%)			Grain size (μm)	Maximum crack on grain boundary (μm)
	Crack	Pore	Total		
Diabase I	0	0.1	0.1	180	30
Gabbro	0	0.2	0.2	2000	180
Raymond granite	0.22	0.4	0.6	2000	180
Katahdin granite I	0.20	0.7	0.9	2000	260
Granite, Westerly	0.20	0.7	0.9	750	100
Granite, Chelmsford	0.16	0.9	1.0	1500	150
Granodiorite	0.29	0.8	1.1	2000	360
Katahdin granite II	0.46	0.8	1.2	3000	600

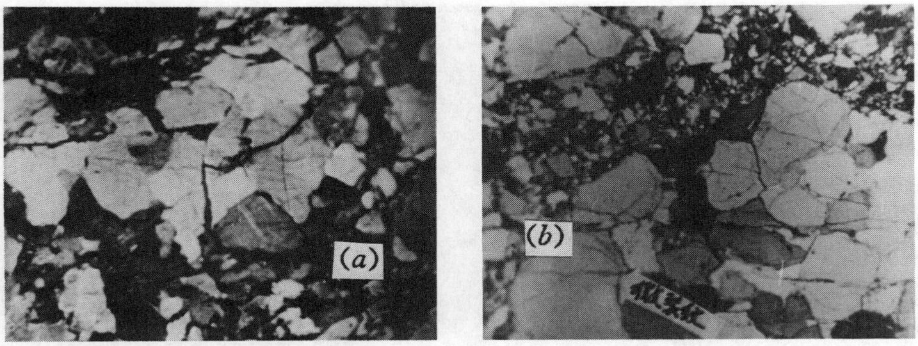

Photo 10.4. Pre-existing microcracks in quartzite (Henan Gonan, China), Optical micrograph × 100.

(3) Frictional sliding on microcracks with accompanying local tensile crack opening. Based on microscopic studies, statements on the relative role of these mechanisms do not always agree with each other. In general, there is a considerable controversy between the results obtained by different observational techniques. However, some general conclusions can be made as follows:

10.3.1 *Size of microcracks*

According to optical observations the typical dimensions of pre-existing microcracks are about the grain size, whereas electron microscope studies show that these dimensions are of the order of $\frac{1}{10}$ of the grain size. It is known that the nucleated cracks usually formed at grain boundaries were nearly parallel to the direction of maximum compression. Studies by Sprunt & Brace indicate that about 30% of the grain boundaries in westerly granite are occupied by low aspect ratio cavities (aspect ratio – ratio of minimum to maximum opening dimension). Nolen-Hoeksema & Gordon (1987) reported their experimental results of marble specimens with a notch under opening-mode loading: both transgranular and intergranular cracking are present in the marble, 82% is formed by separation of the grain boundaries, while 18% is transgranular, and 86% of this is formed on pre-existing cracks within the grains. Twin boundaries do not crack. Thus, since only the largest cracks present in sufficient numbers are to be taken into account (see discussion below), the 'effective' typical size of microcracks can be assumed to be of the order of a grain size.

10.3.2 *Orientation of microcracks*

In uniaxial compression, axial microcracking is dominant, in the first stage of inelasticity, microcracks are increasingly initiated along an orientation of a 15°-30° angle inclining toward the direction of the applied stress. In later stages, however, microcracks tend to initiate and propagate along the direction of axial stress as shown in photo. 10.5.

Photo 10.5. Microcrack growth and coalescence of rock specimen under uniaxial compression. (a) Microcrack coalescence; (b) Final failure.

10.3.3 *Distribution of microcracks*

Microcracks of rock will continually increase in number during loading, and their rate of increase is accelerated as the load increases. Table 10.6 shows the microcrack growth states of rock during loading.

At present, one directly examines the damage growth by using optical microscopes, electron microscopes, and the scanning electron micrograph associated with loading system. These devices can record the entire damage evolution process (microcracks growth process) during loading. Indirect observations of the damage growth are to measure changes of physical quantities caused by damage such as change of strains,

Table 10.6 Microcrack growth states of rock under different load ratio.

Maximum load (%)	Average density of microcracks (Num/mm^2)	Average length of microcracks (mm)	Average width of microcracks (mm)	Microcracks growth states
<0.4	0.5–1	0.08	0.005	No obvious change
0.4–0.6	2.1	0.135	0.043	Obvious increasing occur in the orientation of an angle 15-30° inclining to axial stress direction
0.6–0.85	10.5	0.192	0.015	Microcracks tend to grow in axial stress direction
0.85–1.0	No obvious change	—	A few of cracks reach the length of 0.53 mm	Microcracks interact, coalesce, and lead to failure

change of elastic modulus (see Equation (10.72)), density change (Jonas & Baudelet, 1977), change in sound velocity in acoustic emission (Sondergeld & Estey, 1981), change in the fatigue limit (Bui-Qudc et al., 1971) and so on.

Nolen-Hoeksema & Gordon (1987) used an optical microscope associated loading frame which was designed by themselves to examine the fracture of double cantilever beam test specimen of Stockbridge dolomite marble (Figure 10.43). The length of the newly-formed crack, crack density, and frequency of branching at each level of applied

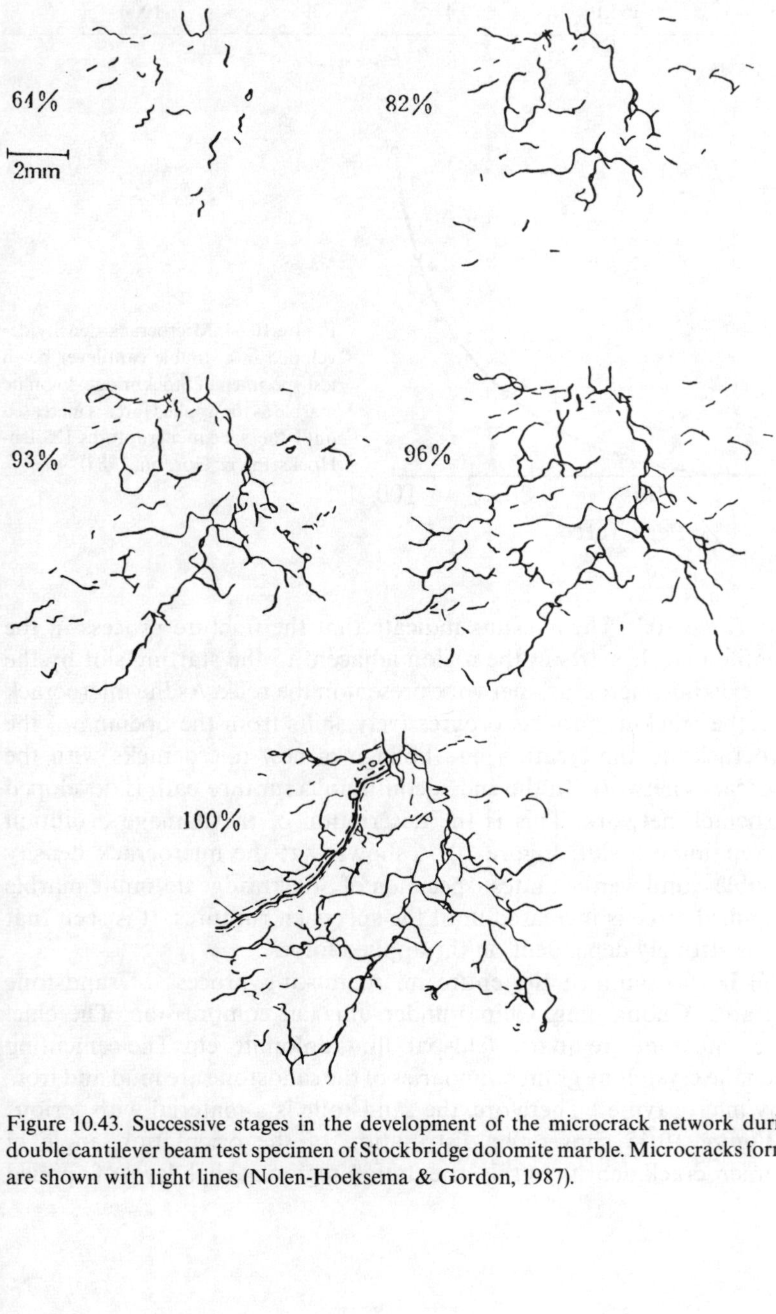

Figure 10.43. Successive stages in the development of the microcrack network during the fracture of a double cantilever beam test specimen of Stockbridge dolomite marble. Microcracks formed in earlier loading are shown with light lines (Nolen-Hoeksema & Gordon, 1987).

Table 10.7. Crack pattern characteristics of Stockbridge dolomite marble specimen (Nolen-Hoeksema & Gordon, 1987).

Maximum load (%)	Crack increment (mm)	Total crack length (mm)	Crack density	% of crack on pre-existing cracks	Frequency of branching (m)
38	0	0	0	0	0
64	9.2	9.2	1.2	55	0
82	30.5	39.7	5.2	49	705
93	35.8	75.5	9.9	41	967
96	30.7	106.2	14.0	32	1083
99	48.9	155.1	20.4	39	1000

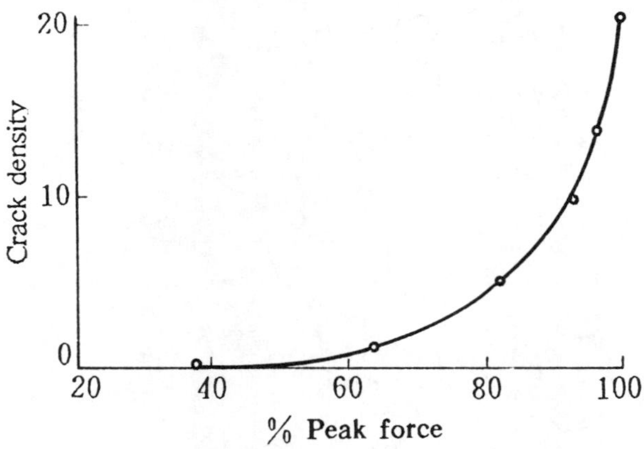

Figure 10.44. Microcrack density developed in a double cantilever beam test specimen of Stockbridge dolomite marble as the applied force is increased until the specimen ruptures (Nolen-Hoeksema & Gordon, 1987).

force are listed in Table 10.7. Their results indicate that the fracture process in the Stockbridge dolomite marble starts in the region adjacent to the starting slot by the opening of the pre-existing microcrack network present in the rock. As the microcrack network develops, the cracking process progressively shifts from the opening of the pre-existing microcracks to the creation and linkage of new microcracks with the pre-existing microcracks network. In the end, a continuous rupture path is developed through the microcrack network. This is the description of the damage evolution process of rock containing a slot. Figure 10.44 shows that the microcrack density developed in a double cantilever beam test specimen of Stockbridge dolomite marble increases as the applied force is increased until the specimen ruptures. It is seen that damage evolution is strongly dependent on the applied forces.

Xu et al. (1986) have examined the entire microcracking process of sandstone (Nantong mining area, Chong qing, China) under uniaxial compression. The chief components of the sandstone are quartz, feldspar, flint, dolomite, etc. The cementing substances between the crystalline grain boundaries of the sandstone are mud and iron, and contain many microcavities. Therefore, the sandstone is a material with serious initial damage. Figure 10.45 shows the distribution of the orientation angle of microcracks and microcrack density in this orientation region at each level of applied

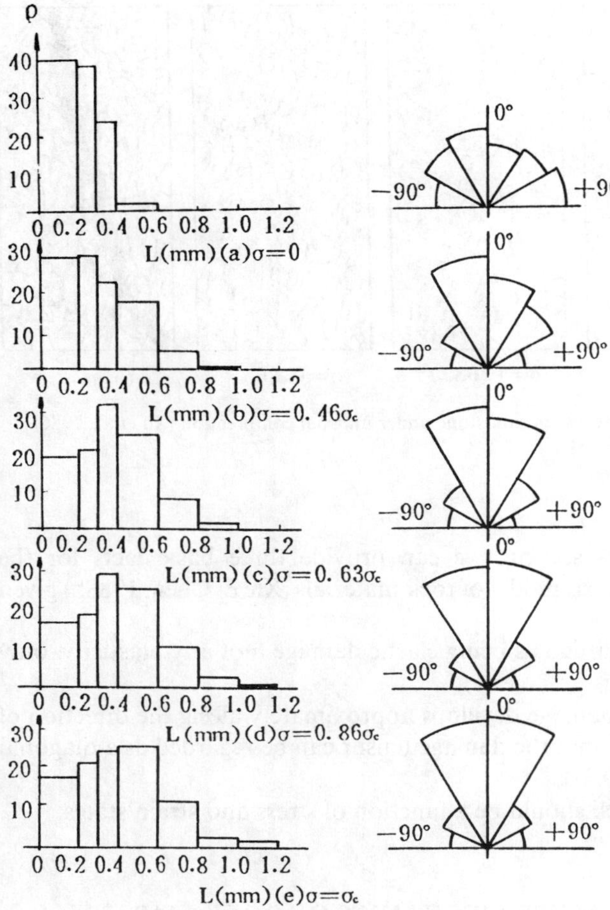

Figure 10.45. Distribution of microcrack orientation and density of sandstone at each level of applied load (Xu et al., 1986).

Table 10.8. Total length L and number N of microcracks of sandstone at each level of applied stress.

σ/σ_c	N_{sum}	L_{sum}(mm)	$N/N_{max} \times 100\%$	$L/L_{max} \times 100\%$
0	125	27.2	26.69	16.69
0.46	236	69.8	56.06	42.82
0.63	259	92.7	61.52	56.87
0.86	311	117.8	73.87	72.27
1.0	421	163.0	100.00	100.00

force. The total length and number of microcracks at each level of applied force are listed in Table 10.8. The entire damage evolution process is recorded as illustrated in Figure 10.46. These experimental results indicate that rock fracture is actually a damage evolution process, and rock damage is induced mainly by applied stresses.

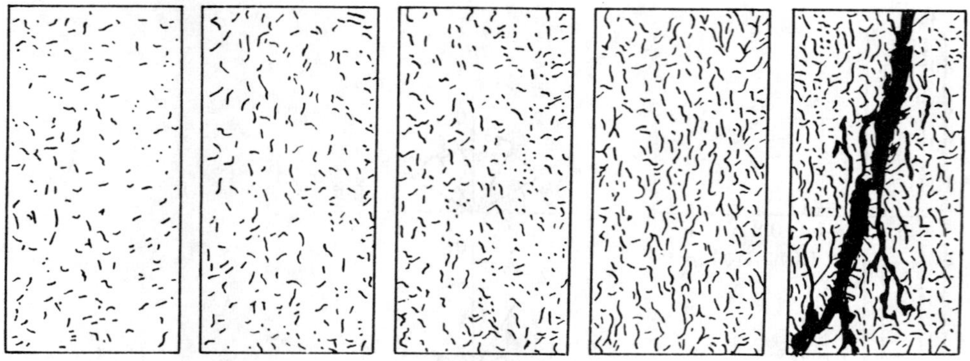

Figure 10.46. The damage evolution process of sandstone under uniaxial compression (Xu et al., 1986).

From the discussion in this section, we can provide three basic facts for the establishment of a damage analysis model of rock materials (Xie & Chen, 1988b) given in the next section:

(1) Rock damage can be regarded as being elastic damage (not any plasticity, only the microcracking process);

(2) The dominate region of damage develops approximately along the direction of a maximum principal stress, so that the damage tensor can be regarded as a diagonal matrix defined by Equation (10.87);

(3) Damage evolution of rock should be a function of stress and strain states.

10.4 THE ANALYSIS OF DAMAGE MECHANICS OF ROCK MASSES

Damage mechanics was first applied to rock materials by Dougill et al. (1976). Dragon & Mroz (1979) have analyzed the continuous damage of rock and concrete materials by using the concept of fracture surfaces, while Krajcinovic et al. (1981) have given descriptions of damage constitutive equations of brittle materials in terms of thermodynamics and void kinematics. Frantziskonis & Desai (1987) simulated the strain-softening phenomenon of brittle materials by using the damage concept. In the last ten years, rock damage has been intensively studied from widely differing viewpoints. However a large amount of research remains in the investigation of the theoretical mechanical behavior of rock damage. In this and next sections, we apply damage mechanics to investigate both the mechanics behavior of rock damage and to solve the problems of rock damage and fracture in underground engineering.

Consider the damage state of an element body in a neighborhood at point O on a co-moving coordinate system (see Figure 10.47). We then define the damage variable as (Xie, 1987a, 1990a):

$$\begin{aligned} \boldsymbol{w} &= (\mathrm{d}\boldsymbol{A} - \mathrm{d}\boldsymbol{A}^c - \mathrm{d}\boldsymbol{A}^D)/(\mathrm{d}A) = (\mathrm{d}A_i - \mathrm{d}A_i^e - \mathrm{d}A_i^D)/(\mathrm{d}A)\cdot \boldsymbol{g}^i \\ &= w_i \cdot \boldsymbol{g}^i = \mathrm{d}A_i^0/\mathrm{d}A\cdot \boldsymbol{g}^i. \end{aligned} \tag{10.120}$$

where $w_i (i = 1, 2, 3)$ are the principal damage values. dA^e is the effective micro-area vector of micro-area vector dA after damage and deformation, and dA^D represents the change of the micro-area vector dA caused only by deformation.

From the experimental results in previous sections, we establish a model of the damage evolution of rock. Using the method of fuzzy probability measuring, we have deduced an equation of the damage evolution of rock (Xie, 1987a; Xie & Chen, 1988b) as follows:

$$w_i = 1 - \exp[-B|Y^i - Y_0^i|^{1/n}]. \tag{10.121}$$

Its rate equation can be derived from Equation (10.121):

$$\dot{w} = B|\dot{Y}|:(I - w)/[n|Y - Y_0|^{(n-1)/n}], \tag{10.122}$$

where B and n are material parameters.

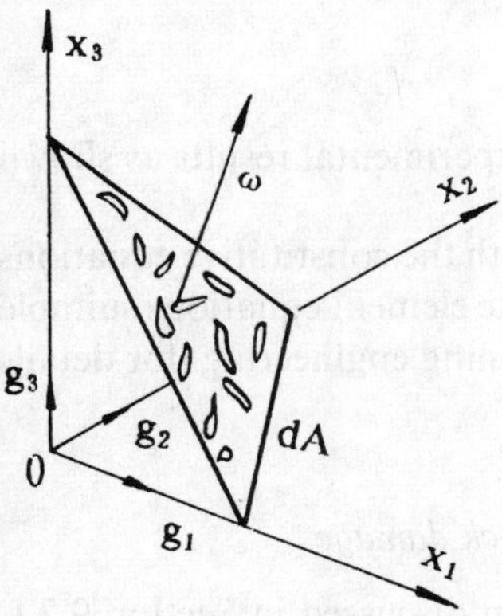

Figure 10.47. Damage state of a point O.

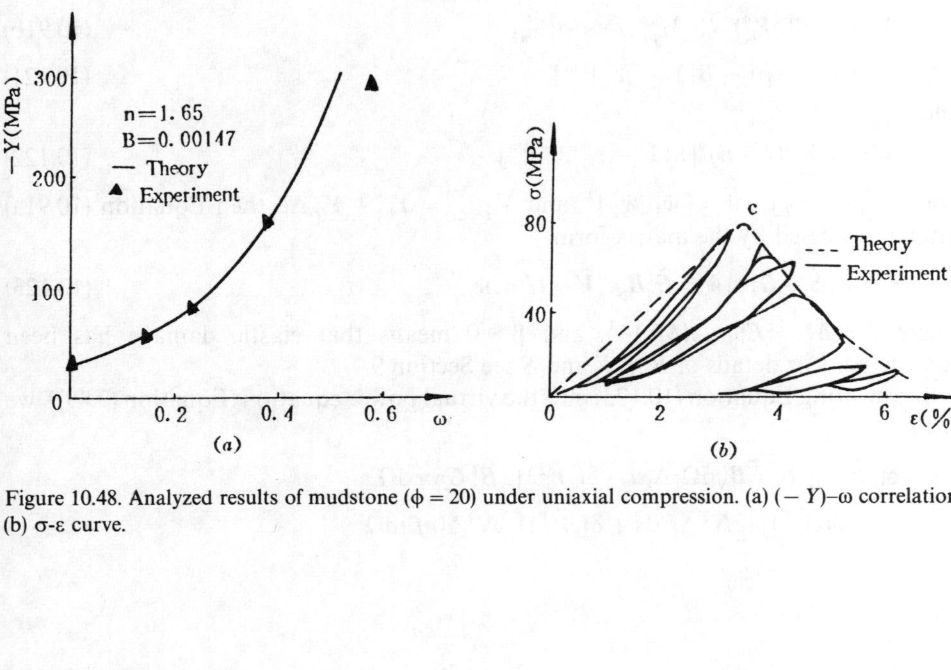

Figure 10.48. Analyzed results of mudstone (ϕ = 20) under uniaxial compression. (a) (− *Y*)–ω correlation; (b) σ-ε curve.

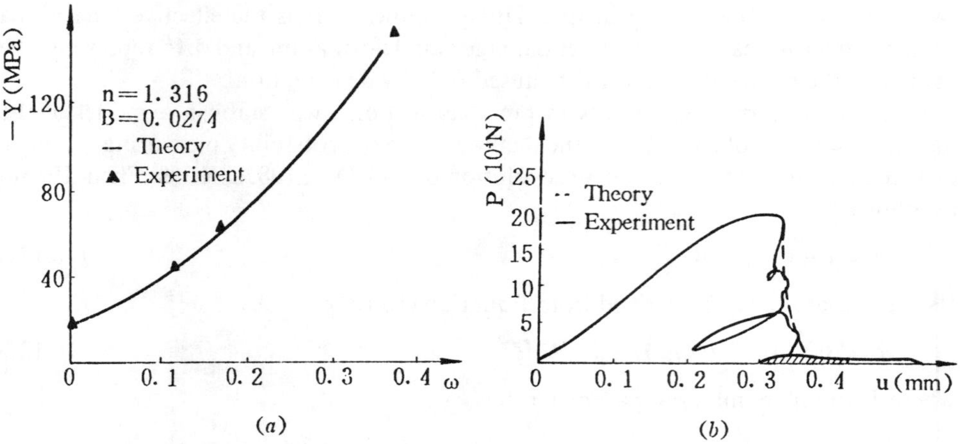

Figure 10.49. Analyzed results of marble under uniaxial compression. (a) $(-Y)$–ω correlation; (b) σ-ω curve.

Here, the theoretical modelling is in agreement with experimental results as shown in Figures 10.48 and 10.49.

Combining the large deformation theory (Chapter 9) with the constitutive equations of damage materials (Section 10.1.2), we have deduced finite element equations suitable for analyses of plane problems in rock mechanics and mining engineering (for details see papers of Xie, 1987a; Xie & Chen, 1988b).

10.4.1 *Finite element equations for large deformation of rock damage*

Finite element equations for large deformation have been discussed in Section 9.7.1. Let us now consider the constitutive equations of rock damage:

$$\dot{\sigma}^i_j = \tilde{E}^{ik}_{jl}\dot{S}^{l(e)}_k - 2(M^{im}_{jn})^{-1}(\partial M^{nk}_{ml}/\partial w_q)\dot{w}_q\sigma^l_k, \tag{10.91a}$$

$$\dot{Y}^k = -2(M^{ip}_{jq})^{-1}(\partial M^{qm}_{pl}/\partial w_k)\sigma^j_i\dot{S}^l_m, \tag{10.91b}$$

$$w_i = 1 - \exp[-B|Y^i - Y^i_0|^{1/n}], \tag{10.121}$$

and

$$\dot{\boldsymbol{w}} = B|\dot{\boldsymbol{Y}}|:(\boldsymbol{I} - \boldsymbol{w})/[n|\boldsymbol{Y} - \boldsymbol{Y}_0|^{(n-1)/n}]. \tag{10.122}$$

Let $\boldsymbol{w} = [w_1, w_2]^T$, $\dot{\boldsymbol{w}} = [\dot{w}_1, \dot{w}_2]^T$, and $\boldsymbol{Y}_{(t+\Delta t)} = \boldsymbol{Y}_{(t)} + \dot{\boldsymbol{Y}}_{(t)}\Delta t$, then Equation (10.91a) can be expressed by the matrix form

$$\dot{\boldsymbol{\sigma}} = \tilde{\boldsymbol{E}}:\dot{\boldsymbol{S}} - \boldsymbol{G}:\boldsymbol{\sigma}:\dot{\boldsymbol{w}} = \tilde{\boldsymbol{E}}:\boldsymbol{B}_N:{}_k\tilde{\boldsymbol{V}} - \boldsymbol{G}:\boldsymbol{\sigma}:\dot{\boldsymbol{w}}, \tag{10.123}$$

where $\tilde{\boldsymbol{E}} = \boldsymbol{M}^{-1}:\boldsymbol{E}|_{\beta=0}:(\boldsymbol{M}^T)^{-1}$, and $\beta = 0$ means that elastic damage has been considered. For details of $\boldsymbol{B}_N$, ${}_k\tilde{\boldsymbol{V}}$ and $\dot{\boldsymbol{S}}$ see Section 9.7.

Substituting Equation (10.123) into the virtual power equation (Equation (9.49)), we have

$$\delta({}_k\tilde{\boldsymbol{V}}^T)\int_\Omega \boldsymbol{B}_N^T\tilde{\boldsymbol{E}}\boldsymbol{B}_N d\Omega\cdot\Delta_k\boldsymbol{u} - \delta({}_k\tilde{\boldsymbol{V}}^T)\int_\Omega \boldsymbol{B}_N^T\boldsymbol{G}\boldsymbol{\sigma}\dot{\boldsymbol{w}}d\Omega$$
$$= \delta({}_k\tilde{\boldsymbol{V}}^T)\int_{\partial\Omega}\boldsymbol{N}^T\Delta\boldsymbol{P}ds + \delta({}_k\tilde{\boldsymbol{V}}^T)\int_\Omega \boldsymbol{N}^T\Delta(\rho\boldsymbol{f})d\Omega.$$

Thus, finite element equations for large deformation of rock damage can be obtained as follows:

$$\begin{aligned} &\boldsymbol{K}_N^w \Delta \boldsymbol{u} = \Delta \boldsymbol{R} + \Delta \boldsymbol{F}_w, \\ &\boldsymbol{K}_N^w = \int_\Omega \boldsymbol{B}_N^T \tilde{\boldsymbol{E}} \boldsymbol{B}_N \mathrm{d}\Omega = \int_\Omega \boldsymbol{B}_N^T \cdot \boldsymbol{M}^{-1} \cdot \boldsymbol{E} \cdot (\boldsymbol{M}^T)^{-1} \cdot \boldsymbol{B}_N \mathrm{d}\Omega, \end{aligned} \tag{10.124}$$

and

$$\Delta \boldsymbol{F}_w = \int_\Omega \boldsymbol{B}_N^T \boldsymbol{G} \boldsymbol{\sigma} \boldsymbol{w} \mathrm{d}\Omega;$$

where $\Delta \boldsymbol{R}$ and $\Delta \boldsymbol{u}$ are the same as definitions given in Section 9.7. $\boldsymbol{K}_N^w$ is the nonlinear stiffness matrix of the large deformation of rock damage. Whereas $\Delta \boldsymbol{F}_w$ is the increment of element equivalent nodal forces caused by rock damage:

$$\boldsymbol{G} = 2\boldsymbol{M}^{-1} : (\partial \boldsymbol{M} / \partial \boldsymbol{w}) = \begin{bmatrix} 2/(1-w_1) & 0 \\ 0 & 2/(1-w_2) \\ 0 & 0 \end{bmatrix}.$$

The stress matrix $\boldsymbol{\sigma}$ associated with $\boldsymbol{G}$ is defined by

$$\boldsymbol{\sigma} = \begin{bmatrix} \sigma_1^1 & \sigma_2^1 \\ \sigma_1^2 & \sigma_2^2 \end{bmatrix}.$$

10.4.2 *Finite element equations for large deformation of rock creep and damage*

When considering an elastic creep state, the total strain rate can be decomposed into two parts:

$$\dot{S}_j^i = \dot{S}_j^{i(e)} + \dot{S}_j^{i(c)},$$

where $\dot{S}_j^{i(e)}$ and $\dot{S}_j^{i(c)}$ are the rate forms of elastic strain tensor and creep strain tensor, respectively. Thus, the constitutive equations of elastic creep damage of rock can be defined by

$$\dot{\sigma}_j^i = \tilde{E}_{jl}^{im} [\dot{S}_m^l - \dot{S}_m^{l(c)}] - G_{jl}^{imk} \sigma_m^l \dot{w}_k. \tag{10.125}$$

The rate of creep strain tensor should be the function of stress state σ_j^i and time t, i.e.,

$$\dot{S}_j^{i(c)} = f(\sigma_j^i, t).$$

In solving practical nonlinear problems, incremental algorithms must be used. This means that the constitutive equations and variational equations should be established on the instantaneous $(t + \Delta t)$ co-moving coordinate system (as discussed in Chapter 9 and the previous section). Thus, we give a Taylor expansion of $\dot{S}_j^{i(c)}$ at $t + \Delta t$:

$$\dot{S}_j^{i(c)}|_{t+\Delta t} = \dot{S}_j^{i(c)}|_t + (\partial \dot{S}_j^{i(c)} / \partial \sigma_q^p)|_t \Delta \sigma_q^p + (\partial \dot{S}_j^{i(c)} / \partial t)|_t \Delta t. \tag{10.126}$$

In each iteration step, we use the algorithm presented by Kiefer & Hilton (1985) in the form:

$$\dot{S}_j^{i(c)} = (1 - \alpha) \dot{S}_j^{i(c)}|_t + \alpha \dot{S}_j^{i(c)}|_{t+\Delta t}, \tag{10.127}$$

where α is an algorithm parameter.

Substituting Equation (10.126) into Equation (10.127), we have

$$\dot{S}_j^{i(c)} = \dot{S}_j^{i(c)}|_t + \alpha(\partial \dot{S}_j^{i(c)}/\partial t)\Delta t + \alpha(\partial \dot{S}_j^{i(c)}/\partial \sigma_q^p)\Delta\sigma_q^p. \tag{10.128}$$

From Equations (10.125) and (10.128), we can obtain the constitutive equations of rock creep and damage established on an instantaneous iteration step at $(t + \Delta t)$, as follows:

$$\dot{\sigma}_l^m = A_{li}^{mj}(w_k, t)[\dot{S}_j^i - \dot{S}_j^{i(c)} - \alpha(\partial \dot{S}_j^{i(c)}/\partial t)\Delta t]$$
$$+ [\alpha \tilde{E}_{li}^{mj}(\partial \dot{S}_j^{i(c)}/\partial \sigma_q^p)\cdot G_{q\beta}^{p\alpha k}\Delta t - G_{l\beta}^{m\alpha k}]\sigma_\alpha^\beta \dot{w}_k,$$
$$A_{li}^{mj}(w_k, t) = \tilde{E}_{li}^{mj}(w_k)[1 + \alpha(\partial \dot{S}_j^{i(c)}/\partial \sigma_q^p)\tilde{E}_{ql}^{pm}(w_k)\delta_m^j \delta_i^l \Delta t]^{-1},$$

and

$$G_{l\beta}^{m\alpha k} = 2(M_{lp}^{mq})^{-1}\partial M_{q\beta}^{p\alpha}/\partial w_k. \tag{10.129}$$

Obviously, Equation (10.129) is the general form of constitutive equations of rock creep and damage. For a practical problem, we must determine the specific form of $\dot{S}_j^{i(c)} = f(\sigma_j^i, t)$ for which the solution of this problem is obtained by solving Equation (10.129).

However, so far there has not been a state equation of creep established by means of rock experiments to be commonly suitable for rock materials. This is also the case for rock creep and damage. Fortunately, the damage variable w_i defined by Equation (10.121) implies the influence of creep, since $Y^i = Y^i(\sigma_j^i, S_j^i, t)$ is a function of creep time t. Thus, we have to determine the specific form of $\dot{S}_j^{i(c)} = f(\sigma_j^i, t)$ only with the help of the following four basic element models of rheological commonly applied to the analyses of rock creep, then deduce their specific constitutive equations of rock creep and damage, which are suitable for various rock materials (where different type of rheological models is chosen for different rock materials):

10.4.2.1 *General Kelvin-substance*

The creep equation for a General Kelvin-substance under uniaxial stress is

$$\varepsilon_c = (\sigma/k_1) + (\sigma/k_2)[1 - \exp(-k_2/\eta \cdot t)],$$

where k_1 and k_2 are elastic constants, and η is a viscosity modulus. Its rate form is

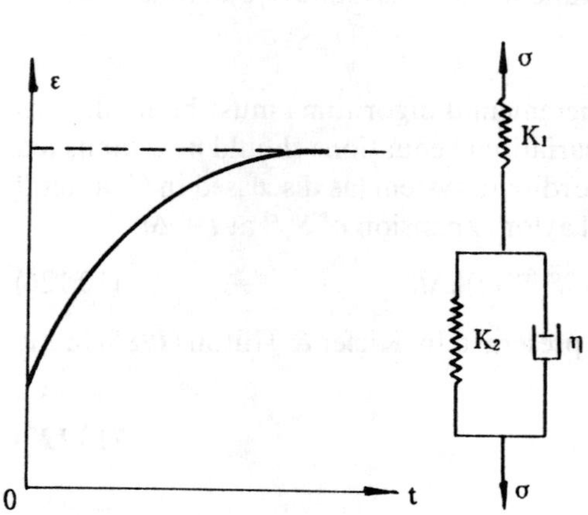

Figure 10.50. General Kelvin-substance of creep.

(σ = constant), thus

$$\dot{\varepsilon}_c = \sigma/\eta \cdot \exp(-k_2/\eta \cdot t).$$

Creep state equations under triaxial stresses can be determined by the normality rule (Boyle & Spence, 1983)

$$\dot{S}_j^{i(c)} = (\tfrac{3}{2})(\bar{\dot{\varepsilon}}_c/\bar{\sigma})T_j^i \tag{10.130}$$

where $\bar{\dot{\varepsilon}}_c$ and $\bar{\sigma}$ are equivalent creep strain rate and equivalent stress, respectively, which are defined by quantities of uniaxial creep states and T_j^i is a deviator stress tensor.

Thus, for a General Kelvin-substance, we have

$$\bar{\dot{\varepsilon}}_c = (\bar{\sigma}/\eta)\exp(-k_2/\eta \cdot t);$$
$$\dot{S}_j^{i(c)} = [3/(2\eta)]\exp(-k_2/\eta \cdot t)[\delta_\alpha^i \delta_j^\beta - (\tfrac{1}{3})\delta_j^i \delta_\alpha^\beta]\sigma_\beta^\alpha;$$
$$\partial \dot{S}_j^{i(c)}/\partial t = -(3k_2/2\eta^2)\exp(-k_2/\eta \cdot t)[\delta_\alpha^i \delta_j^\beta - (\tfrac{1}{3})\delta_j^i \delta_\alpha^\beta]\sigma_\beta^\alpha;$$

and

$$\partial \dot{S}_j^{i(c)}/\partial \sigma_q^p = 3/(2\eta)\exp(-k_2/\eta \cdot t)[\delta_p^i \delta_j^q - (\tfrac{1}{3})\delta_j^i \delta_p^\beta \delta_\beta^\alpha].$$

From Equation (9.37), we express Equation (10.129) in the matrix form:

$$\dot{\boldsymbol{\sigma}} = \boldsymbol{A}[\boldsymbol{B}_N({}_k\tilde{\boldsymbol{V}}) - \dot{\boldsymbol{S}}^{(c)} - \alpha \boldsymbol{L}\Delta t] + \alpha \Delta t \tilde{\boldsymbol{E}}\boldsymbol{H}\boldsymbol{G}\boldsymbol{\sigma}\dot{\boldsymbol{w}} - \boldsymbol{G}\boldsymbol{\sigma}\dot{\boldsymbol{w}},$$

and

$$A = \{\boldsymbol{I} + \alpha \Delta t \tilde{\boldsymbol{E}}\boldsymbol{H}\}^{-1}\tilde{\boldsymbol{E}}. \tag{10.131}$$

where $\boldsymbol{A} = \boldsymbol{A}^T$, $\boldsymbol{H} = \partial \dot{\boldsymbol{S}}^{(c)}/\partial \boldsymbol{\sigma}$, and $\boldsymbol{L} = \partial \dot{\boldsymbol{S}}^{(c)}/\partial t$.

Clearly, for a General Kelvin-substance, specific forms of $\dot{\boldsymbol{S}}^{(c)}$, $\boldsymbol{H}$, and $\boldsymbol{L}$ are defined by

$$\dot{\boldsymbol{S}}^{(c)} = [3/(2\eta)]\exp(-k_2/\eta \cdot t)\begin{bmatrix} \sigma_1^1 - \sigma_k^k/3 \\ \sigma_2^2 - \sigma_k^k/3 \\ 2\sigma_2^1 \end{bmatrix}$$

$$\boldsymbol{L} = -[3k_2/(2\eta^2)]\exp(-k_2/\eta \cdot t)\begin{bmatrix} \sigma_1^1 - \sigma_k^k/3 \\ \sigma_2^2 - \sigma_k^k/3 \\ 2\sigma_2^1 \end{bmatrix}$$

and

$$\boldsymbol{H} = [3/(2\eta)]\exp(-k_2/\eta \cdot t)\begin{bmatrix} \tfrac{2}{3} & -\tfrac{1}{3} & 0 \\ -\tfrac{1}{3} & \tfrac{2}{3} & 0 \\ 0 & 0 & 2 \end{bmatrix}.$$

10.4.2.2 *Maxwell-substance* (*linear model*)

The creep state equation of Maxwell-substance under uniaxial stress is

$$\varepsilon_c = \sigma/\eta \cdot t + \sigma/k.$$

Then,

$$\bar{\dot{\varepsilon}}_c = \bar{\sigma}/\eta;$$
$$\dot{S}_j^{i(c)} = [3/(2\eta)][\delta_\alpha^i \delta_j^\beta - (\tfrac{1}{3})\delta_j^i \delta_\alpha^\beta]\sigma_\beta^\alpha;$$
$$\partial \dot{S}_j^{i(c)}/\partial t = 0;$$

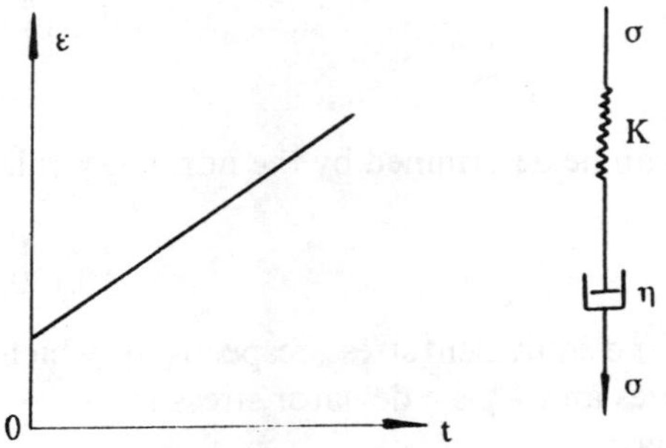

Figure 10.51. Maxwell-substance of creep.

and

$$\partial \dot{S}^{i(c)}_j / \partial \sigma^p_q = 3/(2\eta)[\delta^i_p \delta^q_j - (\tfrac{1}{3})\delta^i_j \delta^q_p].$$

Their matrix forms can be expressed by

$$\dot{\boldsymbol{S}}^{(c)} = 3/(2\eta)\begin{bmatrix} \sigma^1_1 - \sigma^k_k/3 \\ \sigma^2_2 - \sigma^k_k/3 \\ 2\sigma^1_2 \end{bmatrix},$$

$\boldsymbol{L} = 0$, and

$$\boldsymbol{H} = 3/(2\eta)\begin{bmatrix} \frac{2}{3} & -\frac{1}{3} & 0 \\ -\frac{1}{3} & \frac{2}{3} & 0 \\ 0 & 0 & 2 \end{bmatrix}.$$

10.4.2.3 *Burger-substance*

The creep state equation of Burger-substance under uniaxial stress is

$$\varepsilon_c = \sigma/\eta_1 \cdot t + \sigma/k_1 + \sigma/k_2 - \sigma/k_2 \cdot \exp(-k_2/\eta_2 \cdot t);$$

Then,

$$\dot{\bar{\varepsilon}}_c = \bar{\sigma}[(1/\eta_1) + (1/\eta_2)\exp(-k_2/\eta_2 \cdot t)];$$

$$\dot{S}^{i(c)}_j = -(\tfrac{3}{2})[1/\eta_1 + (1/\eta_2)\exp(-k_2/\eta_2 \cdot t)][\delta^i_\alpha \delta^\beta_j - (\tfrac{1}{3})\delta^i_j \delta^\beta_\alpha]\sigma^\alpha_\beta;$$

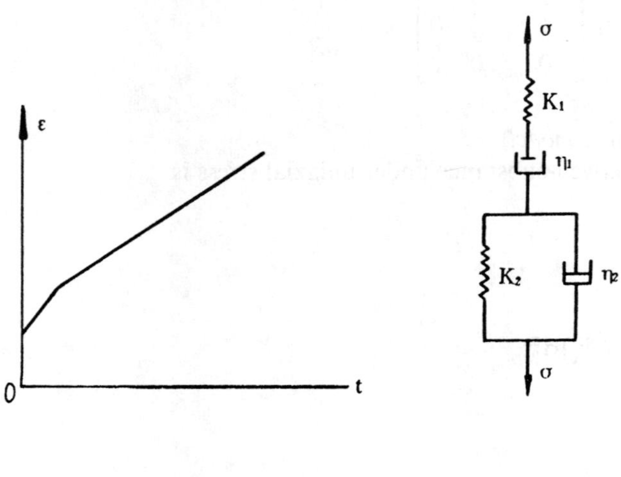

Figure 10.52. Burger-substance of creep.

$$\partial \dot{S}_j^{i(c)}/\partial t = -(3k_2/2\eta_2^2)\exp(-k_2/\eta_2 \cdot t)[\delta_\alpha^i \delta_j^\beta - (\tfrac{1}{3})\delta_j^i \delta_\alpha^\beta]\sigma_\beta^\alpha;$$

and

$$\partial \dot{S}_j^{i(c)}/\partial \sigma_q^p = (\tfrac{3}{2})[1/\eta_1 + (1/\eta_2)\exp(-k_2/\eta_2 \cdot t)][\delta_p^i \delta_j^q - (\tfrac{1}{3})\delta_j^i \delta_p^q].$$

Their relative matrix forms are

$$\dot{\boldsymbol{S}}^{(c)} = (\tfrac{3}{2})[1/\eta_1 + (1/\eta_2)\exp(-k_2/\eta_2 \cdot t)] \begin{bmatrix} \sigma_1^1 - \sigma_k^k/3 \\ \sigma_2^2 - \sigma_k^k/3 \\ 2\sigma_2^1 \end{bmatrix},$$

$$\boldsymbol{L} = -[3k_2/(2\eta_2^2)]\exp(-k_2/\eta_2 \cdot t) \begin{bmatrix} \sigma_1^1 - \sigma_k^k/3 \\ \sigma_2^2 - \sigma_k^k/3 \\ 2\sigma_2^1 \end{bmatrix},$$

and

$$\boldsymbol{H} = (\tfrac{3}{2})[1/\eta_1 + (1/\eta_2)\exp(-k_2/\eta_2 \cdot t)] \begin{bmatrix} \frac{2}{3} & -\frac{1}{3} & 0 \\ -\frac{1}{3} & \frac{2}{3} & 0 \\ 0 & 0 & 2 \end{bmatrix}.$$

10.4.2.4 *Nishihara-substance*

The creep state equation of Nishihara-substance under uniaxial stress is

$$\varepsilon_c = \sigma/k_1 + \sigma/k_2[1 - \exp(-k_2/\eta_2 \cdot t)], \quad \sigma < f_3,$$

and

$$\varepsilon_c = \sigma/k_1 + \sigma/k_2[1 - \exp(-k_2/\eta_2 \cdot t)] + (\sigma - f_3)/\eta_3 \cdot t, \quad \sigma \geq f_3.$$

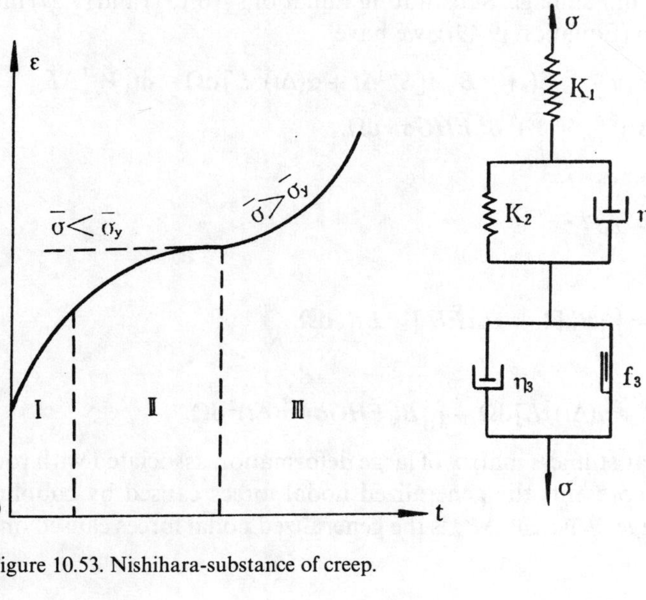

Figure 10.53. Nishihara-substance of creep.

where f_3 is the equivalent yield stress σ_y. Thus we have

$$\dot{\bar{\varepsilon}}_c = \langle \bar{\sigma} - \sigma_y \rangle / \eta_3 + \bar{\sigma}/\eta_2 \exp(-k_2/\eta_2 \cdot t);$$

$$\dot{S}_j^{i(c)} = (\tfrac{3}{2})[\langle \bar{\sigma} - \sigma_y \rangle / (\bar{\sigma}\eta_3) + (1/\eta_2)\exp(-k_2/\eta_2 \cdot t)][\delta_\alpha^i \delta_j^\beta - (\tfrac{1}{3})\delta_j^i \delta_\alpha^\beta]\sigma_\beta^\alpha;$$

$$\partial \dot{S}_j^{i(c)}/\partial t = -[3k_2/(2\eta_2^2)]\exp(-k_2/\eta_2 \cdot t)[\delta_\alpha^i \delta_j^\beta - (\tfrac{1}{3})\delta_j^i \delta_\alpha^\beta]\sigma_\beta^\alpha;$$

and

$$\partial \dot{S}_j^{i(c)}/\partial \sigma_q^p = (\tfrac{3}{2})[\langle \bar{\sigma} - \sigma_y \rangle / (\bar{\sigma}\eta_3) + (1/\eta_2)\exp(-k_2/\eta_2 \cdot t)] \cdot [\delta_p^i \delta_j^q - (\tfrac{1}{3})\delta_j^i \delta_p^q].$$

Their relative matrix forms can be defined by

$$\dot{\boldsymbol{S}}^{(c)} = \begin{bmatrix} \sigma_1^1 - \sigma_k^k/3 \\ \sigma_2^2 - \sigma_k^k/3 \\ 2\sigma_2^1 \end{bmatrix} \times (\tfrac{3}{2})[\langle \bar{\sigma} - \sigma_y \rangle / (\bar{\sigma}\eta_3) + (1/\eta_2)\exp(-k_2/\eta_2 \cdot t)],$$

$$\boldsymbol{L} = -[3k_2/(2\eta_2^2)]\exp(-k_2/\eta_2 \cdot t) \begin{bmatrix} \sigma_1^1 - \sigma_k^k/3 \\ \sigma_2^2 - \sigma_k^k/3 \\ 2\sigma_2^1 \end{bmatrix}$$

and

$$\boldsymbol{H} = \begin{bmatrix} \frac{2}{3} & -\frac{1}{3} & 0 \\ -\frac{1}{3} & \frac{2}{3} & 0 \\ 0 & 0 & 2 \end{bmatrix} \times (\tfrac{3}{2})[\langle \bar{\sigma} - \sigma_y \rangle / (\bar{\sigma}\eta_3) + (1/\eta_2)\exp(-k_2/\eta_2 \cdot t)];$$

where $\langle \bar{\sigma} - \sigma_y \rangle$ is defined as

$$\langle \bar{\sigma} - \sigma_y \rangle = \begin{cases} 0, & \text{if } \bar{\sigma} \le \sigma_y; \\ \bar{\sigma} - \sigma_y, & \text{if } \bar{\sigma} > \sigma_y. \end{cases}$$

From the above discussions, we can deduce the finite element equations for large deformation of rock creep and damage. Substituting Equations (10.131) and (9.37) into the virtual power equation (Equation (9.49)), we have

$$\int_\Omega \delta({}_k\tilde{\boldsymbol{V}})^T \boldsymbol{B}_N^T \boldsymbol{A} \boldsymbol{B}_N \mathrm{d}\Omega \Delta_k \boldsymbol{u} - \int_\Omega \delta({}_k\tilde{\boldsymbol{V}})^T \boldsymbol{B}_N^T \boldsymbol{A}[\dot{\boldsymbol{S}}^{(c)}\Delta t + \alpha(\Delta t)^2 \boldsymbol{L}]\mathrm{d}\Omega - \delta({}_k\tilde{\boldsymbol{V}})^T \Delta \boldsymbol{F}_w = \delta({}_k\tilde{\boldsymbol{V}})^T \Delta \boldsymbol{R} - \alpha(\Delta t)^2 \int_\Omega \delta({}_k\tilde{\boldsymbol{V}})^T \boldsymbol{B}_N^T \tilde{\boldsymbol{E}} \boldsymbol{H} \boldsymbol{G} \boldsymbol{\sigma} \dot{\boldsymbol{w}} \mathrm{d}\Omega.$$

Its matrix expression becomes

$$\boldsymbol{K}_N^{cw} \Delta \boldsymbol{u} = \Delta \boldsymbol{R} + \Delta \boldsymbol{F}_w + \Delta \boldsymbol{F}_w^c;$$

where

$$\boldsymbol{K}_N^{cw} = \int_\Omega \boldsymbol{B}_N^T \boldsymbol{A} \boldsymbol{B}_N \mathrm{d}\Omega = \int_\Omega \boldsymbol{B}_N^T [1 + \alpha \Delta t \tilde{\boldsymbol{E}} \boldsymbol{H}]^{-1} \tilde{\boldsymbol{E}} \boldsymbol{B}_N \mathrm{d}\Omega,$$

and

$$\Delta \boldsymbol{F}_w^c = \int_\Omega \boldsymbol{B}_N^T \boldsymbol{A}[\dot{\boldsymbol{S}}^{(c)}\Delta t + \alpha(\Delta t)^2 \boldsymbol{L}]\mathrm{d}\Omega - \int_\Omega \boldsymbol{B}_N^T \tilde{\boldsymbol{E}} \boldsymbol{H} \boldsymbol{G} \boldsymbol{\sigma} \dot{\boldsymbol{w}} \alpha(\Delta t)^2 \mathrm{d}\Omega.$$

Note that $\boldsymbol{K}_N^{cw}$ is a nonlinear stiffness matrix of large deformation associated with rock creep and damage, $\Delta \boldsymbol{F}_w^c$ represents the generalized nodal forces caused by coupling rock creep with rock damage. Whereas $\Delta \boldsymbol{F}_w$ is the generalized nodal forces caused only by rock damage.

10.4.3 *The numerical simulation for rock creep and damage*

In addition to the finite element program NEPR of the nonlinear large deformation of underground engineering problems discussed in Section 9.7, we have compiled two other two-dimensional finite element programs NEPROD and NOCRED for analyses of the problems of rock creep damage in underground engineering. Their functions are respectively:

Program NEPR – Analyses of elastic large deformation;
Program NEPROD – Analyses for large deformation with rock damage;
Program NOCRED – Analyses for large deformation with rock creep and damage.

These programs are widely used to simulate rock creep and damage phenomena in underground engineering.

10.4.3.1 *The deformation of large displacement caused by excavation*
The model of a subsidence experiment with equivalent materials is shown in Figure 10.54. The numerical simulation is made with the programs NEPR and NEPROD, respectively. Computed results are illustrated in Figures 10.55 and 10.56. It is clear that, in Figures 10.55 and 10.56, the numerical results in which rock damage effect is considered (NEPROD) are in good agreement with, both qualitatively and quantitatively, the measurement values. The phenomenon that displacements computed by classical elasto-plastic theory (NEPR) are always smaller than the measurement values has been improved.

10.4.3.2 *Modelling damage and fracture of a circular tunnel*
For continuous supports, e.g., rock boltsand and shotcrete, it is very important to predict the degree of failure in the region of a circumjacent tunnel. A computational

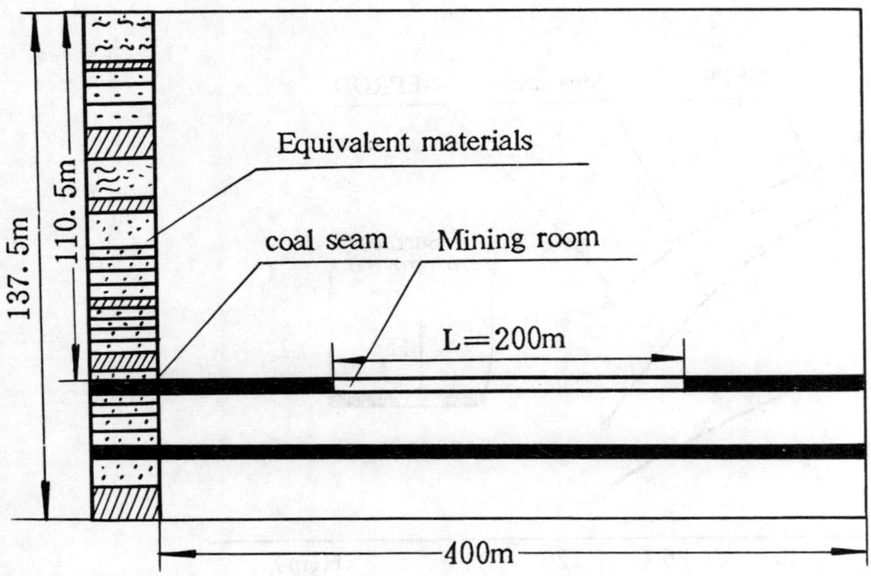

Figure 10.54. Model of the subsidence experiment.

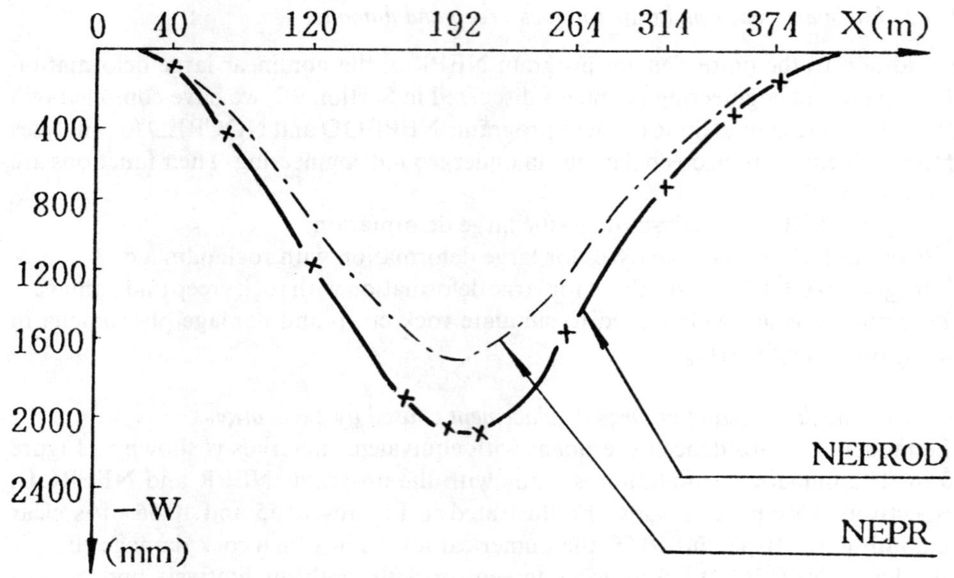

Figure 10.55. Profile of surface movements.

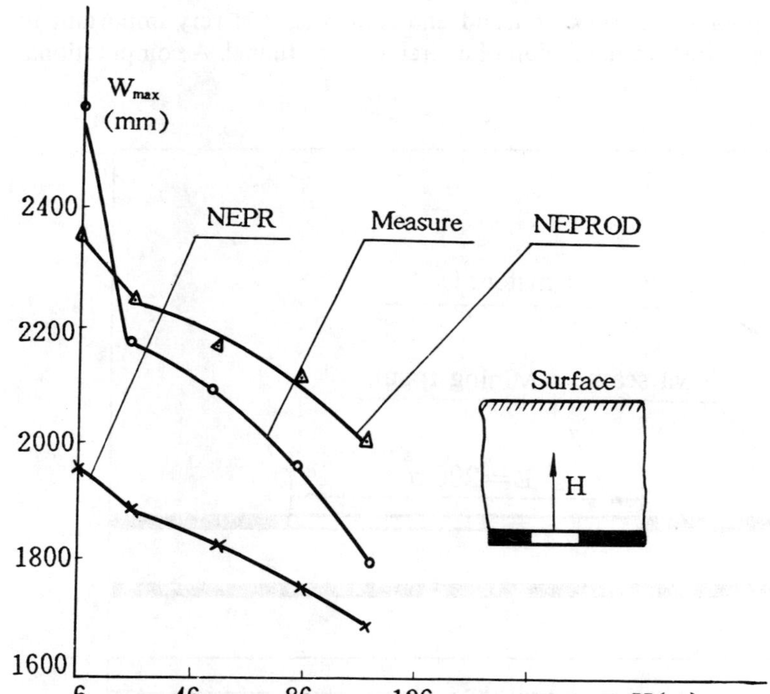

Figure 10.56. Maximum subsidence vs. H.

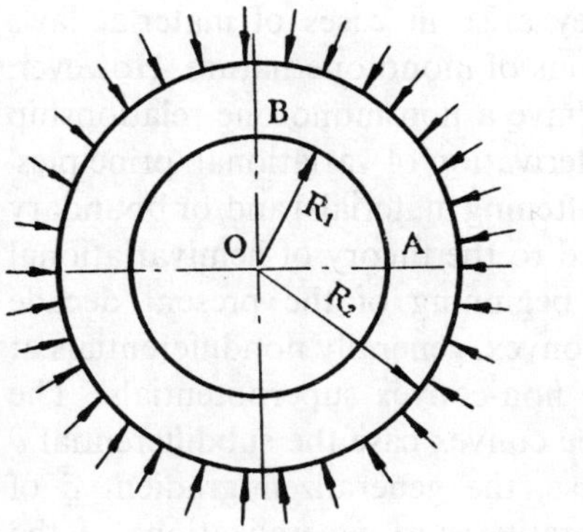

Figure 10.57. Model of the tunnel.

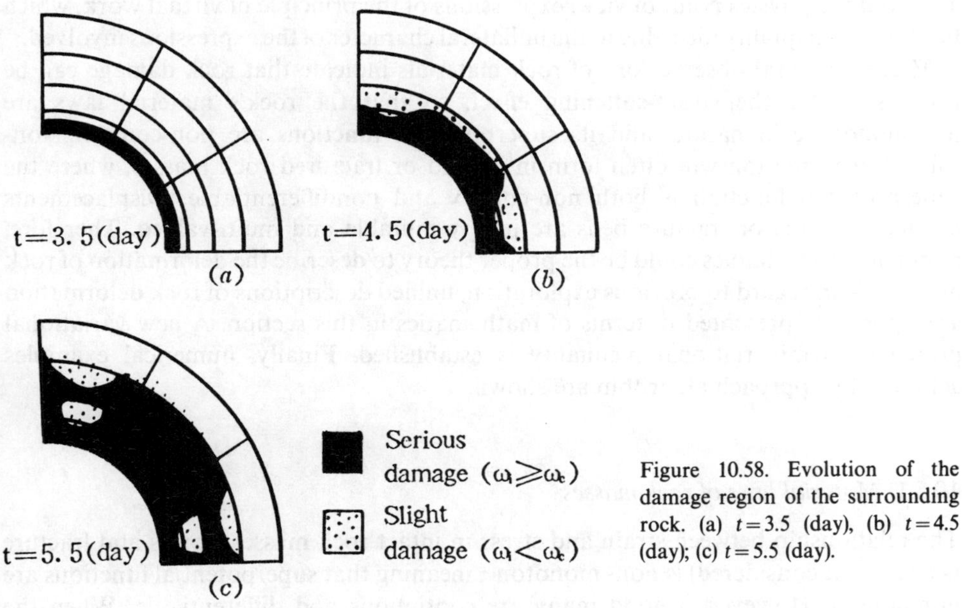

Figure 10.58. Evolution of the damage region of the surrounding rock. (a) $t=3.5$ (day), (b) $t=4.5$ (day); (c) $t=5.5$ (day).

model of a circular tunnel was chosen as indicated in Figure 10.57. The program NOCRED was used to simulate damage and fracture region of the circumjacent tunnel. Figure 10.58 shows the evolution of the damage region of the surrounding rock due to rock creep (here w_c is the critical value for the damage of rock material and is given a value of 0.5 in terms of experimental results of rock damage). Notice that the failure of the surrounding rock of the tunnel will continuously develop with rock creep and damage.

10.5 THE ANALYSIS OF NON-SMOOTH DEFORMATION OF ROCK MASSES

In this section, we discuss the non-smooth deformation of rock materials by means of non-smooth mechanics (Xie & Chen, 1988a). Variational equality and inequality formulations are well known in mechanics. They arise in cases of material laws (constitutive equations) and/or boundary conditions of monotone nature. However, friction, damage and fracture problems in rocks have a non-monotone relationship between their stress and strain. The need for the derivation of variational 'principles' for non-monotone materials laws (such as strain-softening materials) and/or boundary conditions such as joints and fracture beds has led to the theory of hemivariational inequalities. The theory was developed at the beginning of the present decade (Panagiotopoulos, 1981, 1983), arising from non-convex, generally nondifferentials at some points and nonfinite potentials (called here non-convex superpotentials). The differential operator on the superpotential is in the convex case the subdifferential ∂ of convex analysis, and in the non- convex case, the generalized gradient $\bar{\partial}$ of Clarke-Rockafeller (Rockafeller, 1980). Both operatives are generalizations of the classical differential and are multi-valued. Variational and hemivariational inequalities are, from the physics point of view, expressions of the principle of virtual work, which holds to its inequality form due to the unilateral character of the expressions involved.

Microstructural observations of rock materials indicate that rock damage can be responsible for the strain-softening effect, so that the rock's material laws are non-monotone in nature, and its superpotential functions are non-convex. Non-smooth deformation will often form in jointed or fractured rock masses, where the superpotential function is both non-convex and nondifferentiable. Displacements located in joints or fracture beds are jumpy mutable and multi-valued. Therefore, non-smooth mechanics could be the proper theory to describe the deformation of rock masses. With regard to previous exploration, unified descriptions of rock deformation are rigorously presented in terms of mathematics in this section. A new variational principle – hemivariational inequality is established. Finally, numerical examples achieved by approach algorithm are shown.

10.5.1 *Material laws of rock masses*

The relationship between strain and stress in intact rock masses (joints and fracture beds are not considered) is non- monotone, meaning that superpotential functions are non-convex. However, a good many are continuous and differentiable. When the effective stress σ_e is adopted in damage mechanics (discussed in Section 10.1), the superpotential functions about $\sigma_e - \varepsilon$ become smooth and convex. Generally, the superpotential functions about $\sigma - \varepsilon$ are also smooth, but non-convex; its material laws can be formulated one-dimensionally, as follows (Figure 10.59):

$$\sigma \in \bar{\partial}\phi(S).$$

Assuming

$$Z_i = \{S \mid \phi_{i(s)} = \phi_{(s)}\}, \quad (i = 1, 2);$$

where S represents strain. Then,

$$\bar{\partial}\phi_{(S)} = \{\text{grad}\,\phi_{i(s)}\}, \quad S \in Z_i, \quad (i = 1, 2);$$

$$\bar{\partial}\phi_{(S)} = \text{Co}(\text{grad}\,\phi_{1(s)}, \text{grad}\,\phi_{2(s)}), \quad S \in Z_1 \cap Z_2,$$

and 'Co' denotes a convex cone.

For a case of n-dimension, the material laws in every increment are defined by

$$\dot{\boldsymbol{\sigma}} \in \bar{\partial} \dot{\phi}_{(\dot{s})}. \tag{10.133}$$

Jointed rock masses are rock masses which include one or more joints or fracture beds. The global deformation is confined, and discontinuous deformation is formed at these joints or fracture beds, which we call internal boundary restraint. The mechanical properties of jointed rock masses are shown in Figure 10.60 by means of experimental data.

Normal deformation of the joint is in direct proportion to its normal stress:

$$-\sigma_n = K_n \langle u^1 \rangle = K_n[u^2_{+0} - u^2_{-0}]; \quad \text{if} \quad |\sigma_n| \in [\sigma_t, \sigma_c]; \quad \langle u^1 \rangle \leq U^m_n,$$

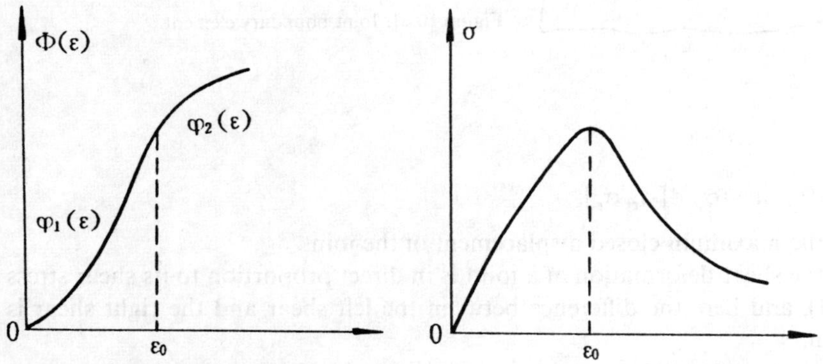

Figure 10.59. Stress-strain curve and superpotential of rock material under uniaxial loading.

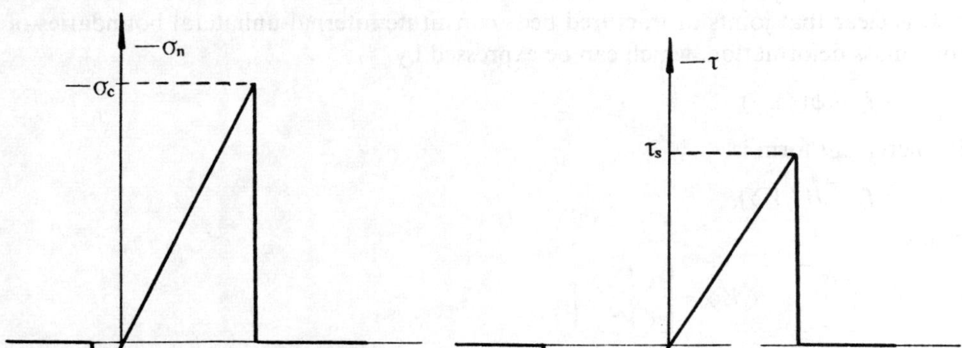

Figure 10.60. Deformation curve of joint.

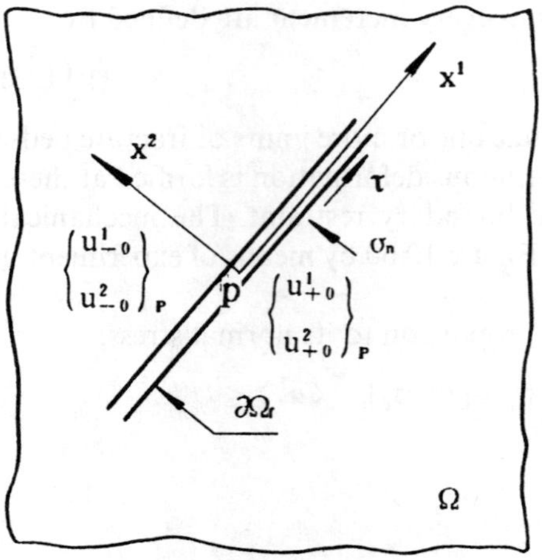

Figure 10.61. Joint boundary element.

and

$$-\sigma_n = 0, \quad \text{if} \quad |\sigma_n| \notin [\sigma_t, \sigma_c];$$

where U_n^m is the maximum closed displacement of the joint.

Similarly, the shear deformation of a joint is in direct proportion to its shear stress (Figure 10.61), and here the difference between the left shear and the right shear is ignored. Then,

$$-\tau = K_s\langle u^2\rangle = K_s[u_{+0}^1 - u_{-0}^1], \quad |\tau| \leq \tau_s, \quad \text{and}$$
$$-\tau = 0, \qquad |\tau| > \tau_s;$$

where K_n and K_s are the normal and shear stiffness, respectively.

It is clear that joints or fractured beds constitute internal unilateral boundaries of rock mass deformation, which can be expressed by

$$-f_c \in \bar{\partial}\phi(\langle u\rangle).$$

Its increment form is

$$-\dot{f}_c \in \bar{\partial}J(\langle V\rangle),$$

where

$$\dot{f}_c = \begin{bmatrix} \dot{\sigma}_n \\ \dot{\tau} \end{bmatrix}, \quad \langle V\rangle = \begin{bmatrix} \langle V^1\rangle \\ \langle V^2\rangle \end{bmatrix};$$

where $\phi(\langle u\rangle)$ and $J(\langle V\rangle)$ are non-convex superpotential functions.

10.5.2 *Hemivariational inequality and the minimal formula of a finite element*

By current configurations, the boundary of body Ω can be defined by $\partial\Omega = \partial\Omega_V + \partial\Omega_p + \partial\Omega_f$ in which $\partial\Omega_V$ is the displacement increment rate boundary, $\partial\Omega_p$ is the tractive boundary, and $\partial\Omega_f$ is the internal unilateral boundary (such as a joint or fracture bed).

Set A_V is defined as the kinematically admissible set of velocity $\boldsymbol{V}^*$ and the strain rate $\dot{\boldsymbol{S}}^*$:

$$A_v = \{\boldsymbol{V}^*, \dot{\boldsymbol{S}}^* \mid \boldsymbol{V}^* = \bar{\boldsymbol{V}}, \text{ on } \partial\Omega_V;\ f(\boldsymbol{V}^*) < \infty, \text{ on } \partial\Omega_f;\ \dot{\phi}(\dot{\boldsymbol{S}}^*) < \infty, \text{ in } \Omega\},$$

where $\bar{\boldsymbol{V}}$ is the known velocity vector.

We adopt the approximate rate form of the equilibrium equation to the current configuration (ref. Section 9.7):

$$\dot{\sigma}^i_j\|_i + \rho\dot{f}_j = 0. \tag{10.134}$$

Multiplying Equation (10.134) by $(V^{*j} - V^j)$, we have

$$\int_\Omega \dot{\sigma}^i_j\|_i (V^{*j} - V^j)\mathrm{d}\Omega + \int_\Omega \rho\dot{f}_j (V^{*j} - V^j)\mathrm{d}\Omega = 0. \tag{10.135}$$

By Green-Gauss's integral theorem,

$$\int_\Omega \dot{\sigma}^i_j\|_i (V^{*j} - V^j)\mathrm{d}\Omega = \int_{\partial\Omega} \dot{\sigma}^i_j n_j (V^{*j} - V^j)\mathrm{d}\Gamma - \int_\Omega \dot{\sigma}^i_j (\dot{S}^{*j}_i - \dot{S}^j_i)\mathrm{d}\Omega. \tag{10.136}$$

Substituting Equation (10.136) into Equation (10.135), then

$$\int_\Omega \dot{\sigma}^i_j (\dot{S}^{*j}_i - \dot{S}^j_i)\mathrm{d}\Omega = \int_\Omega \rho\dot{f}_j (V^{*j} - V^j)\mathrm{d}\Omega + \int_{\partial\Omega} \dot{P}_j (V^{*j} - V^j)\mathrm{d}\Gamma,$$
$$\forall \boldsymbol{V}^*, \boldsymbol{V} \in A_v. \tag{10.137}$$

From Equations (10.133), (10.134) and (10.137), we obtain (Xie, 1987a):

$$\begin{aligned} &\int_\Omega \dot{\phi}\uparrow(\dot{\boldsymbol{S}}, \dot{\boldsymbol{S}}^* - \dot{\boldsymbol{S}})\mathrm{d}\Omega + \int_{\partial\Omega_f} \boldsymbol{J}\uparrow(\langle \boldsymbol{V} \rangle, \boldsymbol{V}^* - \boldsymbol{V})\mathrm{d}\Gamma \\ &\quad \geq \int_{\partial\Omega_p} \dot{\boldsymbol{P}}(\boldsymbol{V}^* - \boldsymbol{V})\mathrm{d}\Gamma + \int_\Omega \rho\dot{\boldsymbol{f}}(\boldsymbol{V}^* - \boldsymbol{V})\mathrm{d}\Omega, \quad \boldsymbol{V}^*, \dot{\boldsymbol{S}}^* \in A_V. \end{aligned} \tag{10.138}$$

Here,

$$A_V = \{\boldsymbol{V}^* \in [W^{1,p}_{(\Omega)}]^3,\ \dot{S}^{*i}_j = [V^{*i}\|_j + V^{*i}\|^T_j]/2 \text{ in } \Omega;\ \boldsymbol{V}^* = \bar{\boldsymbol{V}} \text{ on } \partial\Omega_V\},$$

where '↑' is the upper subderivate (Rockafeller, 1980). $W^{1,p}_{(\Omega)}$ is the Sobolev space of order 1, p.

Domain Ω and its boundary $\partial\Omega$ are divided into a mesh of finite elements and $\partial\Omega_f$ is discretized as thickless Goodman's joint elements. We can express the potential power functional of the system as follows:

$$\pi(\boldsymbol{V}) = \sum_{m=1}^{M} \dot{\phi}_m\uparrow(\dot{\boldsymbol{S}}_m, \dot{\boldsymbol{S}}_m) + \sum_{n=1}^{N} \boldsymbol{J}\uparrow_n(\langle \boldsymbol{V}_n \rangle, \boldsymbol{V}_n) - \dot{\boldsymbol{R}}^T\boldsymbol{V};$$

where M is the number of domain elements, N is the number of joint elements, and $\boldsymbol{R}$ is the vector of the equivalent nodal force rate.

The solution of the problem is now to find the substationary value of $\pi(\boldsymbol{V}^*)$ with respect to A_V at V.

$$\pi(\boldsymbol{V}) = \min_{V \in [W^{1,p}_{(\Omega)}]^3} \{\pi(\boldsymbol{V}^*) \mid \boldsymbol{V}^* \in A_V\}. \tag{10.139}$$

Equation (10.139) is the minimal formula of the finite element, which is valid for the continuous and/or discontinuous deformation of rock masses.

When creep damage and large deformation of rock presented in the previous section is considered, and the initial critical value of the damage is defined as one of the failure criteria of stress, then

$$\dot{w} = 0, \quad w = 0, \quad \text{if } \sigma \in K;$$

and

$$w = I - \exp[-B|Y - Y_0|^{1/n}],$$
$$\dot{w} = B|\dot{Y}|(I - w)/[n|Y - Y_0|^{(n-1)/n}], \quad \text{if } \sigma \notin K. \qquad (10.140)$$

Here

$$K = \{\sigma|(\sigma_1 - \sigma_3) + (\sigma_1 + \sigma_3)\sin 2\phi \leq C\cos\phi\},$$

where C and ϕ are the cohesion and friction angle of the rock, respectively.

Letting

$$\psi(\langle V\rangle, V) = \sum_{n=1}^{N} J_n \uparrow(\langle V\rangle_n, V_n).$$

then Equation (10.138) can be rewritten as follows (Xie, 1987a):

$$({}_kV)^T K_N^{cw} \Delta_k u + \psi(\langle V\rangle \Delta t, V) - ({}_kV)^T(\Delta R + \Delta F_w + \Delta F_w^c) \leq 0. \qquad (10.141)$$

Inequalities in the Equation (10.141) are herein solved by the 'trial and error' method (Panagiotopoulos in 1980). The method proposed by Panagiotopoulos consists of the transformation governed by equalities. Through iteration, with optimality solutions u satisfying certain conditions the solution of the problem can be obtained. These considerations are all compiled in the programs NEPROD and NOCRED.

10.5.3 *Numerical simulations*

10.5.3.1 *Analysis of reverse sliding deformation caused by depression fracture in the Lianyuan area of China*

The Lianyuan area in China is serially depressed by northwest-southeast lateral compression, and can be analyzed by geological data relative to the structural character of the region. Under the conditions specific to the area, the inherent extrusive reverse sliding fracture bed forms are caused by lateral compression of tectonic stress. A model of finite element analysis is established in Figure 10.62. Undeformed and

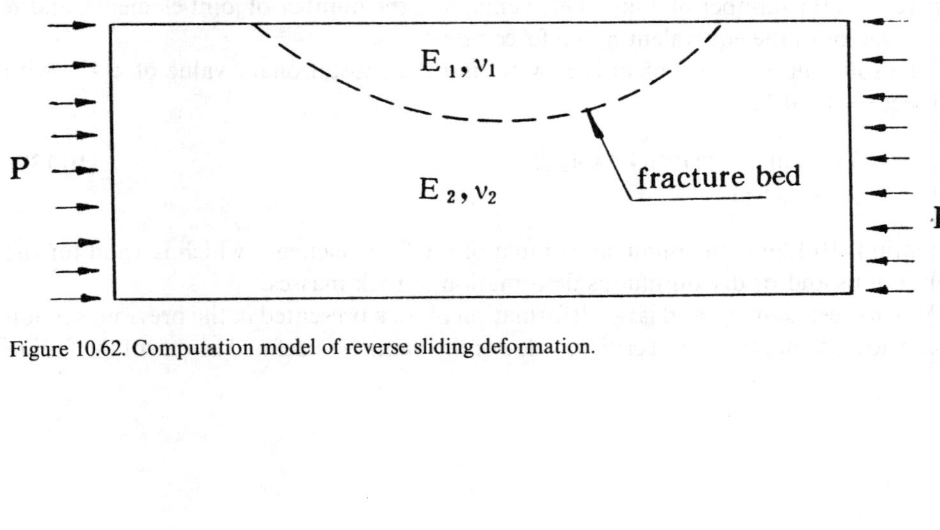

Figure 10.62. Computation model of reverse sliding deformation.

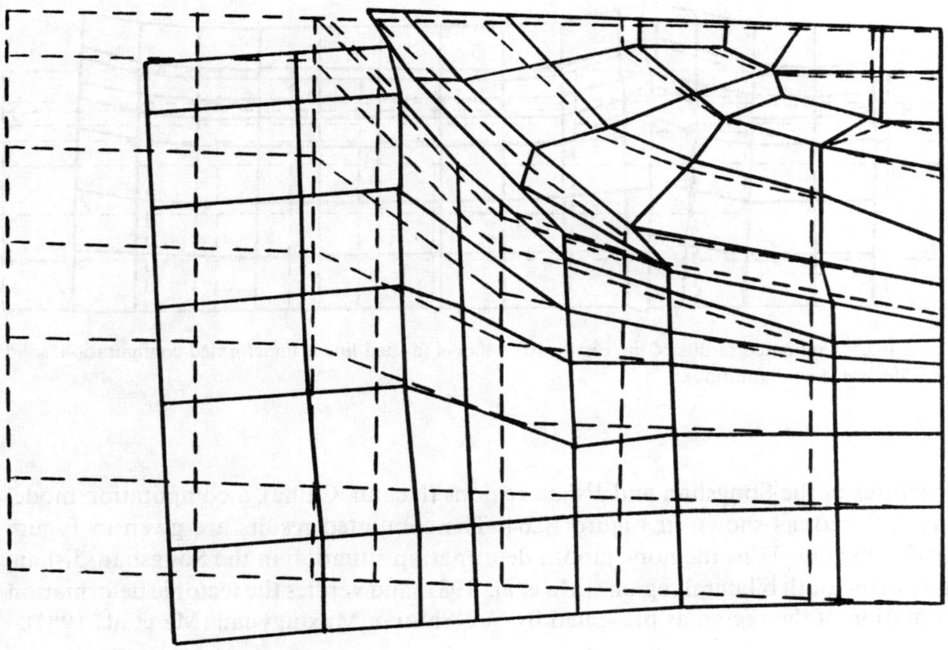

Figure 10.63. Undeformed (dotted line) and deformed (solid line) configuration of Figure 10.62 model.

deformed configurations (symmetrical) obtained by numerical simulations are shown in Figure 10.63. From Figure 10.63, the effect of extrusive structures is clear. The shortened low weak bed is much larger than the upper strong bed. The non-smooth deformation at fracture bed is simulated well.

10.5.3.2 *Modelling non-smooth deformation of shallow beds caused by fracture beds in deep rock masses*

Geological research indicates that fracture structures form early in deep rock masses and strongly influence the deformation of shallow beds over time. To study shallow

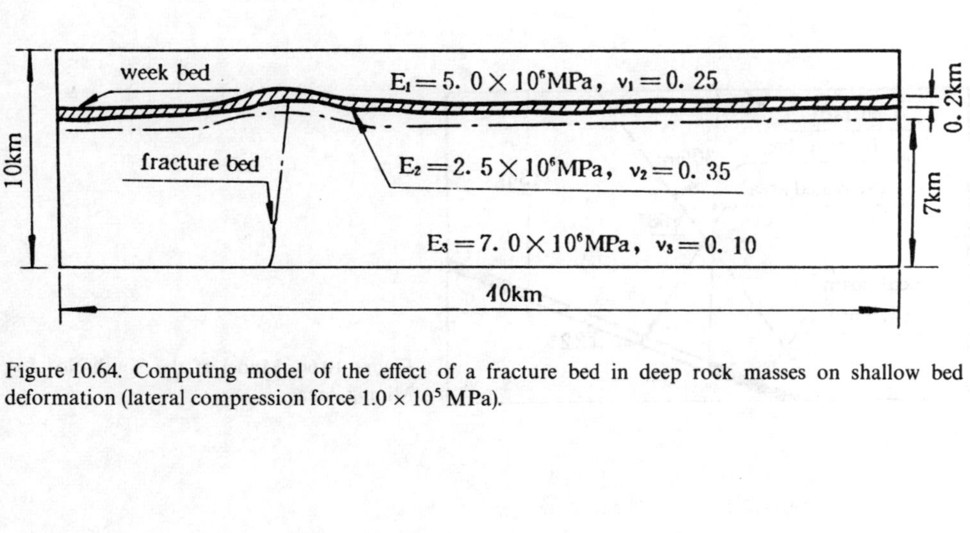

Figure 10.64. Computing model of the effect of a fracture bed in deep rock masses on shallow bed deformation (lateral compression force 1.0×10^5 MPa).

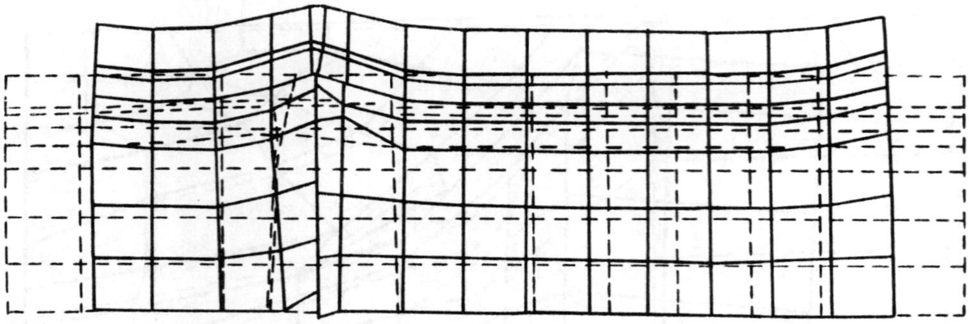

Figure 10.65. Computed results of the Figure 10.64 model (dotted line is undeformed configuration, solid line is deformed configuration).

structures in the Songshan and Jishan regions (Henan, China), a computation model was designed, as shown in Figure 10.64. The computed results are given in Figure 10.65, which explains the non-smooth deformation situation in the Songshan, Jishan, and north-south bilateral regions (Ma et al., 1981) and verifies the tectonic deformation estimation of the region as presented by Suo Shutian Maxingyuan (Ma et al., 1981).

10.5.3.3 *Analysis of non-smooth deformation, creep damage and fracture caused by excavation in jointed rock masses*

In underground engineering, the balance of stress in surrounding rock is disturbed after mining, so that rock movement, deformation, damage and fracture occur, particularly under longwall cavings. Movement, deformation, damage and fracture of the surrounding rock are not only closely connected to active mining, but are also affected by geological structures (such as joints, and fracture beds). In this section, we apply the theories presented in the previous sections to the analyses of non-smooth deformation, creep damage and fracture caused by excavations beneath jointed rock masses in the South No.2 Mining area, Qinghemen Mine, Fuxin, China as shown in Figure 10.66. Using the program NOCRED (in which the Nishihara creep model is considered), the non-smooth deformation, creep damage and fracture of this problem

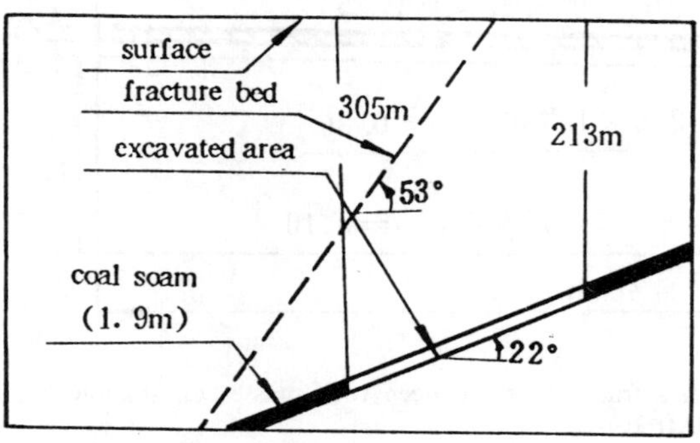

Figure 10.66. Model of the South-No. 2 Mining Area.

have been simulated. Figures 10.67a-c shows the non-smooth deformations near the fracture bed and subsidence caused by excavation.

It is seen that the results of the numerical simulation considering damage and creep damage are much more in agreement with measured values than those of numerical simulation of elasto-plasticity with the same mechanics parameters. The non-smooth deformation at the fracture bed is clearly characterized by the non-smooth mechanics theory presented in this section. The maximum surface movement is strongly

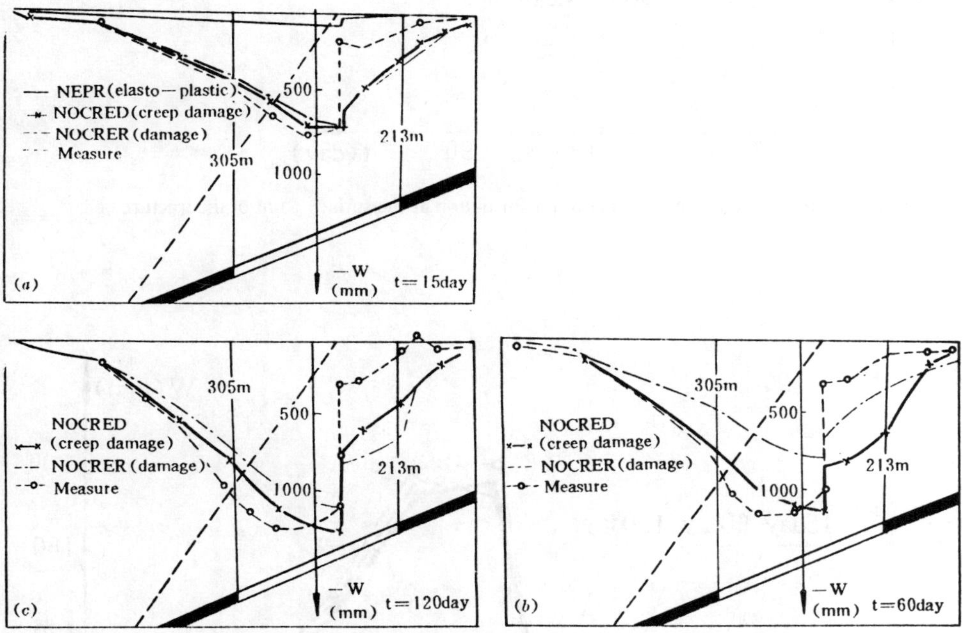

Figure 10.67. Surface movements caused by excavation. (a) $t = 15$ day after mining; (b) $t = 60$ day after mining; and (c) $t = 120$ day after mining.

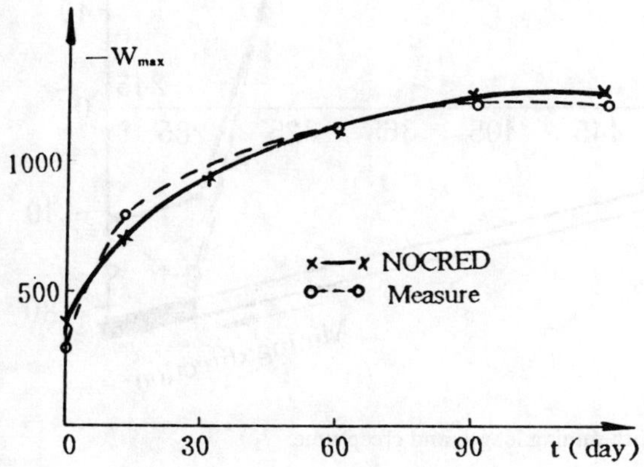

Figure 10.68. The creep curve of maximum value W_{max} of subsidence.

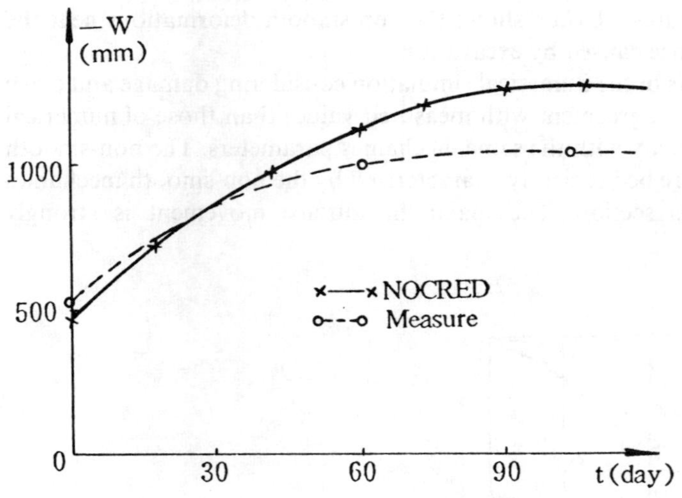

Figure 10.69. The creep curve of non-smooth deformation at the surface point of the fracture bed.

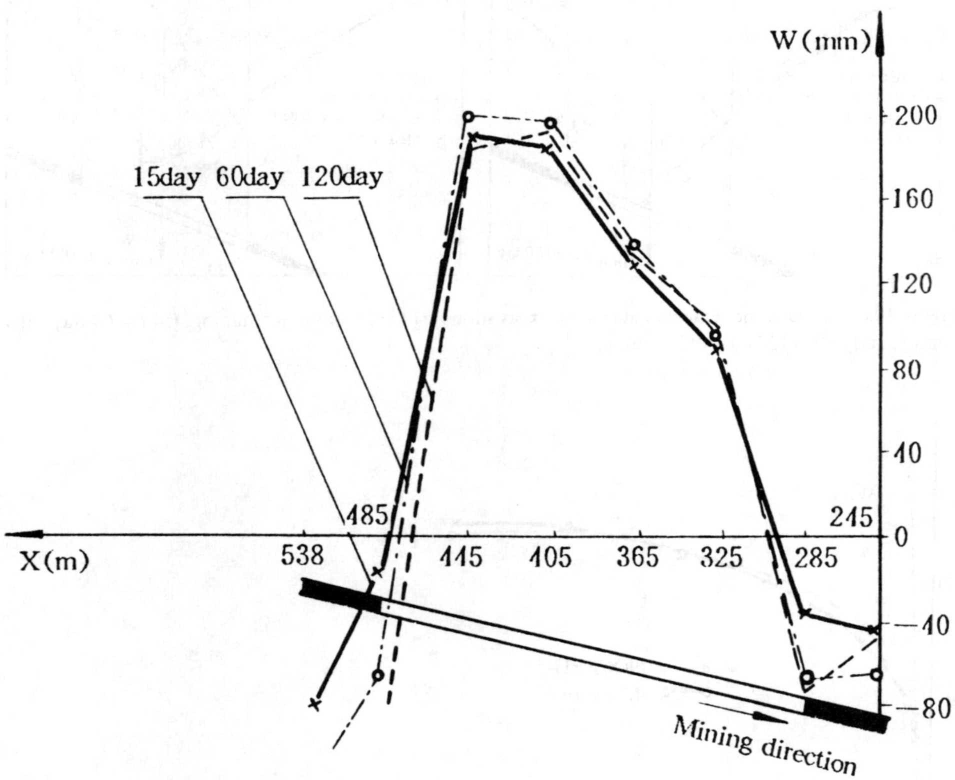

Figure 10.70. The change of bottom swelling with mining length and creep time.

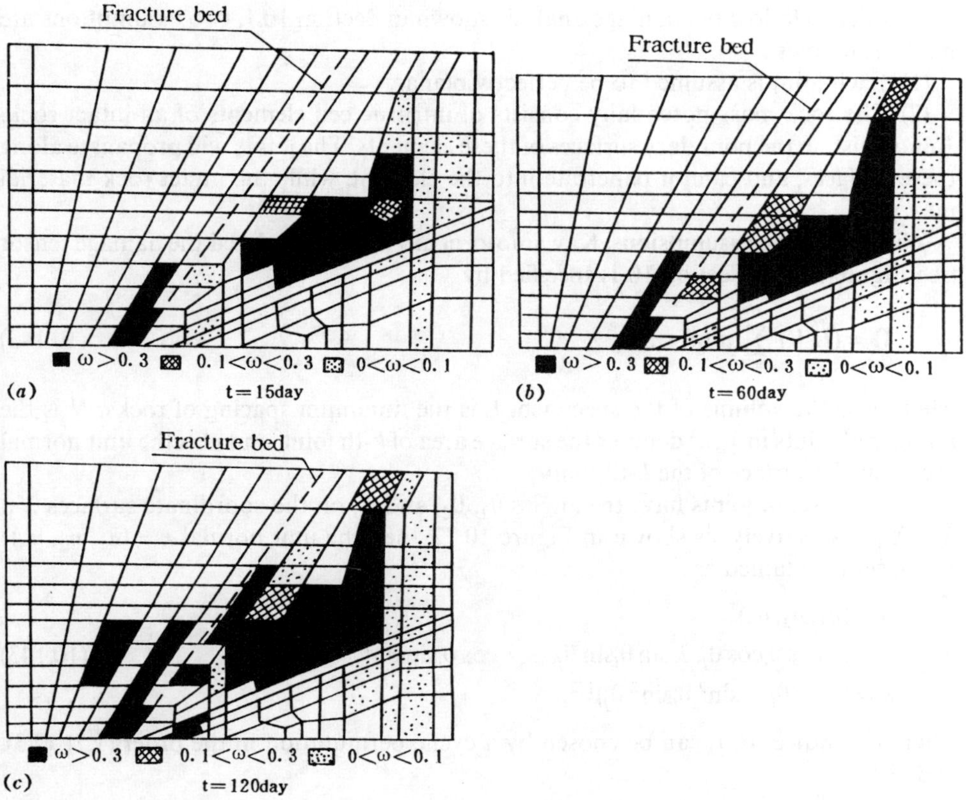

Figure 10.71. The damage evolutional situation of the upper strata caused by excavation: (a) $t = 15$ day after mining; (b) $t = 60$ day after mining; and (c) $t = 120$ day after mining.

dependent on creep time as illustrated in Figure 10.68. The jumpy value of non-smooth deformation at the surface point of the fracture bed also behaves as a creep feature (Figure 10.69). Figure 10.70 shows the bottom swelling phenomenon which changes with mining length and creep time. Figure 10.71 shows the damage evolutional situations caused by excavation. It is obvious that damage over the roof of the workout area is serious, and the damage region increases with creep.

10.6 THE ANALYSIS OF DAMAGE MECHANICS OF DISCONTINUOUS ROCK MASSES

Recently, Kawamoto, et al. (1988), Tao, et al. (1989) have applied a damage mechanics theory to the analyses of discontinuous rock masses. Discontinuities in rock masses are characterized by a damage tensor (a second-order symmetric tensor). The damage tensor is determined from in situ data of joint sets, in which the rock mass itself is considered as an ideal material.

In order to follow the damage analysis shown in Section 10.1, two assumptions are made as follows:

(1) Each joint is assumed to be perfectly planar;

(2) The rock mass potentially consists of intrinsic cell elements of an intact rock. Joints exist on the boundary surfaces of these elements. The joints will propagate along these surfaces, and cannot panetrate into the element, while the intact rock material may yield.

With the above assumpsions, Kawamoto, et al. (1988) considered the damage tensor (see Figure 10.4 in Section 10.1) specified by

$$\mathbf{\Omega} = (L/V) \sum_{k=1}^{N} a^k(n^k * n^k) \tag{10.142}$$

where V is the volume of the specimen, L is the minimum spacing of rocks, N is the number of joints in V, a^k denotes the surface area of k-th joint, and n^k is the unit normal vector at the surface of the k-th joint.

Letting a set of joints have the angles θ_1, θ_2, and θ_3 on the coordinate surfaces X_1, X_2, X_3, respectively, as shown in Figure 10.72, then the unit normal $\boldsymbol{n} = (n_1, n_2, n_3)^{\mathrm{T}}$ of the set is obtained as

$$\begin{aligned} \boldsymbol{n} &= (n_1, n_2, n_3)^{\mathrm{T}} \\ &= (\lambda \cos\theta_i \cos\theta_j,\ \lambda \sin\theta_i \sin\theta_j,\ -\lambda \cos\theta_i \sin\theta_j)^{\mathrm{T}} \\ \lambda &= (\cos^2\theta_i + \sin^2\theta_i \sin^2\theta_j)^{1/2}, \end{aligned} \tag{10.143}$$

where the indices (i, j) can be chosen by a cyclic permutation in the order (1,2), (2,3), and (3,1).

Figure 10.72. Observation of cracks on rock mass surfaces (Kawamoto et al., 1988).

For each angle, we have the following relationship

$$1/[\tan\theta_i \tan\theta_j] = -\tan\theta_k,$$

where $(i, j, k) = (1, 2, 3), (2, 3, 1)$, or $(3, 1, 2)$. Since the unit normal of a single set of joints is specified by Equation (10.143), the damage tensor for this set will be determined if the number and trace length are obtained.

Consider a cube element in local coordinates as indicated in Figure 10.73. The number of joints along the X'_1-axis will be estimated as

$$\bar{N}_{12} = N'_1(V^{1/3}/L'_2),$$

where N'_1 is the number of joints on the X'_1-surface, and L'_2 is the average length of joints appearing on the X'_2-surface. Similarly, along the X'_2-axis, we have

$$\bar{N}_{21} = N'_2(V^{1/3}/L'_1).$$

Thus, the number of joints in this cube is

$$\bar{N} = V^{1/3}(N'_1 N'_2/L'_1 L'_2)^{1/2},$$

and the average surface area of a joint is

$$\bar{a} = L'_1 L'_2.$$

Since $L'_1 L'_2 = L_1 L_2$ and $N'_i = N_i/(1 - n_i^2)^{1/2}$ $(i = 1, 2)$, the number of joints in the cube is estimated as

$$\bar{N} = V^{1/3}\{N_1 N_2/[L_1 L_2((1 - n_1^2)(1 - n_2^2))^{1/2}]\}^{1/2}. \tag{10.144}$$

Substituting Equation (10.144) into Equation (10.142), the damage tensor of a single set of joints is given by

$$\mathbf{\Omega} = \frac{L}{V^{2/3}}\left[\frac{N_1 N_2 L_1 L_2}{\sqrt{[(1 - n_1^2)(1 - n_2^2)]}}\right]^{1/2} (\boldsymbol{n} * \boldsymbol{n}).$$

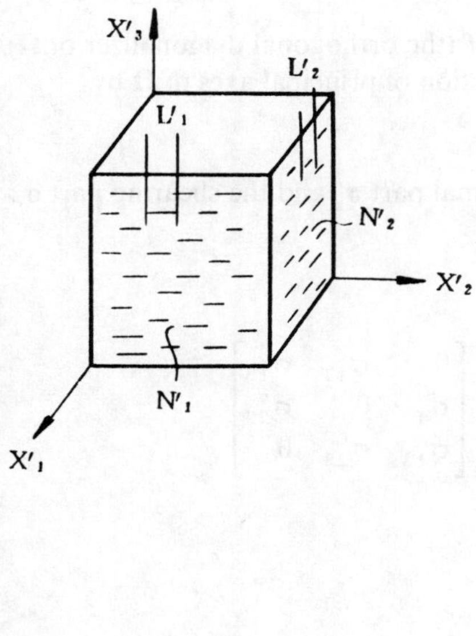

Figure 10.73. Cube element in local coordinates (Kawamoto et al., 1988).

For the general case, if $N_i > N_j > N_k$, the damage tensor is given by

$$\boldsymbol{\Omega} = \frac{L}{V^{2/3}}\left[\frac{N_i N_j L_i L_j}{\sqrt{[(1-n_i^2)(1-n_j^2)]}}\right]^{1/2}(\boldsymbol{n}*\boldsymbol{n}), \quad (i, j\text{:not summed}). \tag{10.145}$$

If a rock mass involves plural sets of joints, we can separately determine the unit normals and the damage tensors for the set. The global damage tensor is then obtained by summing up the sets.

Suppose that on the surface X_1; we have angles θ_{a1}, θ_{b1}, and θ_{c1} and on surface X_2; we have angles θ_{a2}, θ_{b2} and θ_{c2}. The angles θ_{a3}, θ_{b3} and θ_{c3} appearing on surface X_3 are written in the lower row of the table. Computing $1/(\tan\theta_{\alpha 1}\tan\theta_{\beta 2})$ and $-\tan\theta_{\tau 3}$ ($\alpha, \beta, \tau = a, b,$ or c), then, $1/(\tan\theta_{b1}\tan\theta_{a2})$ coincides with $\tan\theta_{c3}$, and the combination (θ_{b1}, θ_{a2}, θ_{c3}) gives one unit normal. The damage tensor of this set is calculated by Equation (10.145)

If there exist N-sets of joints and Ω_i ($i = 1, \ldots, N$) is the damage tensor of the i-th set, the global damage tensor is given as

$$\boldsymbol{\Omega} = \sum_{i=1}^{N} \Omega_i.$$

Note that if the joints are randomly distributed, the global damage tensor is written as

$$\boldsymbol{\Omega} = [1/(3V^{2/3})][N_i N_j L_i L_j)^{1/2}\boldsymbol{I}, \quad (i, j\text{: not summed});$$

and the effect of damage is then isotropic. If joints are parallel to unit vector $\boldsymbol{J} = (J_1, J_2, J_3)$, the damage tensor is expressed as

$$[\Omega_{ij}] = B\begin{bmatrix} 1-J_1^2 & -J_1J_2 & -J_1J_3 \\ & 1-J_2^2 & -J_2J_3 \\ \text{Sym.} & & 1-J_3^2 \end{bmatrix}$$

where

$$B = [1/(2V^{2/3})][N_i N_j L_i L_j]^{1/2}, \quad (i, j\text{: not summed}).$$

The effective stress tensor σ_e in the principal axis of $\boldsymbol{\Omega}$ is defined as

$$\boldsymbol{\sigma}'_e = \boldsymbol{\sigma}'_t(\boldsymbol{I} - C_t\boldsymbol{\Omega}')^{-1} + \boldsymbol{\sigma}'_n[H\langle\boldsymbol{\sigma}'_n\rangle(\boldsymbol{I}-\boldsymbol{\Omega}')^{-1} + H\langle-\boldsymbol{\sigma}'_n\rangle(\boldsymbol{I} - C_n\boldsymbol{\Omega}')^{-1}],$$

where $\boldsymbol{\Omega}'$ is the diagonal tensor of $\boldsymbol{\Omega}$. By using $\boldsymbol{T}$ (the orthogonal diagonalizer of $\boldsymbol{\Omega}$), the Cauchy stress $\boldsymbol{\sigma}$ is transformed into the direction of principal axes of $\boldsymbol{\Omega}$ by

$$\boldsymbol{\sigma}' = \boldsymbol{T}\boldsymbol{\sigma}\boldsymbol{T}^{\mathrm{T}}.$$

Then, we can separate this stress $\boldsymbol{\sigma}'$ into the normal part $\boldsymbol{\sigma}'_n$ and the shearing part $\boldsymbol{\sigma}'_t$:

$$\boldsymbol{\sigma}' = \boldsymbol{\sigma}'_n + \boldsymbol{\sigma}'_t.$$

This is written in a matrix form as

$$\begin{bmatrix} \sigma'_{11} & \sigma'_{12} & \sigma'_{13} \\ \sigma'_{12} & \sigma'_{22} & \sigma'_{23} \\ \sigma'_{13} & \sigma'_{23} & \sigma'_{33} \end{bmatrix} = \begin{bmatrix} \sigma'_{11} & 0 & 0 \\ 0 & \sigma'_{22} & 0 \\ 0 & 0 & \sigma'_{33} \end{bmatrix} + \begin{bmatrix} 0 & \sigma'_{12} & \sigma'_{13} \\ \sigma'_{12} & 0 & \sigma'_{23} \\ \sigma'_{13} & \sigma'_{23} & 0 \end{bmatrix}$$

where $\boldsymbol{H}\langle\cdot\rangle$ is the switching operator:

$$(H\langle\sigma_{ij}\rangle)_{ij} = \begin{cases}\sigma_{ij}, & \text{if } \sigma_{ij} > 0 \\ 0, & \text{if } \sigma_{ij} \leqslant 0\end{cases}; \quad (i, j\text{: not summed}).$$

Thus, the effective stress tensor produced by damage in a rock mass is transformed by

$$\boldsymbol{\sigma}_e = \boldsymbol{T}^{\mathrm{T}}\boldsymbol{\sigma}'_e\boldsymbol{T}.$$

The coefficients C_n and C_t have values between 0 and 1.0, which characterized the rate of transmission of compression and shearing stress components, respectively, throughout a crack or a joint.

This damage model can be solved by following finite element equations:

$$\boldsymbol{Ku} = \boldsymbol{F} + \boldsymbol{F}^*;$$
$$\boldsymbol{K} = \int_{\Omega} \boldsymbol{B}^{\mathrm{T}}\boldsymbol{EB}\,\mathrm{d}\Omega,$$

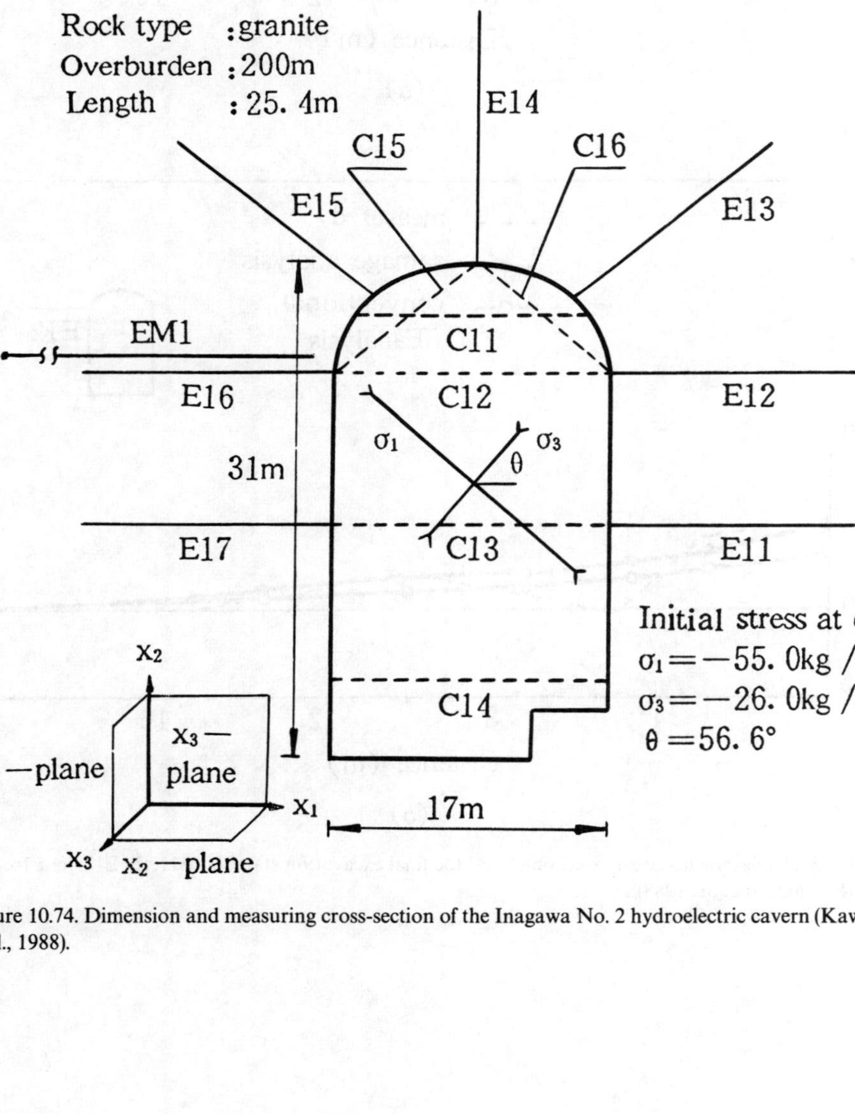

Figure 10.74. Dimension and measuring cross-section of the Inagawa No. 2 hydroelectric cavern (Kawamoto et al., 1988).

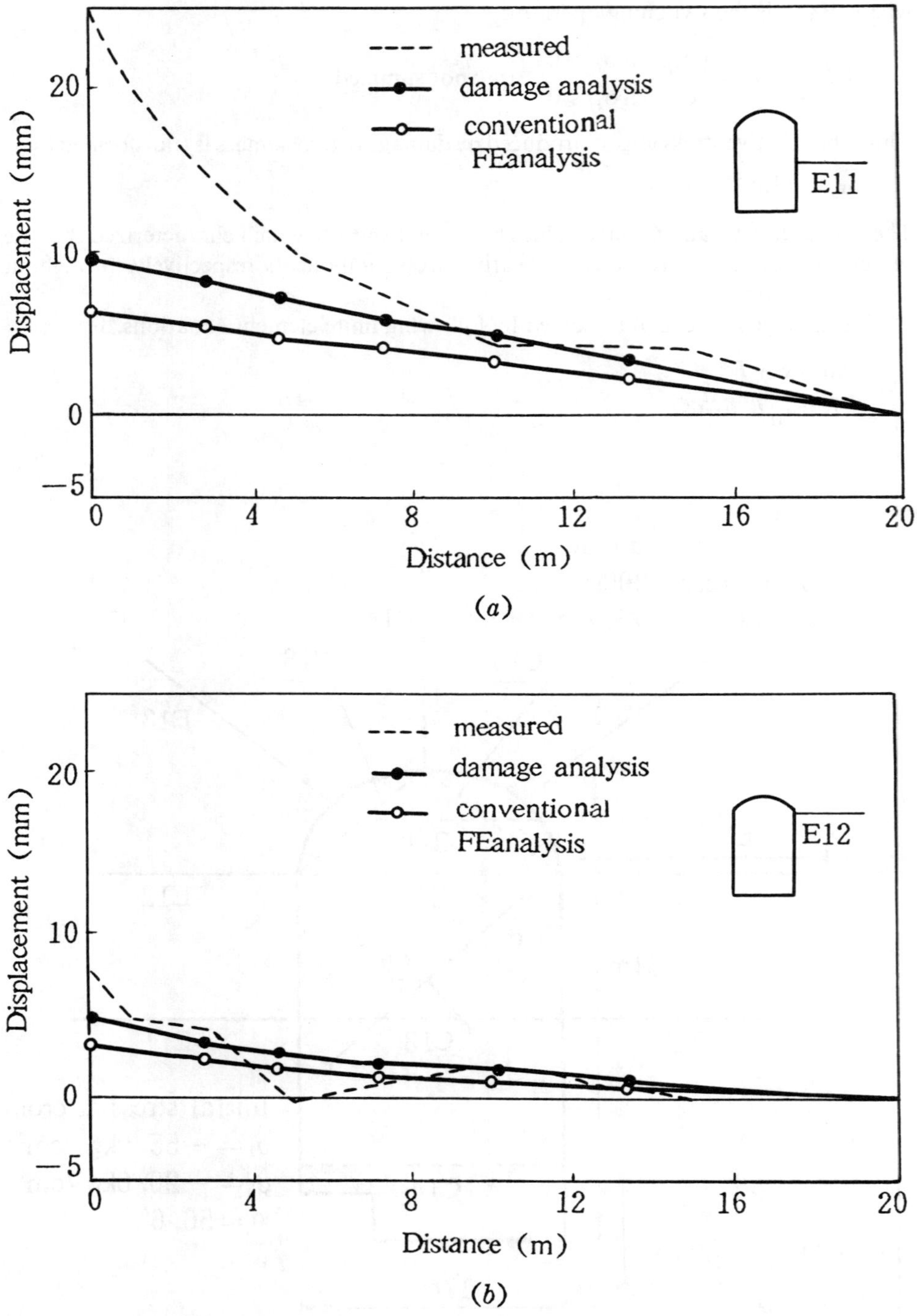

Figure 10.75. Displacements of an extensometer at the final excavation stage: (a) E11; (b) E12; (c) E16; and (d) E17 (Kawamoto et al., 1988).

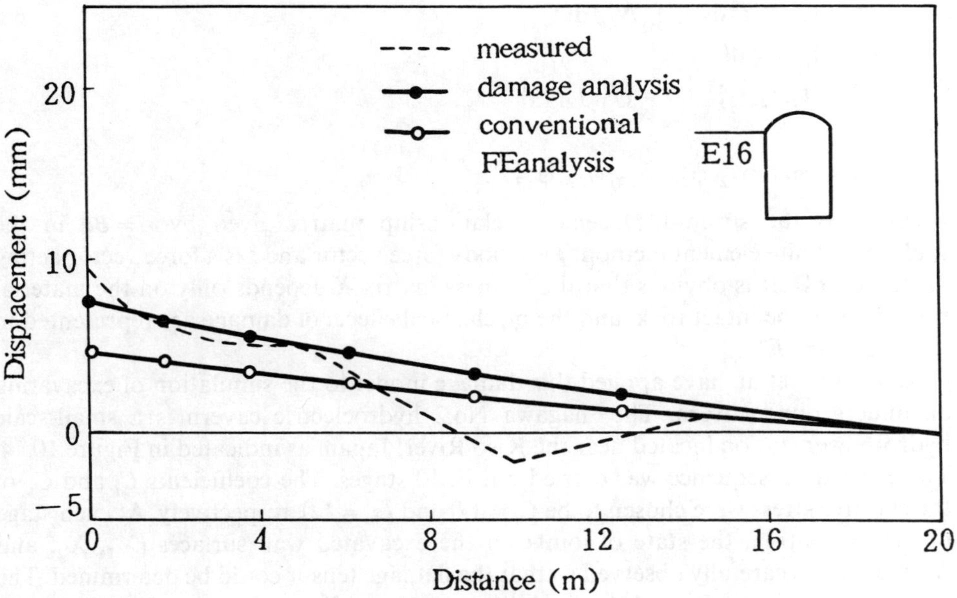

(*c*)

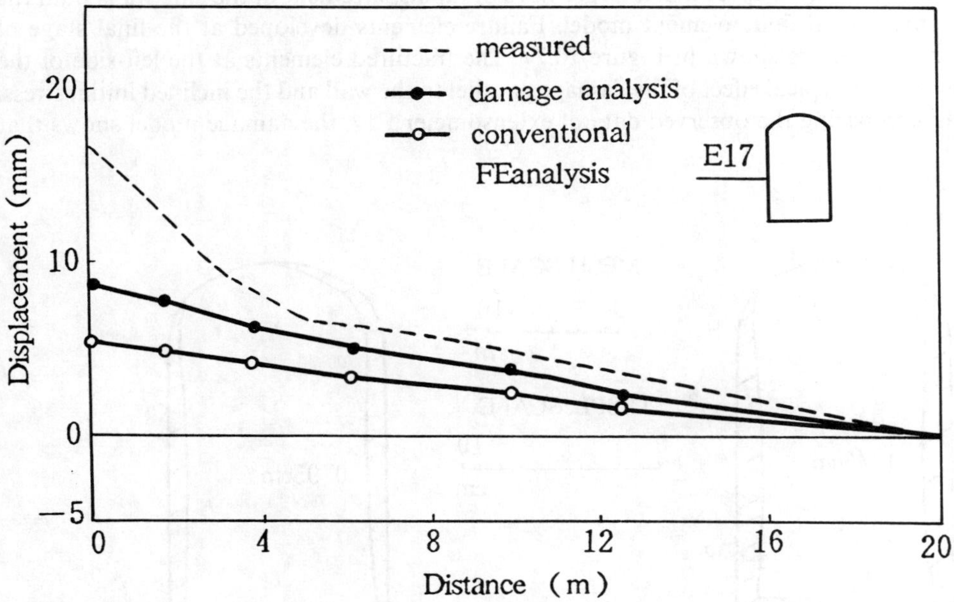

(*d*)

Figure 10.75. continued

$$F = \int_{\partial\Omega} N^{\mathrm{T}} t^0 \mathrm{d}s_t + \int_{\Omega} N^{\mathrm{T}} f \mathrm{d}\Omega,$$
$$F^* = \int_{\Omega} B^{\mathrm{T}} \phi \mathrm{d}\Omega,$$
$$t^0 = \{t_1, t_2, t_3\}^{\mathrm{T}}, \quad f = (f_1, f_2, f_3\}^{\mathrm{T}},$$

and

$$\phi = \{\phi_{11}, \phi_{22}, \phi_{33}, \phi_{12}, \phi_{23}, \phi_{13}\}^{\mathrm{T}},$$

where B is the strain-displacement relationship matrix given by $\varepsilon = Bu$ in the traditional finite element method. f is a body force vector and t is a force vector acting on surface $\partial\Omega$. It is obvious that the stiffness matrix K depends only on the material properties of the intact rock, and the mechanical effects of damage are represented in the force term F^*.

Kawamoto et al. have applied this damage model to the simulation of excavating an underground cavern. The Inagawa No.2 hydroelectric cavern is a small-scale hydro-power station located near the Kiso River, Japan, as indicated in Figure 10.74. The excavation sequence was carried out in 10 stages. The coefficients C_t and C_n of the effective stress were chosen to be $C_t = 1.0$ and $C_n = 1.0$, respectively. At each stage of the excavation, the state of joints on the excavated wall surfaces (X_1, X_2, and X_3-plane) was carefully observed so that the damage tensor could be determined. The damage analysis results and the final displacements of extensometers E11, E12, E16 and E17 are shown in Figure 10.75. Deformation models of the wall surface are computed as shown in Figure 10.76 at the final state of excavation. It is clear that the damage analysis gives much larger horizontal displacement of the side walls than the conventional finite element model. Failure elements developed at the final stage of excavation are shown in Figure 10.77. The fractured elements at the left-side of the wall are a typical effect of the damage parallel to the wall and the inclined initial stress. In comparing the observed data of extensometer E17, the damage model shows that

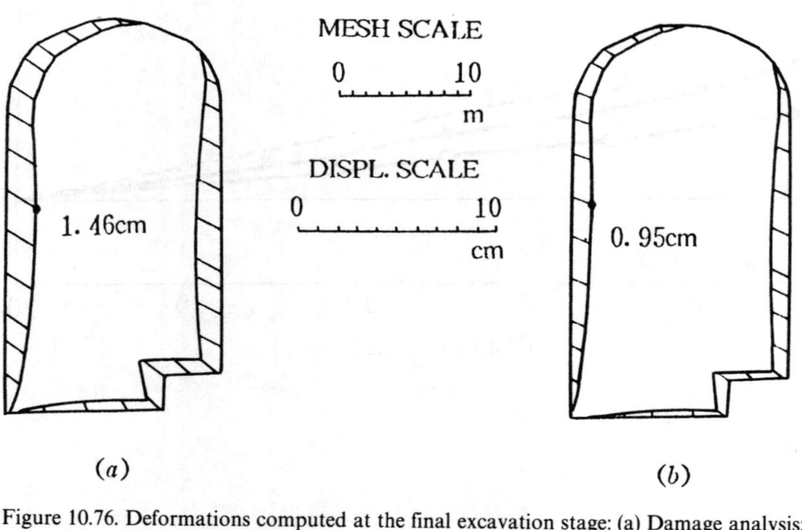

Figure 10.76. Deformations computed at the final excavation stage: (a) Damage analysis; and (b) Conventional finite element analysis (Kawamoto et al., 1988).

Figure 10.77. Fractured elements, predicted by damage analysis (Kawamoto et al., 1988).

a loosening zone is developed with the fractured elements, whereas in conventional finite element analysis, no elements reach failure.

Tao et al. (1989) have also established a damage model of jointed rock mass and simulated the stress-strain relationship of jointed rock mass by this model. They defined the damage tensor as

$$\boldsymbol{\Omega} = (1/V) \sum_{k=1}^{N} \int_{\partial\Omega} \boldsymbol{b}^{(k)} * \boldsymbol{n}^{(k)} \mathrm{d}S_{(k)},$$

where $S_{(k)}$ is the surface area of k-th joint and N is the joint number included in volume V. $\boldsymbol{b}^{(k)}$ is the deformation vector at a point on $S_{(k)}$. $\boldsymbol{n}^{(k)}$ is the unit-normal vector of $S_{(k)}$. Obviously, this second-order tensor is asymmetric. Symmetrizing it,

$$\phi_{ij} = (\Omega_{ij} + \Omega_{ji})/2;$$

that is,

$$\boldsymbol{\phi} = [1/(2V)] \sum_{k=1}^{N} \int_{\partial\Omega} [\boldsymbol{b}^{(k)} * \boldsymbol{n}^{(k)} + \boldsymbol{n}^{(k)} * \boldsymbol{b}^{(k)}] \mathrm{d}S_{(k)}.$$

By visually, inspecting the equation, it is assumed that the shear deformation in the i-direction of the joint in the j-direction equals the shear-deformation in the j-direction of the joint in the i-direction, and this conforms to the deformation characters of joints in the rock mass.

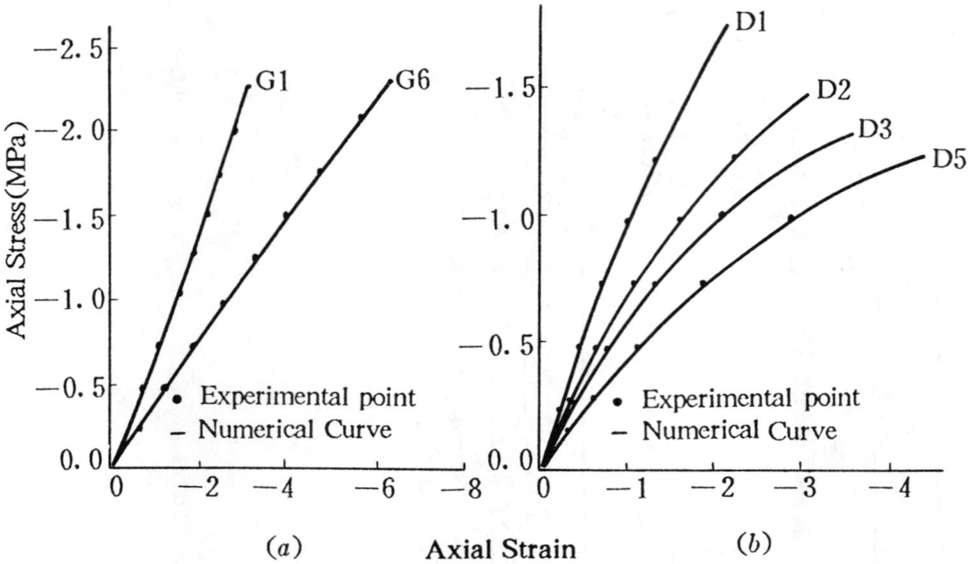

Figure 10.78. The stress-strain relationship of modelling jointed rock mass (tests and numerical); (a) Specimen G1 and G6; (b) Specimen D1–D5 (Tao et al., 1989).

Using the constitutive equations and this damage model, Tao et al. have performed a numerical computation for the unit-compression testing information of a modelled jointed rock mass (plaster, sand, and mica sheets), where the damage is considered as isotropy, the damage tensor is a diagonal tensor, and all elements are the same. The results of these tests and numerical computations are illustrated in Figure 10.78. It is clear to see that the numerical curves are consistent with the test values.

10.7 DYNAMIC DAMAGE ANALYSIS IN BRITTLE ROCK

The wave propagation problem and dynamic loading in a rock mass are of great interest in the fields of underground and geotechnical engineering for the earthquake resistant design of large structures, such as underground power stations, dams, etc. However, dynamic behaviors are very complicatedly refracted and/or reflected at the surface of internal microcracks. Difficulties in solving problems arise from both complex boundary conditions caused by these phenomena and the complicated geological setting of the cracks. It may be said that, there is still no available mechanical model for a body with distributed cracks in the dynamic state.

Recently, the damage mechanics theory has been applied to dynamic analysis in brittle rock. Davison & Stevens (1973) developed a theory of spallation of a brittle thermoelastic body, and of the interaction between propagation waves and accumulated damage. Grady & Kipp (1980) have found that a model coupling fracture, fragmentation, and stress wave propagation and based on the activation and growth of an initial Weibull distribution of fracture-producing flaws effectively described the observed rate-dependent fracture phenomena from static to high strain-rate impulse loading.

They have given continuum modeling of explosive fracture in oil shale by means of damage theory.

When dynamic loading involves the propagation of stress waves, the basic conservation laws of momentum and energy must be incorporated. Neglecting body forces the balance of linear momentum requires that

$$\partial\sigma_{ij}/\partial x_j - \rho\partial V_i/\partial t = 0,$$

where ρ is the mass density and V_i is the particle velocity component. Neglecting thermal and radiative sources, energy conservation yields

$$(\partial\phi/\partial t) - \sigma_{ij}\cdot\partial V_i/\partial x_j = 0,$$

where ϕ is the internal energy. For states of tensile stress and strain, it is assumed that material response will be governed by an isotropic elastic energy function

$$\phi = (\tfrac{1}{2})K(1-w)\varepsilon_{kk}^2 + 2G(1-w)[\varepsilon_{ij}\varepsilon_{ij} - (\tfrac{1}{3})\varepsilon_{kk}^2].$$

Thermodynamic restrictions require the stress to be given by

$$\sigma_{ij} = (\partial\phi/\partial\varepsilon_{ij})|_{w=\text{const.}} = K(1-w)\varepsilon_{kk}\delta_{ij} + 2G(1-w)[\varepsilon_{ij} - (\tfrac{1}{3})\varepsilon_{kk}\delta_{ij}],$$

and further

$$(\partial\phi/\partial w)|_{\varepsilon=\text{const.}}\,\dot{w} \leq 0$$

to satisfy energy dissipation ($Y = \partial\phi/\partial w$).

The activation of flaws could be based on a Griffith criterion due to rapid loading. The total damage at time t is defined by Grady & Kipp (1980) as

$$w = \textstyle\int_0^t \dot{N}(\tau)V(t-\tau)\mathrm{d}\tau,$$

where $\dot{N}$ is the rate of flaw activation, and $V(t-\tau) = (\tfrac{3}{4})\pi C_g^3(t-\tau)^3$. The differential equation governing damage growth has been given by Grady & Kipp in the form

$$\dot{w} = \{8\pi(m+3)^2/[(m+1)(m+2)]\}^{1/3}C_g[n(\varepsilon_{kk})]^{1/3}w^{2/3};$$

where the flaw distribution $n(\varepsilon_{kk})$ is evaluated on the elastic tensile volume strain level, and m and C_g are fracture parameters of rock materials.

Grady & Kipp have applied this coupled stress-wave propagation and damage constitutive model to a one-dimensional cylindrical geometry in which simultaneous loading of an infinitely long, 20 cm dia, cylindrical borehole in oil shale is considered. Dynamic loading is determined by prescribing a pressure-time history on the borehole wall of the form

$$P(t) = P_0(t/t_0)\mathrm{e}^{-t/t_0},$$

where P_0 and t_0 are parameters which characterize the amplitude and duration of the loading pulse, respectively. Material away from the borehole wall is subsequently loaded by the propagation of the compressive stress wave. The outward impulse imparted by the compressive wave leads to both radial and tangential tension at later times in the wave and is the source of tensile fracture damage. Figure 10.79 shows wave profiles for two calculations of the cylindrical problem. The first solution corresponds to the strictly linear elastic response of the medium without damage while, in the second, the constitutive description of damage and fracture for oil shale from the above

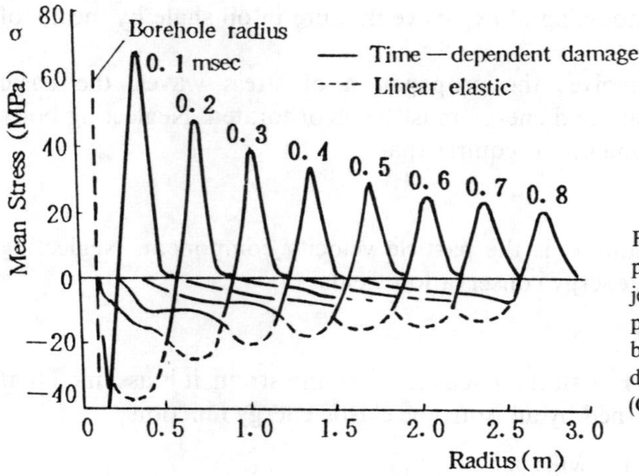

Figure 10.79. Cylindrical wave propagation from 0.2 m dia borehole subjected to 0.2 GPa and 25μsec loading pulse as defined in the text. Calculations based on both linear elastic and time-dependent fracture response are shown (Grady & Kipp, 1980).

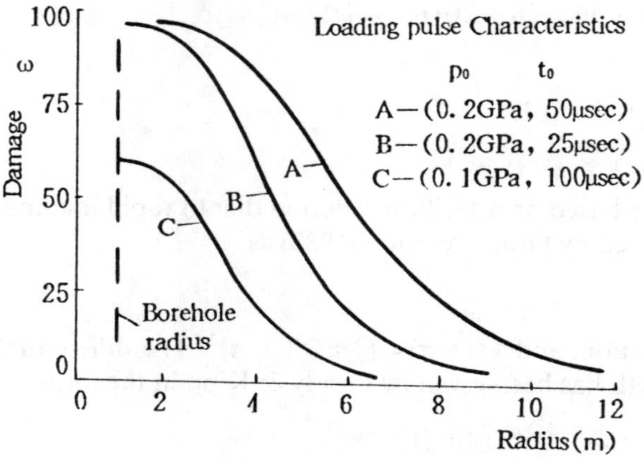

Figure 10.80. Calculated damage distribution around 0.2 m dia cylindrical borehole subjected to different dynamic load conditions (Grady & Kipp, 1980).

model has been applied. In either solution, the initial compressive wave is the same since the fracture signal cannot propagate fast enough to influence the initial outward impulse imparted to the material. In tension, however, the maximum stress calculated in the elastic solution is not achieved in the damage solution due to the onset of fracture. Instead, the tensile wave achieves a peak stress, relaxes slightly, and then remains at a relatively constant level for some characteristic time duration before relaxing rapidly to zero. It is during this characteristic time that crack initiation and growth occurs prior to the onset of catastrophic failure. In Figure 10.80, profiles depicting the calculated radial extent of damage are shown for differing loading pulse conditions.

Grady & Kipp (1980) have also applied the damage model to two-dimensional calculations. Results of a calculation for a 7.6 m long, 0.15 m dia, explosively loaded borehole oriented normal to a free face in oil shale are illustrated in Figure 10.81. The explosive was represented at the borehole end, away from the free face. Damage contours are indicated at a time close to terminal damage growth. The calculation shows a region of extensive, 'near-borehole' damage due to circumferential fracturing,

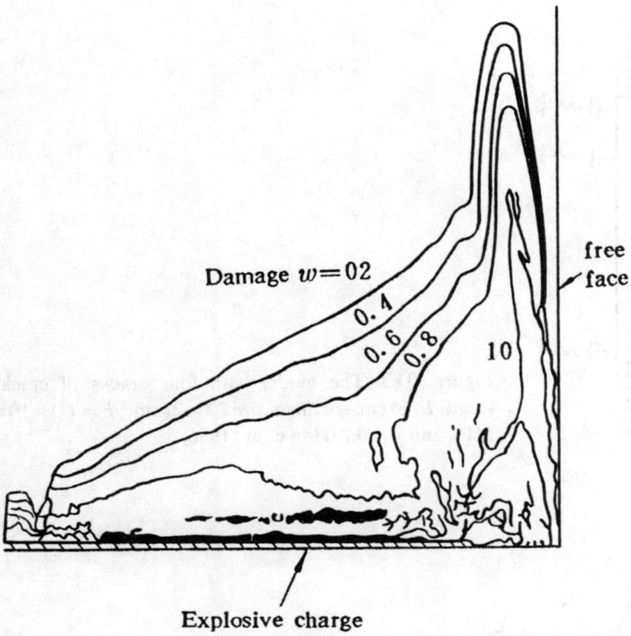

Figure 10.81. Two-dimensional (cylindrical symmetry) fracture damage calculation of an end detonated explosive loaded borehole oriented normal to a free face (Grady & Kipp, 1980).

and a zone of 'spall' tensile damage near the free surface due to the constructive interference of incident and reflected waves.

Taylor et al. (1986) and Seaman et al. (1985) have discussed damage accumulation due to dynamic microcracking in brittle rock and the relationship between dynamic tensile microfracture and fragmentation.

Han et al. (1986) have presented an application of damage mechanics theory to the problem of elastic plane waves travelling in jointed rock mass. For a dynamic problem, they defined the constitutive law as

$$\boldsymbol{\sigma} = \boldsymbol{E}^*\varepsilon + \psi(\varepsilon, \boldsymbol{\Omega}, t),$$
$$\boldsymbol{E}^* = \boldsymbol{P}\boldsymbol{E}$$
$$\boldsymbol{P} = 1/[(\phi_{11}+\phi_{22})(\phi_{11}\phi_{22}-\phi_{12}^2)]$$
$$\times \begin{bmatrix} \phi_{22}(\phi_{11}+\phi_{22})-\phi_{12}^2 & \phi_{12}^2 & -2\phi_{22}\phi_{12} & 0 \\ \phi_{12}^2 & \phi_{11}(\phi_{11}+\phi_{22})-\phi_{12}^2 & -2\phi_{11}\phi_{12} & 0 \\ -\phi_{22}\phi_{12} & -\phi_{11}\phi_{12} & 2\phi_{11}\phi_{22} & 0 \\ 0 & 0 & 0 & 2 \end{bmatrix}$$

where $\boldsymbol{\Phi} = [\phi_{11}, \phi_{12}, \phi_{22}]^{\mathrm{T}}$ is related to damage tensor $\boldsymbol{\Omega}$ defined in the previous section.

$$\boldsymbol{\Phi} = [\boldsymbol{I} - \boldsymbol{\Omega}]^{-1}.$$

Where the term ψ denotes the effect of viscous damping. This may be mainly caused

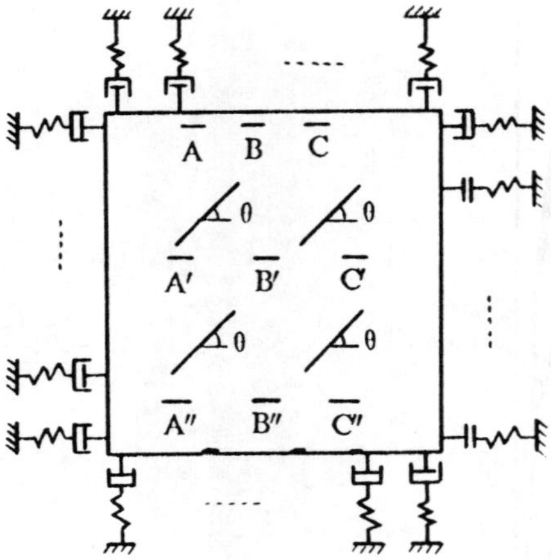

Figure 10.82. The model with four cracks of crack length $L = 5$ cm, specimen size $27 \times 27\,\text{cm}^2$, $E = 1.0 \times 10^2$ MPa and $\nu = 0.2$ (Han et al., 1986).

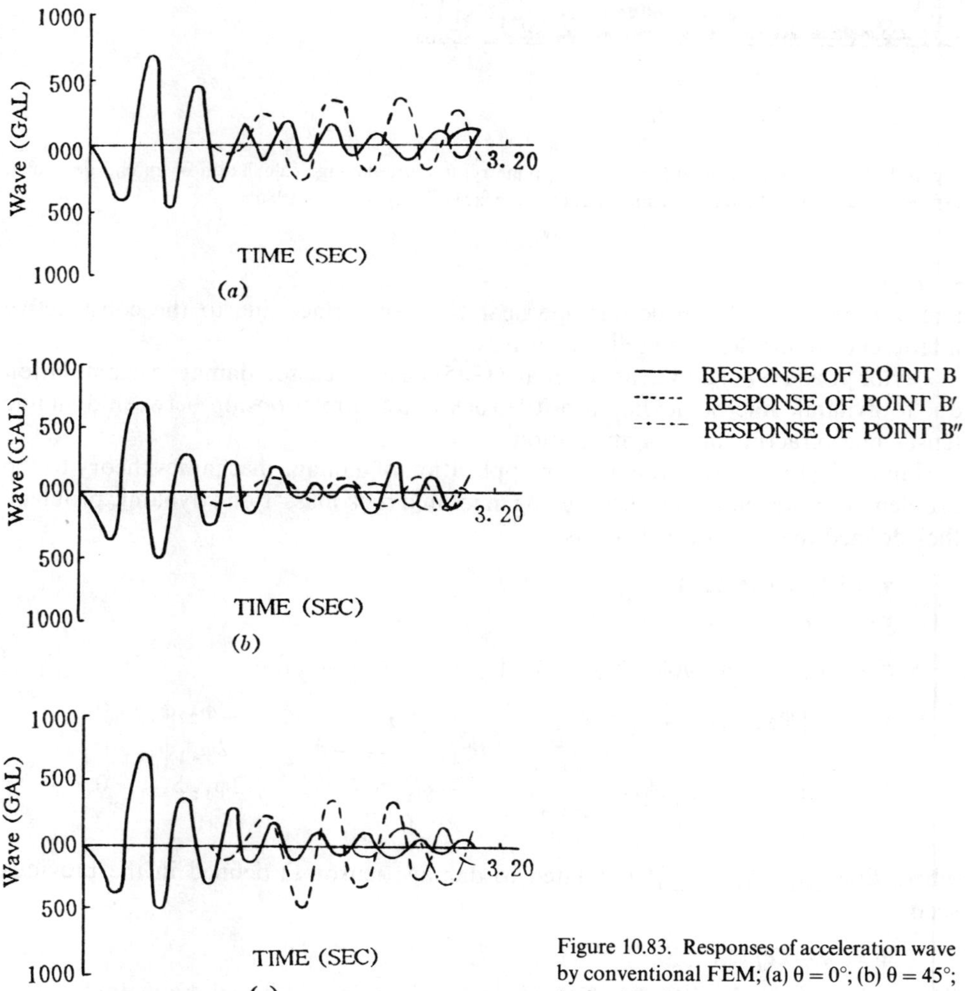

Figure 10.83. Responses of acceleration wave by conventional FEM; (a) $\theta = 0°$; (b) $\theta = 45°$; (c) $\theta = 90°$. (Han et al., 1986).

by internal scattering of the incident wave. If the viscous term ψ is disregarded, then the finite element equation of the problem is written as

$$\boldsymbol{M}\ddot{\boldsymbol{u}} + \boldsymbol{K}\boldsymbol{u} = \boldsymbol{F};$$

where $\boldsymbol{M} = \int_\Omega \rho \boldsymbol{N}^{\mathrm{T}}\boldsymbol{N}\mathrm{d}\Omega$, $\boldsymbol{K} = \int_\Omega \boldsymbol{B}^T\boldsymbol{E}^*\boldsymbol{B}\mathrm{d}\Omega$, and $\boldsymbol{F} = \int_\Omega \boldsymbol{N}^{\mathrm{T}}\boldsymbol{f}\mathrm{d}\Omega + \int_{\partial\Omega}\boldsymbol{N}^{\mathrm{T}}\boldsymbol{t}\mathrm{d}s$.

Han et al. designed an experimental model as shown in Figure 10.82 where the angles of cracks or joints are 0°, 45°, and 90°, and the viscous boundary is introduced at the four sides of the model. Then an acceleration shock wave

$$f(x) = \begin{cases} a_0, & t = 0 \\ 0, & t > 0 \end{cases},$$

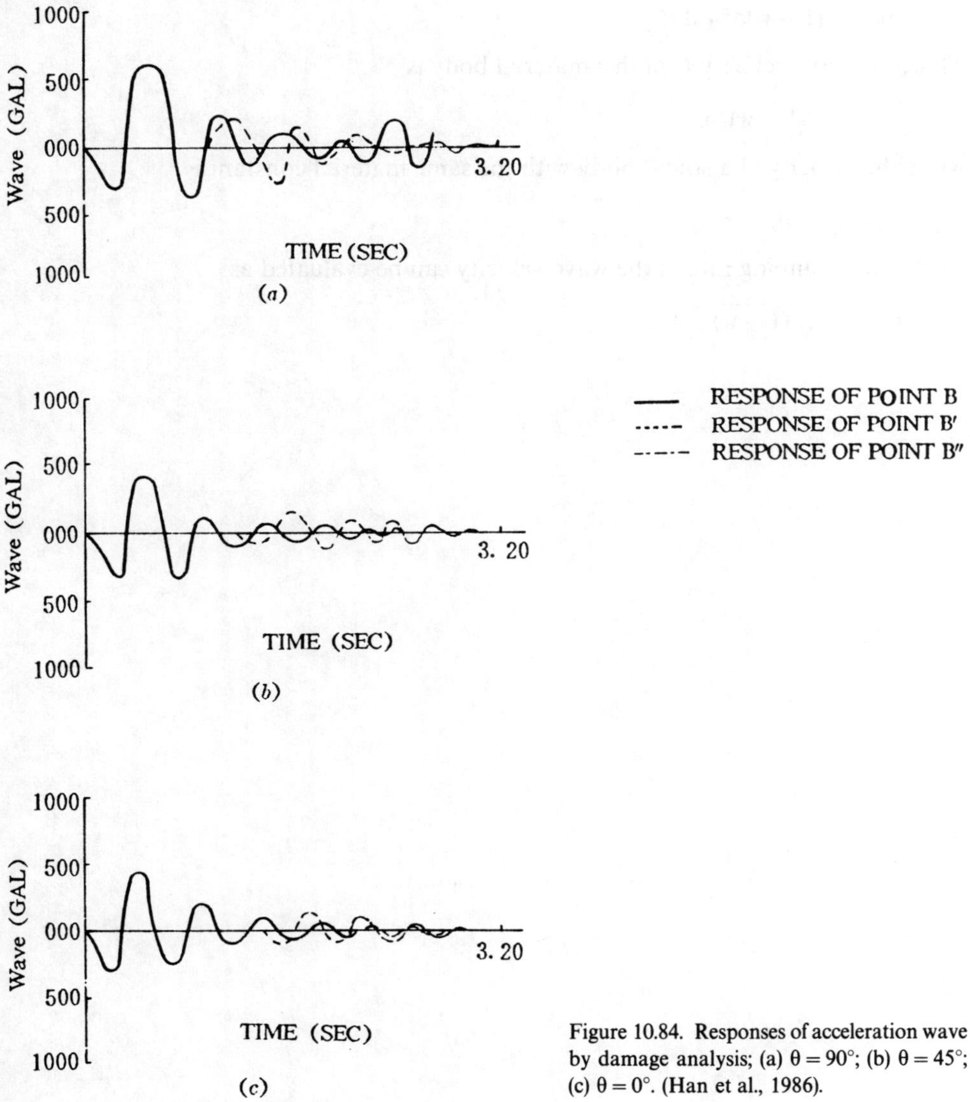

Figure 10.84. Responses of acceleration wave by damage analysis; (a) $\theta = 90°$; (b) $\theta = 45°$; (c) $\theta = 0°$. (Han et al., 1986).

is input at the top surface. In numerical analysis, the time integration scheme of Newmark's β method is used in both cases, and $\beta = \frac{1}{4}$ is chosen. The damage tensor corresponding to the crack angles 0°, 45° and 90° can be respectively obtained as

$$\Omega_{(0^0)} = \begin{bmatrix} 0 & 0 \\ 0 & 0.346 \end{bmatrix}; \quad \Omega_{(45^0)} = \begin{bmatrix} 0.137 & -0.137 \\ -0.137 & 0.137 \end{bmatrix}; \quad \Omega_{(90^0)} = \begin{bmatrix} 0.346 & 0 \\ 0 & 0 \end{bmatrix}.$$

The acceleration responses computed at each point shown in Figure 10.82 are given in Figure 10.83 by conventional FEM and in Figure 10.84 by the damage theory.

Another interesting study presented by Han et al. (1986) is the explanation of why the wave scatters around a crack causing a reduction in the amplitude and velocity of the propagating wave. This may be evaluated as the damping rate.

The one-dimensional wave equation of the elastic damage material can be written as

$$\rho\ddot{u} = E(1 - w)\mathrm{d}^2u/\mathrm{d}x^2.$$

Thus, the wave velocity C of this material body is

$$C = \sqrt{E(1 - w)/\rho},$$

while the velocity of a sound body with the same material constant is

$$C_s = \sqrt{E/\rho},$$

so that the damping rate of the wave velocity can be evaluated as

$$C/C_s = \sqrt{(1 - w)} \leq 1.$$

CHAPTER 11

Fractals and fragmentation of rock materials

As fracture happens because of an external stress system, rock is divided into smaller fragments with arbitrary shapes, and it also absorbs a certain amount of energy. The exploration of energy-size reduction dates back to the late-19th century. Many theories (such as the new surface hypothesis, the similarity hypothesis, and the third comminution theory) were proposed to solve engineering problems such as crushing, grinding, and blasting. These theories are all related to the distribution function of fragmentation which is commonly used in mining, geology, and mineral processing. A wide range of fragmentation phenomena have been discussed by Hartmann (1969). Fragmentation plays an important role in a variety of mining and geological phenomena. Donnison & Sugden (1984) have found that fragmentation controls the determination of the size- frequency relationship of asteroids and meteorites. A variety of statistical relations has been used to correlate data on the size distribution of fragments, and a simple power-law relationship is often used. Recently, Turcotte (1986a) and Zhao et al. (1990) have described these fragmentational distributions with power-law relationships in terms of the fractal theory. We have also proved this description by obtaining much experimental data on rock failure in the laboratory (Xie & Gao, 1991).

From Chapter 2, a definition of a fractal is given by the relationship between number and size. The number N (of objects) with a characteristic linear dimension greater than r is given by

$$N \propto r^{-D}. \tag{11.1}$$

Equation (11.1) is also called the power-law relationship of a fractal distribution (Equation (2.11)). In many cases, fragmentation results in a fractal distribution (Turcotte, 1986a). This is taken as evidence that the fragmentation mechanism is scale invariant. Fragments produced by weathering, explosions, and impacts often satisfy this fractal distribution condition over a wide range of scales. Most correlations for the number versus size for meteorites, asteroids, and interstellar grains also satisfy the fractal condition. In this chapter, we will mainly introduce the fractal distribution of fragmentation, a renormalization group approach to fragmentation presented by Turcotte (1986a), and our work about the description of fractal behavior during the process of rock rupturing (Xie, 1989d). The fractal analyses of the fragmentation of fine particles under different grinding mechanisms (Xie et al., 1993) will be discussed in Section 12.3.

11.1 FRACTAL DISTRIBUTION OF FRAGMENTATION

There are a variety of ways to represent the size-frequency distribution of fragments (Turcotte, 1986a). An exponential dependence can be written as

$$N(m)/N_T = \exp[-(m/\mu)^{\nu}], \qquad (11.2)$$

where $N(m)$ is the number of fragments with a mass greater than m, N_T is the total number of fragments, and μ is related to the average fragment mass. Brown et al. (1983) have provided a derivation of Equation (11.2) based on a Poisson distribution of random cracks. Many experimental results indicate that there is a simple power-law relationship in size-frequency distributions of fragments as given by,

$$N(\alpha \geq m) = Cm^{-b}, \qquad (11.3)$$

where C is a constant. Noting that $m \propto r^3$, from a comparison of Equation (11.3) with the fractal distribution defined in Equation (11.1), it follows that:

$$D = 3b. \qquad (11.4)$$

Thus, the power distribution is equivalent to a fractal distribution.

Figures 11.1 and 11.2 show fractal dimensions related to rock fragments under uniaxial compression and impaction.

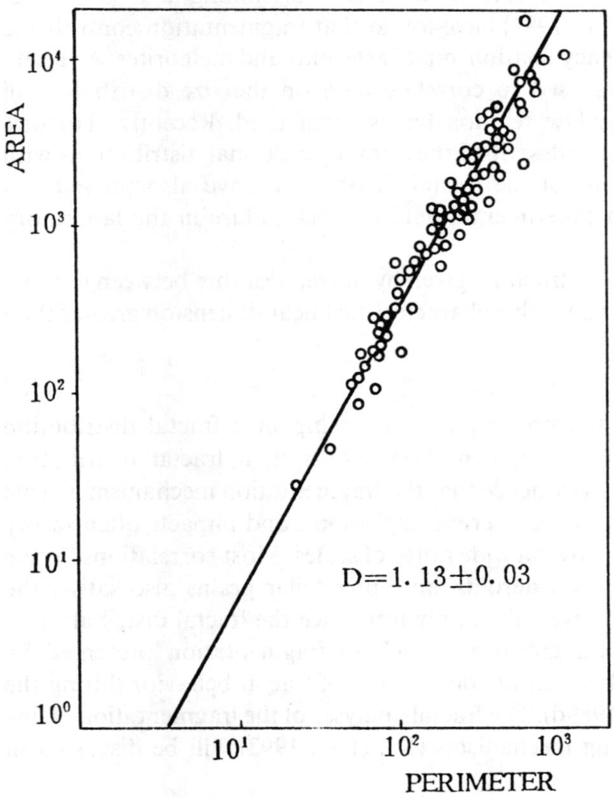

Figure 11.1. The fractal dimension of fragments of basalt under impacting and uniaxial compression (Kawakami et al., 1990).

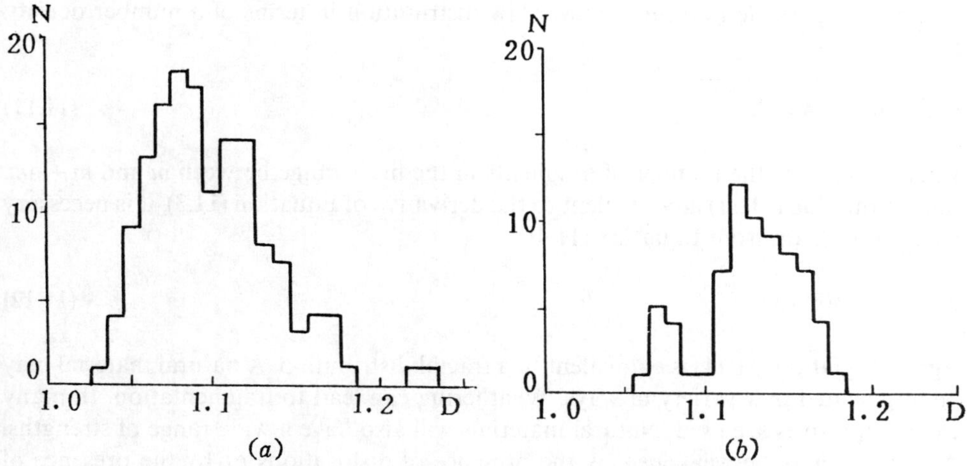

Figure 11.2. Histogram of fractal dimensions of fragments of basalt: (a) Under impacting; (b) Under uniaxial compression (Kawakami et al., 1990).

An alternative representation of the size-frequency distribution is the Weibull dependence, given by

$$M(r)/M_T = 1 - \exp[-(r/\sigma)^{\alpha}], \tag{11.5}$$

where $M(r)$ is the cumulative mass of fragments with a radius (volume $\frac{1}{3}$) less than r, M_T is the total mass, and σ is related to the mean size. If $r/\sigma \ll 1$, Equation (11.5) reduces to the power-law relationship

$$M(r)/M_T = (r/\sigma)^{\alpha}. \tag{11.6}$$

The power-law mass relationship (Equation (11.6)) can be related to the fractal number-size relationship (Equation (11.1)) as follows. Taking the derivative of Equation (11.6) gives

$$dM \propto r^{\alpha - 1}\, dr, \tag{11.7}$$

and taking the derivative of Equation (11.1) gives

$$dN \propto r^{-D-1}\, dr. \tag{11.8}$$

However, the incremental number is related to the incremental mass by

$$dN \propto r^{-3}\, dM. \tag{11.9}$$

Substitution of Equations (11.7) and (11.8) into Equation (11.9) gives

$$r^{-D-1} \propto r^{-3} r^{\alpha - 1} = r^{\alpha - 4},$$

or

$$D = 3 - \alpha. \tag{11.10}$$

The power-law mass distribution is equivalent to a fractal distribution.

It is also possible to write a power-law distribution in terms of a number density function,

$$f(m) = Am^{-s}, \tag{11.11}$$

where $f(m)\mathrm{d}m$ is the number of fragments in the mass range between m and $m + \mathrm{d}m$. Since Equation (11.11) is equivalent to the derivative of Equation (11.3), it is necessary that $b + 1 = s$ or, from Equation (11.4)

$$D = 3(s - 1). \tag{11.12}$$

Again, Equation (11.11) is equivalent to a fractal distribution. A natural material may be fragmented in a variety of ways. Weathering can lead to fragmentation. In many cases, explosives are used. Natural materials will also have a wide range of strengths. A material may be weakened by the presence of dislocations or by the presence of micro-cracks and joints. Hartmann (1969) has summarized a variety of data in terms of the power-law relationship (Equation (11.3)). Using Equation (11.4) to convert b to D, some of his results are summarized by Turcotte in Table 11.1.

A classic study of the size-frequency distribution for broken coal was carried out by Bennett (1936). Data from 13 seams in 6 collieries are averaged and given in Figure 11.3. The number of fragments with linear dimensions (determined from screen size) greater than r is given as a function of r. These data were obtained from coal broken directly from the face of the seam. The scatter in the data between seams was relatively small. It is seen that the distribution is in quite good agreement with Equation (11.1), and that the fractal dimension is $D = 2.50$.

Rock that has been fragmented by an explosion often satisfies a power-law size-frequency distribution. A variety of data have been summarized by Schoutens (1979) and by Okeefe & Ahrens (1985). The number-size distribution in the chimney rubble 43 m above the pulverizer nuclear explosion is also given in Figure 11.3. This

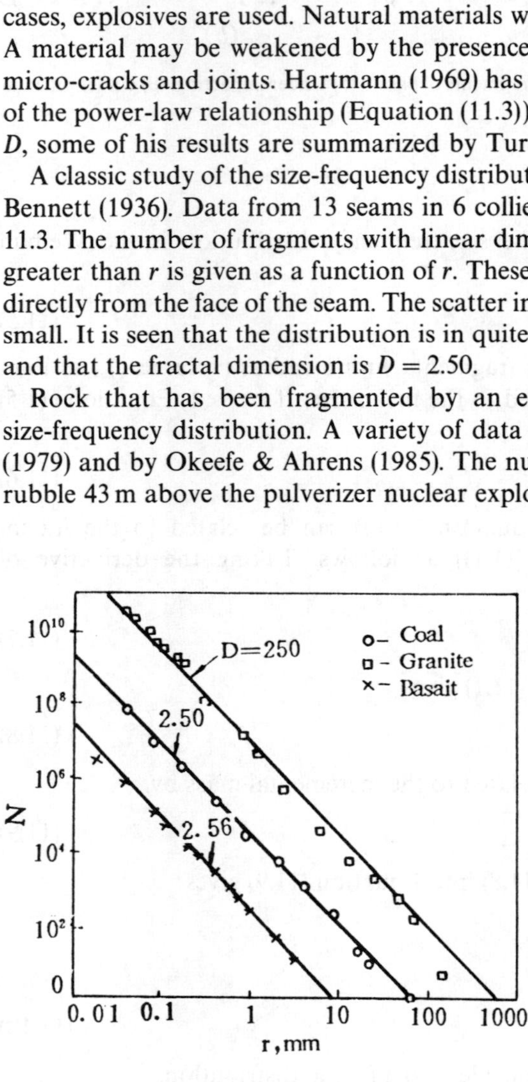

Figure 11.3. Dependence of the number of fragments with a linear dimension greater than r on r. Data for broken coal (Bennett, 1936), granite fragmented in an underground nuclear explosion (Schoutens, 1979), and a basalt fragmented by an impacting projectile (Fujiwara et al., 1977) (Turcotte, 1986a).

Table 11.1. Fractal dimensions for a variety of fragmented objects (Turcotte, 1986a).

Object	Reference	Fractal Dimension
Projectile fragmentation of gabbro with lead	Lange et al. (1984)	1.44
Projectile fragmentation of gabbro with steel	Lange et al. (1984)	1.71
Meteorites (Prairie Network)	McCrosky (1968)	1.86
Artificially crushed quartz	Hartmann (1969)	1.89
Plane of weakness model	Turcotte (1986a)	1.97
Disaggregated gneiss	Hartmann (1969)	2.13
Disaggregated granite	Hartmann (1969)	2.22
FLAT TOP I (chemical explosion, 0.2 kt)	Schoutens (1979)	2.42
Asteroids (theory)	Hellyer (1971)	2.48
PILEDRIVER (nuclear explosion, 61 kt)	Schoutens (1979)	2.50
Broken coal	Bennett (1936)	2.50
Interstellar grains	Mathis (1979)	2.50
Asteroids (theory)	Dohnanyi (1969)	2.51
Projectile fragmentation of basalt	Fujiwara (1977)	2.56
Sandy clays	Hartmann (1969)	2.61
Terrace sands and gravels	Hartmann (1969)	2.82
Pillar of strength model	Allegre et al. (1982)	2.84
Glacial till	Hartmann (1969)	2.88
Stony meteorites	Hawkins (1960)	3.00
Asteroids	Donison & Sugden (1984)	3.05
Ash and pumice	Hartmann (1969)	3.54

was a 61-kt event at a depth of 457 m in granite. The data are in good agreement with Equation (11.1), again taking $D = 2.50$.

Fujiwara et al. (1977) impacted a poly-carbonate projectile with a mass of 0.37 g and a velocity of 2.6 kms^{-1} against the surface of basalt rock. A crater was created, and the frequency-size relationship for the fragments was determined. Their data are also given in Figure 11.3, and a good fit with Equation (11.1) is obtained for $D = 2.56$. Lange et al. (1984) impacted high-velocity projectiles on indurated samples of San Marcos gabbro. For a 0.143 g stainless steel projectile with a velocity of 5.2 kms^{-1}, they found a good number-size correlation using Equation (11.1) with $D = 1.71$; for a 0.195 g lead projectile with a velocity of 4.6 kms^{-1}, they found a good correlation using Equation (11.1) with $D = 1.44$. The fractal dimensions associated with explosion and impact fragmentation are also summarized in Table 11.1. The statistics of the frequency-mass distribution of interplanetary objects has been reviewed by Dohnanyi (1972).

It is generally accepted that the frequency-mass distribution for asteroids and meteorites is due to fragmentation. The problem of multiple collisions with fragmentation has been studied by Dohnanyi (1969). He found that $s = 1.837$ ($D = 2.511$); this result was quite insensitive to the value of s assumed for fragmentation in an individual collision. This result was confirmed by Hellyer (1971), who obtained $s = 1.825$ ($D = 2.475$). The calculated values are slightly less than the observed values. It should be emphasized, however, that the fragmentation models depend on the assumed maximum fragment size and the assumed mass distribution of fragments.

11.2 RENORMALIZATION GROUP APPROACH TO FRAGMENTATION

The renormalization group (RG) theory was proposed by Wilson in 1936 and Wilson & Kogut (1974). The general ideas of the renormalization group approach may be formulated as follows (Wilson & Kogut, 1974; Fisher, 1974):

(1) The given (or initial) Hamiltonian H is transformed or 'renormalized' to obtain a new Hamiltonian H'; formally we write

$$H \rightarrow H' = RH. \tag{11.13}$$

(2) The renormalization group operator R acts to reduce the number of degrees of freedom (i.e., spin variables) from N to

$$N' = N/b^d,$$

where d is the dimensionality, while the 'spatial rescaling factor', b, exceeds unity. Different types of renormalization operator may be constructed. Frequently R is defined by means of 'a partial trace' over $(N - N')$ of the N spin variables, or over suitably 'transformed' variables (denoted by T'_r), so that the renormalized Hamiltonian itself is given by,

$$\exp(H') = T'_{rN-N'}\{\exp H\}. \tag{11.14}$$

(3) The essential condition to be satisfied by R is that the 'partition function' (ref. relative mathematical definition) must be satisfied, that is

$$Z_{N'}[H'] = Z_N[H]. \tag{11.15}$$

This follows from Equation (11.14), for example, merely by taking the remaining part of the trace over N' variables to calculate $Z_{N'}[H']$.

(4) In order to preserve the partial density of degree of freedom (i.e., the partial density of spins), all 'spatial vectors' entering into correlation functions, etc., are rescaled by the factor b according to

$$\boldsymbol{X} \rightarrow \boldsymbol{X}' = \boldsymbol{X}/b. \tag{11.16}$$

(5) Finally, in order to preserve the basic spin fluctuation magnitude, the renormalized 'spin vectors' are rescaled as

$$\boldsymbol{S}_x \rightarrow \boldsymbol{S}'_{x'} = \boldsymbol{S}_x/C, \tag{11.17}$$

where the spin rescaling factor C depends on H, i.e., $C = C(H)$. In more complex cases, different rescaling factors for different classes of spins may be needed.

Evidently the renormalization group operator depends on b and C. For the linear renormalization group (discussed here), one finds that the basic spin-spin correlation function transforms as

$$G[\boldsymbol{X}; H] = C^2 G[\boldsymbol{X}/b; H']. \tag{11.18}$$

Equations (11.15) and (11.18) ultimately lead to scaling properties – scaling invariance.

Although the renormalization group theory was at first independently proposed from physical phenomena for understanding phase transitions, it is currently almost identified with the self-similar fractal theory by the same feature of scaling invariance, and it is widely used as a mathematical tool for the analyses of fractals in nature. As discussed previously in Chapters 3-6, rock transport properties and rock failure (see Chapters 14-16), etc. behave as fractals. Therefore, the renormalization group theory is very useful in modeling rock properties.

Madden (1983) used the RG technique to relate the macroscopic electrical conductivity and fracture of rocks to the micro-crack population. Smalley et al. (1985) applied RG techniques to the failure of asperities under stress transfer in order to model earthquake instability. Allegre et al. (1982) applied RG techniques to study the coalescence of fractures and Turcotte (1986a) directly applied the approach of Allegre et al. (1982) to the fragmentation problem.

In order to model fragmentation, the material is hypothesized to develop micro-fracturing under applied stress. For simplicity, Turcotte considered a cube of material with a linear dimension of h, (as illustrated in Figure 11.4). This cube is referred to as a cell, which is divided into eight cubic elements, each with a dimension of $h/2$. Attention is now focused on one of the cubic elements, and it becomes a cell of dimension $h/2$ at order 1. This cell is then divided into eight first-order elements, each with a dimension of $h/4$, as is illustrated in Figure 11.4. The process is repeated at successively higher orders. Clearly, the division into eight cells is not unique. For example, each cubic cell could be divided into 64 cubic elements and the linear dimension of the elements would be one-quarter the linear dimensions of the cell.

In the application of the renormalization group to fragmentation, the assumption is made that the probability p_c that a cell will fragment into eight elements is the same at all orders. It is also hypothesized that the original body has a linear dimension of h and that the largest fragments have a linear dimension of $h/2$; these in turn fragment to give fragments with a linear dimension of $h/4$, and so forth. Considering m levels of fragmentation, the total number of particles is then

$$N_m = (1 - p_c)[1 + 8p_c + (8p_c)^2 + (8p_c)^3 + \cdots + (8p_c)^m]. \tag{11.19}$$

Similarly, the total number of particles with $m + 1$ levels of fragmentation is

$$N_{m+1} = (1 - p_c)[1 + 8p_c + (8p_c)^2 + (8p_c)^3 + \cdots + (8p_c)^m + (8p_c)^{m+1}]. \tag{11.20}$$

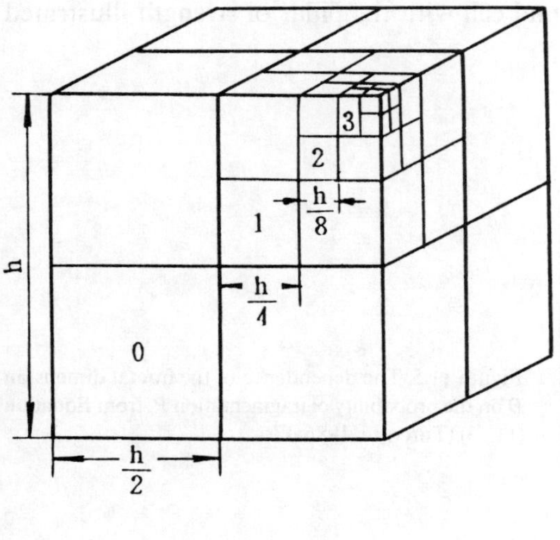

Figure 11.4. Illustration of the renormalization group approach to fragmentation (Turcotte, 1986a).

However, from Equation (11.1) we have

$$N_{m+1}/N_m = 2^D, \tag{11.21}$$

since the change in the linear dimension between levels m and $m+1$ is a factor of two. Combining Equations (11.19), (11.20) and (11.21), and assuming $m \gg 1$ and $8p_c > 1$, a very good approximation can be obtained as

$$8p_c = 2^D, \tag{11.22}$$

or

$$D = \ln(8p_c)/\ln 2. \tag{11.23}$$

In order to define a fractal dimension, the range allowed for the probability of fragmentation is $\frac{1}{8} < p_c < 1$, and the range allowed for the fractal dimension is $0 < D < 3$. The dependence of D on p_c is illustrated in Figure 11.5.

Next, the derivation given above for a cubic cell with dimension h that is divided into 64 cubic elements, each with dimension $h/4$, is repeated. If the probability of fracture at each order is taken to be P_c^2, the above derivation gives

$$64P_c^2 = 4^D, \tag{11.24}$$

and this is simply the square of Equation (11.22) and reduces to Equation (11.23). Thus, Equation (11.23) can be independently derived from the volume ratio chosen, and D is a scale independent measure of the probability of fracture.

Returning to the model illustrated in Figure 11.4 and following Allegre et al. (1982), each element in a cell is hypothesized to be either fragile (f), if it is permeated with microfractures, or sound (s), if it is not. The probability that an element in an nth-order cell is fragile is P_{n+1}. It is necessary to specify a condition for the fragility (soundness) of the cell in terms of the fragility (soundness) of the elements. Allegre et al. (1982) hypothesized that a cell was sound if a 'pillar' of sound elements linked two faces of the cell. In each cell there can be zero to eight fragile elements; there are $2^8 = 256$ possible combinations. Excluding multiplicities, there are 22 topologically different configurations; these are illustrated in Figure 11.6. The fragile elements are indicated by the solid dots at the corners. Configurations 4f, 5c, 6b, 6c, 7, and 8 are fragile and are indicated by solid underlining in Figure 11.6. Examples of sound and fragile cells are illustrated in Figure 11.7; 5b is a sound cell with the pillar of strength illustrated by heavy lines, and 5c is a fragile cell.

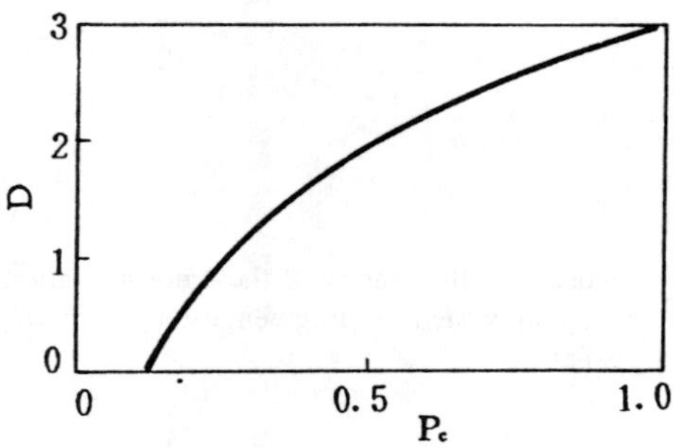

Figure 11.5. The dependence of the fractal dimension D on the probability of fragmentation P_c from Equation (11.13) (Turcotte, 1986a).

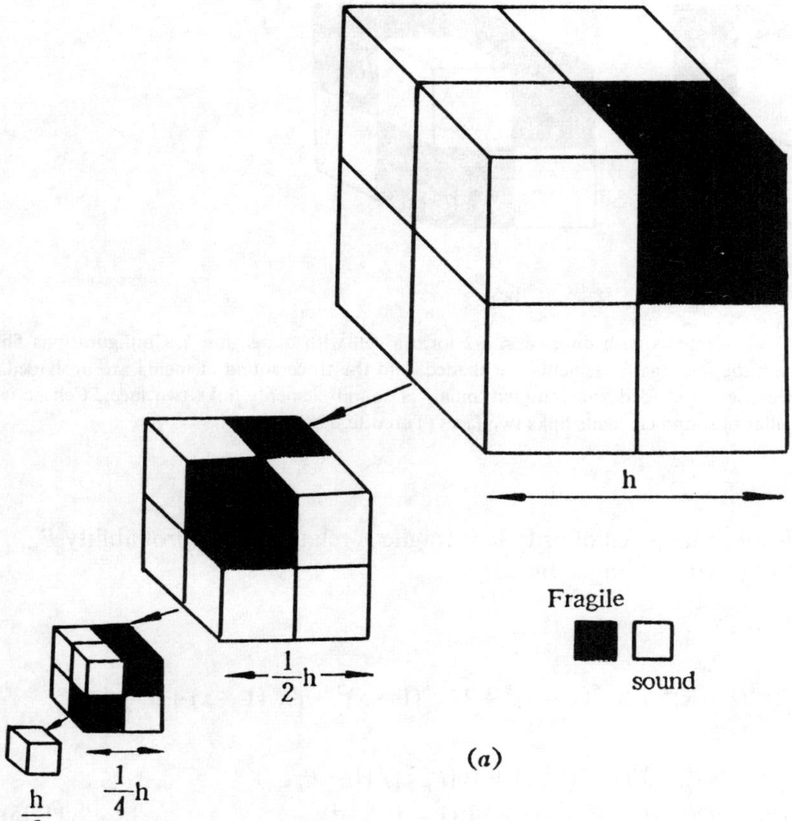

(*a*)

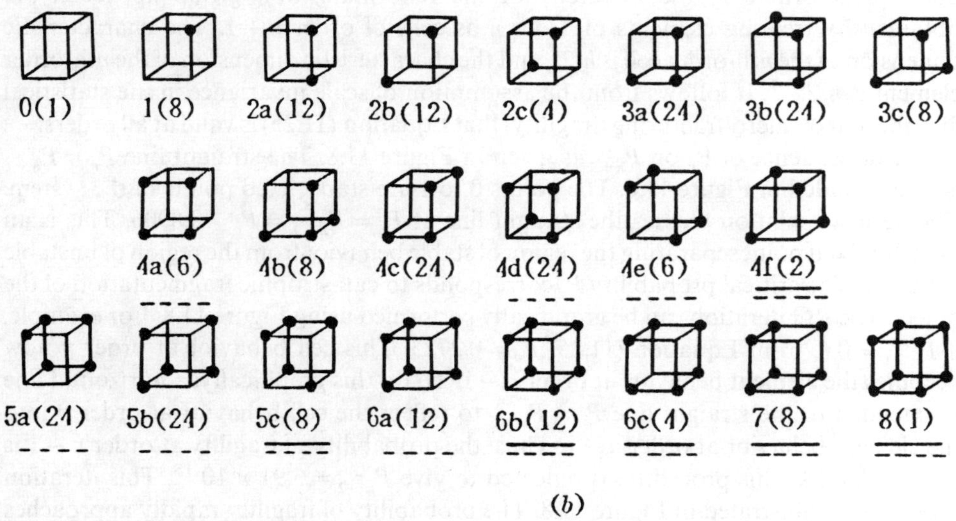

(*b*)

Figure 11.6. Illustration of 22 topologically different configurations. The 'fragile' elements are indicated by the solid dots on the corners. (a) Fragile process; (b) Fragile elements (Turcotte, 1986a).

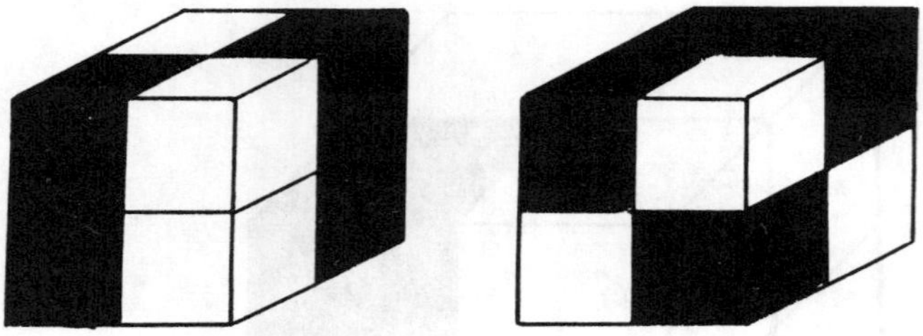

Figure 11.7. Eight cubic elements with dimension $h/2$ form a cell with dimension h. Configurations 5b and 5c are illustrated; the five fragile elements are shaded, and the three sound elements are unshaded. Cell 5b is sound because the shaded and outlined 'pillar' of sound elements links two faces. Cell 5c is fragile because no pillar of sound elements links two faces (Turcotte, 1986a).

The probability P_n that a cell of order n is fragile is related to the probability P_{n+1} that an element of order n is fragile by

$$P_n = R^{-1}(P_{n+1}),$$

with

$$R^{-1}(x) = 2x^4(1-x)^4 + 8x^5(1-x)^3 + 16x^6(1-x)^2 + 8x^7(1-x) + x^8.$$

Thus

$$\begin{aligned} P_n &= (P_{n+1})^8 + 8(P_{n+1})^7(1-P_{n+1}) + 16(P_{n+1})^6(1-P_{n+1})^2 \\ &\quad + 8(P_{n+1})^5(1-P_{n+1})^3 + 2(P_{n+1})^4(1-P_{n+1})^4 \\ &= (P_{n+1})^4[3(P_{n+1})^4 - 8(P_{n+1})^3 + 4(P_{n+1})^2 + 2]. \end{aligned} \tag{11.25}$$

Subsequent orders in the hierarchy of the renormalization group approach are obtained by treating elements of order n as cells of order $n+1$. The characteristic dimension of the nth-order cell is $h/2^n$, and the characteristic dimension of the nth-order element is $h/2^{n+1}$. It follows from the assumption of scale invariance in the statistical distribution of micro-fracturing (fragility) that Equation (11.25) is valid at all orders.

The dependence of P_n on P_{n+1} is given in Figure 11.8. The straight line $P_n = P_{n+1}$ is also included in Figure 11.8. The points 0 to 1 are stable fixed points of the system. The iterative relation crosses the straight line at $P_n = P_{n+1} = P_c = 0.8906$. This is an unstable fixed point separating the region of stable behavior from the region of unstable behavior. The critical probability P_c corresponds to catastrophic fragmentation of the object. The RG iteration can be graphically performed using Figure 11.8. For example, if $P_{n+1} = 0.6$, from Equation (11.25) $P_n = 0.2723$. This cell behavior at order n now becomes the element behavior at order $n-1$. To do this graphically, a horizontal line is extended to the straight line $P_n = P_{n+1}$ to reflect the cell behavior at order n into the element behavior at order $n-1$. Thus, the probability of fragility at order $n-1$ is $P_{n-1} = 0.0118$. This procedure is repeated to give $P_{n-2} = 3.91 \times 10^{-8}$. This iteration procedure is illustrated in Figure 11.8. The probability of fragility rapidly approaches zero as the order is decreased. If $P_{n+1} < P_c$, the iteration approaches the singular point $p = 0$ as n decreases, and the object is sound. If $P_{n-1} > P_c$, the iteration approaches the singular point $p = 1$ as n decreases and the object fragments. Bifurcation of the solution occurs at $P_{n+1} = P_c$; this is the critical value of the probability of microfractures (fragility) that leads to a catastrophic fragmentation. Substituting the critical probability $P_c = 0.896$ into Equation (11.23), Turcotte has found that $D = 2.84$ for this model of fragmentation (Allegre et al., 1982).

The above approach in which fragmentation can be modeled using the RG method is certainly not unique. An alternative approach is now considered using the same basic model. Eight cubic elements with dimension $h/2$ are again combined to form a cubic cell with dimension h. Each element again has a probability P_{n+1} of being fragile. The sides of a fragile element are assumed to form planes of weakness. If the sides of fragile elements form an internal plane through the cell, then the cell is assumed to be fragile. Of the 22 topologically different configurations illustrated in Figure 11.6, 4a, 4d–f, 5a–c, 6a–c, 7, and 8 are fragile and are indicated by dashed underlining. Examples of sound and fragile cells are illustrated in Figure 11.9; 4c is a sound cell, and 4e is a fragile cell with the 'plane' of fragility is illustrated by the heavy lines.

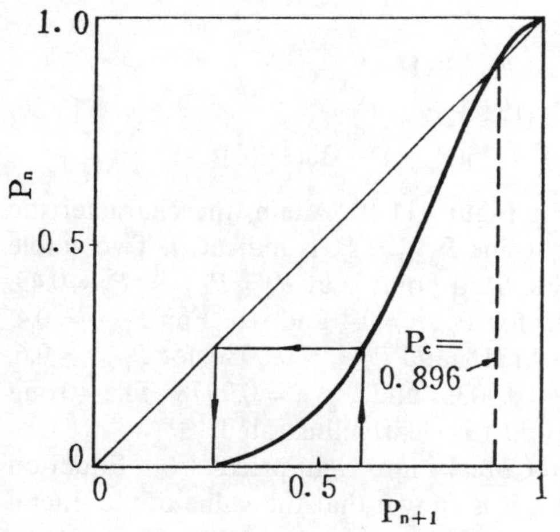

Figure 11.8. Probability of fragility at order n, P_n, as a function of the probability of fragility at order $n+1$, P_{n+1}, from Equation (11.25) (Turcotte, 1986).

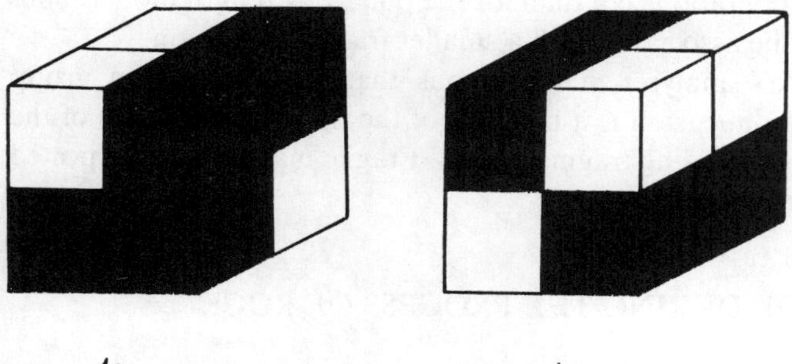

Figure 11.9. Configurations 4c and 4e are illustrated; the four fragile elements are shaded, and the four sound elements are unshaded. Cell 4c is sound because there is no plane of weakness through the cell. Cell 4e is fragile because the shaded plane of weakness penetrates the cell (Turcotte, 1986a).

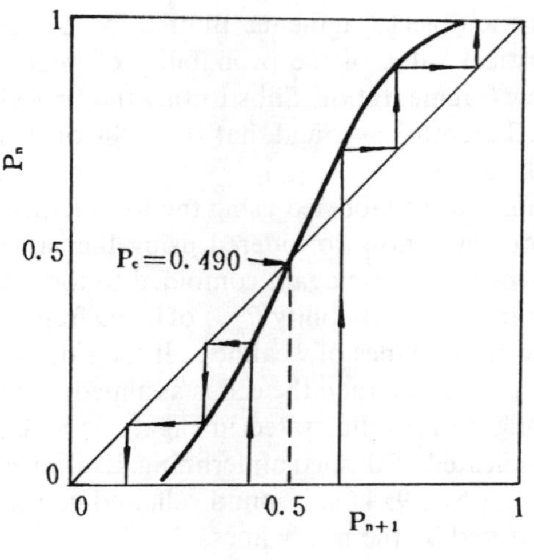

Figure 11.10. Probability of fragility at order n, P_n, as a function of the probability of fragility at order $n+1$, P_{n+1}, from Equation (11.26) (Turcotte, 1986a).

The probability P_n that a cell of order n is fragile is related to the probability P_{n+1} that an element of order n is fragile by,

$$\begin{aligned} P_n &= (P_{n+1})^8 + 8(P_{n+1})^7(1-P_{n+1}) + 28(P_{n+1})^6(1-P_{n+1})^2 \\ &\quad + 56(P_{n+1})^5(1-P_{n+1})^3 + 38(P_{n+1})^4(1-P_{n+1})^4 \\ &= 3(P_{n+1})^8 - 32(P_{n+1})^7 + 88(P_{n+1})^6 - 96(P_{n+1})^5 + 38(P_{n+1})^4. \end{aligned} \tag{11.26}$$

The dependence of P_n on P_{n+1} is given in Figure 11.10. Again, the characteristic S-shaped curve is obtained, and the straight line $P_{n+1} = P_n$ is indicated. Two stable fixed points are at 0 and 1, and the unstable fixed point is at $P_n = P_{n+1} = P_c = 0.49$. The RG iteration is graphically performed for $P_{n+1} = 0.4$ and 0.6. For $P_{n+1} = 0.4$, Equation (11.26) gives $P_n = 0.2997$, $P_{n-1} = 0.1315$, and $P_{n-2} = 0.0080$; for $P_{n+1} = 0.6$, Equation (11.26) gives $P_n = 0.7202$, $P_{n-1} = 0.9035$, and $P_{n-2} = 0.9978$. The strong convergence to the stable fixed points at 0 and 1 is clearly illustrated.

Again assuming that each fragmented cube breaks into eight pieces, from Equation (11.23) with $P_c = 0.490$, then $D = 1.97$. Thus, it is shown that the value of the fractal dimension is sensitive to the particular fragmentation model that is hypothesized.

The plane of weakness model for failure is a model applicable to more fragile materials. That is, more configurations fail than for the pillar of strength model. Thus a more fragile material may be associated with a smaller fractal dimension.

From the above Turcotte's analyses, it is obvious that renormalization group studies show that the fractal dimension is a measure of the fracture resistance of the material relative to the process causing fragmentation. Fragile materials are expected to have a smaller fractal dimension.

11.3 FRACTAL BEHAVIOR DURING THE PROCESS OF ROCK RUPTURE

We can see rock blocks everywhere such as in railways or railroad, mines, and stone pits or quarries. A large block may weigh several hundred tons, and a smaller block may weigh only several micrograms. We discover that there is always a triangular face

Photo 11.1. Rock blocks.

in every large or small rock block produced by blasting, cutting, or hammering unless they are from manmade process as shown in Photo 11.1. From the analyses of physics and mechanics, rock ruptures as blasting or cutting, in which the most stable structural face is a triangle because of the least dissipation of energy. However, from the description of geometry, the process of rock rupture may be considered as follows: A large triangular rock block ruptures to several small triangular ones, then again ruptures to smaller triangular ones. Continuing this process, numerous triangular blocks are formed. Although, large and small rock blocks will be produced at the same time by blasting, a common characteristic is that, in this process, the part is the reduction of the whole, which is a fractal. We can take an equilateral triangle as its initial polygon, and a broken line with angle 60° as a generator. Then a fractal figure of the Sierpinski gasket (also see Figure 2.15) is constructed to perform the fractal characteristics on a specific plane during the procedure of rock rupture. The construct rule is; the largest triangles included in the equilateral triangles are eliminated and by infinitely repeating the same procedure, numerous large and small triangular blocks are obtained. From the discussion given in (Section 2.7),

$$N = 9, \quad 1/r = 4, \quad D = \log 9/\log 4 = 1.5849.$$

On this plane, the fractal dimensions of fractal figures constructed by any other means are all larger than 1.5849. Our research indicates that the fractal dimension on a certain plane during the rock blasting process may be

$$1.5849 \leq D \leq 1.8928.$$

In fact, rock rupture is a cubic problem, it will be between two-and three-dimensional space, its fractal dimension obtained by us will be

$$D \in [2.0, 2.7268].$$

From the principle of critical crack extension force in which fractal effects are considered (see Chapter 13, for details),

$$G_f = 2r_s(1/r)^{D-1},$$

where r_s denotes the surface energy of per unit area. It is obvious that the larger D, and the larger G_f is, i.e., the more energy will be dissipated. Hence, triangular rock blocks will be easily produced in the process of rock rupture by the analyses of fractal geometry.

CHAPTER 12

Fractal pores and particles of rocks and soils

Rock and soil pore (or particle) geometry can be divided into surface geometry, single-pore-volume geometry, and the collective geometry of all of the pore space. The pore geometries of rocks behave as fractals in the range from atomic dimensions of a few angstroms up to grain size (Katz & Thompson 1985, 1986; Krohn & Thompson 1986; Thompson et al., 1987; Thompson, 1991; Wong, 1988; and Roberts, 1986). At present, fractal properties have been widely applied to the investigation of pores, the transport property of rocks and soils, permeability in rock materials, and so on. It may be possible in the future to account for the principal features of rock and soil microgeometry by fractal models of pore evolution (analogous to the models of DLA and the viscous fingering patterns discussed in Chapter 5).

12.1 FRACTAL MODELS OF POROUS MEDIA

From a scanning electron micrograph of some simple 'clean' sandstones, we can see that it is just a random packing of sand grains of comparable sizes – grains of quartz typically tens to hundreds of microns in diameter (Wong, 1988). Such grains are sorted during sedimentation. After burial and compaction with other mineral sediments, they are consolidated under high pressure to form a rigid matrix with a multiply connected network of pores that controls the movement of ions and molecules through the rock. Such ions and molecules, in turn, alter the rock over millions of years – a process called diagenesis. After locally magnifying a portion of a scanning electron micrograph of sandstones, it is seen that a small amount of mineral with a highly irregular shape has grown on the grain surface. Most sandstones are much 'dirtier' and contain many such minerals, the most common of which are clays formed of layered alumina silicate compounds (Wong, 1988).

So far, many fractal models have been developed to describe this random mineral overgrowth. A Koch island is used to model fractal surfaces of rock pores as shown in Figure 12.1 (Wong, 1988). One begins with the initiator of an equilateral triangle of side length L and then successively attaches smaller and smaller equilateral triangles of sizes $L/3$, $L/3^2$, and so on to the middle of every straight segment of the perimeter (for details, see Chapter 2). After n iterations, the perimeter consists of N segments with length $r = L/3^n$,

$$N = A(L/r)^D, \tag{12.1}$$

where A is 3 and D is log 4/log 3. This mathematics can be used to model the pores in a rock because ever-smaller mineral particles can grow or deposit on particles already present. When a pore of size L is observed with resolution r, the number of surface features will be given by Equation (12.1), and the measured surface area S is defined by,

$$S = Br^2(L/r)^D = BL^2(L/r)^{D-2}. \tag{12.2}$$

Whereas the Sierpinski gasket shown in Figure 12.1b can be used to model a volume fractal. Here one starts with a triangle with side length L to represent a smooth pore, then successively cuts out smaller triangles with a side length $L/2$, $L/2^2$, and so on to characterize ever-smaller grains filling up the pore volume and reducing the porosity. After n iterations, the pore space consists only of triangles of size r, given by $L/2^n$, and the number of such triangles is given by Equation (12.1) with the constant A equal to 1, and the fractal dimension $D = \log 3/\log 2$. Because a pore of size r contributes a porosity of the order $(r/L)^3$, the total porosity is

$$\phi = (r/L)^3 N = C(L/r)^{D-3}. \tag{12.3}$$

The pores in actual rocks behave differently from the surface and volume fractals described above because they have random surface and volume elements with a continuous hierarchy of sizes between an upper limit L and a lower limit r. Equations (12.1)-(12.3) are applicable representation of statistical average behavior between these limits (Wong, 1988). The fractal dimension D is always between two and three for either the pore surface or the pore volume. To date, there have been many experiments supporting such a fractal description for sandstones (Wong, 1988; Thompson, 1991).

Frieson & Mikula (1987) have used the Menger sponge (see Chapter 2) as a fractal

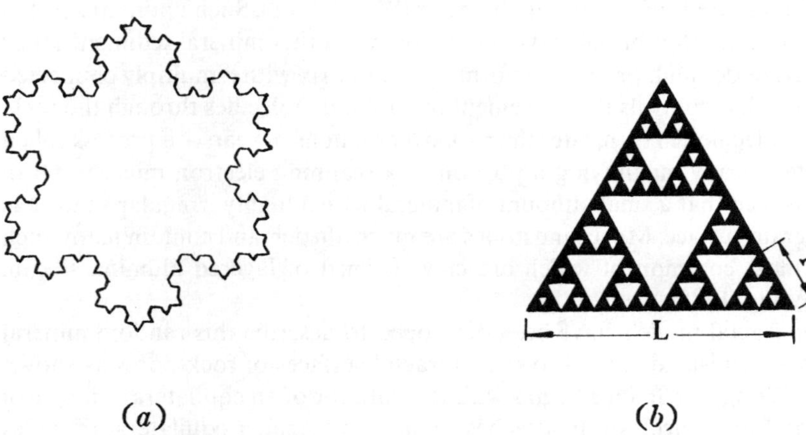

Figure 12.1. Geometrical models for sedimentary rocks: (a) Koch island with a fractal perimeter models a pore with a fractal surface; (b) Sierpinski gasket models a set of tiny pores that link up to form a fractal volume.

model of a bulk porous media for modelling the voids between the particles and pores in the individual particles. Considering the initiator of a cube with edge length R, and dividing the cube into m^3 identical smaller cubes, a number of these cubes can be removed according to a given prescription, leaving a structure of $N_{b1}(m)$ cubes which defines the generator. Then, each subcube is, in turn, replaced by a scaled-down version of the generator. After K iterations of the procedure, the total number of cubes of size $r_k = R/m^k$ remaining in the structure is

$$N_{bk} = N_{b1}^k, \tag{12.4}$$

which can be rewritten as

$$N_{bk} = (r_k/R)^{-D_b}, \tag{12.5}$$

where

$$D_b = \log(N_{b1})/\log(m). \tag{12.6}$$

From Equation (12.5), the volume of the resulting structure is

$$V_k \propto r_k^{3-D_b}. \tag{12.7}$$

Obviously, for $K \to \infty$, or $r_k \to 0$, $V_k \to V(r)$ (a continuous function), in this limit,

$$\mathrm{d}V/\mathrm{d}r \propto r^{2-D_b}.$$

Since the pore volume $V_p(r) = R^3 - V(r)$,

$$-\mathrm{d}V_p/\mathrm{d}r \propto r^{2-D_b}. \tag{12.8}$$

This expression has also been derived by Pfeifer & Avnir (1983) from molecule monolayer adsorbates (see Section 12.2.3).

Note that in order to obtain Equation (12.6), the generator is substituted for each cube throughout the volume at every stage of rescaling. If the generator is taken to be a modified face of the cube, rather than a restructured cube, the surface of the resulting object is a fractal. The number of squares with side length r_k required to cover the surface is

$$N_{sk}(m) \propto r_k^{-D_s}, \tag{12.9}$$

where

$$D_s = \log(N_{s1})/\log(m). \tag{12.10}$$

Hence, $N_{s1}(m)$ is the number of squares of the side R/m contained by the generator, and D_s is the fractal dimension of pore surfaces.

Frieson & Mikula also formulated a model of a particle with a fractal porous surface from a cubic initiator with a generator constructed as follows: consider a unit square which is subdivided into m^2 smaller squares. Remove n subsquares subject to the restriction that they do not have common sides or lie on the square's edges (this requires $n \leq (m^2 - 4m + 5)$). At each square removed, insert a hollow cube which has the top face missing. Thus, the surface area of the square has been increased by $4n$ squares of area $(1/m)^2$, yielding for the generator

$$N_{s1} = m^2 + 4n. \tag{12.11}$$

Applying the generator to each face of the cube gives the fractal dimension for the porous surface:

$$D_s = 2 + \log(1 + 4n/m^2)/\log(m), \tag{12.12}$$

where D_s has a maximum value of 2.37 ($m = 7$, $n = 13$) and decreases to the limiting value of 2 as $m \to \infty$. From this model, we can express the pore volume in terms of D_s. After one reduction in scale, the pore volume on one face is $V_{p1} = n/m^3$, and the number of squares is N_{s1} as given by Equation (12.11). After two steps, the pore volume is

$$V_{p2} = V_{p1} + nN_{s1}/(m^3)^2,$$

and after k steps

$$V_{pk} = V_{p1} + n \cdot \sum_{j=1}^{k-1} N_{sj}/(m^3)^{j+1} = n/m^3 \cdot \sum_{j=0}^{k-1} [N_{s1}/m^3]^j.$$

Solving this geometrical series gives

$$V_{pk} = V_{pt}[1 - (N_{s1}/m^3)^k], \tag{12.13}$$

which can be rewritten as

$$V_{pk} = V_{pt}[1 - r_k^{3-D_s}], \tag{12.14}$$

where $V_{pt} = n/(m^3 - N_{s1})$ is the total pore volume for the infinitely iterated structure ($k \to \infty$). Differentiating the continuum limit of Equation (12.14) yields

$$\log[-\mathrm{d}V_p/\mathrm{d}r] \propto (2 - D_s)\log(r). \tag{12.15}$$

This means that the fractal dimension of the surface can be determined by measurement of the pore volume as a function of the pore radius. Note that Equation (12.15) differs from Equation (12.8) only in that D_s replaces D_b. This does not imply that $D_s = D_b$, (although there have been some experimental results of $D_s = D_b$, Katz & Thompson, 1985), since the latter equation describes a fractal porous surface, but the former corresponds to a void between particles by construction. Thus, a measurement of $V_p(r)$ alone can not distinguish between a fractal surface and a fractal volume if the observed $D > 2$. In this case, additional information is required (Frieson & Mikula, 1987; and Thompson et al., 1987, 1991).

12.2 FRACTAL PORES AND THEIR FRACTAL MEASUREMENTS IN ROCK MATERIALS

After understanding the concept of fractal pores in rocks, one tries to design some methods to measure the fractal dimensions of porous rocks. There have been many reported measurements of fractal pores over the last few years to support the theoretical analysis of fractal pores discussed in the previous section. These methods of fractal measurements in porous rock material can be classified as three types: discrete methods, scattering methods, and adsorption methods. In this section we will discuss these methods, respectively, and some interesting results of fractal pores associated with these methods.

12.2.1 *Discrete methods*

The discrete methods for fractal measurements are based on the earliest description of curves with nonintegral dimension (see Chapter 2). In these methods, the fractal object is measured by directly using rulers with different lengths (purely geometrical measurements).

In the earliest direct measurements of fractal dimension, Kay (1986) characterized the shape of fine particles of carbon black using two-dimensional projection and obtained the fractal dimension to be 1.32 from data covering one-half an order of magnitude in length scale. Orford & Whalley (1983) attempted to discriminate between various sedimentary particles based on their fractal characters. They measured the roughness of the individual grains by analyzing the outline of the grains imaged from a grain mount. They examined fractal dimensions in a range from 1.0 to 1.30 by using SEM (Scanning Electron Microscopy) measurement of grain roughness. Thus, different particle types were distinguished on the basis of both the fractal dimension and the length scales over which fractal behavior was observed. Clark (1986) has described several approximate divider methods for a more rapid determination of an approximate fractal dimension on particle shapes.

Mandelbrot (1982) introduced the correlation measurement of fractal pores in thin sections as shown in Figure 12.2. So-called correlation measurement is the pore-space density-density correlation measurement. The origin of the coordinate system is located in pore space, and progressively larger circles are constructed about the origin. If the pore space is a fractal, then the probability of finding pore space at distance r falls off as r^{D-2}.

This method is simple and convenient. A rock-sample is first saturated with epoxy, then sliced to a thickness of 1 mm or less, polished to a final thickness of 10-100 μm, and mounted on a glass microscope slide. Pore-space-grain-space high-contrast images are optically made using a dye in the pore-filling epoxy or by using SEM images. However, this method has some limitations: the range of length scale that can be measured is too limited, and uncorrelated pores are overlapped to distort the

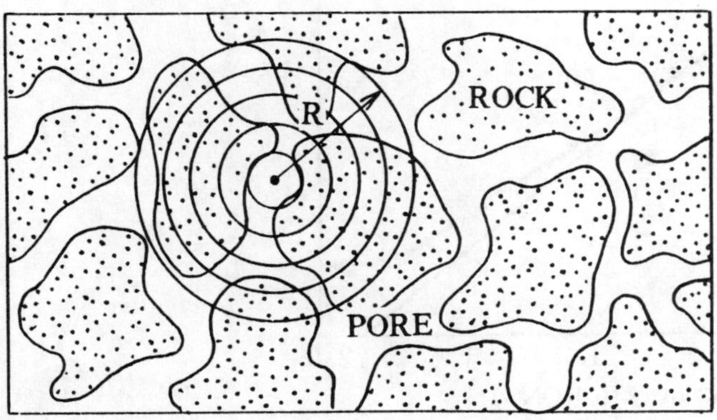

Figure 12.2. Schematic diagram of the pore-space density-density correlation-function measurement. The probability of finding pore space at distance R falls off as R^{D-2} if the pore space is a fractal (Thompson et al., 1987).

power-law result of fractals. For these reasons, Katz & Thompson (1985) have generalized the correlation measurement into the autocorrelation measurement of fractal dimension. In the autocorrelation measurement, backscattered electron micrographs of polished thin sections are photographically enhanced with high-contrast black and white film to produce a binary image. Two identical negatives are then made on 35 mm-film format. The two films are placed in an optical microscope to measure the transmitted light through both films. The transmitted intensity is measured with a chopper – stabilized photomultiplier detector. The autocorrelation function is directly given by the transmitted intensity as a function of the distance one film is translated relative to the other (Thompson et al., 1987). It is clear that the autocorrelation measurement makes the fractal measurement of pores more accurate and faster than the correlation measurement. Figures 12.3 and 12.4 show the typical autocorrelation data obtained by Katz & Thompson (1985) and Thompson et al. (1987). Especially, Katz & Thompson (1985) have used scanning electron and optical data to show that the pore spaces of several sandstones are fractals, and they used fractal statistics to predict the correct porosity of rocks. Their results indicate that the density-density correlation function of the pore space is constant over length scales greater than a characteristic length L_2, which is approximately the size of sand grains in a sandstone (typically 100 μm across as an upper limit of self-similarity). At length scales $L_1 < L < L_2$, the pore-space-rock interface is a self-similar manifold with a well-defined fractal dimension, D, and a lower limit of self-similarity L_1. In fact, that the sandstone pore space has the fractal properties of the interface implies a simple relationship between the fractal dimension and the porosity of the rock. In a volume element of size L_2, the

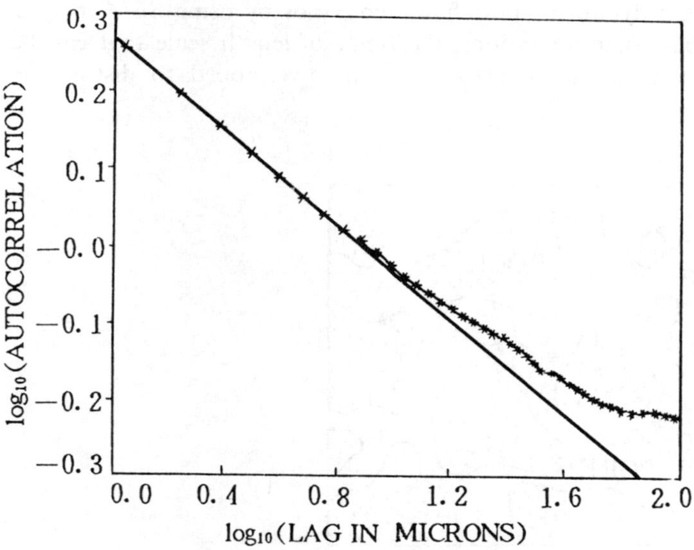

Figure 12.3. The pore density-density correlation or auto-correlation for a sample of sandstone, with a slope of −0.27. The linear region at small lag has the same slope as determined from other fractal measurements (Katz and Thompson, 1985).

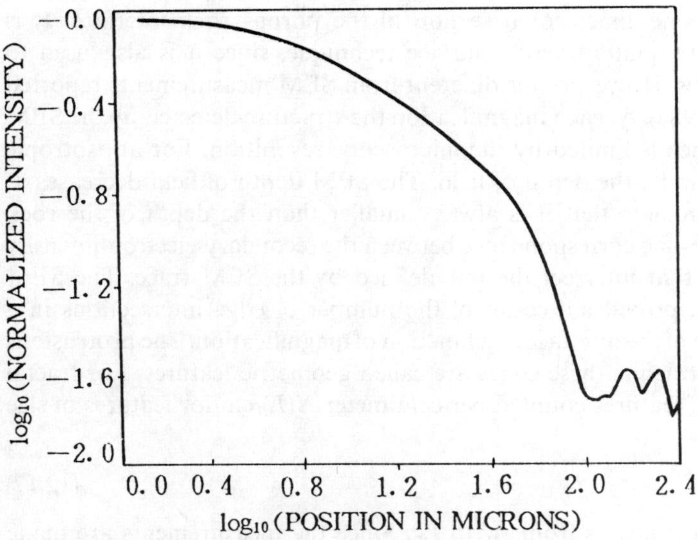

Figure 12.4. The pore-space autocorrelation function for Coconino sandstone (Thompson et al., 1987).

Table 12.1. Results of SEM measurements on various sandstones (Katz & Thompson, 1985).

Sample	Fractal dimension	L_2 (μm)	Porosity (%)	
			Calculated	Measured
Tight gas sand #965	2.57	2.5	4.7	5.3-5.6
Tight gas sand #466	2.68	6	7.6	6.9-7.6
Coconino	2.78	98	10	11-12.5
Navajo	2.81	50	15	16.4
St. Peter's	2.87	50	27	24-28

'volume' of the pore filling surface is given in units of L_1 by $A(L_2/L_1)^D$, where A is a constant. Hence, the porosity ϕ can be defined by

$$\phi = A(L_1/L_2)^{3-D}. \qquad (12.16)$$

This has the same form as Equation (12.3). For all the sandstones, assuming that L_1 is a constant on the order of 20 Angstroms (Å), the minimum size of a crystal nucleus in the pore space, and $A = 1$, Katz & Thompson have found that the porosities of the five sandstones calculated by terms of Equation (12.16) are in good agreement with the values measured by using SEM measurements (Table 12.1). This successful prediction of the porosities from the fractal parameters verifies that the pore surface and volume may be fractals with the same fractal dimension.

The SEM measurement here presented by Katz & Thompson (1985) and Krohn & Thompson (1986) is an automatic measurement method. It is used to measure the fractal dimension of a one-dimensional section of the porous rock interface. It is sometimes called the automatic fracture surface technique, since it is also used to measure fracture surfaces. However, it is different from SEM measurements reported by Orford & Whalley (1983). At each magnification the structure detected by the SEM beam along a linear trace is limited by the microscope resolution. For an isotropic fractal, this resolution limits the depth of field. The SEM depth of field decreases at increasing magnification such that it is always smaller than the depth of the rock interface. There is one-to-one correspondence between the secondary-electron-intensity extreme and the edges that intersect the line defined by the SEM trace. The SEM measurement, therefore, provides a count of the number of edge intersections in a one-dimensional section of the interface as a function of magnification. The protrusions into the pore space that define these edges are called geometric features. For fractal behavior, the number of features counted per centimeter $N(L)/\mathrm{cm}$ for features of size L can be expressed as

$$N(L)/\mathrm{cm} \propto L^{2-D}, \tag{12.17}$$

where L includes all feature sizes from L_1 to L_2. Since the measurements are made from images, a convenient measure of feature size is in terms of pixels, with a pixel being $\frac{1}{512}$ of the image (512×512). The digitized data are denoted by an intensity sequence $I(J)$ (where J is a pixel ranging from 1 to 512). The feature size L for a feature with an edge at J_1 and at J_2 is $J_2 - J_1$. The width in centimeters of the field of view for each image is 12 divided by the magnification or $12/M$. Thus, the number of features per centimeter from Equation (12.17) becomes

$$N(L)/\mathrm{cm} = A(12L/512M)^{2-D}, \tag{12.18}$$

where L is the feature size in pixels, A is a constant, and M is the magnification.

Since edges of a feature are defined to be a local maximum in the secondary-electron-intensity of the SEM, features may pertain to grains, clay particles, and pits along the surface of grains and clay particles. In performing the analysis, a digital filter is convolved with the data in order to establish a resolution which is uniform at all magnifications and independent of the optics of the SEM. Thus, counting features gives a histogram for each magnification. The number of features actually counted $N(L)$ in the feature histogram can be defined by

$$N(L) = N(L)/\mathrm{cm}\, P(L)R(L)\,(\mathrm{cm}), \tag{12.19}$$

where $P(L)$ is the probability of detecting a feature in the feature histogram, and $R(L)$ is the distance in centimeters over which the features are counted. The probability $P(L)$ of detecting a feature of size L is set by the digital filter, and it can be explicitly determined by performing the Fourier transform of the impulse response and expressing the amplitude as a function of distance rather than frequency. The distance $R(L)$ of counting features of size L is the product of the width of the field of view times the fraction of that field of view which is unoccupied by smaller features, i.e., large features are not seen if small features on top of them are resolved. Thus, $R(L)$ should be a function of the number of features counted for sizes less than L. The fraction of the field of view $F(L-1)$ occupied by features of sizes less than L is

$$F(L-1) = \frac{1}{512} \cdot \sum_{k=0}^{L-1} KN(k), \tag{12.20}$$

where K is a feature size smaller than L. The function $F(L)$ is called the cumulative feature distribution defined by Krohn & Thompson. As L goes larger, $F(L)$ must become equal to 1.0. Since $R(L)$ is the width of the field of view in centimeters times the fraction of the field of view unoccupied by smaller features, then

$$R(L) = [(12\,\mathrm{cm})/M][1 - F(L-1)]. \tag{12.21}$$

From Equations (12.18) and (12.19), a feature histogram can be rewritten as

$$N(L) = A(12L/512M)^{2-D}P(L)(12/M)[1 - F(L-1)], \tag{12.22}$$

where A and D are two parameters to be determined together. Equation (12.22) is used to fit the measured feature histograms of digitized SEM data. If the measured feature distribution can be described by a power-law over a certain range of length, the fractal dimension is determined from Equation (12.22). Thus, SEM and image digitization provide a means to automatically measure the fractal dimension of porous rocks. Krohn & Thompson reported their measured fractal dimensions ranging from 2.55 to 2.85, as illustrated in Figures 12.5-12.8. The samples include standard outcrop rocks such as Boise, Berea, and Coconina as well as sandstones from the Blackhawk formation from a well near Price River, Utah. The properties of samples used in the fractal measurements are listed in Table 12.2. The long-length limit L_2 represents the end of the fractal range which is marked by a definitive transition from fractal to nonfractal behavior. In general, smaller values of the long-length limit correlate with lower values of the fractal dimension. It is seen from Table 12.2 that neither the fractal dimension nor the long-length limit is related to the averaged grain size of the sandstones (Krohn & Thompson, 1986).

It should be pointed out that the pore volume of some rocks is not a fractal. For example, the L_2 value for some rocks is less than the average pore size (Krohn, 1988a). Fractal pores of rocks are associated with diagenesis. The structures including euhedral quartz overgrowths, druse quartz, calcite, dolomite, and clays which give rise to fractal properties are shown to be diagenetic in the number of the samples. Uniform authigenic minerals, such as calcite cement or quartz which are irregularly arranged are shown to have fractal distributions, whereas uniform crystals which are uniformly arranged do not generate a fractal (Krohn, 1988b). The fractal structures can be regarded as inhomogeneous distributions within the rock, particularly for sandstones. Sandstone formation involves burial of the sand grains, and compaction and alteration of the pore structure by the flow of fluids through the pore space. Such alteration involves crystal growth and nucleation on the pore surface. Recent studies show that the flow rate of fluid through the rock strongly influences the extent of mineral formation: a slow flow rate leads to a greatly reduced porosity and permeability. The rate of diagenesis, therefore, is a function of time (Thompson, et al., 1987).

Katz & Thompson (1985) and Thompson et al. (1987) have described a simple fractal crystal growth in rocks. They formulated the competing processes of crystal growth and crystal nucleation in terms of two kinetic equations:

$$\mathrm{d}R/\mathrm{d}t = KC^n, \quad \text{and} \quad \mathrm{d}N/\mathrm{d}t = K_1 C^p,$$

where K and K_1 are equilibrium constants, and C is the concentration of surface molecules. The first equation describes the radial growth of a crystal of size R. The second equation describes the heterogeneous nucleation of a new crystal on a unit path

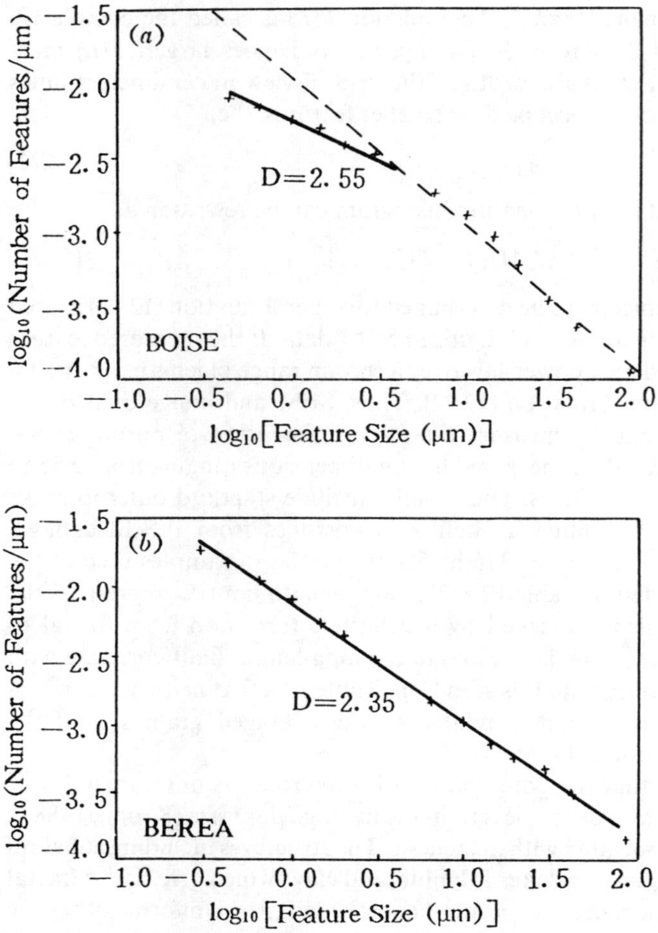

Figure 12.5. Results comparing two sandstones with extreme values of the fractal dimension. Plus signs represent fits to individual magnifications, solid lines are the result of simultaneous fits to a range of magnification, and dashed lines are lines with a slope of − 1.0 which match data beyond the fractal regime; (a) Log-log plot of the number of features per micrometer; (b) Log-log plot of the number of feature per image (Krohn & Thompson, 1986).

of the original unaltered surface. Since growth involves the addition of a single molecule to a growth site, n should be 1; on the other hand $p > 1$, since nucleation requires the coincidence of two or more molecules at a nucleation site. Once diagenesis begins and mineral crystals form, the porosity and permeability of the rock continually decrease, and concentration C increases with time. This variation in ion concentration is sufficiently slow so that the growth rate and nucleation rate for a particular set of particles is well described by steady-state conditions. The ultimate size R of a particular crystallite is limited by nucleation of new crystals on the rock surface: a newly formed particle grows at the expense of the underlying surface. Under these conditions,

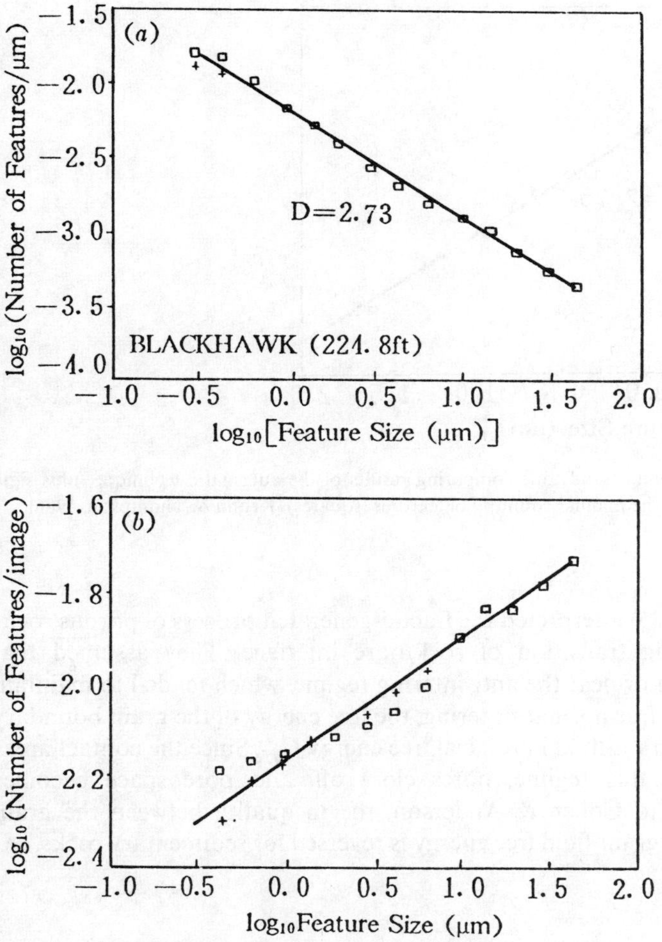

Figure 12.6. Results for a Blackhawk sandstone showing the behavior for a single position (squares) and for the average of four positions (plus signs). (a) Log-log plot of the number of features per micrometer with a slope of $2 - D$; (b) Log-log plot of the number of features per image with a slope of $3 - D$ (Krohn & Thompson, 1986).

$R \propto KC^n/N$ and $N \propto K_1 C^p$. Combining these expressions gives

$$N(R) \propto R^{-p/(p-1)}.$$

Classical nucleation theory requires that $p > 2$ so that $2 \leq p(p-1)$. Moreover, the imposition that the total volume of particles cannot exceed the available volume in the pore space requires that $2 \leq p/(p-1) \leq 3$. It is obvious that this model describes one possible mechanism for fractal crystal growth in rocks and is a mechanism capable of producing fractal pore spaces with a continuous range of fractal dimensions. Because the model is dependent only on the orders of the nucleation and growth reaction which determine the value of the exponent, it may be useful in understanding the influence of nucleation on diagenesis.

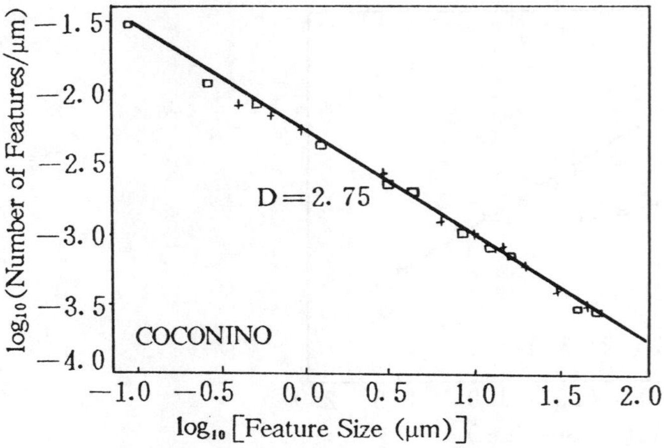

Figure 12.7. Results from Coconino sandstone comparing results of the automatic technique (plus signs and solid lines) to the results of the manual counting of features (squares) (Krohn & Thompson, 1986).

Cohen & Anderson (1985) interpreted the fractal-generated process of porous rocks in terms of a roughening transition of rock/pore interface. They assumed that sedimentary rocks were at or near the antisintering regime, which tended to maintain small intergrain contacts. For normal sintering, the free energy of the grain boundary G_{gb} is less than twice the grain-fluid interfacial free energy G_{gf}. Since the contact areas between grains grow in this regime, pores close off, and pore space becomes disconnected. According to Cohen & Anderson, the inequality between the grain boundary free energy and grain-fluid free energy is reversed for sedimentary rocks, i.e.,

Table 12.2. Properties of samples used in the fractal measurements (Krohn & Thompson, 1986).

Sample	Porosity (%)	Permeability (mD)	Grain size (µm)	Fractal *D*	L_2 (µm)
Berea	20	123	190	2.85	>32
Boise (Table)	35	4933	330	2.55	5
Coconino	9.9	0.037	120	2.75	>50
Navajo	18	246	210	2.74	>50
St. Peter's	9.3	5.15	140	2.61	18
Blackhawk sandstones from a well near Price River, Utah.					
137.3 ft	12	1.8	120	2.60	18
154.8 ft	4.3	0.02	100	2.59	20
161.7 ft	11.5	21.7	180	2.66	11
169.3 ft	14.3	13.6	200	2.72	22
224.8 ft	7.5	0.02	70	2.73	>50
Austin chalk				2.59*	6.3
Carboniferous shale	17			2.74	>10

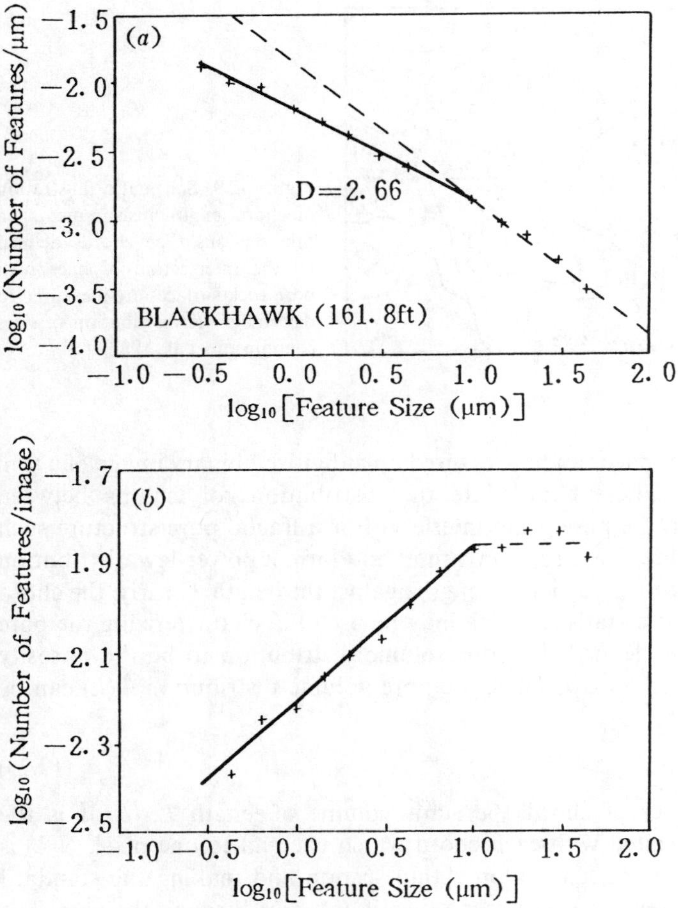

Figure 12.8. Results for a Blackhawk sandstone showing the transition to nonfractal behavior at the long-length limit; (a) Boise with $D = 2.55$; (b) Berea with $D = 2.35$ (Krohn & Thompson, 1986).

$G_{gb} > 2G_{gf}$. Under these conditions (antisintering) the grain contact areas do not grow, and the pore space remains connected. Further increase of the ratio $G_{gb}/2G_{gf}$ leads first to a roughening transition, which gives a rough but not fractal rock/pore interface (corresponding to the first prefractal rock/pore interface), and eventually to a crossover to a fractal rock-pore interface. Ultimately the interface becomes sufficiently rough and results in a volumetric fractal pore space (Thompson et al., 1987).

Krohn (1988a) and Thompson et al. (1987) have developed the thin section technique to provide data complimentary to the SEM measurement discussed above. Unlike the fracture surface measurements discussed in Chapter 3, the thin section technique requires delineating pore space from grain-pore structures and making measurements on the pores themselves. Thus, it can be used to measure the pore volume distributions for porous rocks. A typical measurement in the thin section technique may be a chord-length measurement as shown in Figure 12.9. The chord-length distributions

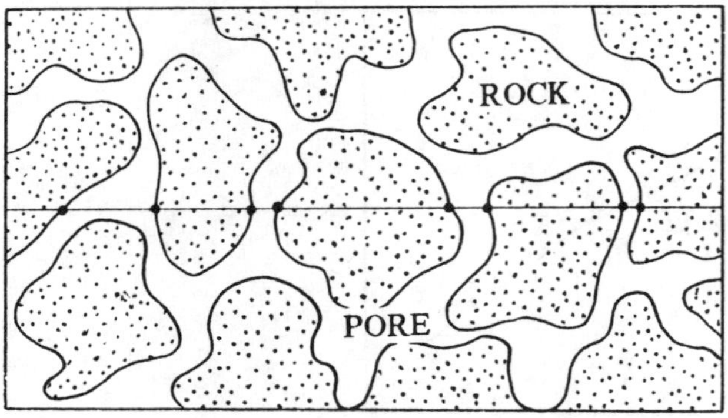

Figure 12.9. Schematic illustration of chord-length measurements on thin sections. The chords defined by the intersection of lines with pore-rock surfaces are measured to determine the distribution of sizes (Thompson et al., 1987).

across the grain and pore space can be measured on a digitized binary image of a thin section (Krohn, 1988a). These chords are the distributions of lengths between intersections of lines with the pore-grain interface. For a fractal pore structure such chord distributions should behave scale invariant and form a power-law distribution of the number of chords with a particular length against the length. Clearly, the chord length of the pore space is a statistic which may be useful in characterizing the pore space. Krohn (1988a) has defined the pore volume distribution to be the porosity associated with each chord length. Thus, the pore volume distribution $\Phi(L)$ can be expressed as

$$\Phi(L) = N(L)L(\Delta L)^2, \tag{12.23}$$

where $N(L)$ is the number of chords per unit volume of length L; $(\Delta L)^2$ is the cross-sectional area associated with each chord which is equal to one pixel. $N(L)$ is obtained from counting chord lengths on a thin section and making the standard assumption that the thin section is representative of any slice through the core. The total porosity is the sum of the pore volume distribution for all chord lengths.

The thin section technique is quite simple. SEM images of the thin section are digitized, and pore size is delineated wherever the intensity is less than a set gray level. Then, using horizontal lines to scan the images, chord lengths are measured from the interception of the line with the pore surface. A histogram of the number of chords of each chord length is then compiled using a logarithmic bin size. It should be noted that the choice of a gray level to delineate the pore space requires consistency from image to image and from magnification to magnification. Examination of gray level histograms of the images is easy to identify the pore distribution from the high contrast achieved with the SEM as shown in Figure 12.10 (Krohn, 1988a). Figures 12.11-12.12 show measurements using the thin section technique obtained by Krohn. In these plots, data from different magnifications are numerically designated. It is seen that the thin section data have slopes in agreement with the automatic fracture surface measurement results below L_2.

The chord-length technique is also used to measure fracture surface (Thompson et al.,1987). The principle of measurement is shown in Figure 12.13. A hypothetical cross-section is through a fracture surface. Thus, a measurement of chord length on a fracture surface should give the fractal distribution of the lengths if the fracture surface

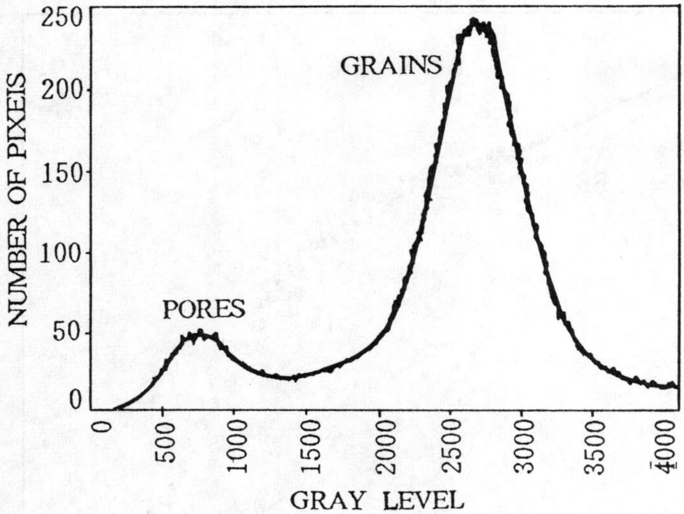

Figure 12.10. Gray level histogram for a thin section. Grain space is brighter than the pore space, and the distributions are well separated (Krohn, 1988a).

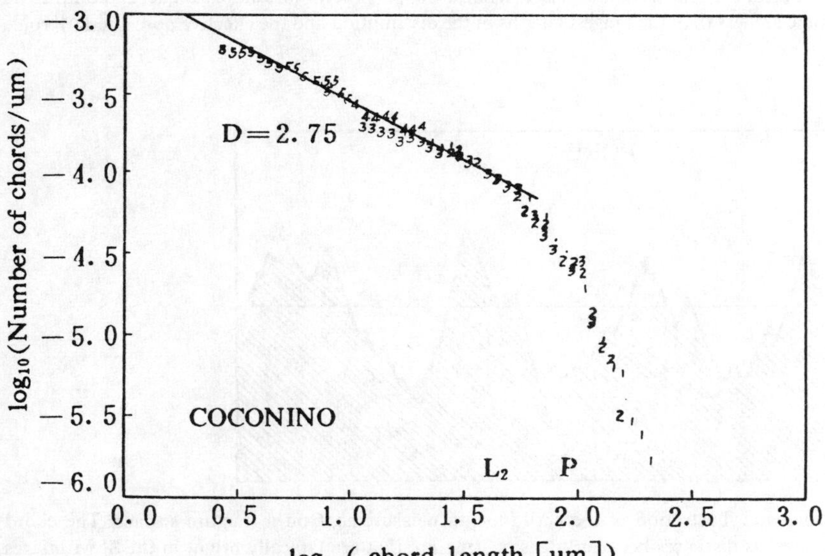

Figure 12.11. Results of the thin section measurement for the Coconino sandstone. L^2 and the end of the linear regime are close to the longest lengths in the distribution and to the effective pore size P (Krohn, 1988a).

is a fractal. Obviously, the advantage of application of this method to a fracture surface is that the polishing of the surface is not needed. However, a carefully designed experiment is necessary to ensure that a section of the surface and not a projection of the surface is measured.

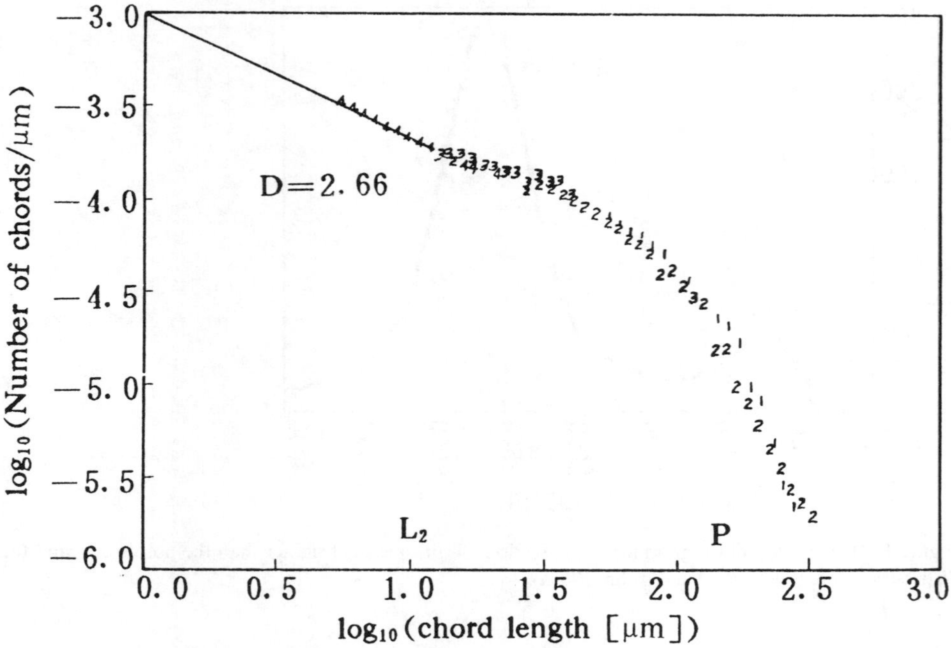

Figure 12.12. Results of the thin section measurement for the Blackhawk foreshore sample. L^2 and the end of the linear regime are less than the longest lengths in the distribution and the effective pore size P (Krohn, 1988a).

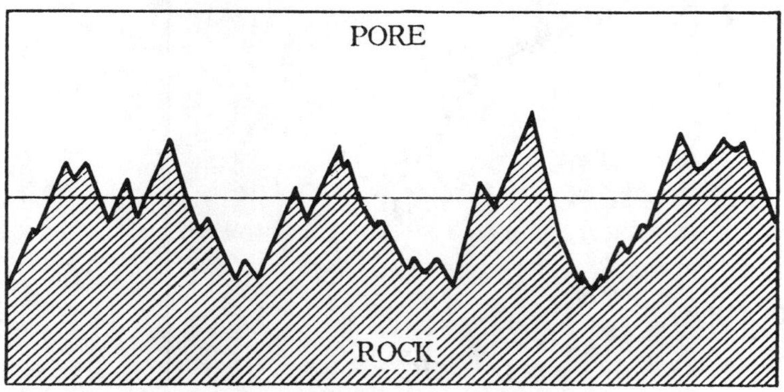

Figure 12.13. Schematic illustration of the chord-length measurement on a fracture surface. The chord lengths are measured as distances between edges of structure that are typically bright in the SEM images (Thompson et al., 1987).

12.2.2 *Scattering methods*

Small-angle scattering of thermal neutrons or X-rays has been used in recent years to study the wide variety of disordered systems such as polymer gels, aggregated colloids, oil-water emulsions, and cements. It provides a measurement of fractal properties at length scales between 0.5 nm and 50 nm.

In a typical scattering experiment, the sample is irradiated by a collimated beam of neutrons or X-rays with a fixed incoming wave-vector $\boldsymbol{K}_i$. The scattered radiations have wave-vectors $\boldsymbol{K}_f$. The change in the wave-vector is

$$\boldsymbol{q} = \boldsymbol{K}_f - \boldsymbol{K}_i; \quad |\boldsymbol{q}| = 2K \sin\theta, \tag{12.24}$$

where K is the magnitude of vector $\boldsymbol{q}$, and 2θ is the scattering angle between these two vectors. According to the Born approximation (in physics), the scattered intensity for any $\boldsymbol{q}$ is just the Fourier transform of the density-density correlation that characterizes the structure. If the local density $\rho(x)$ in a porous material is defined to be zero in the pore and one in the solid, then the scattered intensity $I(\boldsymbol{q})$ can be expressed by

$$I(\boldsymbol{q}) = V \int g(r) e^{i\boldsymbol{q}r}\, dr, \tag{12.25}$$

where V is the sample volume, and $g(r)$ is defined approximately as (Porod, 1982):

$$g(r) \propto 1 - \phi - rS/(4V), \tag{12.26}$$

for small r, where S is the surface area and ϕ is the porosity. The constant term $(1 - \phi)$ gives a $\delta(\boldsymbol{q})$ term in $I(\boldsymbol{q})$. The linear term in r gives a q^{-4} behavior. It applies to systems with random surfaces which are smooth on the length scales of q^{-1}, i.e., area S is independent of resolution r. Thus, the length scale r is given by q^{-1} so that one has to go to a smaller q as r increases. Therefore, a smaller θ is required in the experiment, and for this reason it is named small-angle scattering.

For self-similar surfaces on rock grains, $S \propto r^{2-D}$ from Equation (12.2) or (2.7b), and substituting it into Equations (12.26) and (12.25) yields

$$I(q) \propto q^{-(6-D)}. \tag{12.27}$$

For a volume self-similar fractal, however,

$$I(q) \propto q^{-D}. \tag{12.28}$$

The different powers of the wave vector $\boldsymbol{q}$ in Equations (12.27) and (12.28) readily distinguish volume fractals from surface fractals, because $D \leq 3$ and $6 - D \geq 3$. Experiments on many sandstones and shales have found the behaviors described by Equations (12.27) and (12.28) (Wong, 1988). A log-log plot of the wave-vector transform function $I(q)$ (Figure 12.14) shows slopes between -3 and -4, indicating fractal dimensions between 2 and 3. The length scale $r = 1/q$ probed by these experiments is under about 500 Angstroms (Å).

One of the advantages of the small-angle scattering method is being able to distinguish self-affine fractal surfaces from fractal surfaces. For self-affine fractal surfaces, the surface area S in Equation (12.26) is estimated by (Wong, 1985, 1988; Wong & Lin, 1988);

$$S \propto S_0[1 + (b/r)^{1-\alpha}]. \tag{12.29}$$

Thus, substituting Equation (12.29) into Equations (12.26) and (12.25), we have

$$I(q) \propto A/q^4 + B/q^{3+\alpha}, \tag{12.30}$$

where $B/A = b^{1-\alpha}$ determines crossover length b. Because $\alpha = 3 - D$, the $q^{-(3+\alpha)}$ is the

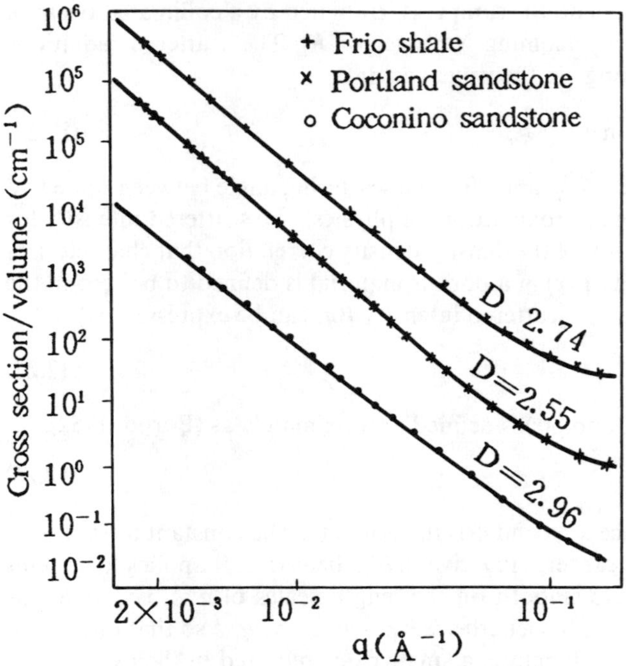

Figure 12.14. Neutron scattering data for sandstone. The intensity of neutrons scattered at small angles shows the power-law behavior described by Equation (12.27) plus a constant background. This behavior supports a fractal pore-surface picture below 500 Angstroms (Wong, 1988).

same as the $q^{-(6-D)}$ in Equation (12.27). The additional term q^{-4} reflects the fact that self-affine surfaces become effectively smooth at large length scales. Wong et al. (1986) have shown such a behavior (Figure 12.15) from small-angle neutron scattering data on limestones.

Recently, Hurd et al. (1989) have studied the surface area of silica and carbon samples by using small-angle scattering. The samples have fractal surfaces that behave as though the surface area increases as the fractal dimension approaches 3. Their results again support the surface fractal models for scattering intensity.

12.2.3 *Adsorption methods*

In addition to small-angle scattering, the fractal structure of surfaces at molecular length scales can be investigated by molecular adsorption. As one of the earliest experiments on fractal surfaces, Pfeifer & Avnir (1983) and Avnir et al. (1983) reported data on surface coverage as a function of molecule size. They used an absorption technique to cover the pore surface with a monolayer of molecules. The number of molecules N varied with their size r. Subsequent studies indeed indicate that, at the molecular-size range, the pore surfaces of most materials are fractals, that is, at this range, surface geometric irregularities and defects are characteristically self-similar upon variations of resolution (Avnir & Farin, 1984). When Equation (12.1) is used to measure pore surfaces of different materials at the molecular-size range, then the yardsticks are molecules. There exists a series of relative equations for determination of fractal dimension as listed in Table 12.3. In these equations, n is the monolayer coverage, for example, moles of adsorbate per gram absorbent; σ is the effective

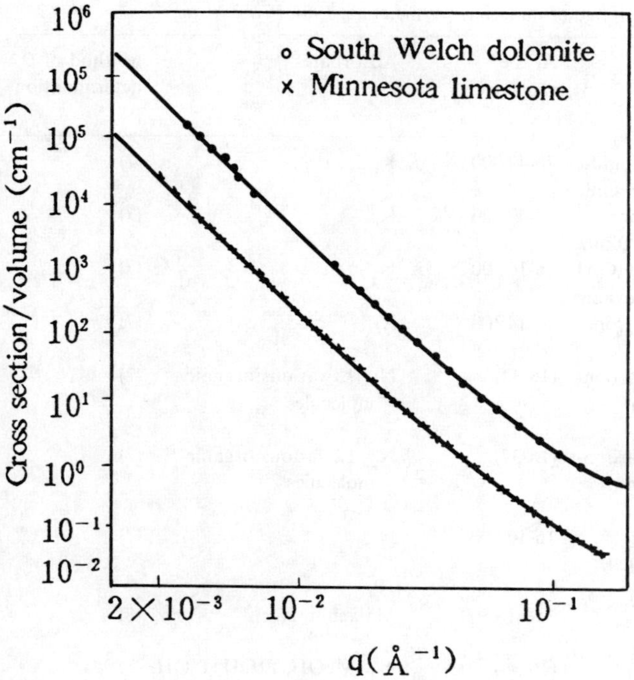

Figure 12.15. The data on carbonate rocks are well-fitted by Equation (12.30) for self-affine fractal surfaces. Least-squares fit of the data gives $b \propto 20$ Angstroms (Wong & Lin, 1988).

Table 12.3. Equations derived for determination of fractal dimensions (Avnir et al., 1984).

Equation No.	
(1)	$n \propto r^{-D}$
(2)	$n \propto \sigma^{-D/2}$
(3)	$A \propto \sigma^{(2-D)/2}$
(4)	$n \propto R_g^{-D}$
(5)	$n \propto V^{-D+1}$
(6)	$V_m \propto \sigma^{-D/2}$
(7)	$A \propto R^{D-3}$
(8)	$n \propto R^{D-3}$
(9)	$\sigma_0 \leq \sigma \leq \sigma_0(R_{max}/R_{min})^2$

cross-sectional area of adsorbate; A is the apparent surface area (M^2g^{-1}); R_g is the radius of gyration; V is the molar volume; V_m is the monolayer coverage; R, R_{max} and R_{min} are values for particle radius; and σ_0 is the cross-sectional area of adsorbate, as used for Equations (7) and (8) in Table 12.3. Pore surface fractal dimensions of some common rock materials measured by these equations are listed in Table 12.4 (Avnir et al., 1984).

Another interesting work is the study on fractal surfaces of coal pores. Since the porous nature oi coal is important in its handling, preparation, and utilization, much

Table 12.4. Fractal dimensions in a range of materials (Avnir et al., 1984).

Fractal dimension*	Found in	At the range (Å^2)↑	Adsorbates	Method of *D* determination
High				
2.91 ± 0.02	Upper Columbus dolomitic rock from Bellevue, Ohio	20-47 000	K_r	(7)
2.97 ± 0.01	Goodland high calcium rock, from Idabel, Okln.	20-47 000	K_r	(7)
2.88 ± 0.02	Granitic rock from SHOAL nuclear test site, Nevada	16-16 500	N_2	(7)
2.73 ± 0.05	Igneous rock sample from SHOAL slit	14-14 300	A_r	(7)
2.71 ± 0.14	Granular activated carbon-Tsurumi Hc.8, from coconut shell	(16-37)	N_2, 12 various organic molecules	(2)
2.80 ± 0.16	Granular activated carbon-Fujisawa B-CG, from coconut shell	(16-37)	N_2, 12 various organic molecules	(2)
2.90 ± 0.01	Carbonate rock from groundwater test well, Yucca Flat Nevada	16-16 500	N_2	(7)
2.92 ± 0.02	Soil (Kaolinite, trace halloysite)	150-16 500	Malachite green	(8)
2.94 ± 0.04	Porous silicic acid	(16-34)	N_2, MeOH, EtOH, P_rOH, iso-P_rOH, BuOH	(3)
2.79 ± 0.03	Activated alumina grade F-20 (Alcoa Corp.)	16-45 100	Polystyrene fractions, N_2	(4)
2.78 ± 0.21	Charcoal (BDH) of animal origin	1400-180 000	Styrene-methyl-methacrylate random copolymers	(4)
2.67 ± 0.16	Porous coconut charcoal (Standard Chemical Co., Montreal)	(16-47)	$N_2, C_2H_2, C_2H_4, CH_4$, $C_2H_6C_3H_8$, n-C_4H_{10}, iso-C_4H_4	(2)
Medium				
2.57 ± 0.04	Porous a-FeOOH pigment for magnetic tapes	16-980	N_2	(7)
2.35 ± 0.11	Crushed Corning 0010 lead glass	21-14 900	Hexadccyl-l-^{14}C-trimethyl ammonium bromide	(7)
2.52 ± 0.07	Coal mine dust from Western Pennsylvania	16-180	N_2	(7)
2.33 ± 0.08	Coal mine dust from Western Pennsylvania	16-270	N_2	(7)
2.55 ± 0.09	Carbon black	(16-71)	N_2, benzene, naphthalene, anthracene, phenanthrene	(2)
2.54 ± 0.12	Slightly porous coconut charcoal (Standard) Chemical Co., Montreal)	(16-47)	N_2, C_2H_2, C_2H_4, CH_4, C_2H_6, C_3H_8, C_4H_{10}, iso-C_4H_{10}	(2)
2.30 ± 0.07	Slightly porous coconut charcoal (Standard Chemical Co., Montreal)	(16-47)	N_2, C_2H_2, C_2H_4, CH_4, C_2H_6, C_3H_8, C_4H_{10}, iso-C_4H_{10}	(2)
2.63 ± 0.03	Mosheim gigh calcium from Stephens City, Va.	20-47 000	K_r	(7)

Table 12.4. (Continued)

Fractal dimension* High	Found in	At the range (Å^2)†	Adsorbates	Method of D determination
2.58 ± 0.01	Niagara (Guelph) dolomite from Woodville, Ohio	20-47 000	K_r	(7)
2.46 ± 0.11	Glassy melted rock from Rainier nuclear zone, Nevada	14-14 300	A_r	(7)
2.29 ± 0.06	Soil (mainly feldspar quartz and limonite)	150-21 800	Malachite green	(8)

effort is expended in the measurement of pore size distribution. Frieson & Mikula (1987) described measurements of pore volume V_p of a number of coal and char samples by mercury intrusion porosimetry (MIP). In the MIP experiment, the mercury intrudes only into the voids between the particles in the sample at very low pressures, and at higher pressures the mercury is forced into the pores in the individual particles. To overcome the interfacial tension between the mercury and the solid, a pressure $P(r)$ must be exerted on the mercury before it will fill a pore of size r. For a cylindrical pore, P and r are related by the well-known Washburn equation

$$P = 2\sigma \cos\theta / r, \tag{12.31}$$

where σ is the surface tension of mercury and θ is the mercury-solid contact angle. In the experiment, the total pore volume at a given pressure is obtained by the volume of mercury forced into the pores. Frieson & Mikula have used the fractal model defined by Equation (12.15) to measure the total volume by means of MIP. Combining Equations (12.15) and (12.31), then

$$\log[\mathrm{d}V_p/\mathrm{d}P] \propto (D-4)\log(P). \tag{12.32}$$

Thus the fractal dimension of the surface of the porous medium can be simply measured from the slope of log-log plot of $\mathrm{d}V_p/\mathrm{d}P$ versus P. Figures 12.16 and 12.17 show the data plotted in the form $\log(\mathrm{d}V_p/\mathrm{d}P)$ versus $\log(P)$ for the various samples (Frieson & Mikula, 1987). From these results, it is possible to distinguish three regions of pressure – low ($P \leq 0.1$ MPa), intermediate ($1 \leq P \leq 10$ MPa) and high ($P \geq 10$ MPa) – which are characterized, respectively, by D_1, D_2, and D_3 (as listed in Table 12.5). In all cases, the low and intermediate pressure data are similar, giving $1 < D_1 < 2$ and $2 < D_2 < 3$, respectively. However, the high pressure data have two different regimes; $D_3 = D_2$ for the chars (samples 5-9), and $3 < D_3 < 4$ for the coals (samples 1-4). Frieson & Mikula explained these results. They thought D_1 as the fractal dimension for a bulk porous medium. At low pressures, the increase in pore volume was primarily due to mercury intrusion into the interparticle voids and the surface pores could not be observed at these pressures. In the region of intermediate pressures, all sample showed fractal behavior as described by the model equation (Equation (12.15)). Different types of samples had different values of $D_2 \in (2,3)$, for coal A, $D_2 = 2.84$ (untreated) and $D_2 = 2.92$ (treated), and so on. Note that the nominal pore diameters determined from

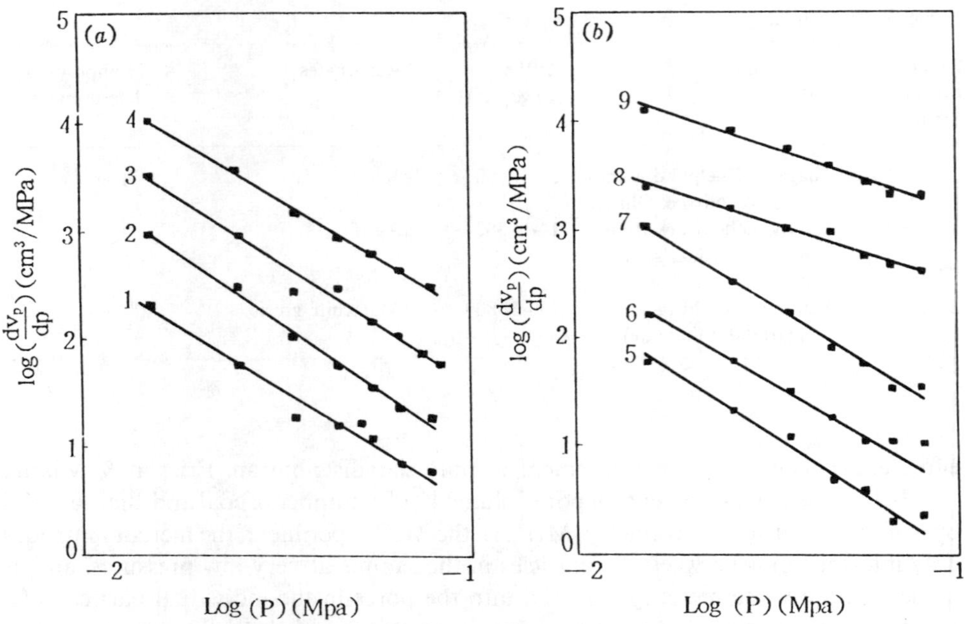

Figure 12.16. Mercury porosimetry data in the pressure range 0.01-0.1 MPa: (a) Coal samples: 1-coal A, 2-coal A (treated), 3-coal B, 4-coal B (treated); (b) Char samples: 5-coal A old char (oxidized), 6-coal A new char (oxidized), 7-coal A old char, 8-coal A old char (uncrushed), 9-coal A new char (uncrushed) (Friesen & Mikula, 1987).

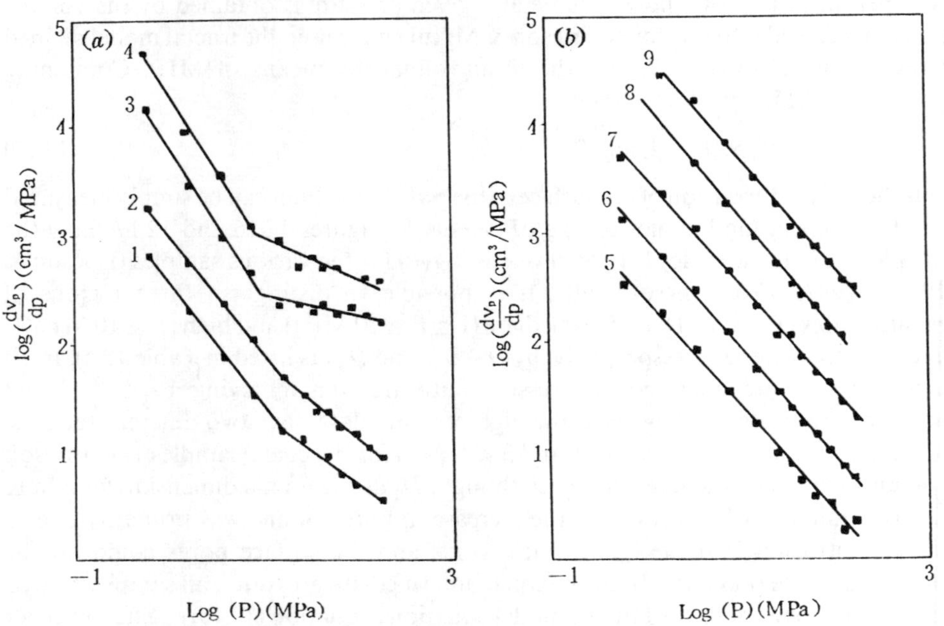

Figure 12.17. Mercury porosimetry data in the pressure range 1-200 MPa: (a) Coal samples; (b) Char samples. The samples are numbered as in Figure 12.16.

Table 12.5. Fractal dimension of coal obtained by Friesen & Mikula (1987).

Sample	Curve No.	D_1	D_2	D_3
coal A	1	1.84	2.84	3.37
coal A treated	2	1.54	2.92	3.33
coal B	3	1.70	2.64	3.87
coal B treated	4	1.82	2.62	3.67
coal A old char (Oxidized)	5	1.73	2.95	2.95
coal A new char (Oxidized)	6	1.93	2.97	2.97
coal A old char	7	1.68	3.00	3.00
coal A old char (uncrushed)	8	2.84	2.94	2.94
coal A new char (uncrushed)	9	2.80	2.99	2.99

Equation (12.31) ranged from 1.2 μm (0.5 MPa) down to 0.05 μm (12 MPa), and the size range of the crushed samples was 0.5×0.25 mm, while that of the uncrushed samples was approximately 5×10 mm. At high pressures, the values of $D_3 > 3$ were nonphysical from a geometric point of view since $D = 3$ corresponded to volume filling. It may be attributed to the compressibility of the coal and failure of coal particles under high pressures.

Recently, Pfeifer et al. (1989) and Albano & Martin (1989) and Drake & Klafter (1990), respectively, have quantitatively measured the absorbed gas at one monolayer, adsorption isotherms and molecular dynamics to determine surface or pore properties at molecular dimensions. However, we should realize that absorption methods may, in some circumstances, yield biased values of the fractal dimension if chemical disorder on the surfaces is a dominant factor, or if molecular orientation or conformation is a function of coverage or surface structure (Thompson, 1991). In addition, Fripiat et al. (1986) have found that molecular adsorption does not ideally cover the interface. However, these concerns have not proven to be of a general significance in studies on geological materials in which measured values of D are in approximate agreement with those obtained by other means.

In this section we introduced some methods of fractal measurements in porous rocks. It is seen that numerous methods have been successfully used to determine the microscopic fractal properties of porous rocks. However, it should be pointed out that each of the reported methods has its advantages and also some limitations. The choice of these methods depends on the questions being asked and the length scales of relevance. A systematical review of these methods has been shown in the paper of Thompson (1991).

12.3 FRACTAL PARTICLES

On the question of why crushed stones (particles) should have a fractal surface, Mandelbrot wrote (1982, pp. 262-263): 'A stone is made of grain stuck together into domains organized hierarchically into bigger domains sticking less strongly together than their smaller components. The energy generated when a stone is hit would dissipate itself very easily by separating big domains, but there is no reason to expect such separation to be allowable geometrically, therefore the break is likely to combine portions belonging to inter-domain walls of diverse hierarchical level'. The following examples presented by Avnir et al. (1985) show that this is indeed the case. Whittaker & Love in 1954 have studied the surface properties of crushed agricultural limestones and dolomite, as summarized by Avnir, 'When particle size was varied, the total surface area showed relations with geometric surface (i.e., average particle radii) varying from nearly linear to markedly curvelinear'. Fractal analysis certifies the situation: all six limestones have self-similar surfaces (Table 12.6). The carbonate rock was collected from different sites indicating how common in nature self-similarity occurs.

Soils are characterized, in general, by surface area and particle size distribution. Cohen & Knight (1974) have measured these two parameters for seven soils. The soils were fractionated into 12 grain sizes, and the surface area of each fraction was measured by malachite green absorption. A priori, one may perhaps expect not to find self-similarity in these composite agglomerated materials, yet, the fractal nature from Mandelbrot again proved to be correct (Avnir et al., 1985). Over the entire wide range of particle sizes, two well-defined regions of fractality could be identified (Table 12.7), corresponding to fine particles as one group and coarse particles as a second group. In fact, Serra (1982) analyzed magnified pictures of clays in order to determine the surface fractal dimension and, interestingly, obtained D values in the range from 2.54 to 2.80, very close to the range of D for the clay-rich fractions (2.42-2.99 in Table 12.7) obtained by Avnir et al. (1985). It should be pointed out here that the magnifications used by Serra are far from the molecular length scale range.

Particles of coal dusts can be regarded as the fragmentation with fine size. They are characterized, in general, by surface areas and particle size distributions. Recently, Xie, et al. (1993) have discussed fractal nature on fine particles under different grinding mechanisms, and found that the fractal dimensions D (for particle-size distribution) and D_s (for particle surfaces) are dependent upon grinding mechanisms, sample material and probably residence time to a certain degree.

Table 12.6. Surface areas, particle sizes and fractal dimensions of various carbonate rocks (Avnir et al., 1985).

Average particle diameter (μm)	Surface area (m^2/g)						
	Sphere	Iceland spar	Niagara	Mosheim	Ranport	Goodland	Columbus
2605	0.00085	0.008	0.136	0.149	1.73	11.7	0.303
1300	0.0017		0.175	0.237			0.293
651.0	0.0034	0.031	0.238	0.277	2.38	11.54	0.315
325.5	0.0069		0.301	0.332		12.23	0.354
163.0	0.0137	0.079	0.442	0.479	3.31	12.10	0.400
81.0	0.0275	0.171	0.586	0.506	3.09	12.63	0.375
53.5	0.0389	0.213	0.679	0.765		12.85	0.425
Fractal dimension	2.01 ± 0.01	2.16 ± 0.04	2.58 ± 0.01	2.63 ± 0.03	2.81 ± 0.04	2.97 ± 0.01	2.91 ± 0.02

Note: Range of self-similarity for all samples: 20-47000 (Angstrom)^2.

Table 12.7. Surface areas, particle sizes, and fractal dimensions of various soils (Avnir, et al., 1985).

	Adsorption of malachite green (mg/g)			
Particle size (μm)	Soil 1	Soil 2	Soil 3	Soil 4
325.5	3.6	3.6	3.6	3.9
223	4.6	4.6	4.1	4.0
111.5	(2.0)	6.8	5.4	4.2
53.5	13.1	9.6	10.7	4.8
		$D = 2.46 \pm 0.02$		$D = 2.89 \pm 0.02$
40	13.6	(17.0)	12.8	(9.5)
33	15.9	(17.9)	14.1	13.1
	$D = 2.34 \pm 0.03$		$D = 2.37 \pm 0.05$	
27	15.8	21.6	18.6	17.6
21	18.1	22.2	22.7	18.5
15	18.3	22.6	23.0	21.0
8.5	18.4	22.7	23.4	24.9
3.5	16.9	26.4	22.9	50.5
2	17.8	25.7	28.2	71.5
	$D = 2.99 \pm 0.03$		$D = 2.90 \pm 0.04$	
		$D = 2.92 \pm 0.01$		$D = 2.43 \pm 0.04$
Particle size (μm)	Soil 5	Soil 6	Soil 7	
325.5	3.1	7.3	2.7	
223	4.1	9.5	3.3	
111.5	10.8	16.0	6.1	
53.5	13.4	15.5	6.3	
40	20.0	45.0	(11.9)	
33	20.9	48.1	17.9	
27	22.5	66.6	19.6	
	$D = 2.19 \pm 0.07$			
21	46.4	67.4	20.4	
15	52.8	85.9	24.0	
8.5	56.0	94.7	26.7	
3.5	60.8	97.4	52.0	
2	62.4	157.3	51.0	
	$D = 2.89 \pm 0.02$		$D = 2.59 \pm 0.04$	
		$D = 2.65 \pm 0.05$		

In particle technology, particle size distribution functions are defined as follows:

$Y_{n(x)}$ = Fraction of total number of particles which are finer than size x; and
$Y_{v(x)}$ = Fraction of total volume of particles which are finer than size x.

From the definition of the number distribution function $Y_{n(x)}$, the actual number $\mathrm{d}N$ of particles whose size lies between x and $x + \mathrm{d}x$ is given by

$$\mathrm{d}N = N_t \mathrm{d}Y_{n(x)}, \tag{12.33}$$

where N_t is the total number of particles in the system. The volume dV of particles in this size interval is obtained by multiplying dN by the volume of a single particle, i.e.,

$$dV = K_v x^3 N_t \, dY_{n(x)}, \tag{12.34}$$

where K_v is the volume shape factor. Note that dV can also be determined directly from the volume distribution function $Y_{v(x)}$:

$$dV = V_t \, dY_{v(x)}. \tag{12.35}$$

Where V_t is the total volume of particles in the system. From Equations (12.34) and (12.35), it follows that

$$dY_{v(x)} = N_t (K_v / K_t) x^3 \, dY_{n(x)}. \tag{12.36}$$

From the basic definition of the number distribution function, $Y_{n(x)}$ can be written by

$$Y_{n(x)} = (N_t - N)/N_t \% = (1 - N/N_t)\%, \tag{12.37}$$

where N is defined as the number of particles with a characteristic size greater than x. If particle size distribution is a fractal, then from Equation (2.11) or (11.1), we have

$$Y_{n(x)} \propto x^{-D}, \tag{12.38a}$$

and

$$dY_{n(x)} \propto x^{-1-D}. \tag{12.38b}$$

Substituting Equation (12.38b) into Equation (12.36), Equation (12.36) becomes

$$dY_{v(x)} \propto x^{2-D}. \tag{12.39a}$$

Integrating Equation (12.39a), we have

$$Y_{v(x)} \propto x^{3-D}. \tag{12.39b}$$

This is the fractal expression of the volume distribution functions. The fractal dimension D can be obtained from the slope of $\log Y_{v(x)} - \log x$ plot ($D = 3$-slope).

In fact, in classical particle technology, the Gaudin-Schuhmann distribution and the Rosin-Rammler distribution are strongly related to the fractal distribution. The Gaudin-Schuhmann distribution is expressed by a simple power law:

$$Y_{v(x)} = (x/K)^{\alpha}. \tag{12.40}$$

Where K is known as the size modules and α as the distribution modules. Comparing Equation (12.39b) with Equation (12.40), we have

$$D = 3 - \alpha. \tag{12.41a}$$

Whereas the Rosin-Rammler distribution is expressed in exponential form as:

$$Y_{v(x)} = 1 - \exp[-(x/K_r)^m], \tag{12.42}$$

where m and K_r are experimentally determined parameters. By expanding the exponential term as a power series, for $x \ll K_r$, Equation (12.42) can be approximated in the form:

$$Y_{v(x)} = (x/K_r)^m; \quad (x \ll K_r). \tag{12.43}$$

Thus, the same result as Equation (12.41a) is obtained, given by

$$D = 3 - m. \tag{12.41b}$$

Above conclusions are the same as those discussed in Section 11.1. This means that the Gaudin-Schuhmann and Rosin-Rammler distributions are actually fractal distributions, and the fractal dimension of particle size distribution can be obtained directly from the two empirical parameters α and m.

Specific surface area of particles (area per unit mass or volume) is a useful measure of particle size. Specific surface areas of fine powders are very large indeed if surface roughness is considered. This surface roughness of particles can be regarded as a fractal similar to a quadratic Koch island. From Equation (2.7b), a fractal surface area can be expressed by

$$S = S_0 x^{2-D_s}. \tag{12.44}$$

Where D_s is the surface fractal dimension ranging from 2.0 to 3.0. S_0 is the surface area of particles with smooth surface, i.e., $S_0 = K_s x^2$ where K_s is the surface shape factor. It is clear that if the surfaces of the particles are smoother, then $D_s \to 2.0$ and Equation (12.44) reduces to its Euclidean surface area formulation. Thus, the volume specific surface S_v is given by

$$S_v = S/(K_v x^3) = K_s x^{1-D_s}/K_v = K_{sv} x^{1-D_s}, \tag{12.45}$$

where K_{sv} is the specific surface-volume shape factor. In practice, the mass specific surface S_m is more commonly used. This is simply related to S_v through

$$S_m = S_v/\rho = K_{sv} x^{1-D_s}/\rho, \tag{12.46}$$

where ρ is the true density of the solid. It is seen that Equations (12.44) and (12.45) reduce to the classical results in particle technology as $D_s = 2.0$.

The mass specific surface area is also obtained from the particle size distribution. In a system containing a range of particle sizes, the contribution to the total surface area of a mass W of powder by particle size x to $x + \mathrm{d}x$ is simply determined by the number of such particles in the form

$$\mathrm{d}A = K_s x^2 x^{2-D_s} \mathrm{d}n = K_s x^{4-D_s} \mathrm{d}n, \tag{12.47}$$

but

$$\mathrm{d}n = \{\text{mass of particles in the interval } x \text{ to } x + \mathrm{d}x\}/\{\text{mass of one particle}\},$$

i.e.,

$$\mathrm{d}n = W \mathrm{d}Y_v/(\rho K_v x^3). \tag{12.48}$$

Substituting Equation (12.48) into Equation (12.47), we have

$$\mathrm{d}A = W K_s x^{1-D_s} \mathrm{d}Y_v/(\rho K_v). \tag{12.49}$$

Thus, the mass specific surface area is given by

$$S_m = (K_{sv}/\rho) \int_0^\infty x^{1-D_s} \mathrm{d}Y_v. \tag{12.50}$$

If the material is assumed to follow the Gaudin-Schuhmann distribution, Equation (12.50) can then be integrated to give

$$S_m = \{K_{sv}(3 - D) K^{1-D_s}\}/\{\rho(4 - D - D_s)\}, \tag{12.51}$$

where K is considered as the size in microns of the largest particle.

Equation (12.51) indicates that the actual specific surface area is not only dependent on the fractal dimension of particle size distribution, but also dependent on the surface fractal dimension D_s. The more rough the particle surface is, the larger is the specific surface area of particles that can be calculated from Equation (12.51). Also it is seen that if surfaces of particles are smooth, i.e., $D_s = 2.0$ and $D = 3 - \alpha$ is considered (see Equation (12.41a)), Equation (12.51) reduces to its classical result in particle technology as shown below:

$$S_m = K_{sv}\alpha/\{\rho K(\alpha - 1)\}. \tag{12.52}$$

The theoretical analyses discussed above were tested by the experimental data as shown in Table 12.8 (Li, 1990). A brief description about sample ID and sample preparation process is given in Table 12.9. Figures 12.18 and 12.19 show the plots of $\log Y_{v(x)}$ vs. $\log x$ for the Holmes pulverizer (HP) and mining-generated dust samples. It is seen that good linear correlations between $\log Y_{v(x)}$ and $\log(x)$ exist. This implies that particle size distribution for coal dust is a fractal distribution.

Based on the above conclusion, the fractal dimension D (size) and D_s (surface area) were calculated using Equations (12.41) and (12.51), respectively. The computation of surface fractal dimension D_s from Equation (12.51), where D is determined from Equation (12.41a), is more difficult than the calculation of D. It requires solving a

Table 12.8. Fine particle size distributions (Li, 1990).

Sample size, μm	CG (%)	HP (%)	DP (%)	HM (%)	RM (%)	SB5 (%)	SB30 (%)	ZB5 (%)	ZB30 (%)	JM (%)	MGD (%)
42.21	100	100	100	100	100	100	100	100	100	100	100
29.85	100	100	100	100	100	100	100	100	100	100	100
21.10	100	100	100	100	100	100	100	100	100	100	100
14.92	100	100	100	100	100	100	100	100	100	100	100
10.55	87.8	89.5	86.3	85.3	89.5	96.0	96.0	92.8	96.5	87.2	92.9
7.46	75.8	73.6	71.6	72.1	74.3	84.2	85.7	77.4	85.3	64.5	76.5
5.27	55.6	51.9	51.8	53.7	51.7	64.1	65.5	53.4	69.0	37.5	57.3
3.73	36.7	34.9	32.5	31.7	33.2	42.1	43.1	34.1	47.5	21.0	38.3
2.63	23.4	22.5	20.8	19.4	21.6	27.7	27.7	20.8	30.9	12.4	25.0
1.69	15.8	14.1	13.0	13.6	13.7	15.7	15.1	10.6	19.4	7.5	15.7
1.01	7.7	6.3	5.7	6.5	5.9	8.1	7.6	5.3	10.1	2.8	8.0
0.66	3.3	2.8	2.3	2.6	2.5	3.5	3.0	2.0	4.1	1.2	3.5
0.43	2.8	2.3	1.6	1.8	1.3	0.8	0.6	0.4	0.9	0.9	1.3
0.24	0.9	0.6	0.5	0.6	0.7	0.3	0.1	0.1	0.3	0.3	0.7
0.17	0.4	0.3	0.1	0.2	0.2	0.1	0	0	0.1	0.0	0.3
Distribution parameters											
16%, μm	1.71	1.90	2.05	2.09	1.97	1.71	1.76	2.19	1.45	3.09	1.72
50%, μm	4.82	5.11	5.13	5.02	5.13	4.28	4.20	5.00	3.91	6.28	4.67
84%, μm	9.52	9.52	10.1	10.3	9.43	7.44	7.28	8.78	7.29	10.1	8.87
MA, μm	2.44	2.64	2.87	2.78	2.74	2.45	2.52	3.04	2.23	3.75	2.44
MV, μm	5.43	5.55	5.78	5.78	5.57	4.64	4.58	5.37	4.36	6.49	5.16
S_m	2.46	2.26	2.09	2.16	2.20	2.44	3.73	1.97	2.69	1.60	2.46

Note: MA – mean area diameter; MV – mean volume diameter; S_m – surface area per unit volume (m^2/cm^3).

Table 12.9. Sample preparations and codings (Li, 1990).

No.	Prepared by equipment	Sample coding
1	Coffee grinder	CG
2	CG ⇒ Holmers Pulverizer (HP)	HP
3	Disk Pulverizer	DP
4	Hammer mill	HM
5	Roller mill	RM
6	Stainless steel ball mill, 5 minutes	SB5
7	Stainless steel ball mill, 30 minutes	SB30
8	Zirconium oxide ball mill, 5 minutes	ZB5
9	Zirconium oxide ball mill, 30 minutes	ZB30
10	CG ⇒ HP ï Jet mill	JM
11	Mining-generated dust	MGD

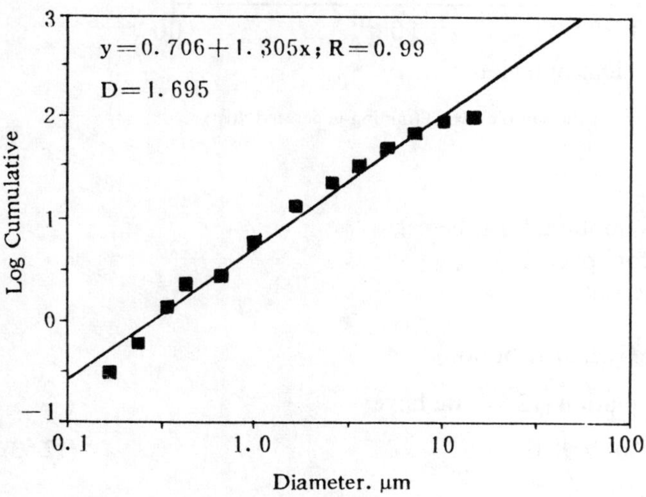

Figure 12.18. Plot of log $Y_{v(x)}$ vs. log x for the sample HP (Holmes pulverizer).

nonlinear equation. A computer program using a finite difference method was written to solve Equation (12.51). The values and units used for the parameters in Equation (12.51) are shown in the following example using data for the mining-generated dust sample (Table 12.8 and Figure 12.19). It should be pointed out that keeping the units in the equation consistent is very important.

For the mining-generated dust sample, the surface fractal dimension D_s is calculated as follows:

$\alpha = 1.450 \times 10^6\ \mathrm{g/m^3}$;

$S_m = 2.458\ \mathrm{m^2/cm^2}$ (surface area per unit volume)

$= 1.695\ \mathrm{m^2/g}$ (surface area per unit mass);

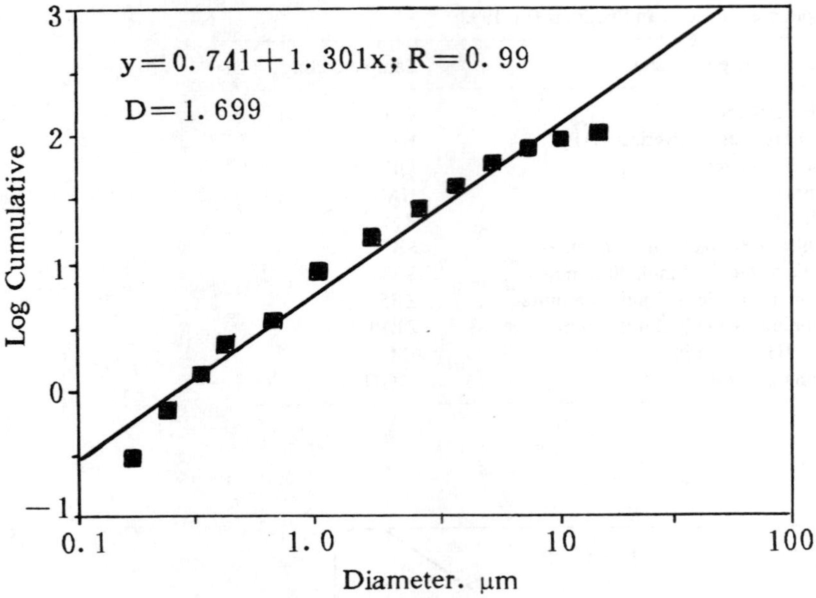

Figure 12.19. Plot of log $Y_{v(x)}$ vs. log x for the sample MGD (mining-generated dust).

$K_s = (\pi d^2)/(\pi d^2) = 1$ (assumption: for spheres);
$K_v = (\pi d^3)/(\pi d^3/b) = 6$ (for spheres);
$K_{sv} = K_v/K_s = 6$ (for spheres);
$K = 14.92\ \mu m$; and
$D = 1.699$ (from particle size distribution).

Substituting the data into Equation (12.51), we have

$$(14.92)^{1-D_s} = 31.485(2.301 - D_s). \qquad (12.53)$$

Solving Equation (12.53), we obtain:

$$D_s = 2.164.$$

The D_s values for other samples were calculated in a similar manner. The results of D and D_s for all samples are collectively shown in Table 12.10 and graphically in Figure 12.20. A different grinding mechanism corresponds to a different value of fractal dimension, which is equivalent to a different fractal distribution. It is found that all of D_s values are larger than 2.0 and smaller than 3.0. It is expected that the fractal dimensions D and D_s are dependent on the grinding mechanism and particle material. It is interesting to see in Figure 12.20 that the trends of D and D_s are opposite. As D_s increases, the corresponding D for the same sample decreases. This is because of that as the amount of fine dust particles present in a sample increases, the fractal dimension D decreases, and D_s increases because of the increase of particle surface area. Furthermore, the fineness of dust particles is proportional to the amount of grinding that the dust sample has been subjected to. The amount of grinding is generally dependent on the effectiveness of a grinding mechanism and residence time. Therefore,

Table 12.10. The calculated fractal dimensions D and D_s.

Grinding mechanism	Slope	R	D	D_s
Coffee grinder (CG)	1.230	0.99	1.770	2.058
Holmes pulverizer (HP)	1.305	0.99	1.695	2.151
Disk pulverizer (DP)	1.445	0.98	1.555	2.232
Hammer mill (HM)	1.357	0.99	1.643	2.216
Roller mill (RM)	1.341	0.99	1.659	2.197
Stainless steel ball mill (SB5)	1.508	0.97	1.492	2.431
Stainless steel ball mill (SB30)	1.619	0.96	1.381	2.584
Zirconium oxide ball mill (ZB5)	1.650	0.98	1.350	2.579
Zirconium oxide ball mill (ZB30)	1.496	0.96	1.504	2.425
Jet mill (JM)	1.489	0.99	1.511	2.340
Mining-generated dust (MGD)	1.301	0.99	1.699	2.164

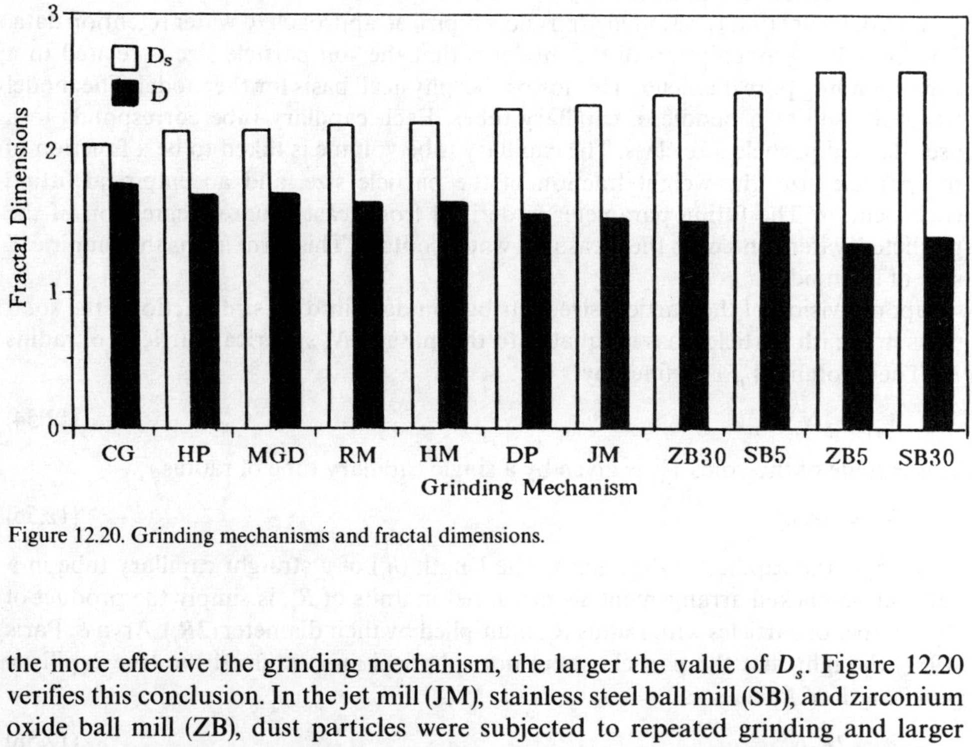

Figure 12.20. Grinding mechanisms and fractal dimensions.

the more effective the grinding mechanism, the larger the value of D_s. Figure 12.20 verifies this conclusion. In the jet mill (JM), stainless steel ball mill (SB), and zirconium oxide ball mill (ZB), dust particles were subjected to repeated grinding and larger residence time than in the coffee grinder, continuous miner (mining dust), Holmes pulverizer (HP), hammer mill (HM), and disk pulverizer (DP). Thus, these samples have relatively larger D_s and smaller D values. However, dust particles cannot be ground down to infinitely small sizes. This is because after a certain period of residence time, particle agglomeration occurs. This phenomenon is also indicated in Figure 12.20. Samples ZB5 and ZB30 were prepared by the zirconium oxide ball mill with a residence time of 5 and 30 minutes, respectively. The values of D and D_s do not follow the regular pattern.

12.4 FRACTAL CAPILLARY TUBE MODEL IN SOIL WATER RETENTION ESTIMATION

Numerous attempts have been made to relate particle-size distribution as discussed in the previous section and Chapter 11 to soil water retention data. The relative ease with which this data may be attained, as well as the similarity in shape of the retention and cumulative distribution curve for soils, suggested such efforts were justified. In recent years, two approaches, Arya & Paris (1981) and Haverkamp & Parlange (1986), have been presented that show significant promise in relating the particle-size distribution data to the retention curve in nonswelling soils. These simplified techniques provide a variable method for bridging the data availability gap. Traditional soil surveys provide a wealth of spatially variable particle-size distribution data, but are often lacking in retention data. However, studies of flow and transport in field soils require input data on the variability of hydraulic properties (retention and conductivity data). Techniques that can bridge this data gap are, therefore, critical in estimating the impacts of agriculture and industry on soil and groundwater quality.

Arya & Paris (1981) presented a 'physicoempirical' approach to water retention data. The underlying assumption of the model is that the soil particle size is related to a corresponding pore diameter. This forms the 'physical' basis for the model. The model treats the soil as a bundle of capillary tubes. Each capillary tube corresponds to a user-defined particle-size class. The capillary tube volume is taken to be a function of the particle size, the weight fraction of the particle size, and an empirical fitting coefficient, α. The fitting parameter is derived from least squares regression of the predicted water content to the measured water content. This term forms the 'empirical' side of the model.

Upon division of the particle-size distribution data into M size fractions, the solid mass in the ith particle class is equated to the mass of N_i spherical particles of radius R_i. Their volume V_{pi} is defined by

$$V_{pi} = \tfrac{4}{3} N_i \pi R_i^3. \tag{12.54}$$

The volume of the voids V_{vi} is given by a single capillary tube of radius r_i,

$$V_{vi} = \pi r_i^2 h_i, \tag{12.55}$$

where h_i is the capillary tube length. The length (h_i) of a straight capillary tube in a cubic close-packed arrangement, as measured in units of R_i, is simply the product of the number of particles with radius R_i, multiplied by their diameter ($2R_i$). Arya & Paris (1981) thought that the particles were not spherical and assumed the true capillary pore length of R_i class as

$$h_i = 2R_i N_i^{\alpha}, \tag{12.56}$$

where α is an empirical constant ranging from 1 to 2. It is clear that Equation (12.56) accounts for the nonspherical nature of the particles.

By combining Equations (12.54)-(12.56), Tyler & Wheatcraft (1989) arrived at an expression for the capillary tube radius r_i in the form

$$r_i = R_i[\tfrac{2}{3} e N_i^{(1-\alpha)}]^{1/2}, \tag{12.57}$$

where e is the sample void ratio. The void ratio for each pore class is assumed constant and equal to the bulk sample void ratio. The tube radius is then related to the capillary pressure ψ_i via the equation of rise in a capillary tube. The water content corresponding to the capillary pressure ψ_i is evaluated by summing the available pore space contained in all particle classes from the smallest class to the *i*th class.

Equation (12.57) does not account for hysterisis nor entrapped air. The latter condition may be overcome by assuming that any entrapped air is equally divided among all pore classes (Haverkamp & Parlange, 1986). By the application of the equation of capillary rise (and assuming a zero contact angle), Equation (12.57) closely models the main drying curve for a nonswelling soil.

The power law relationship given by Equation (12.56) indicates that as R_i decreases (assuming N_i increases proportionally) h_i will grow exponentially. This apparent length-increase behavior actually is a fractal effect. In fact, instead of the straight

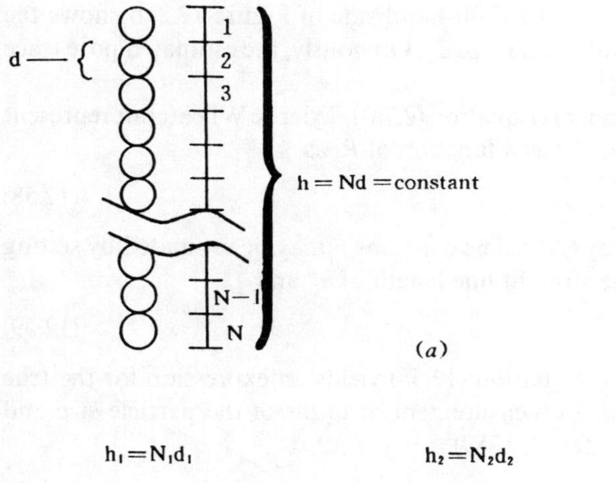

Figure 12.21. (a) Straight capillary tube comprised of uniform spheres; (b) Fractal capillary tube model (Tyler and Wheatcraft, 1989).

capillary tube approach of Arya & Paris (1981), fractal curve measures as discussed in Section 2.2 can be used to evaluate the pore length as a function of measuring scale. Tyler & Wheatcraft (1989) proposed a fractal capillary tube model as shown in Figure 12.21. In their model, smaller and larger particles line the pore wall. Figure 12.21a represents a Euclidean (nonfractal) pore wall whose length is independent of the scale at which it is measured. The volume of the pore space is easily measured and is constant. Figure 12.21b, however, represents a móre realistic soil pore. Both large and small grains line the pore. If any portion of the pore is magnified, more detailed grains lining the pore wall will be observed. Thus, the volume of such a pore is a function of the scale at which it is measured, and its length is also a function of the scale. The pore length and radius are 'measured' in a retention experiment by varying the tension applied to the soil water. An equivalent radius may be estimated from Laplace's equation, while pore length may be calculated from the change in water content. The left-hand trace in Figure 12.21b is shown measured with a ruler length of d_1, and the total pore length h_1 is simply $N_1 d_1$. The right-hand side of Figure 12.21b shows the same pore as measured with a smaller ruler size d_2. Obviously, the estimated pore trace length h_2 will be larger than h_1.

From the fractal curve measurement (Equation (2.7a)), Tyler & Wheatcraft represent the true length of the pore channel h_i^* as a function of R_i as

$$h_i^* = F(2R_i)^{1-D}, \tag{12.58}$$

where the yardstick δ is replaced by $(2R_i)$. The constant F may be estimated by setting measuring length $2R_i$ equal to the straight line length of h_i^* and

$$F = h_i^D. \tag{12.59}$$

Substituting Equation (12.59) into Equation (12.58) yields an expression for the true fractal pore length at the ith scale of measurement in terms of the particle size and straight line length (Tyler & Wheatcraft, 1989)

$$h_i^* = h_i^D(2R_i)^{1-D}. \tag{12.60}$$

Note that the estimated length h_i is always equal to $2R_iN_i$ or d_iN_i (Figure 12.21a) because the nonfractal is considered here. Thus, Equation (12.60) can rewritten as

$$h_i^* = 2R_iN_i^D. \tag{12.61}$$

This relationship actually is equivalent to Equation (12.56) obtained by Arya & Paris (1981). The exponent D in Equation (12.61) is the fractal dimension describing the tortuosity of the pore channels and is equivalent to Arya & Paris's α (Tyler & Wheatcraft, 1989).

The volume of the fractal pore channel V_{vi} can now be expressed as

$$V_{vi} = \pi r_i^2 N_i^D 2R_i. \tag{12.62}$$

After replacing Equation (12.55) with Equation (12.62), Tyler & Wheatcraft arrived at a similar equation for the effective pore radius

$$r_i = R_i[\tfrac{2}{3}eN_i^{1-D}]^{1/2}. \tag{12.63}$$

The difference, however, is that the exponent D is a true measure of scale-dependent tortuosity of the pore trace.

The capillary pressure ψ_i in terms of equivalent height of water for the corresponding pore radius r_i is obtained using the equation of rise in a capillary tube and is given by Tyler & Wheatcraft in the form

$$\psi_i = (2\gamma\cos\beta/\rho_w g R_i)[\tfrac{2}{3}eN_i^{1-D}]^{-1/2}, \qquad (12.64)$$

where γ is the surface tension of water, β is the contact angle, ρ_w is the density of water, and g is the acceleration due to gravity.

In order to apply Equation (12.64), it is necessary to assume that the bulk sample void ratio is equivalent to the void ratio of each particle class. This assumption, in light of the power-law scaling used to estimate the pore length, has interesting ramifications. Most pore-size/grain-size relationships (for example, see Haverkamp & Parlange, 1986) assume a linear correlation of the form: $r_i = CR_i$, where C is a constant indicative of the grain packing. Such linear behavior is not preserved in porous media exhibiting fractal pores.

Assuming a constant void ratio for all pore classes, Equations (12.54) and (12.62) may be used to show the relationship of pore size to grain size. The void ratio is given as

$$e = (\pi r_i^2 N_i^D 2R_i)/(\tfrac{4}{3}\pi N_i R_i^3). \qquad (12.65)$$

Rearranging terms yields

$$r_i = C(N_i^{1-D})^{1/2} R_i, \qquad (12.66)$$

where $C = (\tfrac{2}{3}e)^{1/2}$.

It is clear that the relationship (Equation (12.66)) between pore and grain size derived by Tyler & Wheatcraft (1989) is a function of the distribution and number of soil grains. Equation (12.66) shows several interesting features. For soils exhibiting nonfractal pores ($D = 1$), Equation (12.66) reduces to the traditional linear proportionality of grain size to pore size. The same is also true if the soil exhibits a constant number of grains in each particle-class size. Such a soil would be dominated (on a weight basis), however, by a few large grains.

Most soils, however, show an increasing number of grains with decreasing grain size. For such soils, Equation (12.66) indicates a more rapid decrease in effective pore radius with decreasing grain size than a purely linear decrease. Such behavior is consistent with fractal pore traces. Such traces will show increased tortuosity as the scale of observation is decreased. This increased tortuosity effectively lengthens the pore, resulting in a smaller cross-sectional area (pore radius) necessary to maintain the assumption of a constant void ratio.

Tyler & Wheatcraft (1989) also tested above fractal analysis by using a set of the particle-size distribution and water retention data for five soils (soils B-F) presented by Arya & Paris (1981). The particle fractal dimension D_p is derived from the particle-size fractal distribution defined by Equation (11.1). Their results are listed in Table 12.11. The fractal dimensions of the soils analyzed by Tyler & Wheatcraft are generally larger than those reported by Turcotte (1986a) (see Table 11.1). It is not clear if these differences are related to the choice of the mean particle radius or the more coarse texture of soils investigated by Turcotte (1986a). The fractal dimension of pore traces D is estimated from the additive law of fractal dimension (see Section 3.6) where the topological dimension of particles is equal to 3.0, i.e., $D = 1 + (D_p - 3)$, ranging from 1.011 to 1.485 for all but one soil tested, as listed in Table 12.11. The pore trace

Table 12.11. Fractal dimensions from Particle-size data (Tyler & Wheatcraft, 1989).

Soil type	Particle fractal dimension	Pore fractal dimension
Sand		
Sable de Riviere	2.70	0.70
Oakley sand	3.138	1.138
Sandy loam		
Arya and Paris soil F	3.011	1.011
Gilat sandy loam	3.160	1.160
Clay loam		
Yolo light clay	3.071	1.071
Silty clay loam		
Arya and Paris soil B	3.404	1.404
Loam		
Arya and Paris soil D	3.264	1.264
Arya and Paris soil E	3.163	1.163
Silty loam		
Arya and Paris soil C	3.419	1.419
Silt		
Columbia silt	3.485	1.485

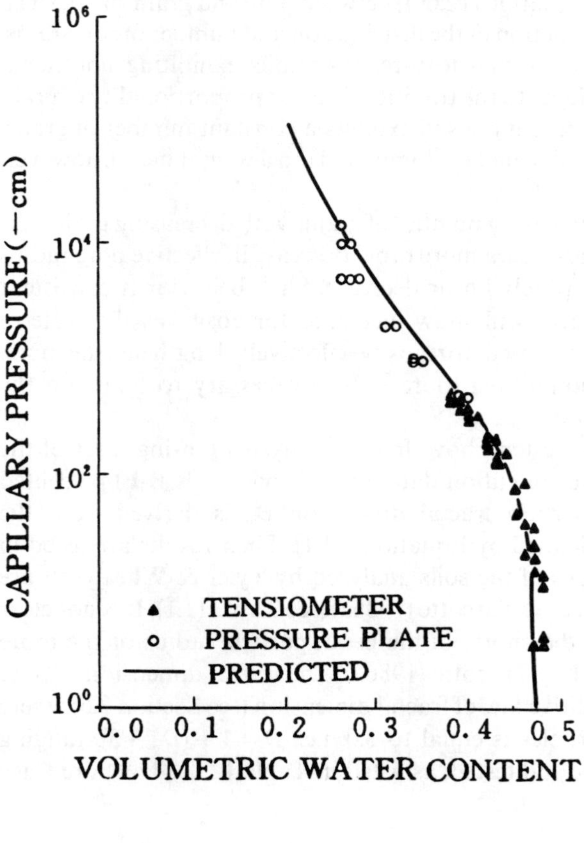

Figure 12.22. Predicted and measured retention data for Arya & Paris soil B (Tyler & Wheatcraft, 1989).

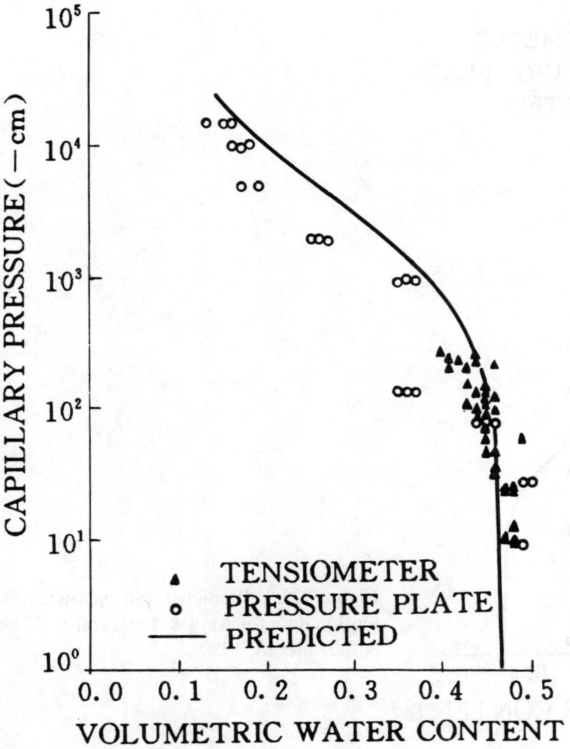

Figure 12.23. Predicted and measured retention data for Arya & Paris soil C (Tyler & Wheatcraft, 1989).

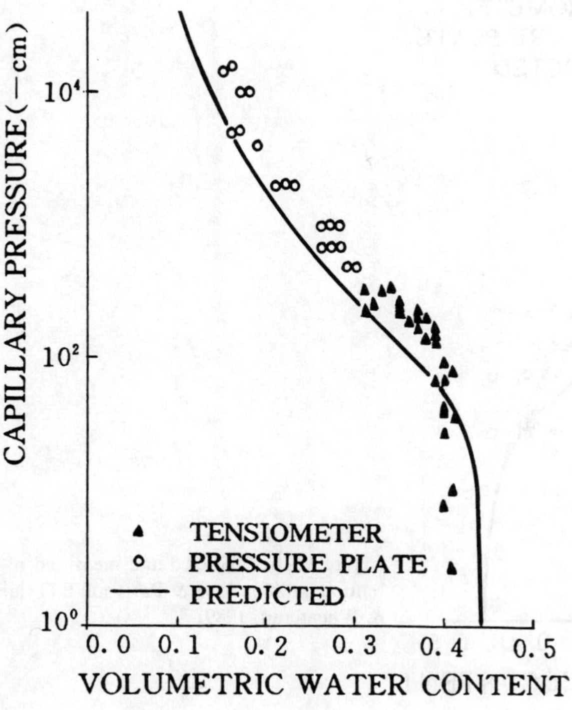

Figure 12.24. Predicted and measured retention data for Arya & Paris soil D (Tyler & Wheatcraft, 1989).

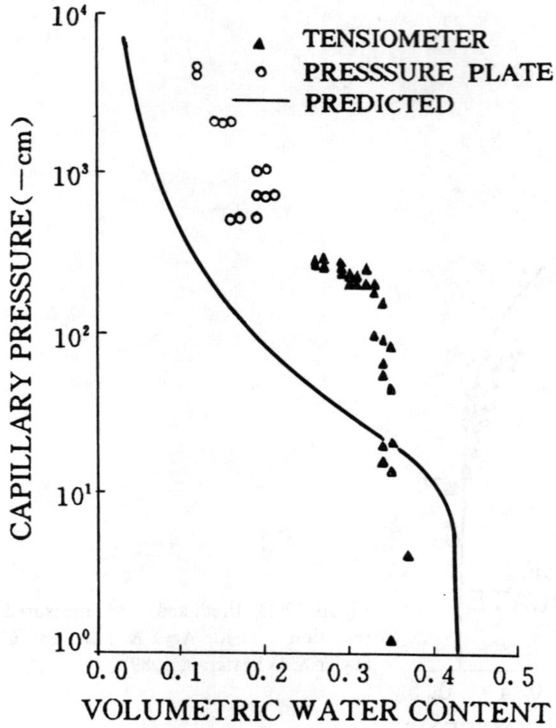

Figure 12.25. Predicted and measured retention data for Arya & Paris soil F (Tyler & Wheatcraft, 1989).

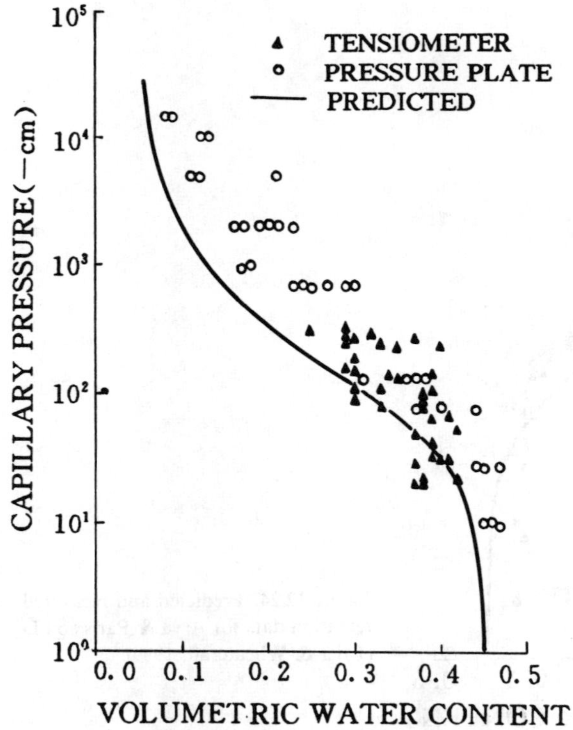

Figure 12.26. Predicted and measured retention data for Arya & Paris soil E (Tyler & Wheatcraft, 1989).

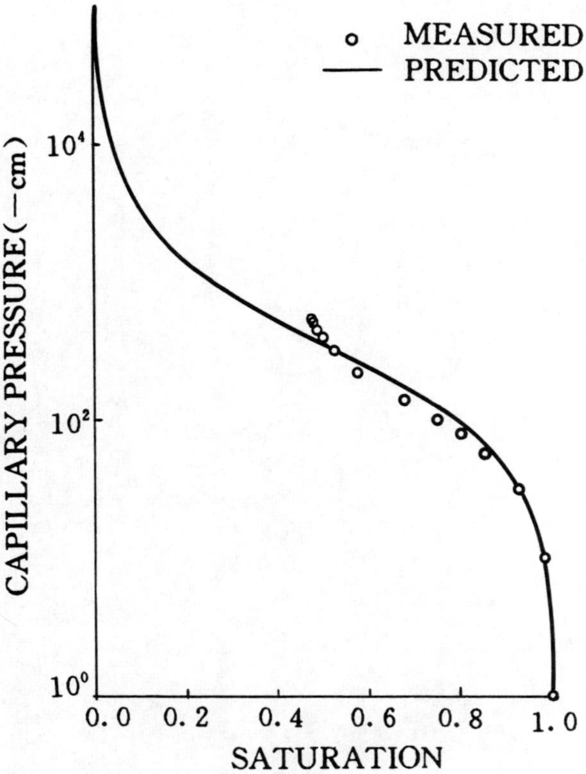

Figure 12.27. Predicted and measured retention data for Columbia silt (Tyler & Wheatcraft, 1989).

dimension of the Sable de Riviere sand (see Table 12.11) is less than 1.0, implying nonfractal behavior. The capillary pressure for the *i*th particle-size class was calculated by Tyler & Wheatcraft using Equation (12.64) assuming a zero contact angle and fluid properties at 25°C. The soil water content θ_i for each ψ_i was calculated as in Arya & Paris (1981) by assuming the available pore space from the smallest size class up to the *i*th class. The void ratios were estimated from bulk density or saturated water content data depending upon the data source. The resulting pairs of capillary pressure and water content (ψ_i, θ_i) were fitted using van Genuchten's retention model (van Genuchten, 1980)

$$\theta = \theta_r + (\theta_s - \theta_r)/[1 + (a\psi)^n]^m, \tag{12.67}$$

where a, n and m, and θ_r were taken to be fitting parameters. Figures 12.22-12.27 show a comparison of calculated and measured water retention data for the soils analyzed by Tyler & Wheatcraft. These results indicate that water retention data may be estimated with reasonable accuracy for soils in which the particle-size data shows power-law scaling with a fractal dimension of > 3.0. Such soils are those with a wide range of particle sizes. The results also indicate that a nonlinear relationship between pore and grain size is appropriate for such soils and should be incorporated into retention and conductivity models of porous media.

CHAPTER 13

Fractal models of rock micro-fractures

One of the difficult problems in geological science and mining engineering is the fracture modelling of rock-material. Macro-fracture mechanics is very difficult to apply to rock-material, because it does not take into account the complexity of microstructures of the rock-material itself nor the variability of conditions affecting the rock-material. One of the research projects on geological materials listed in the report on the research trends in solid mechanics (recently proposed by the American Society of Mechanical Engineering) is to determine parameters which can describe micro-structures of materials and their evolutions and also to develop experimental techniques for measuring these parameters. Therefore, it is very important both in theory and in engineering practice to establish a model for the brittle micro-fracture of rock-material and to uncover the relationship between intrinsic parameters at the microstructural level and macro-mechanics quantities. Our studies indicate that fractal dimension D may play the role of the intrinsic parameter at the microstructural level. In this Chapter, we will introduce our studies on fractal models of rock micro-fractures (Xie & Chen, 1988c; Xie, 1989d).

In Irwin's approach to fracture mechanics for brittle fractures, the critical crack extension force may be written in the form

$$G_{\text{crit}} = 2r_s \quad \text{(for details, see Section 10.1.3)},$$

where r_s represents the surface energy per unit apparent macro-metric area. Here, only the crack extension along a straight line is considered. In fact, the crack in brittle materials frequently propagates along one or more irregular paths, (e.g., in a zigzag where a large zigzag is composed of smaller zigzags, as shown in Photo 13.1, especially in the case of crack branching). Hence, the true area of the fracture surface is actually larger than the data obtained by measurements in which this irregularity is not considered. The area of the fracture surface per unit thickness of a specimen should be (Lung, 1985):

$$A_{\text{true}} = (L(\delta)/L_0(\delta))A, \tag{13.1}$$

where $L_0(\delta)$ is the length of the fractal curve which is straight. It follows that the critical crack extension force should be generalized by,

$$G_{\text{crit}} = 2(L(\delta)/L_0(\delta))r_s. \tag{13.2}$$

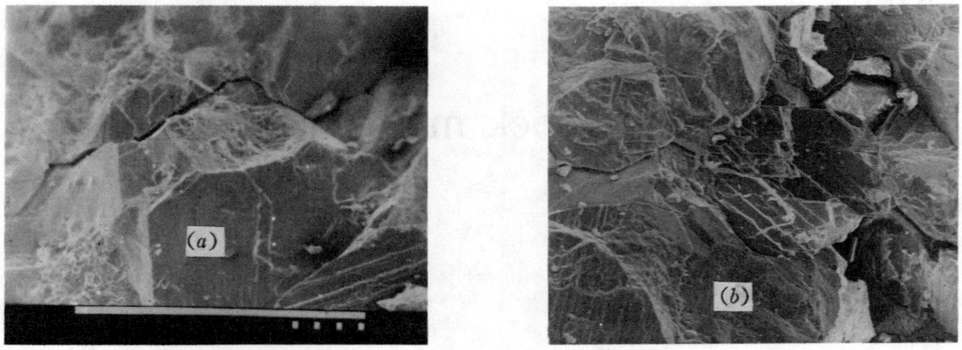

Photo 13.1. Submicroscopic crack extension path of marble under compressional failure. (a) Scanning electron micrograph $\times 63$; (b) Scanning electron micrograph $\times 135$.

It is clear that the smaller the grain size in a certain range, the larger the true area of fractal surface will become. According to Mandelbrot (1982), a crack curve length is defined by;

$$L_i(\delta_i) = L_0^D \delta_i^{(1-D)}.$$

Choosing L_0 as a unit length (say 1 cm) and l as the grain size which is used as the yardstick, we find that

$$G_{crit} = 2r_s l^{(1-D)}. \tag{13.3}$$

The microstructure of a rock consists of large grains. Its fracture is brittle, which is mainly characterized by intergranular brittle fracture, transgranular brittle fracture, and their combinations (for details, see Section 10.2). The crack extension path of these fractures is irregular, can be considered as statistically self-similar, and can then be modelled by a fractal.

13.1 A FRACTAL MODEL OF INTERGRANULAR BRITTLE FRACTURE

In the intergranular fracture case, a crack would propagate along grain boundaries as shown in Photo 13.2. Lung (1985) proposed two fractal models of intergranular brittle fracture (Figure 13.1a, b). From Section 2.7.5, we have

(a) $N = 2$, $r = 1/1.732$, $D = \log(2)/\log(1.732) = 1.26$, and
(b) $N = 4$, $r = 1/3$, $D = \log(4)/\log(3) = 1.26$,

where N is the number of fragments and r represents the ratio of similarity ($N = L_i/\varepsilon_{0i}$, $r = \varepsilon_{0i}/L_{0i}$). Let $\delta_i = \varepsilon_{0i}$, then $L_i/L_{0i} = (L_{0i}/\varepsilon_{0i})^{D-1}$.

From Equation (13.2), we have

$$G_{crit} = 2r_s(L_{0i}/\varepsilon_{0i})^{D-1} = 2r_s(1/r)^{D-1}; \tag{13.4}$$

(a) $G_{crit} = 1.73^{0.26} \times 2r_s = 1.15 \times 2r_s$, and
(b) $G_{crit} = 3^{0.26} \times 2r_s = 1.33 \times 2r_s$.

We can see that case (a) will take place more frequently, for it consumes less energy than case (b).

Photo 13.2. Intergranular brittle fracture of marble under three point bending loading (scanning electron micrograph × 45).

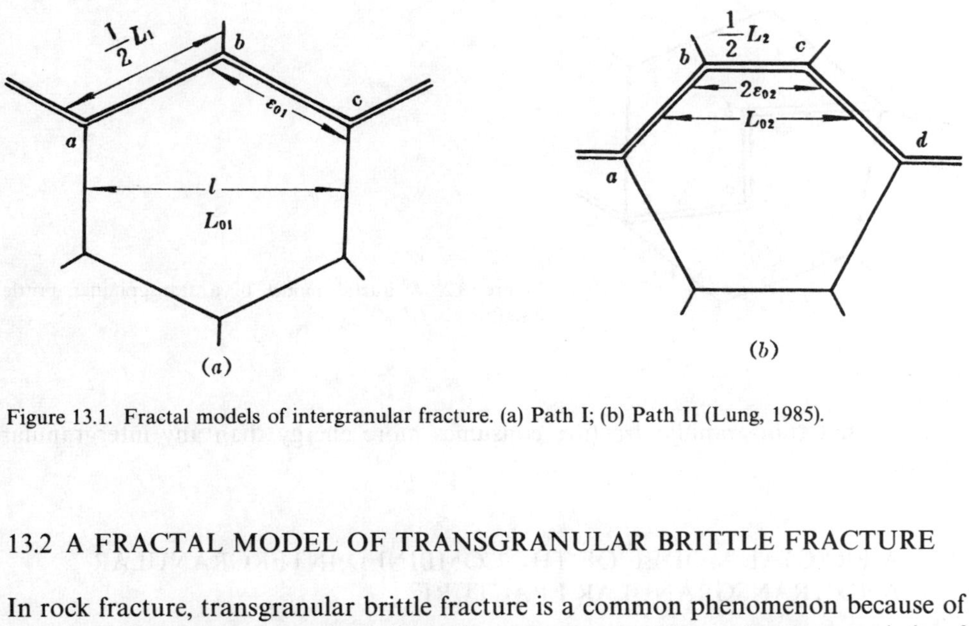

Figure 13.1. Fractal models of intergranular fracture. (a) Path I; (b) Path II (Lung, 1985).

13.2 A FRACTAL MODEL OF TRANSGRANULAR BRITTLE FRACTURE

In rock fracture, transgranular brittle fracture is a common phenomenon because of micro-voids and micro-cracks existing in rock masses. The main characteristic of transgranular fracture is the formation of opposite, non-flat facets of cleavage fracture, i.e. formation of the cleavage step in crystallographic plane as shown in Photo 13.3. A fractal model of transgranular fracture is shown in Figure 13.2,

$$N = 3; \quad r = 1/2.236,$$

and

$$D = \log(3)/\log(2.236) = 1.365.$$

Its critical crack extension force is

$$G_{\text{crit}} = (2.236)^{0.365} \times 2r_s = 1.34 \times 2r_s.$$

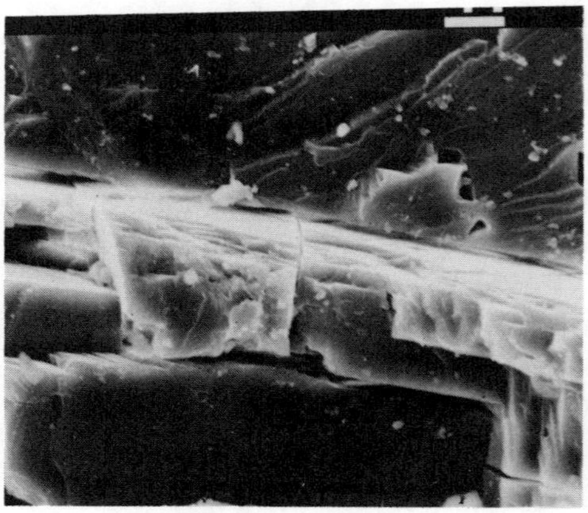

Photo 13.3. Cleavage step in the crystallographic plane of marble (three point bending loading). Arrows indicate the direction of crack pro: pagation (scanning electron micrograph × 1000).

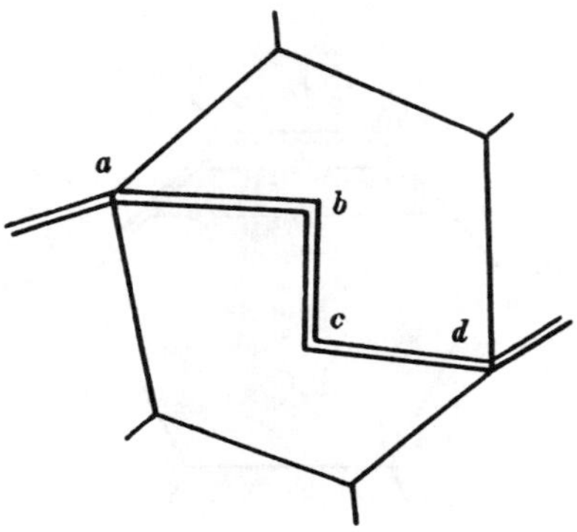

Figure 13.2. A fractal model of a transgranular brittle fracture.

We see that transgranular fracture consumes more energy than any intergranular fracture.

13.3 A FRACTAL MODEL OF THE COMBINED INTERGRANULAR AND TRANSGRANULAR FRACTURE

The fracture profiles of intergranular and transgranular fractures are often found on fracture surfaces of a rock. Local stress concentrations at micro-voids, micro-cracks, and inclusions etc. are the causes of transgranular fracture. Generally, the fracture will be intergranular because less energy is needed. Hence, a fracture surface of rock is that of combined transgranular and intergranular fracture as shown in Photo 13.4. Its fractal model is pictured in Figure 13.3,

$$N = 5; \quad r = 1/3.605; \quad D = \log(5)/\log(3.605) = 1.255.$$

Its critical crack extension force becomes

$$G_{\text{crit}} = (3.605)^{0.255} \times 2r_s = 1.38 \times 2r_s.$$

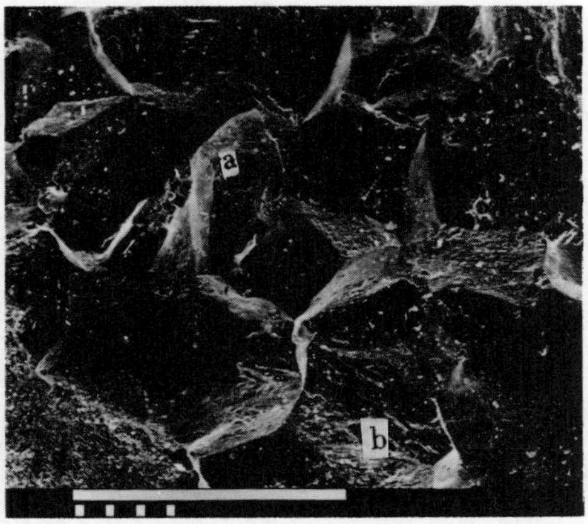

Photo 13.4. Combined (a) intergranular and (b) transgranular fractures of marble (three point bending loading, scanning electron micrograph ×45).

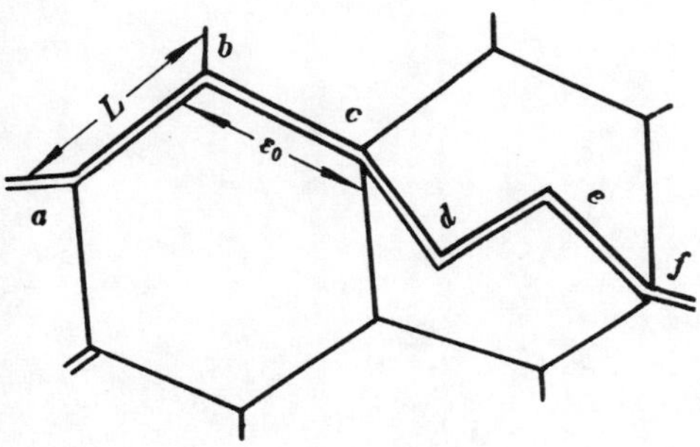

Figure 13.3. A fractal model of a combined intergranular and transgranular fracture.

The analytical results and measured values are listed in Table 13.1. It is clear that calculated results are in good agreement with measured ones. The reason why the intergranular brittle fracture and the combining fracture forms take place more easily for the same grain size can be seen quantitatively in Table 13.1.

Table 13.1. Fractal dimension and G_{crit} under different type of brittle fracture

Fracture type	Calculated value of fractal dimension	Measured fractal dimension of Marble under tensile failure	G_{crit} values for the grain size of 10^{-2} cm $G_{crit} = 2r_s \times l^{(1-D)}$	Liability of rock fracture under the same r_s value
Intergranular fracture (a)	1.26	1.18	$3.31 \times 2r_s$	easy
Intergranular fracture (b)	1.26	1.18	$3.31 \times 2r_s$	easy
Transgranular fracture	1.365	1.31	$5.37 \times 2r_s$	difficult
Combined of intergranular and transgranular fracture	1.255	1.29	$3.25 \times 2r_s$	easy

Figure 14.[illegible] Contours [illegible] and [illegible] fracture [illegible] (section [illegible])

Figure 14.[illegible] [illegible] model of [illegible] intergranular and transgranular [illegible]

The analytical [illegible] and measured values are listed in Table 14.1. It is clear that calculated [illegible] are in good agreement with measured [illegible] intergranular [illegible] and the [illegible] take place [illegible] for the same [illegible] can be seen [illegible] in Table 14.1.

Table 14.1 [illegible]

Fracture type	[illegible]	Measured fracture toughness, Marble under [illegible]	[illegible]	[illegible]
Intergranular	[illegible]	[illegible]	[illegible]	[illegible]
Transgranular	[illegible]	[illegible]	[illegible]	[illegible]
[illegible]	[illegible]	[illegible]	[illegible]	[illegible]

CHAPTER 14

Fractal analyses of rock damage and fracture

Rock damage and fracture behave as good self-similarity in range from a micro-cracking with size $L_1 \leq 1$ µm to an earthquake with size $L_2 \geq 10^6$ kilometers. One can say that fracturing with self-similarity occurs in rocks at all scales, from the microscale (microcracks) to the continental scale (megafaults) and the geologist can equally well observe embrittlement and rupture phenomena under the microscope as on satellite photographs. In the last ten years, fractal geometry has been widely applied to describe microcracking, damage evolution, and the fracture of rocks as well as faults in geology. In this chapter, we will mainly discuss fractal analyses of rock damage and fracture. Fractal characteristics of faults in geology and earthquakes in seismology will be discussed in detail in Chapter 16.

14.1 FRACTAL ANALYSIS OF FRACTURE SURFACES OF ROCK

In recent years, quantitative analyses of fractured surfaces have become an important feature in the process of obtaining a better knowledge of the microstructural processes involved during the deformation and rupture of materials. Coster & Cherment (1983) have given an excellent review of recent developments in quantitative fractography. One of the newest methods, described briefly in the above-mentioned review, is the use of fractals to characterize fractured surface. Fractal analyses of fractured surfaces of metal materials were discussed in detail in Chapter 3. A set of literature about them can be found in various natural scientific journals. Here, we mainly introduce our work on fractal analyses of the fracture surface of rocks (Xie & Chen, 1988c; Xie, 1987b).

14.1.1 *Macroscopic experiments*

Experiments of rock under tensile loading and three point bending are examined. The test pieces are standard specimens. The typical experimental curve of three point bending loading is given in Figure 14.1. When stress reaches its peak value, sufficient energy becomes stored in the specimen. Hence, unsteady crack extension starts, and after failure, stress finally decreases to zero. Let us draw a straight line CD parallel to the elastic loading segment OA as given in Figure 14.1, and assume that the material has the property of elastic unloading at point C_0. The area surrounded by OABCDO

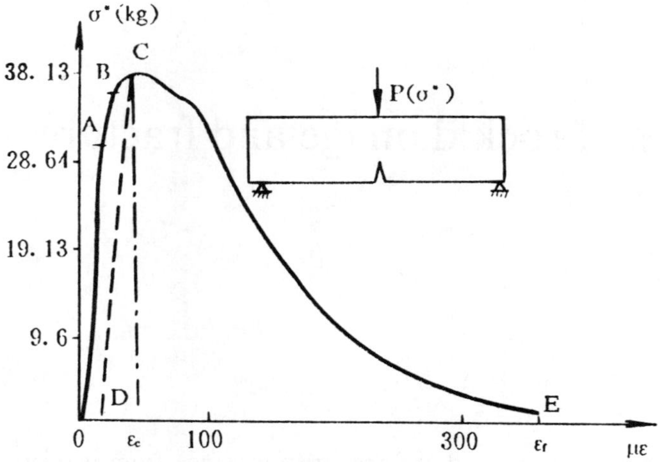

Figure 14.1. Typical experimental curve of marble under three point bending.

then represents the plastic dissipation work and the area under arc DCE is the energy G_f expended in damage and fracture of the material. The expression G_f can be obtained by

$$G_f = \tfrac{1}{2}(\sigma_c^2/E) + \int_{\varepsilon_c}^{\varepsilon_f} \sigma \, d\varepsilon, \tag{14.1}$$

where ε_c and ε_f denote strain at the stress peak value and at failure, respectively.

14.1.2 *Estimation and measurement of fractal dimensions of rock fracture surfaces*

In Chapter 3, we discussed the $P-A$ relationship of fractal sets. Equation (3.2) represented the relationship between non-standard circumferential length P and the area of standard facet A. By using Equation (3.2), we determine that

$$1/D \log P = \text{const.} + (\tfrac{1}{2}) \log A, \tag{14.2}$$

and

$$D = 2\Delta \log(P)/\Delta \log(A). \tag{14.3}$$

The fracture surfaces of rock obtained by macro-fracture experiments are rough and irregular. After statistical self-similarity of these fracture surfaces is tested by the method discussed in Section 3.5, they really are fractals and their fractal dimensions can be estimated by Equation (14.3).

In a series of tests, a fractured rock specimen was plated with gold (or carbon) and the specimen was then continuously polished parallel to the plane of fracture. 'Islands' of rock surrounded by gold (or carbon) became rock slit islands as shown in Photo 14.1. The non-standard circumferential length of rock slit islands were accurately measured with a fixed yardstick length by electron scanning and optical fractographic analyses. Their areas were also obtained. We found that there is a linear relationship between $\log(P)$ and $\log(A)$ (Figure 14.2). For this reason, the data of n groups of $\log(P)$ and $\log(A)$ are analyzed by the linear regression method. An estimation of the fractal

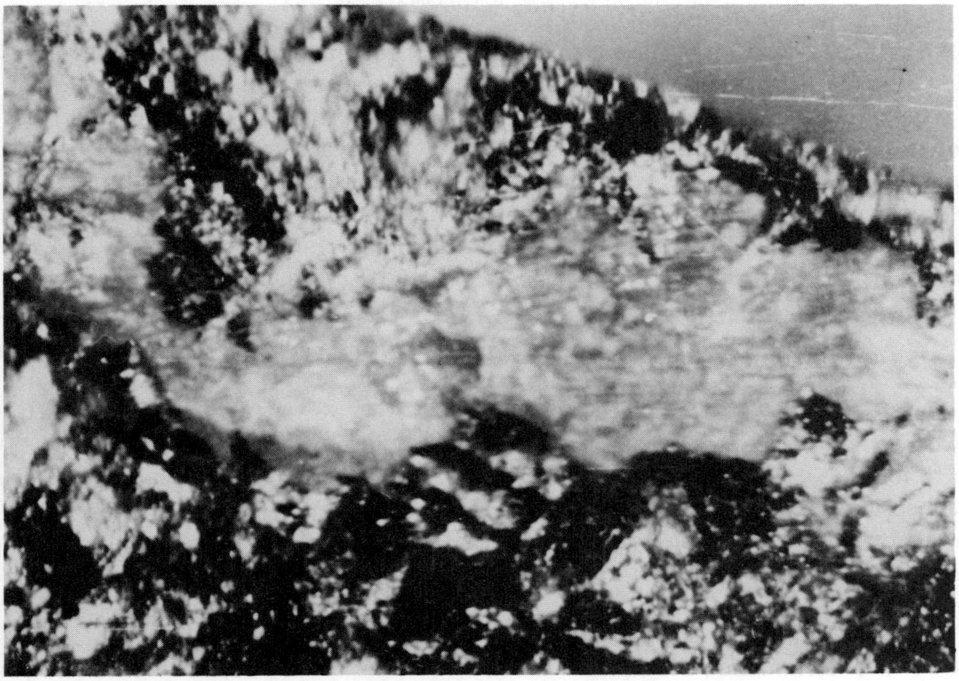

Photo 14.1. Rock slit islands similar to clouds (optical micrograph).

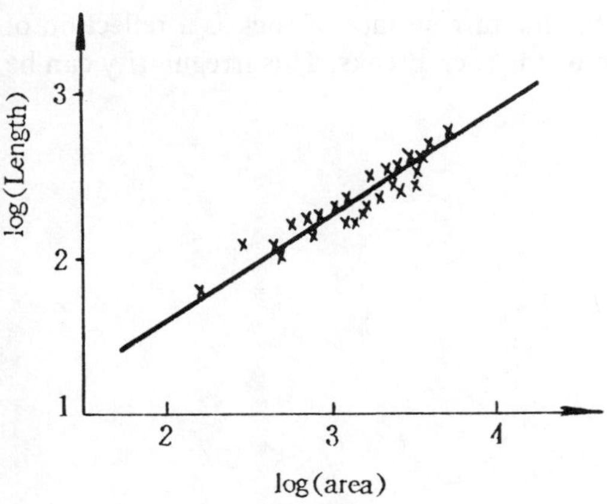

Figure 14.2. The relationship between log (length) and log (area) on a fracture surface of marble under tensile failure.

dimension is then defined by,

$$D = 2\frac{\sum_{i=1}^{n} \log(P)_i \log(A)_i - (1/n)\left(\sum_{i=1}^{n} \log(A)_i \sum_{i=1}^{n} \log(P)_i\right)}{\sum_{i=1}^{n} (\log(A)_i)^2 - (1/n)\left(\sum_{i=1}^{n} \log(A)_i\right)^2} \tag{14.4}$$

14.1.3 *Fractal characteristics of fracture surfaces of rock*

Fractal characteristics of the fracture surfaces of rock under tensile and three point bending are examined and the corresponding fractal dimensions thus obtained. We then have to find the relationship between macroscopic mechanical quantities and these microscopic geometrical intrinsic ones. In the tensile case, macroscopic mechanical quantities obtained in the experiment are only the tensile strength σ_t and the elastic strain energy $U = (\frac{1}{2})(\sigma_t^2/E)$. The relationship between them and the fractal dimension is given in Figure 14.3. In the three point bending case, we first try to find the relationships between fractal dimensions and load P_{max}, the critical stress strength factor K_{Ic}, and the geometric size factor

$$f(\alpha/w) = (a/w)^{1/2}(2.9 - 4.6(a/w) + 21.8(a/w)^2 - 37.8(a/w)^3 + 38.7(a/w)^4),$$

where a is the initial crack length and w is the specimen width.

It looks as if a certain relationship exists (Figure 14.4), but it is not ideal. The fractal dimensions are then connected with the dissipation energy G_f of damage and fracture evaluated by macroscopic experiments. The result is very good (Figure 14.5). The relationship can be expressed by

$$D = 2.49 - 0.045 \times 10^3[\sigma_c^2/(2E) + \textstyle\int_{\varepsilon_c}^{\varepsilon_f} \sigma d\varepsilon]. \tag{14.5}$$

Its general formulation becomes

$$D = K_1 - K_2 G_f, \tag{14.6}$$

where K_1 and K_2 are material parameters dependent on the micro-structure of materials.

It can be seen that the irregularity of the fracture surface of rock is a reflection of the damage and fracture energy expended as the rock breaks. This irregularity can be

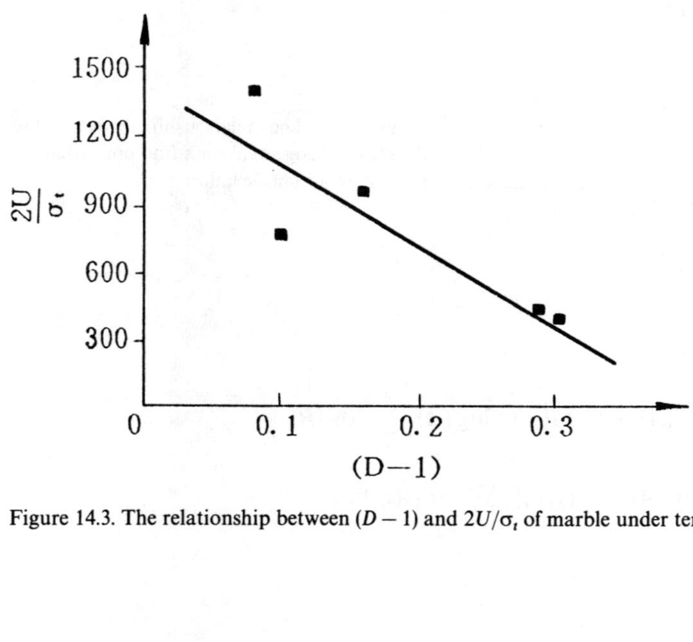

Figure 14.3. The relationship between $(D - 1)$ and $2U/\sigma_t$ of marble under tensile loading.

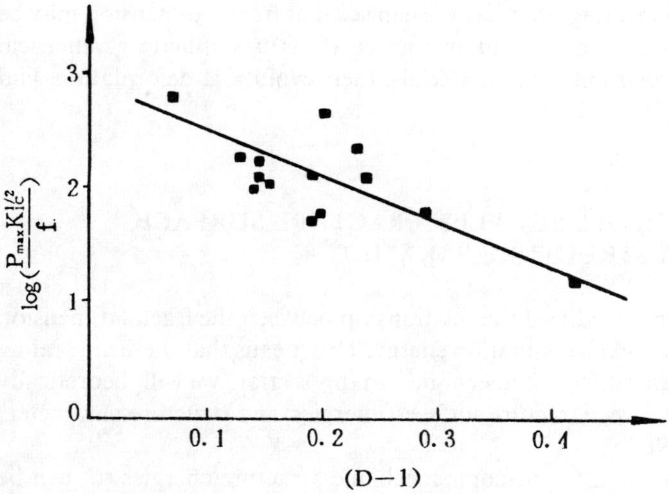

Figure 14.4. The relationship between $(D-1)$ and $\log\{[\mathrm{P}_{\max}(K_{\mathrm{Ic}})^{1/2}]/f\}$ of marble under three point bending.

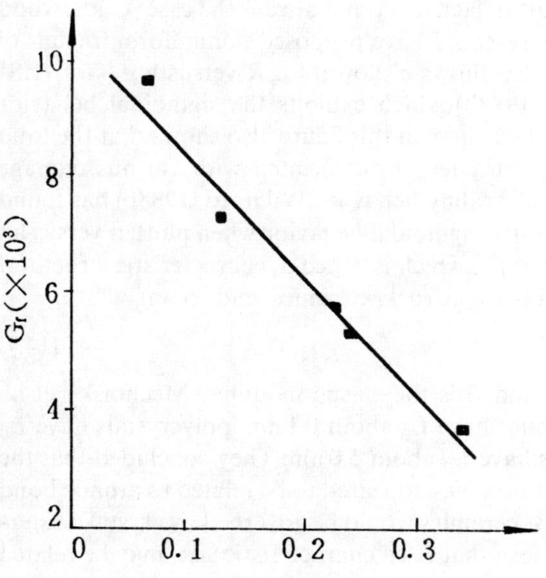

Figure 14.5. The relationship between $(D-1)$ and G_f of marble under three point bending.

modelled by fractal geometry. Therefore, it is possible to retrace the fracture mechanics behavior of rock by means of fractal analyses of the fracture surface. Although G_f (energy) is a synthetical quantity of mechanics behavior, it is a new way to retrace the origin of fracture properties of failed materials. It is clear that the research method given in this section may find very important applications in the analyses of material fracture and in mining engineering practice. We can see that fractal dimension may be useful for fracture analyses of rock, and it may serve as a synthetic geometrical parameter related to microstructures of materials, their evolution, deformation, and fracture properties.

14.2 SCALING SIMILARITIES BETWEEN FRACTURE SURFACE, ENERGIES AND A STRUCTURE PARAMETER

In the previous section, we introduced the relationship between the fractal dimension of fracture surfaces of rock and its dissipation energy. This means that there are scaling similarities between fracture surfaces and energies. In this section, we will theoretically discuss the relationships between fracture surfaces, energies, and structure parameters presented by Williford (1988b).

From fracture mechanics and microscopic mechanics, fracture energies (G) can be quantified according to the scale of observation (L), e.g., J integrals on the macroscopic scale in fracture mechanics or interatomic potentials on the lattice scale in microscopic mechanics. Fracture surface geometries have been described by the roughness parameters (R) or by the fracture surface dimensionality $D \propto \log R/\log L$ (Underwood & Banerji, 1986). A frequently employed assumption is that D is the same for all scales, or that a plot of $\log R$ vs. $\log L$ is linear. In fact, this is not always the case (Underwood & Banerji, 1986, Xie, 1989c). For this reason, I have proposed nonuniform fractals of fracture surfaces (Xie, 1989c). Figure 6.6 shows a plot of $\log R$ versus $\log L$ for AISI 4340 tempered at 200°C (Williford, 1988b) which exhibits the sigmoidal behavior illustrated by the solid curve. The dashed curve in this figure also shows that the total fracture energies ($G_t = G \cdot a$, a is the crack length) associated with various damage mechanisms exhibit the same sigmoidal scaling behavior. Williford (1988b) has found that the fracture energies exhibit the same sigmoidal behavior when plotted versus an apparent material structure parameter, L_0, which is called a 'characteristic structural unit for fracture'. For brittle material (such as rock, ceramics, and so on),

$$L_0 = 2G/E(D-2), \tag{14.7}$$

where $G = 2r_s$ is the surface energy, and E is the elastic modulus. Mecholsky et al. (1986) found that single-crystal ceramics have L_0 about 0.1 nm, polycrystals have L_0 about 0.3–0.4 nm, and glass ceramics have L_0 about 5.0 nm. They concluded that the smaller L_0's for single and polycrystal ceramics appeared to be related to atomic bond breaking, while the larger L_0's for glass ceramics would be governed by crystal clusters surrounded by glassy phases. This is seen that L_0, a characteristic size, may be related to microcracking or damage of materials.

When the microcracks at a crack tip are assumed to be geometrically self-similar to the macrocrack, and material damage is assumed to be distributed in a fractal manner (in Section 14.4, we will certify that the microcracks at a crack tip are self-similar and damage evolution is a fractal, and give fractal analyses of damage of rock), Williford (1988b) has derived the relationship between L_0, energies, and crack length by using the fractal concept. Letting V_f be the fractal 'area' or 'volume' of damage, then

$$V_f = L^D, \tag{14.8}$$

where L is the characteristic length of observation. If the energy density associated with the creation of damage is assumed to be a material constant C, then the total dissipation energy causing damage per unit fractal 'volume' can be written as

$$G_t/L^D = C. \tag{14.9}$$

By rearranging a version of Irwin's equation, a rough estimation of the material constant C corresponding to $D = 3$ is given by

$$C = G_c/a = \pi\sigma_y^2/E, \tag{14.10}$$

where G_c is the crack driving force, a is the crack length, and σ_y is the yield strength instead of the useful fracture stress, whereas $C = G_c$ when $D = 2$. From Equation (14.9), the fractal dimension can be expressed as

$$D = \log(G_t/C)/\log L. \tag{14.11}$$

Let L_0 be the characteristic length of an elemental component of damage, i.e., a microcrack. If there are N such microcracks, their total length (L_t) is

$$L_t = NL_0. \tag{14.12}$$

If the damage field is assumed to be a fractal, then

$$N = b^D, \tag{14.13}$$

where the scale factor b can be defined by the self-similarity between the macrocrack length a and its growth increment Δa;

$$b = a/\Delta a. \tag{14.14}$$

Combining Equation (14.12) with Equation (14.8) yields

$$V_f/L^D = 1 = L_t/(NL_0). \tag{14.15}$$

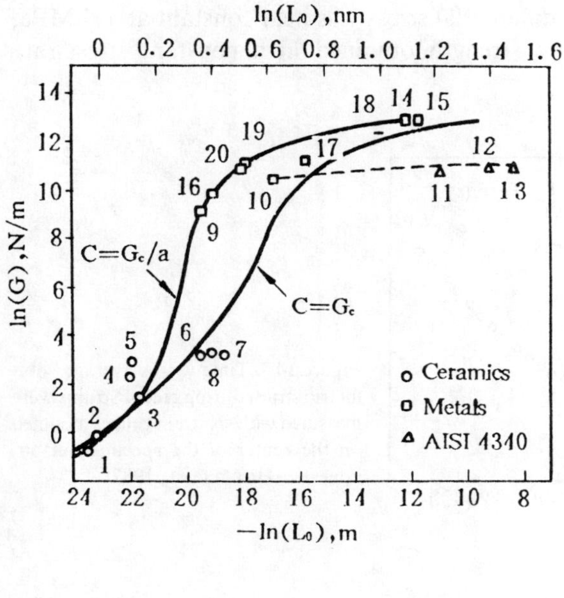

Figure 14.6. Sigmoidal scaling for G and L_0 (Williford, 1988b).

Substituting Equations (14.10) and (14.11) into Equations (14.13) and (14.14) with the scale of observation L taken as Δa and solving for L_0 gives

$$L_0 = L_t(G_t/C)[1 - \log(a)/\log(\Delta a)], \tag{14.16}$$

where L_t is approximated as the sample thickness (W), on the basis that the connection of the microcracks across the crack front is what causes the crack to advance by Δa. G_t is approximated as the total energy at fracture (J_{Ic}) times the projected crack area produced at the scale of observation $L = \Delta a$ (i.e. $G_t = J_{Ic}(\Delta a)^2$). Figure 14.6 shows the simulation results of Williford (1988b). The two solid curves in this figure bound the expected ranges of L_0 for ceramic and metal materials and AISI 4340. The sigmoidal scaling behavior of $\log G$ versus $\log(L_0)$ is also shown in Figure 14.6.

14.3 FRACTAL CHARACTERS OF MICROFRACTURING DURING CREEP OF ROCK

In the previous section we discussed the idea that scaling similarities between fracture surface energies and structure parameters under the assumption that microcracking is a fractal. In this and the next sections, we will give some experimental results to show that microcracking is indeed a fractal, not only during loading process, but also during creep.

A microcracking occurs accompanying a sudden release of elastic strain energy stored in a material. This process generates an elastic stress wave which travels from the point of origin (hypocenter) within the material to a boundary where, using the acoustic technique, the position of the hypocenter (called the acoustic emission hypocenter) can be measured. Hirata et al. (1987) have found that the spatial distribution of hypocenters of acoustic emission in rocks is a fractal, and that the fractal dimension decreases with the evolution of rock fracturing from the rock fracture experiments under a constant-stress.

Their constant stress fracture experiment on Oshima granite was carried out under a confining pressure of 40 MPa. The differential stress was held constant at 547 MPa, which was about 85% of the failure strength obtained in a constant stress-rate

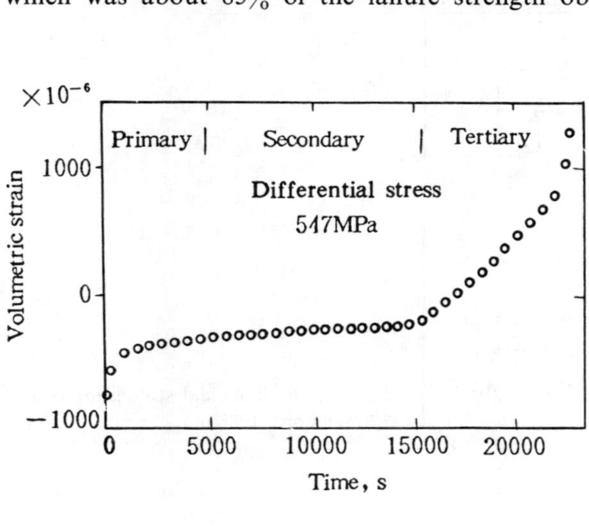

Figure 14.7. Time versus average volumetric strain during creep. Strains were measured with six strain gauges mounted on the center of the specimen at 60° intervals (Hirata et al., 1987).

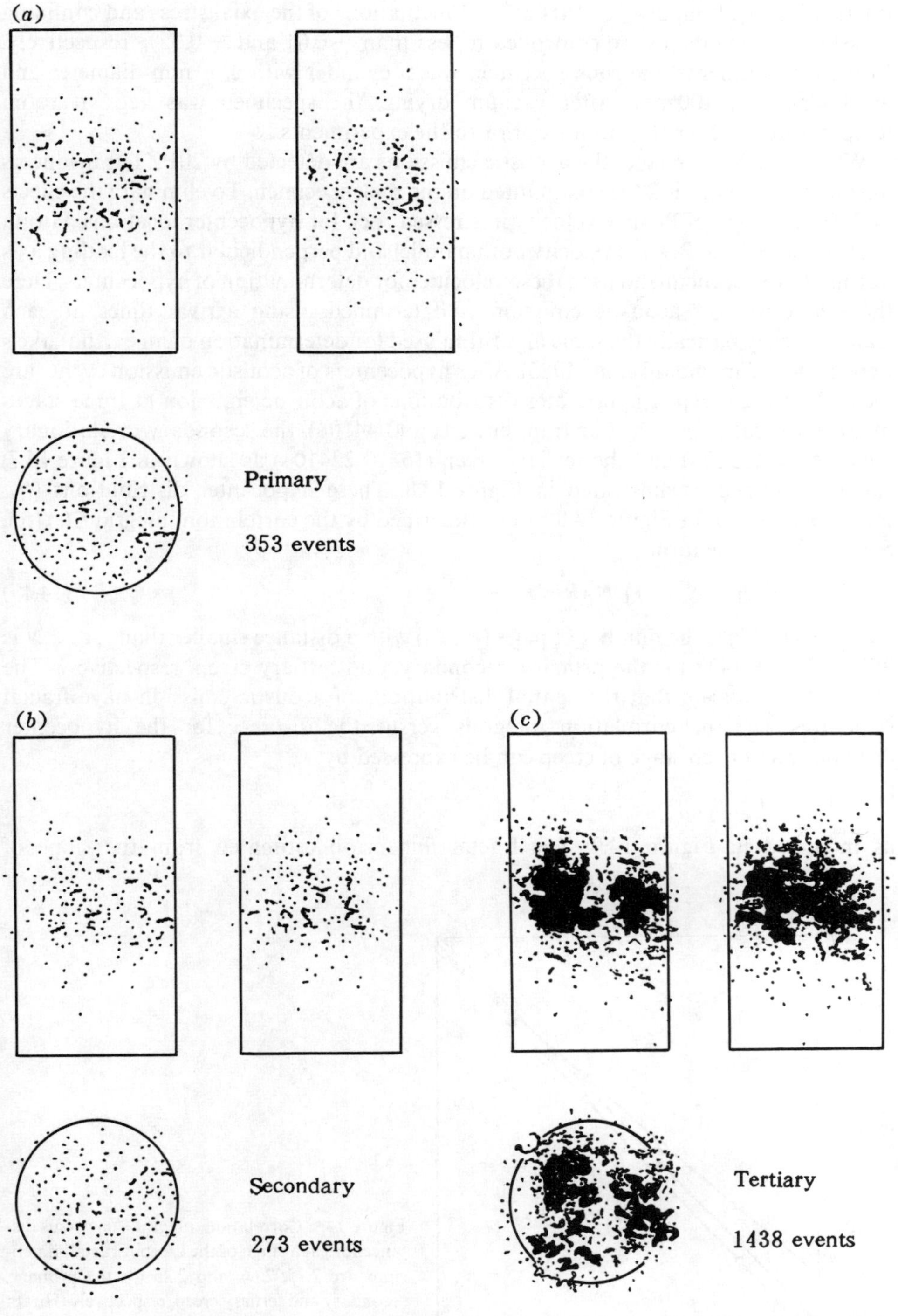

Figure 14.8. Orthographic projections of acoustic emission hypocenters. Hypocenters of (a) 353, (b) 273, and (c) 1438 events were determined during the primary, secondary, and tertiary creep, respectively (Hirata, et al., 1987).

fracture test of about 2×10^{-1} MPaS^{-1}. Fluctuations of the axial stress and confining pressure during creep were controlled to less than ± 0.01 and $\pm 0.02\%$ respectively. In their experiments, the rock specimen was a cylinder with a 50 mm diameter and axial length of 100 mm. After vacuum drying, the specimen was kept at room temperature for over two months prior to the experiments.

When a rock is stressed, the acoustic emissions are detected by 20PZT transducers (resonant frequency is 2 MHz) mounted on the rock specimen. To eliminate the effects of the anisotropy of P-wave velocity in stressed rock for hypocenter locations, Hirata et al. measured the P-wave velocity both parallel and perpendicular to the loading axis during the experiment and used these velocities for determination of hypocenters. Here the hypocenter of acoustic emission is determined, using arrival times at each transducer, by basically the same algorithm used for determination of an earthquake's hypocenter (Yanagidani et al., 1985). After hypocenters of acoustic emission events are located during creep, a hypocenter distributions of acoustic emission at three stages of creep (so-called primary or transient creep (0-4410 s), the secondary or stationary creep (4410-15210 s) and the tertiary creep (15210-22410 s) as shown in Figure 14.7) can be obtained as illustrated in Figure 14.8. These hypocenter distributions (P_1, $P_2, \ldots, P_N$) shown in Figure 14.8 can be described by the correlation integral $C(r)$ (ref. Section 2.7) in the form

$$C(r) = 2/[N(N-1)] \cdot N_f(R < r), \tag{14.17}$$

where $N_r(R < r)$ is the number of pairs (P_i, P_j) with a distance smaller than r, and N is 353, 273, and 1438 for the primary, secondary, and tertiary creep, respectively. The actual data indicate that the spatial distributions of acoustic emission have fractal structures, i.e., the correlation integrals versus the distance for the hypocenter distribution at each stage of creep can be expressed by

$$C(r) \sim r^D,$$

as indicated in Figure 14.9. The fractal dimension estimated from the slope of

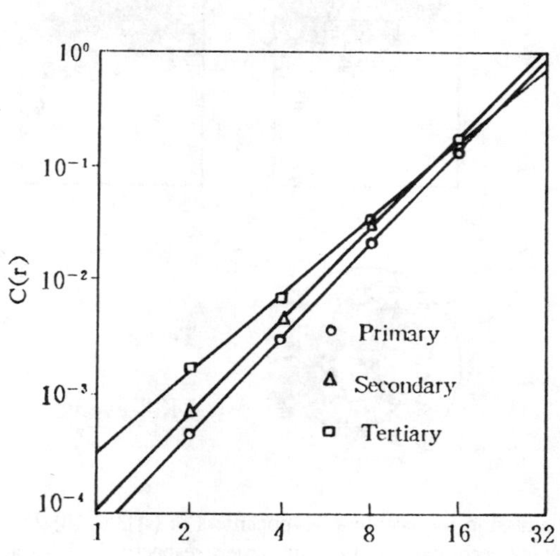

Figure 14.9. Correlation integral $C(r)$ versus distance for each stage of the creep. Fractal dimensions are 2.75, 2.66, and 2.25 for the primary, secondary and tertiary creep, respectively (Hirata et al., 1987).

$\log C(r) - \log r$ plots are 2.75, 2.66, and 2.25 at the stages of primary, secondary, and tertiary creep, respectively.

In this section, the experimental results obtained by Hirata et al., in which the hypocenters of acoustic emission tend to cluster with a self-similar structure in space, directly demonstrate that microcracking in rock is indeed a fractal.

14.4 FRACTAL NATURE ON THE DAMAGE EVOLUTION OF ROCK MATERIALS

The final macroscopic fracture of materials is closely related to the growth and coalescence of existing microcracks. Recently, there has been a great deal of interest in studying how the damage evolves and damage evolution processes in different kinds of materials. A large number of experiments have been performed (Costin, 1983, Botsis & Kunin, 1987; and Nolen-Hoeksema & Gorden, 1987). The experimental results (Chudnovsky, 1984) indicate that under simple loading, the shape and range of the damage zone near the tip of a macrocrack for any kind of material vary with a similar rate as a function of time and space. Chudnovsky regards this property as an important assumption or definition in his *CL* theoretical system, i.e., damage around a macrocrack evolves by the way of self-similarity.

The experimental results of damage evolution in rock materials also show a statistical self-similarity that had been neglected, prior to the development of fractal geometry. Returning to Chapter 10, let us again examine Figures 10.43 and 10.46. The statistical self-similarity of damage evolution really exists. In this section, we will again analyze and describe them by using fractal geometrical methods (Xie & Gao, 1991; Xie, 1991).

A fractal quantity differs from the classical geometrical quantities in that it has a fractional dimension. How do we image a fractal structure or a fractal body which has a fractal dimension? The most classical and visual example is the Menger sponge (Chapter 2). A Menger sponge is a highly ordered structure with holes in it. It is shaped like a cube (three dimension) and has a certain structural rule; that is, the filled original cube becomes a partially filled three-dimensional structure in which its dimension varies in the range of 2.0 to 3.0. In fact, the most natural phenomena rarely possess integer dimensions (La-Point, 1988). For example, a mountain is considered as fractals because its range can only fractionally fill a volume of its space. Although mountain peaks and valleys are located more randomly, the mountain's range is neither a cube [it does not fill all of the (x, y, z) space in a cubical volume] nor is confined to a two-dimensional plane. If it were approximated by an array of appropriately located and length-wise stacked blocks, and then placed within a large rectangular box, it would not fill the entire space and unfilled space would remain. Its dimension must also vary between 2.0 and 3.0, the same as with the Menger sponge (La-Point, 1988). It may be said that the more the substance filling the space, the higher the fractal dimension is. According to this concept, we develop the approach proposed by La-Point (Section 14.9) to measure the fractal dimension of the damage evolution of materials as follows.

Imagine that a grid of squares with an edge length L is overlayed on a damage region (for example, Figure 14.10a, corresponding to the damage region under $\sigma/\sigma_c = 93\%$

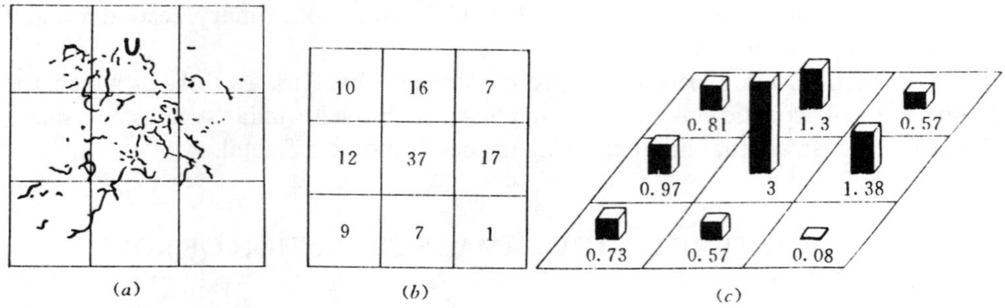

Figure 14.10. An estimation of the fractal dimension of the damage region. (a) A grid to cover the damage region; (b) The number of microcracks in the grid cells; (c) The histogram of a damage field.

in Figure 10.43). These squares divide this grid into nine grid cells. Thus, the number of microcracks intersecting or contained within each square is counted (Figure 14.10b). The distribution of these microcracks can be characterized by a three-dimensional histogram, in which the total size of squares needed to cover the damage region in the $X - Y$ plane indicates the damage range at each step (it is noted that the size of the grid should be able to cover the maximum damage region). The Z-coordinate reflects the damage density which is proportional to the number of the microcracks. The Z-coordinate is then normalized: If the $X - Y$ grid is composed of nine squares (a 3×3 grid), then, the vertical dimension should also be discretized to three divisions. Each division is equal to one-third of the maximum number of microcracks found in any one of the nine squares. For example, as shown in Figure 14.10b, 37 are the most found in the nine squares, and the unit of the Z-dimension will characterize 37/3 microcracks. Thus, the scale occupied by each square of the grid overlying the damage region can be determined (or drawn) by the number of microcracks in each grid cell. This scale reflects only the statistical degree of damage. Figure 14.10c shows the histogram of the damage field. From this histogram, we can figure out the number of the cubes with side length L, $N(L)$. Here, $N(L)$ is not required to be an integer. For example, $N(L) = 9.40$ in Figure 14.10c. By successively overlying the same damage region by the same grid with cells of different size L, a set of data of $N(L)$ and L can be obtained. According to the basic definition of fractal dimension given by Mandelbrot, the fractal dimension of the damage field can be estimated by the following equation:

$$D = \log N(L)/\log(1/L). \qquad (14.18)$$

It is clear that, by using this method, we can determine the fractal dimension of the damage field at each stage in the process of damage evolution of materials.

We have examined the experimental results of damage evolution obtained by Nolen-Hoeksema & Gordon (1987) as shown in Figure 10.43 and determined the plots of $\log(1/L)$ vs. $\log N(L)$ as shown in Figure 14.11 by using the method discussed above. Figure 14.11 shows that plots of $\log(1/L)$ vs. $\log N(L)$ all have a good linear correlation; their slopes are just the fractal dimensions corresponding to the damage fields at different loading levels. This indicates that the process of damage evolution really is a fractal and has good self-similarity (the slopes are basically constants within the

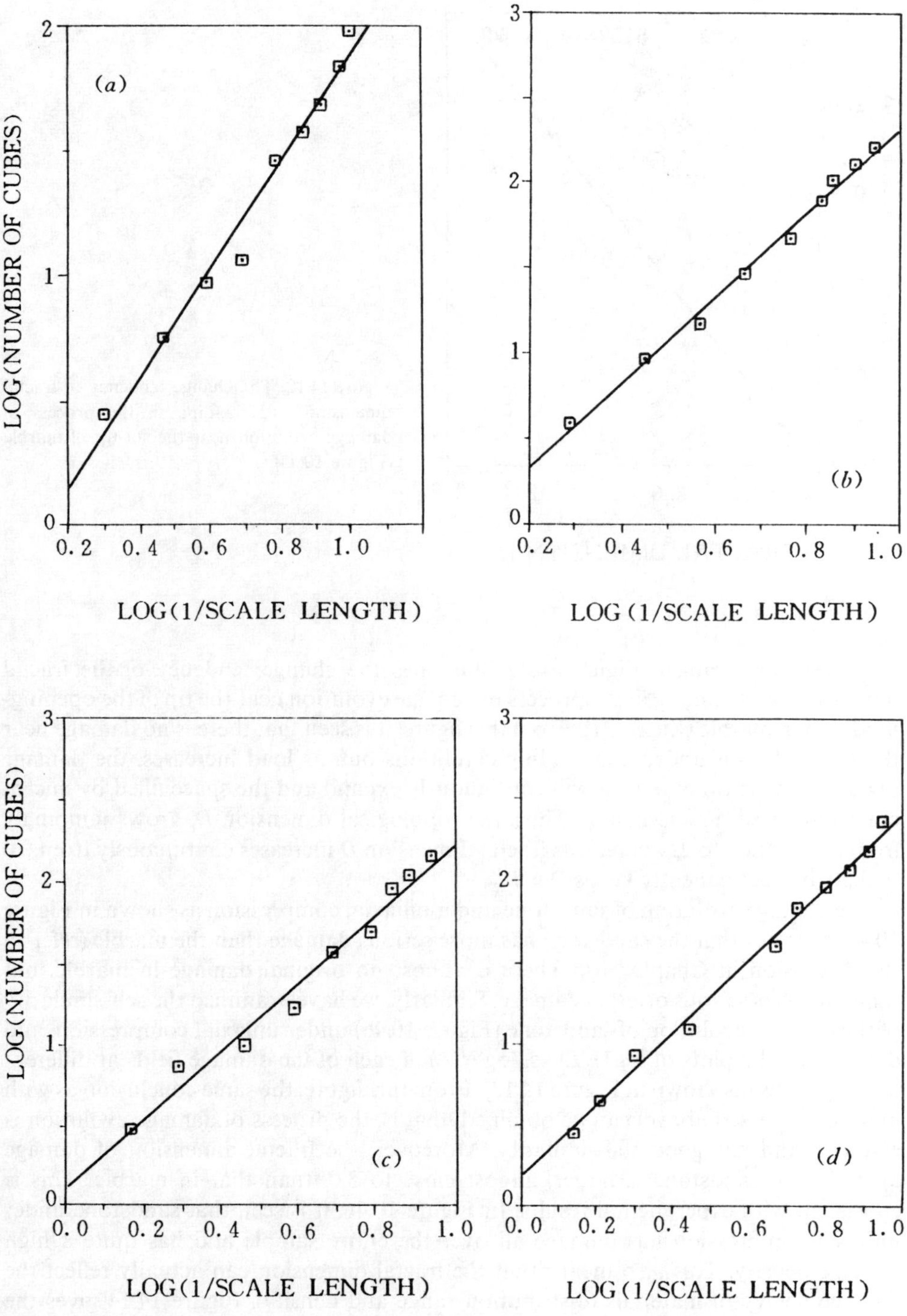

Figure 14.11. The plots of $\log(1/L)$ vs. $\log N(L)$ for the damage region near the tip of the slit in marble (Figure 10.43) at different loading levels in the process of damage evolution. (a) $\sigma/\sigma_c = 64\%$ and $D = 1.70$; (b) $\sigma/\sigma_c = 82\%$ and $D = 2.04$; (c) $\sigma/\sigma_c = 93\%$ and $D = 2.19$; (d) $\sigma/\sigma_c = 96\%$ and $D = 2.32$.

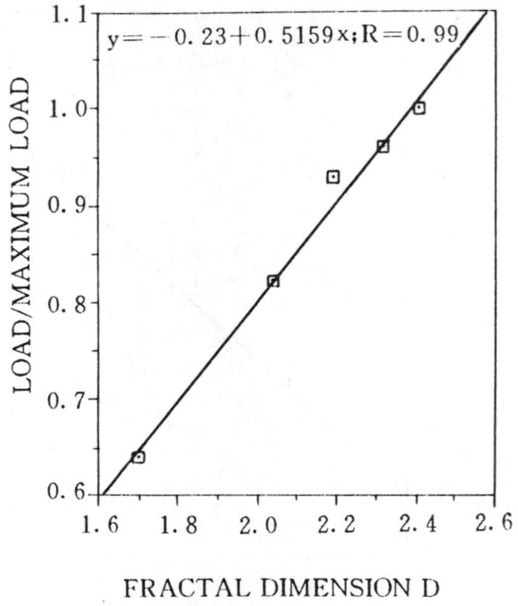

Figure 14.12. The change tendency of fractal dimension with loading in the process of damage evolution near the slit-tip of marble (Figure 10.43).

measured scale range). Figure 14.12 illustrates the change tendency of the fractal dimension with loading in the process of damage evolution near the tip of the opening-mode slit in marble (Xie, 1991). From this figure, it is seen that there is no damage near the tip of the slit under non-loading conditions but, as load increases, the damage region near the tip of the slit will continuously expand and the space filled by micro-cracks increases in succession. Thus, the topological dimension D_t grows jumpingly from 0 to 1, then to 2, whereas its fractal dimension D increases continuously from 0.0 to 2.50, but permanently keeps $D > D_T$.

The damage evolution of sandstone under uniaxial compression as shown in Figure 10.46 indicates that the sandstone has more serious damage than the marble (refer to the discussion in Chapter 10). There is almost no original damage in marble, but sandstone has serious original damage. Similarly, we have examined the self-similarity of the damage evolution of sandstone (Figure 10.46) under uniaxial compression, and determined the plots of $\log(1/L)$ vs. $\log N(L)$ of each of the damage fields at different loading levels (as shown in Figure 14.13). From this figure, the same conclusion as with marble (discussed above) can be obtained; that is, the process of damage evolution is a fractal and has good self-similarity. Moreover, the fractal dimension of damage evolution in sandstone is larger, almost close to 3.0, than that in marble. This is consistent with experimental results. In Figure 10.46, it is seen that sandstone under uniaxial compression has damage all over the entire sample and has quite a high damage density. This also means that the fractal dimension can actually reflect the damage degree of materials (distribution range and density). Figure 14.14 gives the relative curve of the fractal dimension varying with loading for sandstone under uniaxial compression. The fractal dimension increases linearly with the applied load (Xie, 1991). We can obtain this linear correlation between applied load and fractal

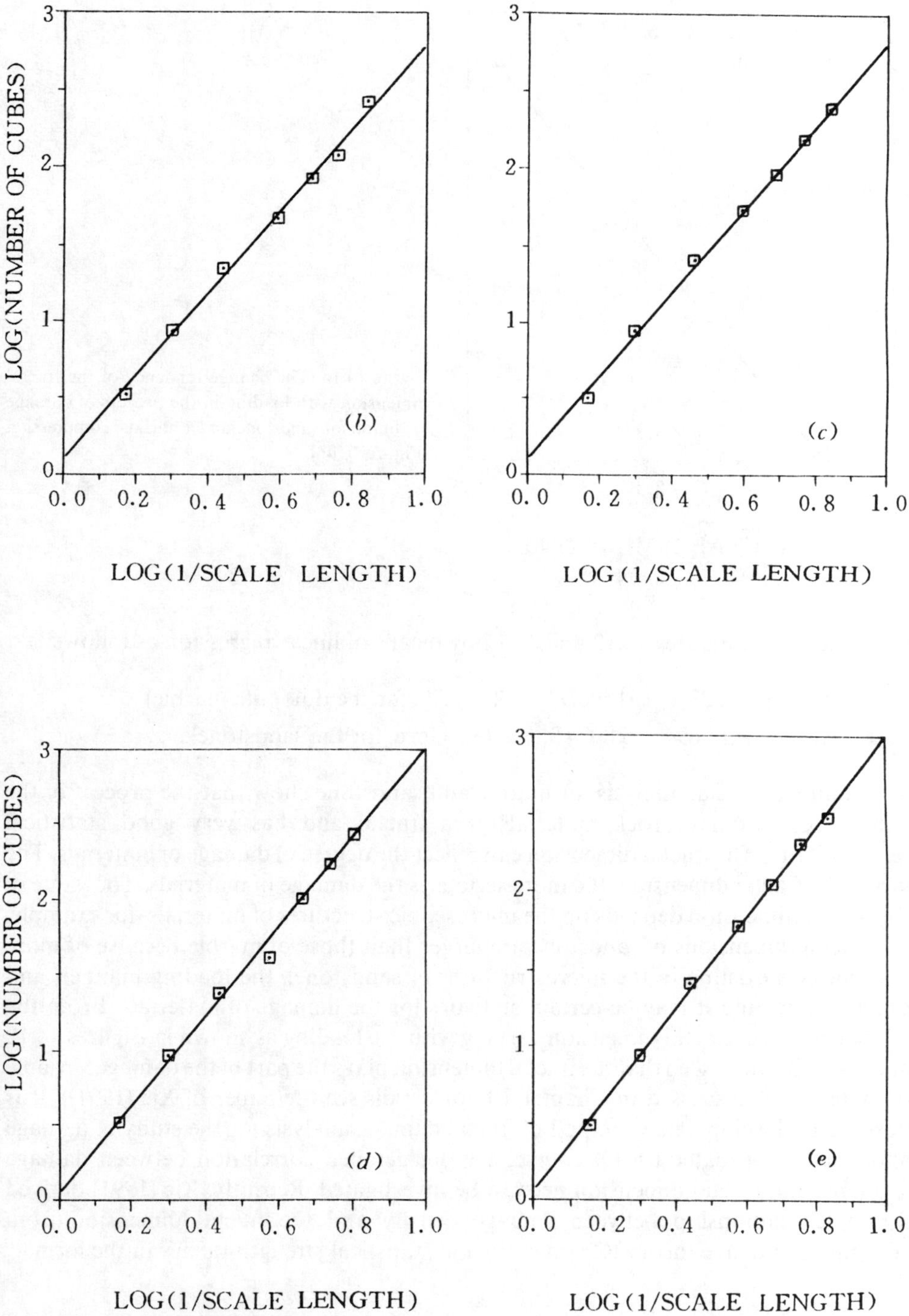

Figure 14.13. The plots of $\log(1/L)$ vs. $\log N(L)$ for the damage region of sandstone under uniaxial compression (Figure 10.46) at different loadings in the process of damage evolution. (a) $\sigma/\sigma_c = 46\%$ and $D = 2.70$; (b) $\sigma/\sigma_c = 63\%$ and $D = 2.81$; (c) $\sigma/\sigma_c = 86\%$ and $D = 2.82$; (d) $\sigma/\sigma_c = 100\%$ and $D = 2.89$.

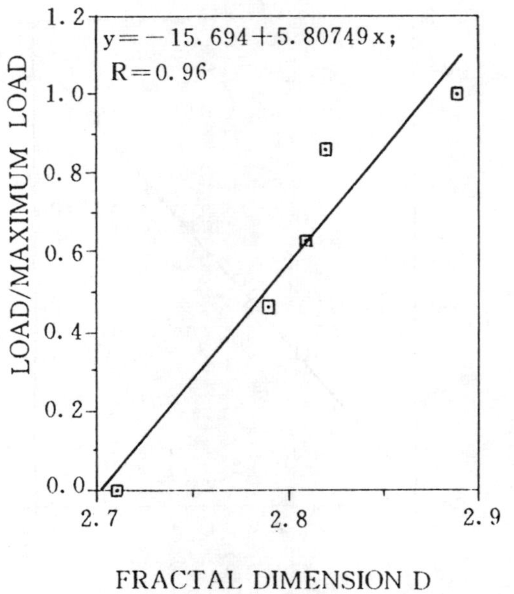

Figure 14.14. The change tendency of the fractal dimension with loading in the process of damage evolution for sandstone under uniaxial compression (Figure 10.46).

dimension from Figures 14.12 and 14.14 by means of linear regression as follows:

$$(\sigma/\sigma_c) = -0.23 + 0.51590D; \quad (R = 0.99, \text{ for the dolomite marble});$$
$$(\sigma/\sigma_c) = -15.694 + 5.8074D; \quad (R = 0.96, \text{ for the sandstone}).$$

The results of fractal analysis for marble and sandstone show that the process of the damage evolution of rock materials is a fractal and has very good statistical self-similarity. The fractal dimension can reflect the degree of damage of materials. The larger the fractal dimension, the more serious is the damage of materials. The value of the fractal dimension depends on the microscopic structures of materials (for example, the fractal dimensions of sandstone are larger than those of marble because of more microdefects existing in the microstructures of sandstone), the loading manner, and other factors, and it may be certain statistics for the damage of materials. From the related curves of fractal dimension varying with the loading as shown in Figures 14.12 and 14.14, it can be seen that the fractal dimension plays the part of the damage variable of materials as discussed in Chapter 10 (for details see the paper of Xie (1991)). It is possible to develop a new method of fractal damage analysis for the study of damage and fracture of materials. Of course, a more detailed correlation between damage available and fractal dimension need to be investigated. Recently, Xie (1991) derived a simple relationship between damage density and the fractal dimension from continuous damage theory (Chapter 10) and statistical strength theory in the form:

$$w/w_c \propto (D)^m;$$

where the damage density is defined by $w/w_c = N/N_{max}$, and N is the number of microcracks. This relation is tested by experimental results (Figures 10.43 and 10.46)

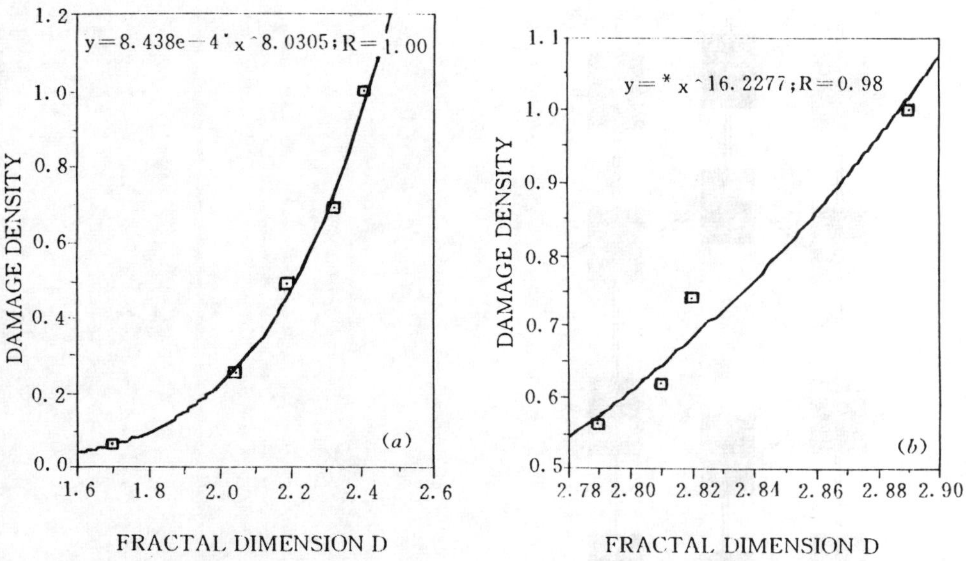

Figure 14.15. The relationship between damage density and fractal dimension. (a) For the dolomite marble (Figure 10.43); (b) For the sandstone (Figure 10.46).

as shown in Figure 14.15:

$$w/w_c \propto (D)^{8.0305}; \quad (R = 1.0, \text{ for Figure 10.43});$$
$$w/w_c \propto (D)^{16.2277}; \quad (R = 0.98, \text{ for Figure 10.46}).$$

This means that there may be a nonlinear relation between fractal dimension and damage density.

14.5 FORMATION OF FRACTAL CRACKS IN A KINETIC FRACTAL MODEL

Making use of a molecular model for material failure, Termonia & Meakin (1986) have studied surface cracks which lead to fractal-like structures based on the kinetic theory of fracture.

The surface is represented by a two-dimensional $(x - y)$ network of nodes representing the difficult-to-deform strong parts of the material (Figure 14.16). The nodes are joined in the y and x directions by elastic bonds with force constants K_y and K_x representing the easily deformable and easily broken regions. In the case of a fibre that is perfectly ordered and oriented along the y-axis, the nodes represent the elementary repetition units of the macromolecules, and K_y and K_x are the elastic constants for the primary (for example, $C - C$) and secondary (van der waals or hydrogen) bonds, respectively. The network is stretched along the y-axis by a constant strain, and basic events in the network are controlled by thermally activated bond breakages; accumulation of these events leads to crack formation and, ultimately, to network breakdown. The bond breakage process is examined with the help of the following stochastic model. The

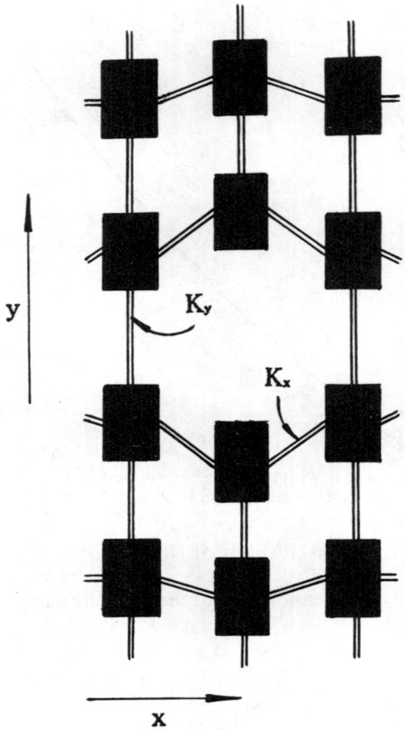

Figure 14.16. Two-dimensional network of nodes with one broken bond along the y-axis. K_y and K_x are the elastic constants for the bonds (Termonia & Meakin, 1986).

(unbroken) bonds are subjected at random to a Monte-Carlo lottery which breaks a bond with probability

$$P_i = \exp\{(-U_i + E_i)/kT\}/P_{\max}, \tag{14.19}$$

where U_i is an activation energy, E_i is the elastic energy stored in the bond, k is Boltzmann's constant, T is the absolute temperature and $P_{\max}$ is a normalization constant which ensures that the probability of breaking the most stressed bond in the array is unity. E_i is defined by

$$E_i = (\tfrac{1}{2})K_i(\Delta l_i)^2; \tag{14.20}$$

in which K_i is the force constant for bond i, and Δl_i is the elongation. As a bond breaks, the two nodes connected through it will move apart (see below). For simplicity, Termonia & Meakin assumed that the motions of the nodes along the x- and y-axes were mutually independent and focused on their displacements along the y-axis, where the network was strained. Thus, the Δl_i values (see Equation 14.20) are for elongations along the y-axis and correspond either to an axial tensile force (for a bond along the y-axis) or to a shear force (for a bond along the x-axis). After a very small number of bonds have been broken, the network is relaxed to its minimum energy configuration in a series of steps. At each step, a node i is visited and its y-ordinate is adjusted so as to reduce to zero the net residual force (R_i)

$$R_i = \sum_{\substack{[\text{intact} \\ \text{bonds } j]}} K_j(\Delta l_j) + \sum_{\substack{[\text{nodes } i' \\ \text{in contact}]}} R_i, \tag{14.21}$$

acting on that node. The first term on the right-hand side of Equation (14.21) is due to the elastic force acting on the bonds j originating from node i, with $K_j = K_y$ and $K_j = K_x$ for a bond along the y- and x-axis, respectively. Δl_j is the elongation of bond j, measured by the difference in the y-ordinates of the two ends of the bond. The second term on the right-hand side of Equation (14.21) accounts for the external forces acting on node i, when it contacts with any of its two neighbors i' along the y-axis.

It is important to note that if the residual force for a node is reduced to zero, it often results in an automatic increase of (the absolute value of) the residuals for neighboring nodes. Therefore, the liquidation of all the residuals in the array is a very computer-time-consuming process. For that reason, the liquidation of the residuals has been performed using two well known computational devices, called over-relaxation and block relaxation, which considerably speed up the convergence of the calculations (Termonia & Meakin, 1986). Over-relaxation moves a node not simply to the position at which the residual R is zero, but rather to a position at which the sign of the residual is opposite to that of the average for neighboring nodes. The second, and more important, device is known as block relaxation and consists of adjusting the y-ordinates of more than one node at a time (Avnir et al., 1983).

Termonia & Meakin started with an intact network (no broken bonds), stretched by a constant strain of 20%. The stochastic model then proceeds in a series of steps. At each step, 0.1% of the total number of bonds are broken according to Equations (14.19) and (14.20), after which the network is relaxed towards mechanical equilibrium, using Equation (14.21). These steps are repeated until the network breaks. Typically, a network consists of 60 nodes in the y-direction and 100 nodes in x-direction, with periodic boundary continues. The small size of the network is dictated by the mechanical relaxation steps which represent the most computer-time-consuming processes in their simulations. A relaxation step is considered to be complete when the net forces measured at the two ends of the network along the y-axis are equal to within a few per cent. A typical simulation takes 6-8 hours of CPU time on an IBM 3081D. For the calculation of the elongations Δl_i (Equation (14.20)), the length of a node along the y-axis is set to be equal to 1 unit length (its width being set to 0.2 units). For simplicity, equal activation energies (Equation (14.19)) is chosen, and $kT = 1$.

Figure 14.17 shows a two-dimensional network with $K_y = K_x = 20$ at different stages of the simulations (Termonia & Meakin, 1986). Long before failure (Figure 14.17a), small cracks initiate and grow. As failure is approached (Figure 14.17b), these cracks coalesce together, giving rise to a smaller number of larger cracks. Note that all the cracks at the top of Figure 14.17a and b have the same 'dog' shape and already suggest self-similarity properties. At failure (Figure 14.17c), only one macrocrack is present, which extends throughout the network along the x-axis. All the other cracks have either been incorporated into the macrocrack or have been closed through relaxation of the remaining elastic bonds towards their un-stretched state. The fraction of broken bonds in Figure 14.17c is about 30%.

The degree of irregularity of a crack perimeter is given by its fractal dimension D, obtained by measuring the perimeter length using yardsticks of different sizes. The perimeter is a fractal if Equation (2.11) is satisfied here. Because, in general, one cannot cover the entire perimeter with an integral number of yardsticks, the termination is handled by taking the straight line segment between the current point and end point of the perimeter. The ratio of the segment length to the yardstick length is then

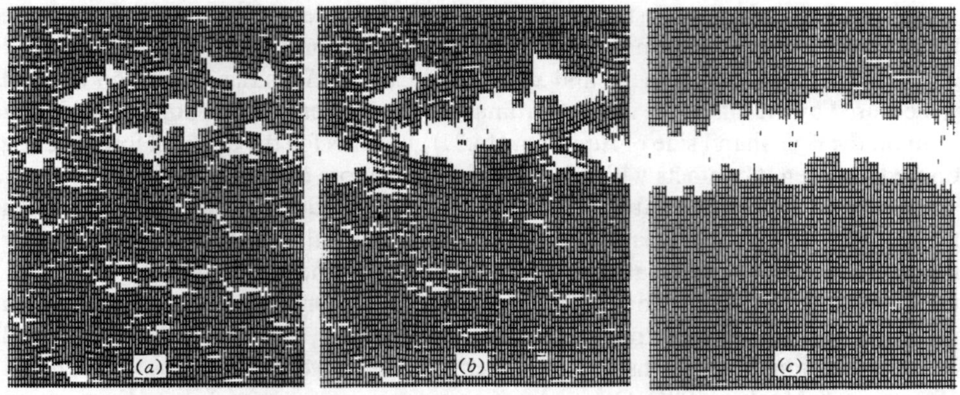

Figure 14.17. Results of the simulations for $K_x = K_y = 20$ ($kT = 1$, U_i = constant for all the bonds). (a) Long before failure; (b) Close to failure; (c) At failure (Termonia & Meakin, 1986).

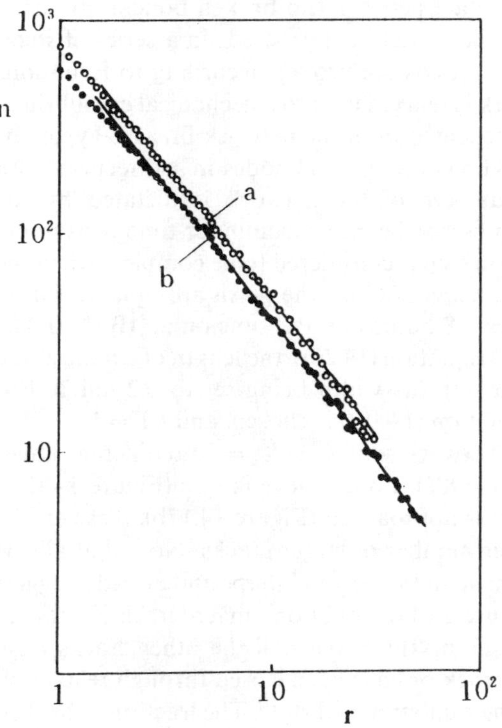

Figure 14.18. Log-log dependence of n on r (see Equation (2.11)), where n is the number of yardsticks of r needed to cover the perimeter of a crack and; r is in units of width of a node along the x-axis. Curve a: $K_x = K_y = 20$; Curve b: $K_x = 200$, $K_y = 20$ (Termonia & Meakin, 1986).

added to the yardstick count. Curve a of Figure 14.18 shows the log-log dependence for the perimeter of the 'infinite' crack in Figure 14.17c. Fitting a straight line in the range $2 < r < 30$ leads to a slope $D = 1.27 \pm 0.02$ (Termonia & Meakin, 1986). A similar value of D, although for a smaller size range of r, has been obtained for the perimeters of the finite cracks in Figure 14.17b.

An alternative statistical measure of D for an ensemble of finite cracks examined with the same yardstick size r defined by the basic perimeter-area relation (Equation (3.2)):

$$P(r) \propto [A(r)]^{D/2} \quad \text{(see Chapter 3)}, \tag{14.22}$$

(for a given finite crack, the area A includes all nodes and groups of nodes completely surrounded by the crack in question). The $P-A$ relation has been applied to the cracks in Figure 14.17a and b by determining the value of D which, for a given r, minimizes the standard deviation for the ratios $P(r)/A(r)^{D/2}$. In the range $1 < r < 8$, Termonia & Meakin found $D = 1.31 \pm 0.15$. Although the statistical error is large because of the small number of cracks investigated, the latter value of D is consistent with that obtained using Equation (2.11).

The range of D values achievable by the model can be easily determined. In the limit of large kT, p_t is constant (see Equation (14.19)) and the crack corresponds to a two-dimensional percolation cluster whose exterior perimeter (hull) has a fractal dimension of $D \approx 1.75$. In the limit $kT = 0$, the crack develops horizontally from a small nucleus, and $D = 1$. For finite kT, D also depends on K_x and K_y. Computer simulation results for $kT = 1$, with $K_x = K_y = 10$ and $K_x = 200$, $K_y = 20$ (see Figure 14.18, curve b) leading to D values which are identical, within experimental error, to those ($1.27 \pm 0.0.2$) found for the initial choice $K_x = K_y = 20$. This indicates that for intermediate values of kT, the fractal dimensional D is insensitive to model details and may take on a universal value for all finite, non-zero values of kT.

It is clear that above model presented by Termonia & Meakin may be only a crude representation of real experimental systems. For example, the model in its present form does not allow for out-of-plane buckling phenomenon or in-plane distortion effects. However, this idea and method can be generalized, modified, and applied to model the formation of fractal cracks in rock masses.

14.6 THE FRACTAL EFFECT OF IRREGULARITY OF CRACK BRANCHING ON THE FRACTURE TOUGHNESS OF ROCK MATERIALS

The bifurcation of a running crack is an instability phenomenon. Dependent on differing loading and geometrical conditions, bifurcation events occur in a variety of materials. This phenomenon, however, is most frequent in essentially brittle fractures and at high crack-tip speeds.

The basic reason for crack to branch in non-homogeneous materials (e.g., specimens containing inclusions, or several other types of interfaces), to date is not well understood. Research results show that the size of the fracture mirror zone and the branching event can be correlated by fracture surface topography and surface roughness. The degree of irregularity increases with crack bifurcation. Smith (1986) made a study of the increase of the fracture toughness of brittle materials caused by crack branching which may be considered to be the result of irregularity in terms of theoretical analysis. Using fractal geometry, in this section we analyze this irregularity. That is to say, a new geometrical method is applied to the analysis of crack branching problems (Xie, 1989d, 1990b).

14.6.1 *A fractal model for crack branching*

In a brittle material, a propagating crack may either depart from its original straight trajectory to curve or split into two or more branches or, under a high state of stress, divide into a river delta crack pattern. This fragmentation may often be a succession of multiple branchings of what was initially a single crack. Typical crack curves and crack branches (microscopic and macroscopic fracture systems) in rock materials are shown in Figures 14.19-14.22 and Photos 14.2 and 14.3.

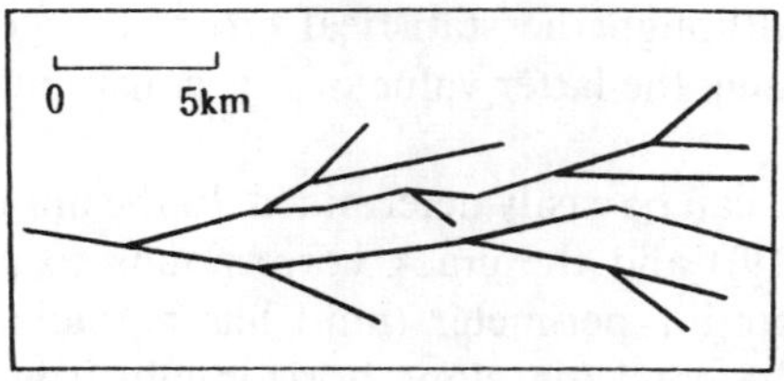

Figure 14.19. Macrofracture system in the Qin-ling area, Shanxi, China.

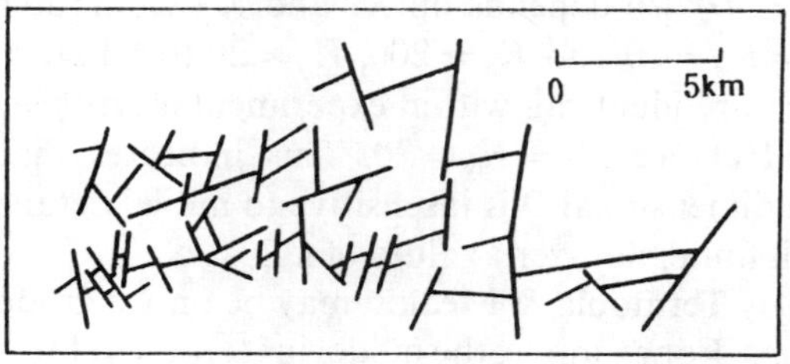

Figure 14.20. Macrofracture system in the Jishan area, Henan, China.

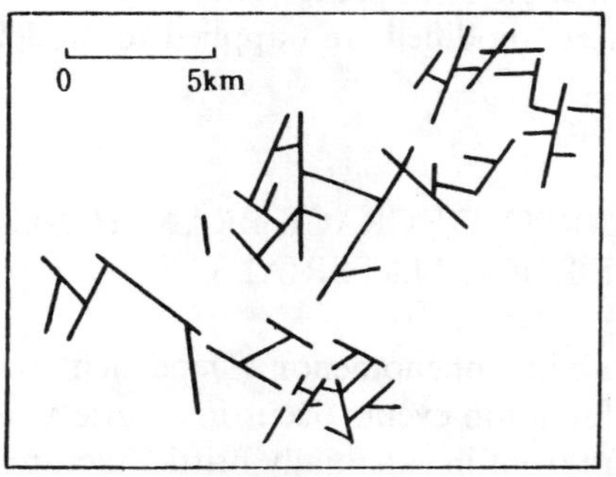

Figure 14.21. Macrofracture system in the Malou area, Henan, China.

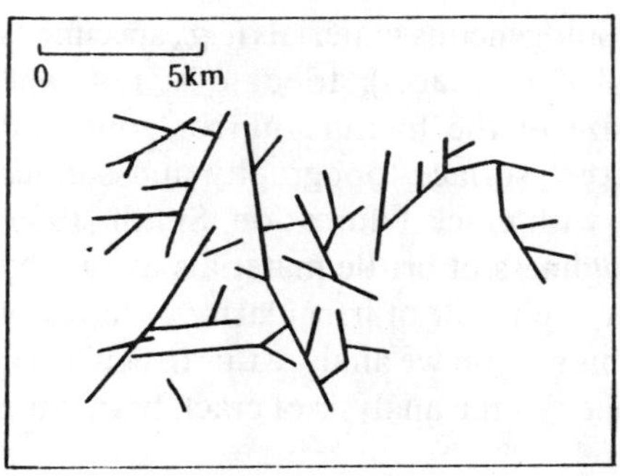

Figure 14.22. Macrofracture system in the Wuyang area, Henan, China.

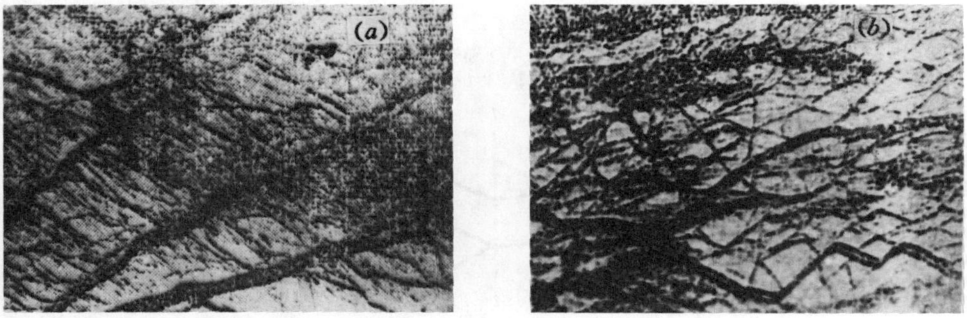

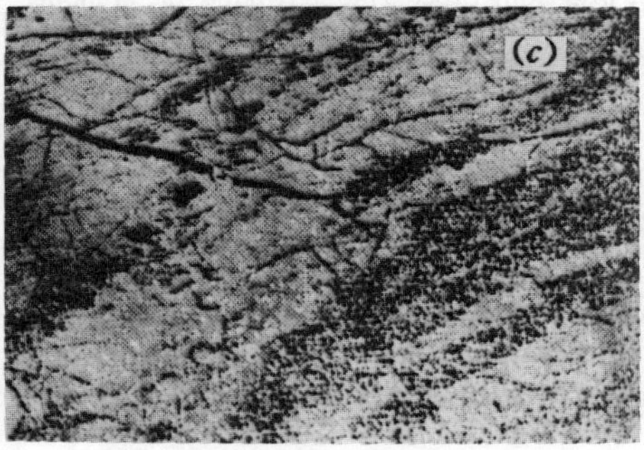

Photo 14.2. Microcrack branching of outburst coal (optical micrograph) in Hunan, China, (a) Coal I; (b) Coal II; (c) Coal III.

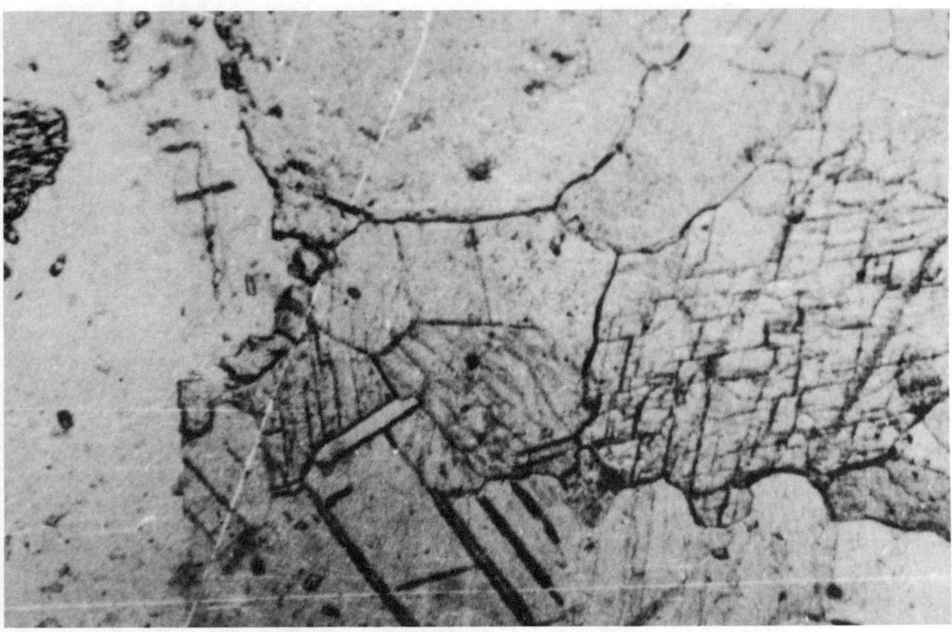

Photo 14.3. Microfracture system of marble (optical micrograph × 50) of Xuzhou, China.

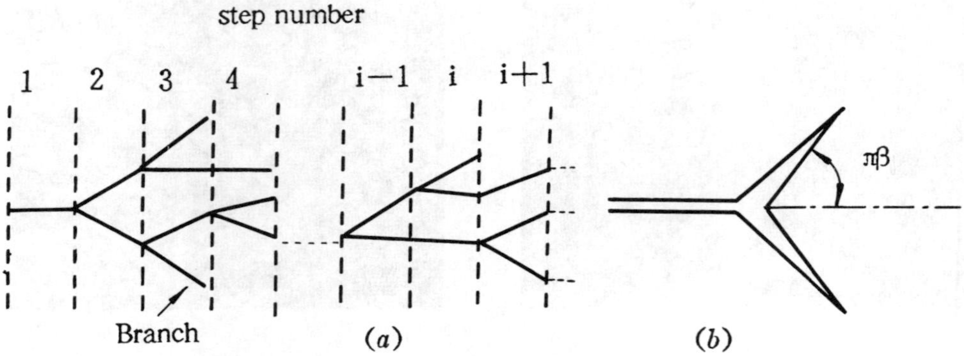

Figure 14.23. The fractal model of crack branching: (a) the branching process model; (b) the fractal model.

From Watanabe (1986), the fracture system as shown in Figures 14.19-14.22 and Photos 14.2 and 14.3 may be considered to be a cluster of branches propagating in such a way that new branches in the $i+1$-th step are successively created from a former branch at the i-th step (as shown in Figure 14.23a). In other words, the fracture system can be considered as statistically self-similar, and thus can be modelled as a fractal. Therefore, the fractal model of crack branching may be established as shown in Figure 14.23b.

The fractal dimension in Figure 14.23b can be estimated by

$$N = 3;\ 1/r = 2\cos(\pi\beta/2), \qquad (15.23)$$

$$D = \log(3)/\log[2\cos(\pi\beta/2)]; \qquad (14.24)$$

where N is the number of fragments, and r represents the ratio of similarity.

14.6.2 *The correlation between fractal dimension and fracture toughness*

According to Equation (2.7a),

$$L_i(\delta_i) = L_0\delta_i^{(1-D)},$$

when choosing L_0 as a unit length, l as grain size, and the yardstick

$$\delta_i = \begin{cases} l & \text{(microscopy)} \\ r & \text{(macroscopy)} \end{cases}. \qquad (14.25)$$

From Equation (13.3), we have

$$G = \begin{cases} 2r_s l^{(1-D)} & \text{(microscopy)} \\ 2r_s(1/r)^{(D-1)} & \text{(macroscopy)} \end{cases}. \qquad (14.26)$$

Applying the principles of fracture mechanics (see Section 10.1.3), the relationship between G and fracture toughness K (valid for model I, II, III) is

Table 14.1. The effects of geometric irregularity of crack branching ($\pi\beta$) and grain size l on the fracture toughness.

	K/K_0								
$\pi\beta$ l	10°	20°	30°	40°	50°	60°	70°	80°	90°
10^{-2} cm	3.93	4.18	4.66	5.51	7.03	10.0	16.79	38.02	149.91
10^{-4} cm	15.42	17.46	21.68	30.34	49.43	100.0	281.83	1445.43	21877.6
10^{-6} cm	60.53	72.95	100.9	167.1	347.5	1000.0	4731.5	54954.1	3235936.0

$$K = AG^{1/2} = \begin{cases} K_0 l^{(1-D)/2} & \text{(microscopy)} \\ K_0(1/r)^{(D-1)/2} & \text{(macroscopy)} \end{cases}, \tag{14.27}$$

where $K_0 = AG_0^{1/2} = A(2r_s)^{1/2}$, and

$$A = \begin{cases} [E/(1-v^2)]^{1/2} & \text{(plane strain)} \\ (E)^{1/2} & \text{(plane stress)} \end{cases} \quad \text{(for model I, II)}. \tag{14.28}$$

K_0 is the fracture toughness in which the irregularity is not considered.

14.6.3 *Microscope fractal effects of the irregularity caused by crack branching*

In microscopic analyses, we use grain size l as a gauge of the fractal metric for modelling the irregularity of crack branching because sizes of micro-cracks in rock materials are of approximately the same order as grain size (Section 10.3). For the fractal model of crack branching shown in Figure 14.23b, by substituting Equation (14.24) into Equation (14.27), we have

$$K/K_0 = l^{\{1-\log 3/\log[2\cos(\pi\beta/2)]\}/2}. \tag{14.29}$$

Values of K/K_0 for different ($\pi\beta$) and grain size are listed in Table 14.1.

Obviously, the fracture toughness increases exponentially in relationship to the crack branching angle $\pi\beta$, (i.e., the larger $\pi\beta$, the more dissipation energy that is required for crack branching).

In the meantime, microstructural effects on the fracture toughness of crack branching are very serious; furthermore, it appears difficult to form the crack of branching angle $\pi\beta \geq 60°$ in a material of grain size $l \leq 10^{-4}$ cm.

14.6.4 *Macroscopic fractal effects of the irregularity caused by crack branching*

Macroscopic crack branching produces an increase in the fracture toughness of brittle materials. Smith (1986) has modelled this effect by theoretical analysis. For the crack branching model shown in Figure 14.23b, the equation of estimating K derived by Smith is given by

$$K_{\text{Smith}}/K_0 = 2[(1-\beta)/\beta]^{-\beta/2}. \tag{14.30}$$

In our macroscopic analyses, from Lung (1985), we take r (ratio of similarity) as the gauge of fractal metric for modelling this effect by fractal geometry. Then Equation (14.29) becomes:

$$K/K_0 = [2\cos(\pi\beta/2)]^{\{\log 3/\log[2\cos(\pi\beta/2)] - 1\}/2}. \tag{14.31}$$

or

$$\begin{aligned}\log(K/K_0) &= \log[2\cos(\pi\beta/2)]\cdot\{\log 3/\log[2\cos(\pi\beta/2)] - 1\}/2 \\ &= \{\log 3 - \log[2\cos(\pi\beta/2)]\}/2.\end{aligned} \tag{14.32}$$

Values of K/K_0 given by Equations (14.30) and (14.31) are listed in Table 14.2.

The tendency of K/K_0 to vary in relationship to the crack branching angle ($\pi\beta$) is shown in Figure 14.24. Results obtained by using Equation (14.31) have good monotonicity. This indicates that the smaller the crack branching angle ($\pi\beta$), the smaller is the amount of strain energy dissipated by crack branching (i.e., the easier crack branching is). It must be pointed out that there is a minimum value in Smith's equation when $\pi\beta = 39.24°$. It seems easiest to yield crack branches along $\pi\beta = 39.24°$; in general however, this is impossible. In fact Sih (1977) predicts, using the minimum

Table 14.2. The effects of geometrical irregularity of macrocrack branching on the fracture toughness.

$\pi\beta$	10°	20°	30°	40°	50°	60°	70°	80°	90°
D	1.574	1.621	1.668	1.741	1.847	2.00	2.225	2.58	3.17
K_{smith}/K_0	1.306	1.260	1.236	1.230	1.238	1.260	1.295	1.346	1.414
K/K_0	1.227	1.234	1.246	1.263	1.286	1.316	1.353	1.407	1.456

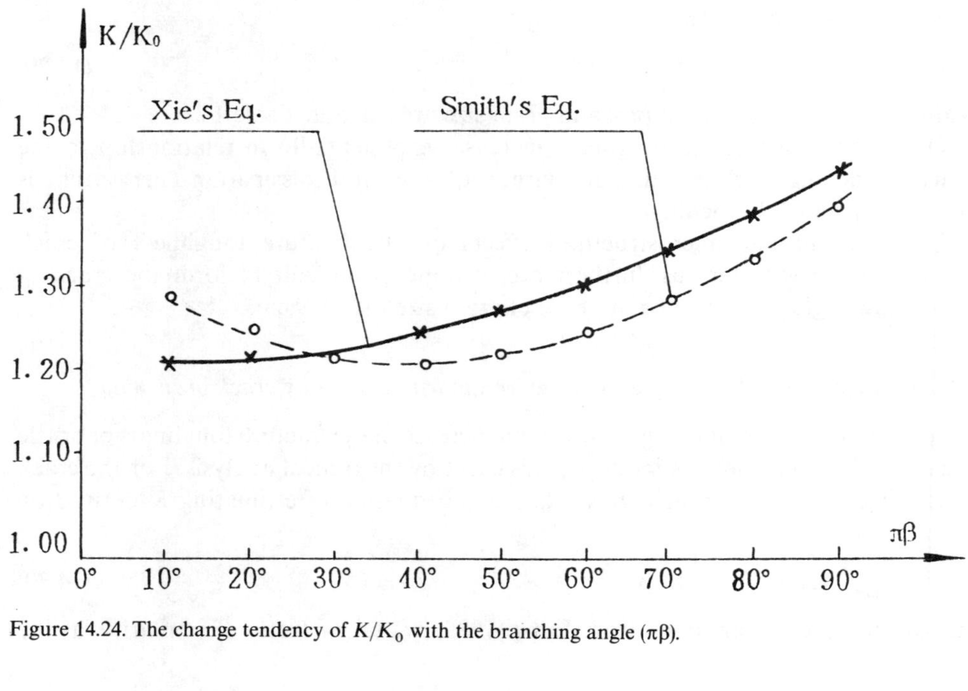

Figure 14.24. The change tendency of K/K_0 with the branching angle ($\pi\beta$).

strain energy density criterion, a branching angle between 15-18 degrees (i.e., $\pi\beta = 7.5\text{-}9°$) under the dynamic load conditions of model I. In general, theoretically predicted branching angles are larger than those experimentally observed. On the other hand, Table 14.2 shows that, except for $\pi\beta = 39.24°$, the values obtained by Equation (14.31) approach those predicted by Smith's equation at other points. Therefore, it is clear that the analysis in this section is available, and that the phenomenon that fracture toughness can be raised by crack branching may be summarized as fractal effects of the irregularity.

From the above analyzed results, we can obtain conclusions as follows:

Crack branching irregularity depends on the physical and mechanical behavior, microstructural events in materials, the specimens and loading conditions. This irregularity can be modelled well by fractal geometry. The essence of such a complicated phenomenon as branching can be revealed by the properties of the fractal dimensional space. Because of the difficulties in analyzing problems in Euclid's space, even when analyzing simple crack branching phenomenon, the mathematically analytical process is laborious (such as Smith's analyses), and is even impossible. In this case, fractal geometry exhibits the evident advantages of conciseness and clarity. Thus, we say that fractal geometry provides a new approach for the investigation of crack irregularity, and for ascertaining the relation between macro-fracture and microfracture parameters.

14.7 FRACTAL EFFECTS ON THE CRACK EXTENSION RATE

We now consider the energy equilibrium of the object, including a coin crack with length a under loading, as follows (Xie, 1989d):

$$\dot{W} = \dot{U}_e + \dot{K} + \dot{\mathfrak{C}} + \dot{F}_s,$$

where $\dot{W}$ represents the power of external loads, $\dot{U}_e$, $\dot{K}$ and $\dot{\mathfrak{C}}$, respectively, are the rates of elastic deformational energy, dynamic energy, and damage dissipation energy of materials, and $\dot{F}_s$ denotes the rate of surface energy needed to form the new surface of the crack in which fractal effects are considered.

$$\dot{F}_s = 2r_s \cdot h \cdot L(\varepsilon)/L_0(\varepsilon) \cdot \dot{a},$$

where h is the thickness of a specimen. We then obtain

$$\dot{a} = [\dot{W} - \dot{U}_e - \dot{K} - \dot{\mathfrak{C}}]/[2r_s hL(\varepsilon)/L_0(\varepsilon)] = \dot{a}_0 l^{(D-1)};$$

where $\dot{a}_0 = (\dot{W} - \dot{U}_e - \dot{K} - \dot{\mathfrak{C}})/(2r_s h)$ is the crack extension rate of the damage material in the macroscopic sense, and $\dot{a}$ represents the crack extension rate in which fractal effects are considered in microscopic sense. The above equation implies that the crack extension rate will increase with material damage ($\mathfrak{C} \leq 0$) and decrease with the crack irregular extension (microstructure effects). From the previous analysis (Chapter 13), its equivalent alternative expression is written by

$$\dot{a} = \begin{cases} \dot{a}_0 \cdot l^{0.26} & \text{(intergranular brittle fracture),} \\ \dot{a}_0 \cdot l^{0.365} & \text{(transgranular brittle fracture),} \\ \dot{a}_0 \cdot l^{0,255} & \text{(the combination).} \end{cases}$$

Table 14.3. Effects on crack extension rate caused by grain size and modes of microfracture.

		$\dot{a}$ for different Grain size l		
Fracture mode	$D-1$	$l=10^{-4}$ cm	$l=10^{-3}$ cm	$l=10^{-2}$ cm
Intergranular	0.26	$\dot{a}_0/10.98$	$\dot{a}_0/6.03$	$\dot{a}_0/3.31$
Transgranular	0.365	$\dot{a}_0/28.84$	$\dot{a}_0/12.45$	$\dot{a}_0/5.37$
Combination of intergranular and transgranular fractures	0.255	$\dot{a}_0/10.47$	$\dot{a}_0/5.82$	$\dot{a}_0/3.24$

The fractal effects of various grain sizes on the crack extension rate in the materials are listed in Table 14.3. Obviously, under the condition of the same grain size, intergranular and combination brittle fracture will produce a fast crack extension rate; under the condition of the same fracture mode, a fast crack extension rate will be produced in the materials with larger grain size. These results are in agreement with the known physical phenomena, but if fractal effects are not considered, these natural principles will not be quantitatively characterized (Xie, 1989d).

14.8 FRACTAL PHENOMENA OF ROOF FALL CAUSED BY EXCAVATION

After excavation of the underlying seams, the roof of the workout area bends and moves under gravity and reaction of upper strata. The roof will fracture, resulting in failure and fall of the entire upper strata if its internal tension stress exceeds the limit strength of rock. It is a very complicated mechanical procedure correlated with mechanics properties of the upper strata, geological conditions, mining velocity, and so on. We find that there is a fractal geological phenomenon of the fracture system of roof fall. The fracture system of roof fall obtained by experiments of equivalent materials (Photo 14.4) behaves as self-similarity. We may take a straight line as an

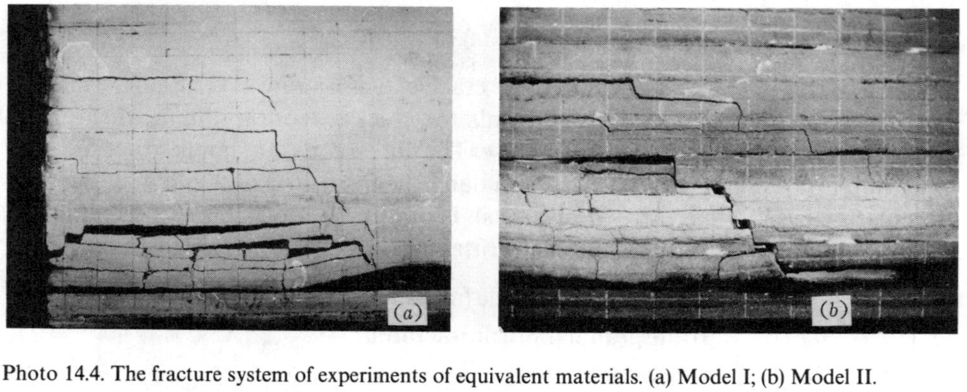

Photo 14.4. The fracture system of experiments of equivalent materials. (a) Model I; (b) Model II.

initial polygon generator

Figure 14.25. A fractal model of roof fall

Table 14.4. The relationship between fracture angle α and fractal dimension D of roof fall caused by excavation. $G_f = 2r_s(1/r)^{D-1}$.

α	10°	20°	30°	40°	50°	60°	70°	80°	90°
D	1.003	1.013	1.029	1.053	1.085	1.129	1.187	1.263	1.365
$G_f/2r_s$	1.003	1.014	1.031	1.057	1.090	1.134	1.189	1.257	1.341

initial polygon to establish its generator as shown in Figure 14.25. Its fractal dimension is

$$N = 3, \quad 1/r = (5 + 4\cos\alpha)^{1/2},$$
$$D = \log 3/(1/2 \log(5 + 4\cos\alpha)).$$

It is clear that D and G_f vary with angle α (Table 14.4). Data at the site indicate that α depends on the range of the workout area, mining velocity, and mechanics properties of the roof strata, that is, D also depends on these factors. Therefore, it becomes possible to explore the complicated mechanical problems of roof fall caused by excavation in terms of fractal geometry (Xie, 1989b).

14.9 FRACTAL SIMULATION OF FRACTURE NETWORKS IN ROCK MASSES

Fractures form three-dimensional interconnected networks in nearly all rocks at or near the earth's surface. The measurement and prediction of the density and concreteness of natural fracture networks are important, yet elusive, tasks. The amount and degree of interconnection strongly affect, and may even dominate, deformability, failure, and permeability of the rock mass (La-Point, 1988).

Very few civil engineering and mining projects require accurate estimations of fracture density or connectedness; the cost and performance of the excavation can be constrained sufficiently by simpler parameters, such as fracture orientation, openness, filling materials and average spacing. These conventional geo-technical aspects provide good three-dimensional information on fracturing from the drifts, tunnels, and benches that are common in such work. However, this is not the case for deep mining and projects such as the underground storage of nuclear and toxic wastes or the production of hydrocarbons from naturally-fractured reservoirs. In these undertakings, the misjudgment of permeability or stability can lead to sizeable financial loss or social risk. There have been several methods proposed to predict the mechanical or hydrological response of naturally-fractured rock masses, but all require an estimation of fracture size or assume it to be at least greater than the scale of interest. Thus, it is

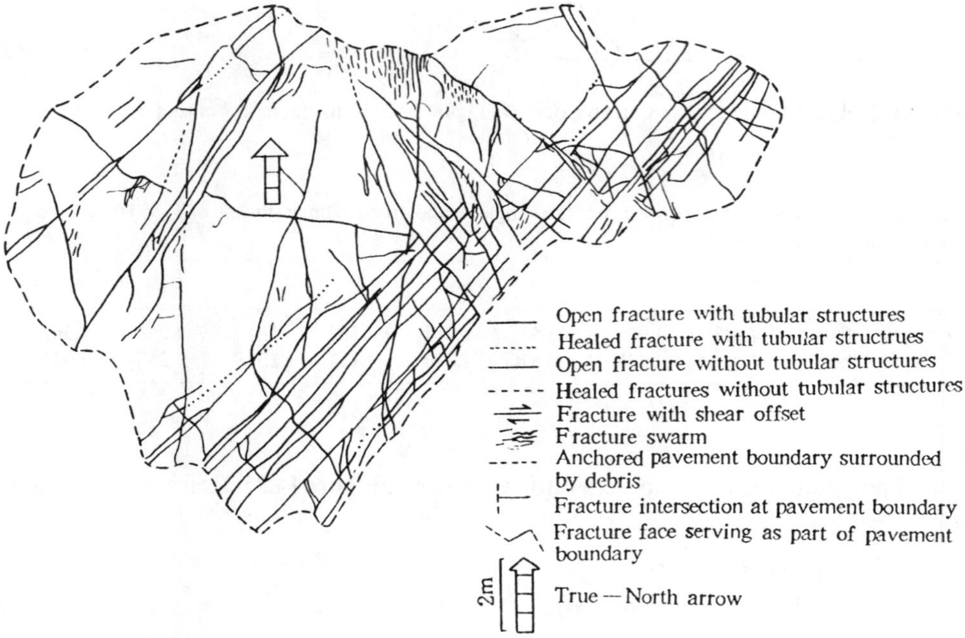

Figure 14.26. Fracture-trace map of pavement 100 (area = 214 m^2) (Barton & Larsen, 1985).

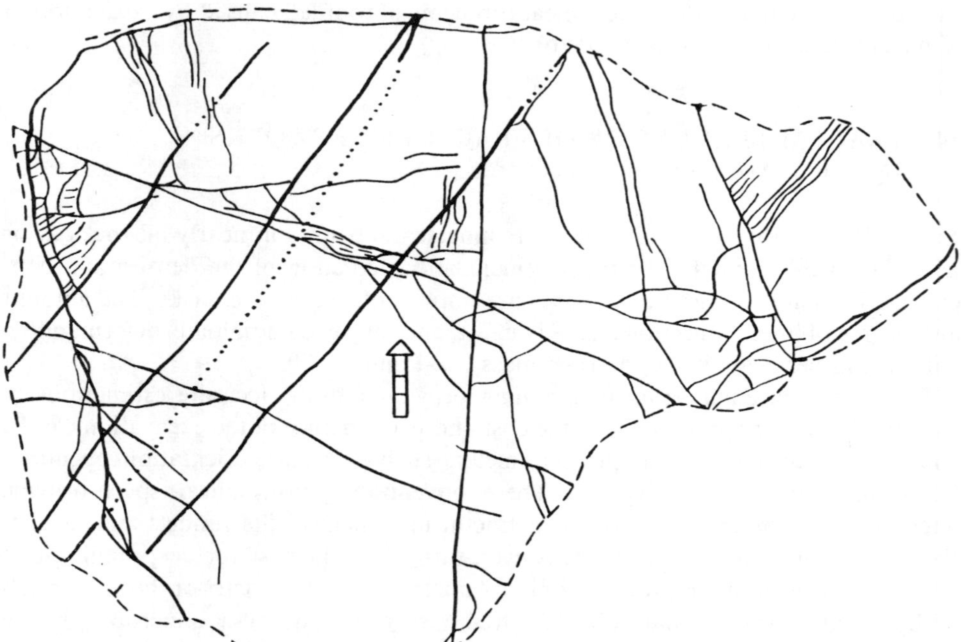

Figure 14.27. Fracture-trace map of pavement 200 (area = 260 m^2) (Barton & Larsen, 1985). See Figure 14.26 for explanation of symbols.

difficult to apply these methods where only borehole data are available, since it is impossible to measure fracture dimensions. The new methods used for this problem are fractals and the geostatistical method. Barton & Larsen (1985) has been the first to characterize maps of fracture-trace networks on two-dimensional sections through the three-dimensional fracture network by using fractal geometry. Velde et al. (1990, 1991) and Chiles (1988) have studied the fracture networks by Cantor-fractals and geostatistical methods (see Chapter 4), whereas La Pointe (1988) has given a method to compute an index of fracture density using fractal geometry. The fracture networks generated by La Pointe's method looks the same as natural rock fracture patterns. In this section, we mainly introduce La Pointe's and Barton's works in this field.

Barton & Larsen (1985) examined fracture traces exposed on three 214 to 260 m^2 pavements designated by the numbers 100, 200, and 300 in the same Miocene ash-flow tuff at Yucca Mountain, southwestern Nevada and mapped them at a scale of 1:50 as shown in Figures 14.26-14.28. These maps are two-dimensional sections through the

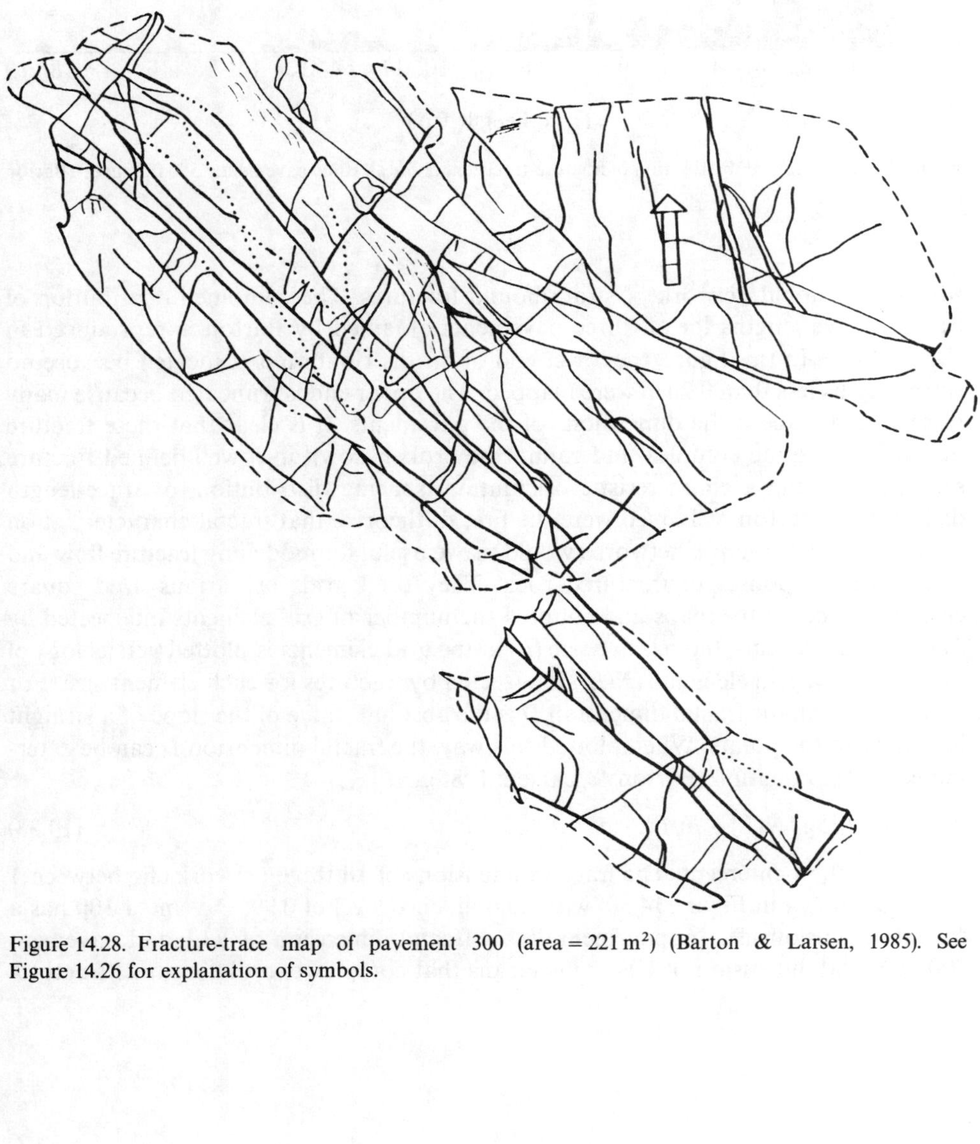

Figure 14.28. Fracture-trace map of pavement 300 (area = 221 m^2) (Barton & Larsen, 1985). See Figure 14.26 for explanation of symbols.

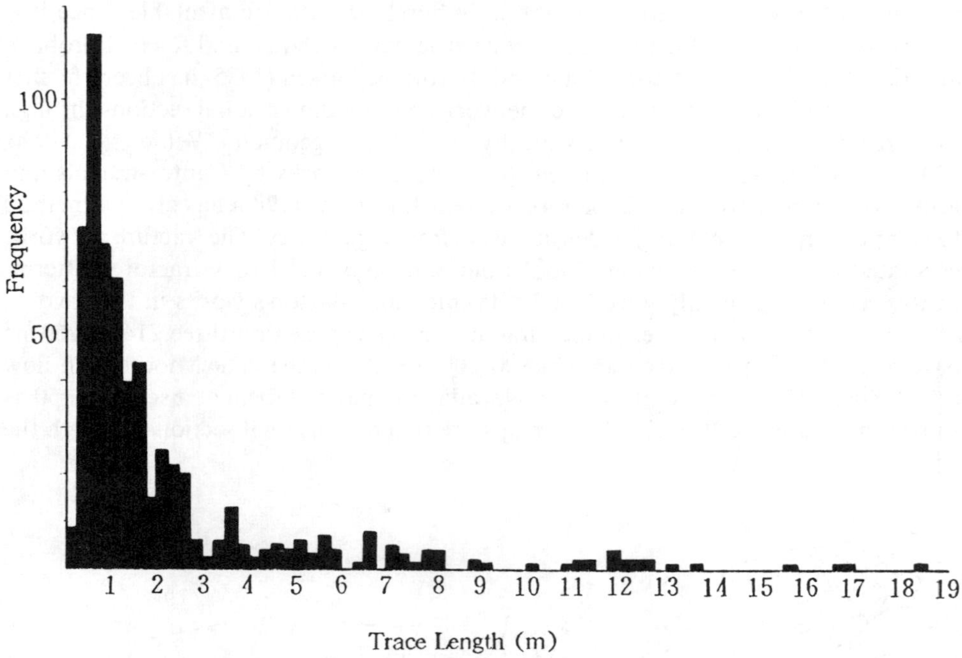

Figure 14.29. Combined distribution of fracture-trace length for all three pavements (580 fractures) (Barton & Larsen, 1985).

three-dimensional network of strata-bound fractures. The combined distribution of fracture-trace lengths for all three pavements obtained by Barton is reproduced in Figure 14.29. In this figure, the lower end of the distribution is truncated because no trace lengths less than 0.20 m were mapped. The upper end is truncated because many trace lengths exceed the dimensions of the pavements. It is clear that these fracture networks are quite complex and cannot be broken down into well-defined fracture sets, each having a characteristic orientation, spacing distribution, or trace-length distribution. Barton & Larsen were the first to theorize that fractal characterization of such complex fracture networks would prove useful for modelling fracture flow and mechanical responses of fractured rock. They used grids of various-sized square elements to cover the maps and counted the number of grid elements intersected by fracture traces. Thus, the relative size (L) of the grid elements is plotted versus $\log_2$ of the number of grid elements ($N(L)$) intersected by fractures for each element size. For each pavement, the fractal dimension D is the absolute value of the slope of a straight line fitted to the points. When plotted this way, the fractal dimension D can be determined by the equation (Barton & Larsen, 1985),

$$D = |(\log_2 N(L) - b)/L|, \tag{14.33}$$

where b is the y-intercept. The fractal dimensions of all three networks lie between 1 and 2 (as shown in Figure 14.30) with a confidence level of 0.99. Pavement 100 has a fractal dimension of 1.16; pavement 200, a fractal dimension of 1.12; and pavement 300, a fractal dimension of 1.14. This means that complex two-dimensional fracture-

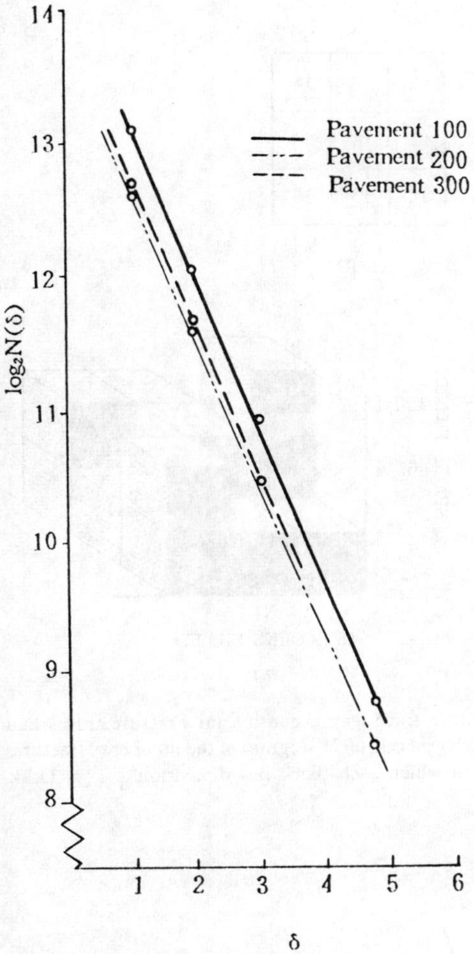

Figure 14.30. Fractal plot of fracture networks (Barton & Larsen, 1985).

trace networks can be quantitatively described using fractal geometry. The fractal dimensions D are tightly clustered (1.12-1.16) for three laterally separated pavements in the same stratigraphic subunit, even through the fracture network visually appears quite different. Over two orders of magnitude (0.2 m-25 m), the networks are scale independent, even when the trace-length distribution is truncated.

La Pointe (1988) proposed fractal formulation of fracture density by using grid covering fracture networks as discussed in Section 14.4 (Figure 14.31). After obtaining the histogram of the fracture density, by referring to the discussion about fractal description of damage evolution in Section 14.4, La Pointe further suggested normalization of the histogram as indicated in Figure 14.31(c) into the filled blocks (Figure 14.31(d)) where cells are to be mapped as either filled or unfilled. For example, if the number of fractures intersected with or contained within the grid of any cell were 18 (Figure 14.31(b)), and the unit z-dimension was 14 fractures (here the fracture number of unit z-dimension = (maximum number of fractures in all cells)/3), then the dimension of the z-direction on this cell is 2, that is, the two bottom cells would be

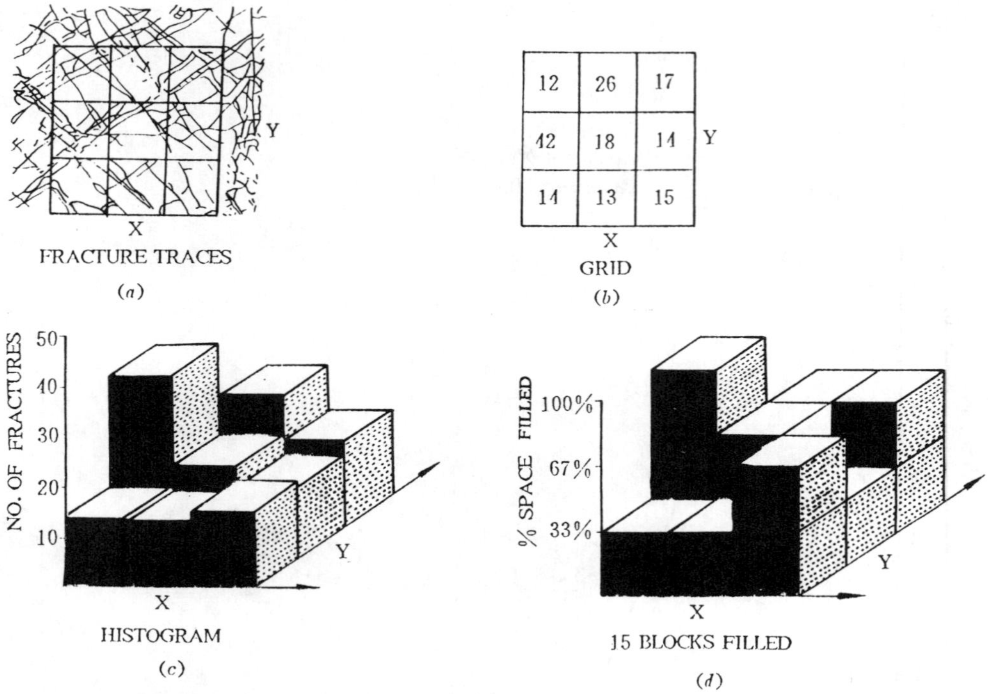

Figure 14.31. Example of fractal formulation for fracture trace length density. (a) Fracture traces and superimposed grid, $R = \frac{1}{3}$; (b) Number of fractures in each grid cell; (c) Histogram of the number of fractures in each cell; and (d) Filled blocks in $3 \times 3 \times 3$ cube, in which each block has dimension $\frac{1}{3} \times \frac{1}{3} \times 33.3\%$ (La Pointe, 1988).

Figure 14.32. Typical synthetic line pattern with overlying grid. The grid squares are of edge length 1/5. There are two line sets, each set having 100 lines (La Pointe, 1988).

counted as being filled (as shown in Figure 14.31(d)). Thus there will be $N(R)$ filled cells for a given $x - y$ cell with edge length R (here $N(R)$ is integral). The fractal dimension of the fracture networks can be determined by means of Equation (14.18).

La Pointe generated the patterns of lines by a computer, based on their calculated fractal dimension, (as shown in Figure 14.32). These patterns have much in common with the pattern of fracture traces that might be found on a bedding plane in a sedimentary rock unit. They differ from a real rock fracture pattern in two important ways: natural fractures have more spatial structure in their location; and common termination styles, such as hooking and splaying, are not reproduced in the fractal simulations. Four geometric factors were examined: the number of lines/unit area, line lengths, orientations, and spatial distribution.

La Pointe simulated six different cases successively; 1 line, 2 lines, 10 lines, 50 lines, 100 lines and 500 lines, in each set. Values of D varied from 2.06 for the simulation with 2 lines total, to 2.75 for the case with 1000 lines total, in which the fractal dimension is determined by Equation (14.18) (as illustrated in Figure 14.33 for a simulation with 2 sets of lines). A plot of the number of lines vs. the resulting fractal dimension is reproduced in Figure 14.34. In each simulation of La Pointe, there were two sets of lines making an acute angle of 60°, and each set had the same number of lines in it. Lengths of the lines were selected at random from the same negative exponential trace length distribution. This graph shows that the greater the number of lines, the higher the fractal dimension. Moreover, there is not a linear relationship between the number of lines and D. After a certain point, equivalent to a dimension of about 2.5, very large increases in the number of lines are necessary for D to increase. Fortunately, the typical joint densities found in outcrop are probably on the lower end of the curve and, thus, the fractal dimension may prove to be a sensitive indicator of the number of joints.

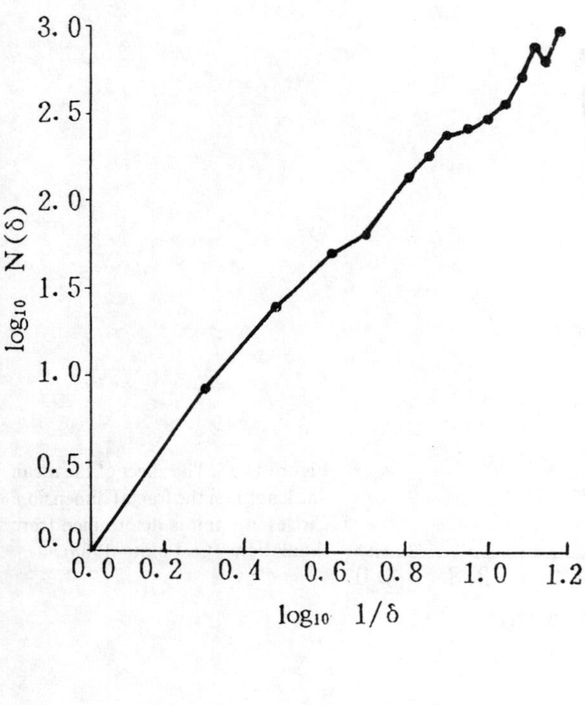

Figure 14.33. Typical plot of the log of the number of filled cells vs. the log of 1/cell edge length. This plot is for a simulation with 2 sets of lines, 50 lines per set, and a mean line length of $\frac{1}{5}$. The fractal dimension (D) is equal to the slope of the line. For this pattern, $D =$ 2.48. Circles are points determined from simulation (La Pointe, 1988).

Holding the number of lines constant at 50 lines per set and, instead, varying the mean line length, yields the results summarized in Figure 14.35. As the mean length increases from $\frac{1}{10}$ to $\frac{1}{5}$, to $\frac{2}{5}$, and finally to $\frac{1}{2}$, the fractal dimension increases, but not markedly so. The 10-fold increase in mean line length only led to an increase in D from

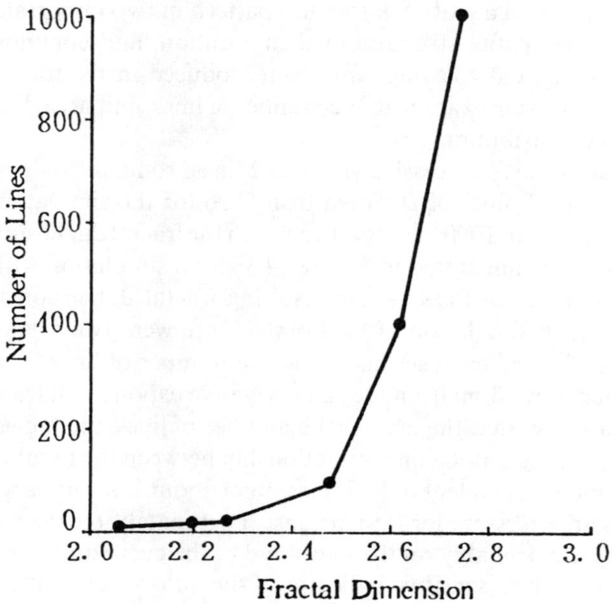

Figure 14.34. The effect of the number of lines on the fractal dimension. Circles are points determined from simulations (La Pointe, 1988).

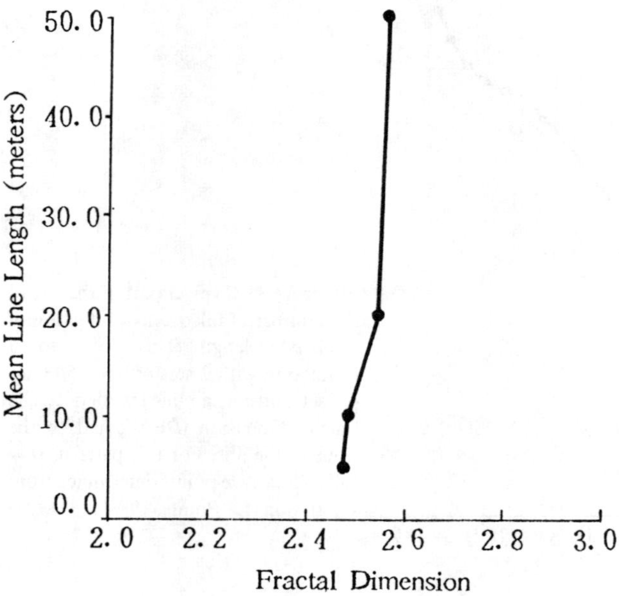

Figure 14.35. The effect of the mean line length on the fractal dimension. Circles are points determined from simulations (La Pointe, 1988).

2.47 to 2.56, a range of variation of 9%. An interesting effect is produced by reducing the number of sets from two to one. The dimension for the simulation employing two sets, each with 50 lines, was 2.48. For the case with a single set consisting of 100 lines, *D* equalled 2.56, virtually the same effect as increasing the length 10 times.

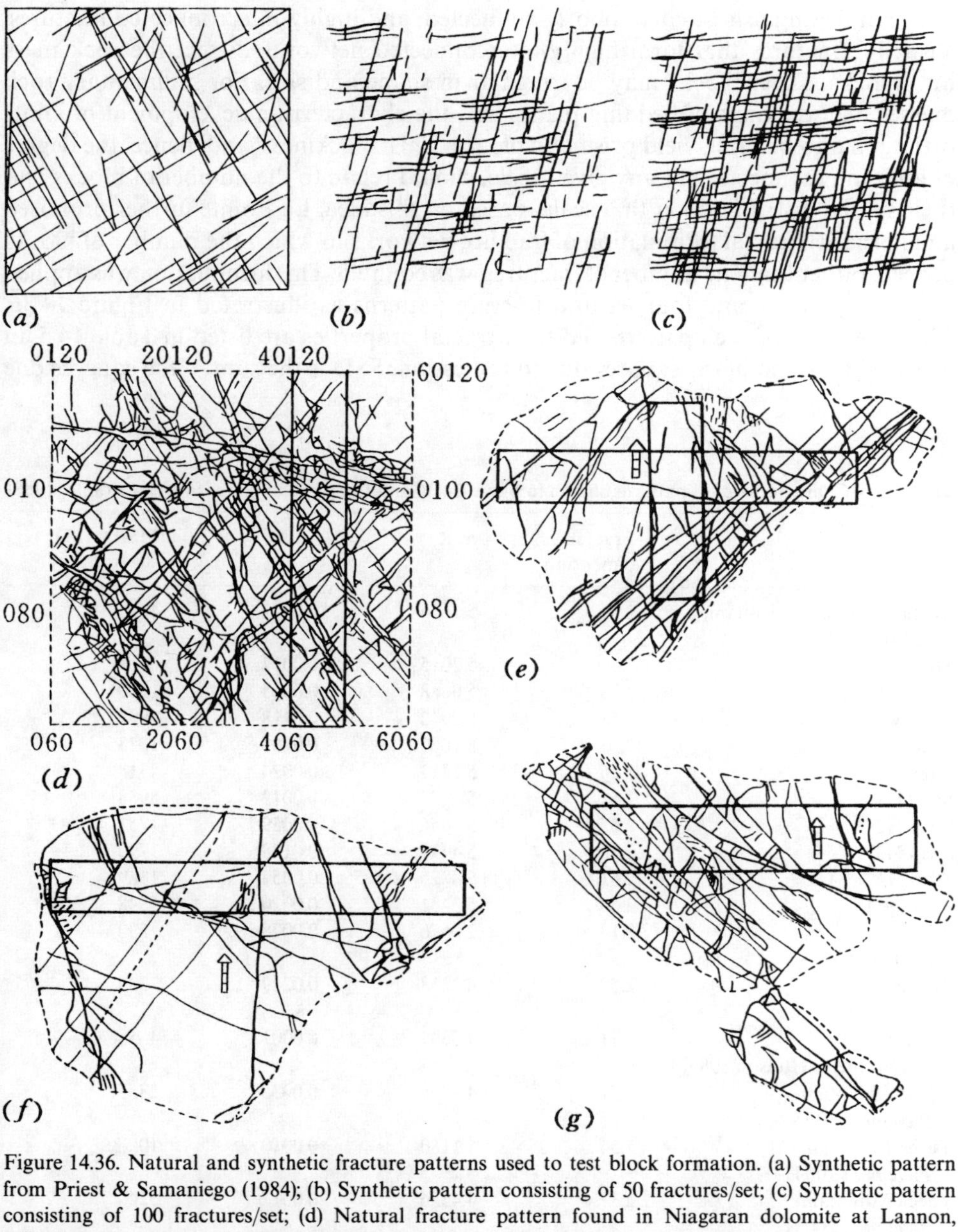

Figure 14.36. Natural and synthetic fracture patterns used to test block formation. (a) Synthetic pattern from Priest & Samaniego (1984); (b) Synthetic pattern consisting of 50 fractures/set; (c) Synthetic pattern consisting of 100 fractures/set; (d) Natural fracture pattern found in Niagaran dolomite at Lannon, Wisconsin (La Pointe & Hudson, 1985); (e) Natural fracture pattern from pavement 100 (see Figure 14.26); (f) Natural fracture pattern from pavement 200 (see Figure 14.27); and (g) Natural fracture pattern from pavement 300 (see Figure 14.28) (La Pointe, 1988).

It should be pointed out that the fractal formulation of fracture networks proposed by La Pointe differs from the fractal measurement by using the grids of squares with side L to cover the fracture networks as used by Barton & Larsen. The later is to cover the fracture-trace (in a section), and its fractal dimension is in the range from 1.0 to 2.0. The former is to cover the fracture region, its fractal dimension ranges between 2.0 and 3.0. Moreover, the fractal formulation of fracture networks proposed by La Pointe does not distinguish isolated, poorly connected, and highly-interconnected fractures. Visually, fractures that form highly-interconnected networks dissect the rock mass into blocks. These blocks may be irregular in shape and size. For sedimentary rock masses, the weakness of bedding planes and the characteristic development of joints orthogonal to bedding yield prismatic blocks. This 'blockiness' and, hence, the degree of joint interconnection at any given scale, should relate to the number of blocks and their spatial distribution at that scale. Based on this idea, La Pointe further proposed an alternative fractal formulation of fracture networks in which the number of blocks per cell, rather than the number of fractures, was counted. This formulation was applied to three synthetic and four natural fracture patterns as illustrated in Figure 14.36. Characteristics of each pattern and their fractal properties are listed in Table 14.5, in which SSR is sum of the squares due to regression, SSD is the sum of the squares due

Table 14.5. Characteristics of patterns shown in figure 14.36 and their fractal dimension (La Pointe, 1988).

Pattern	Fractal dimension	SSR	SSD	MSR/MSD
Synthetic (Priest & Samaniego 1984)	2.55	5.1676	0.0022	4698
Strip V1	2.56	5.2015	0.0017	6119
Strip V2	2.53	5.0688	0.0065	1560
Strip V3	2.58	5.2942	0.0118	897
Strip V4	2.55	5.1667	0.0100	1033
Average V	2.56	5.1813	0.0024	4318
Strip H1	2.61	5.3927	0.0012	8988
Strip H2	2.55	5.1656	0.0059	1751
Strip H3	2.62	5.4475	0.0138	790
Strip H4	2.61	5.3825	0.0057	1889
Average H	2.60	5.3481	0.0020	5348
Synthetic (100 joints)	2.14	2.0676	0.0078	530
Synthetic (200 joints)	2.58	4.2258	0.0249	339
Lannon (La Pointe & Hudson,1985)	2.71	4.5547	0.0001	455 470
Yucca Mountain Pavement 100 V	2.51	4.9968	0.0035	2855
Yucca Mountain Pavement 100 H	2.54	5.1180	0.0010	10 236
Yucca Mountain Pavement 200 H	2.37	4.4536	0.0017	5240
Yucca Mountain Pavement 300 H	2.69	5.7222	0.0064	1788

to deviation, MSR is the mean sum of the squares due to regression, and MSD is the mean sum of the squares due to deviation.

Before discussing the results of this formulation, the method used to determine the fractal dimension for the more complicated patterns needs to be explained. In regions where there may be several thousand blocks, it is impractical to try to incorporate the entire region into the calculations. The pattern shown in Figure 14.36d is an example of a pattern with too many blocks to identify and count by hand. Although sophisticated

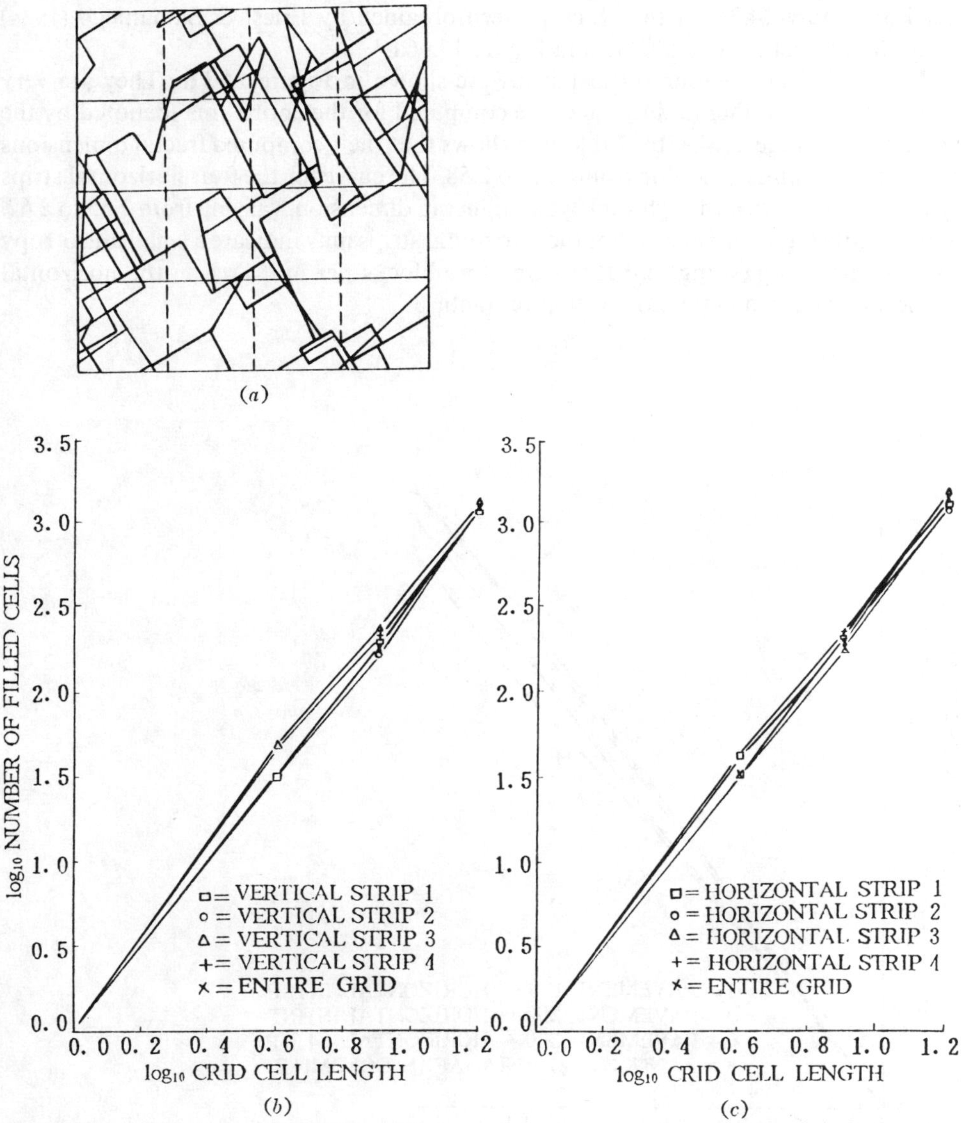

Figure 14.37. Strip locations and computed fractal curves for strips. (a) Location of strips; (b) Fractal curves for four vertical strips; and (c) Fractal curves for four horizontal strips. Fractal curve for the entire grid is denoted by symbol 'X'. The pattern was reproduced from Priest & Samaniego (1984) (La Pointe, 1988).

image scanning techniques could be used to identify and digitally represent the blocks, this is an expensive procedure rarely available except through research laboratories. There exists an alternative, however, that utilizes a smaller portion of the region and makes the block-counting procedure manageable. Moreover, it may indicate a method to scale-up core data to the surrounding rock mass.

Some experiments on the synthetic pattern published by Priest & Samaniego (1984) suggest a way to scale up the core data. Rather than incorporate the entire area, the fractal dimension of strips in different locations and orientations was computed. The location of the strips are outlined in Figure 14.37a. Four are horizontal and four are vertical. Figure 14.37a is the block pattern obtained by Priest & Samaniego (1984) from the synthetic pattern shown in Figure 14.36a.

The results for the four vertical strips are shown in Figure 14.37b. They are very similar to one another and to the curve computed for the entire area (denoted by the symbol 'X' in Figure 14.37b). Table 14.5 shows that their computed fractal dimensions are also very similar, varying from 2.52 to 2.58. The curve for the four horizontal strips is also very similar, although somewhat higher in dimension, varying from 2.55 to 2.62. This slightly higher dimension for the horizontal strips may indicate a small anisotropy in blockiness, suggesting that there are more blocks per unit area in the horizontal strips and, hence, a better connection of fractures.

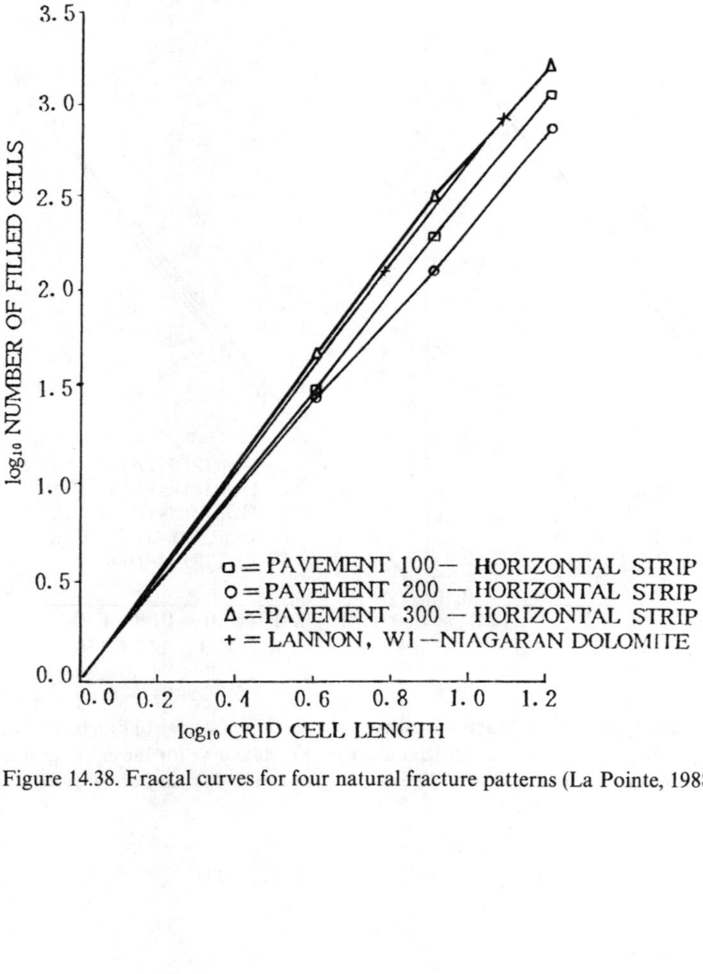

Figure 14.38. Fractal curves for four natural fracture patterns (La Pointe, 1988).

All four natural fracture patterns (Figure 14.36d-g) are well-approximated by least-squares straight lines fit to the data (Figure 14.38 and Table 14.5). The previous synthetic pattern (Figure 14.36a) is also fairly linear (Figure 14.39). The other two synthetic patterns (Figure 14.36b and c) are less linear; they appear more parabolic (Figure 14.39) and their sums of squares due to deviations (SSD) are much higher, while their ratios of mean sums of squares due to regression to those due to deviation (MSR/MSD) are lower (Table 14.5).

It is seen that, in La Pointe's fractal simulations, the fractal dimensions span a good portion of the interval between 2.0 and 3.0. The pattern with the lowest dimension, 2.14, corresponds to the simulation having the fewest blocks, but not necessarily the fewest fractures. The low dimension originates from the 'blockiness' of the pattern, in which the few blocks formed are distributed sporadically throughout the region. This is similar to the behavior seen in the previous computer-generated line experiments. The pattern having the highest fractal dimension is the natural fracture pattern from Lannon, Wisconsin in its Niagaran dolomite (see Table 14.5 and Figure 14.38). Its dimension is 2.71, and it is very well-approximated by a straight line. Fracturing in the dolomite is really more intense than in the other patterns presented, leading to the formation of many blocks. This, in turn, gives rise to a lower variance in block density among the cells and is responsible for the higher dimension. The other synthetic and natural patterns fall nearly midway between being two- and three-dimensional.

Figure 14.39. Fractal curves for three synthetic patterns (La Pointe, 1988).

We can conclude that, from the discussion given by La Pointe, geometric properties that tend to produce a more uniform spatial density of lines lead to a higher fractal dimension. For the fracture density formulation (Equation (14.18)), the greater the number of lines, the larger or longer and the more randomly distributed in space they are, and the higher will be their fractal dimension. For the block formulation, a higher fractal dimension is brought about by a greater number of blocks and, perhaps, a lower block density variance. The computer simulations made by La Pointe suggest that, for natural fracture patterns, the fractal dimension for either formulation may be more sensitive to the number of fractures or block and their spatial distribution, rather than their size and orientation. Of course, the fractal formulations of fracture networks proposed by La Pointe need to be tested in the field in a variety of different settings, and the relationship between the fractal dimension of the fracture network and its stability or permeability needs to be further investigated. However, these fractal formulations are very useful for the analysis of fractures in the range from microscopic fracture networks (such as damage evolution discussed in Section 14.4) to macroscopic joint networks. It may be particularly interesting to apply these methods to fracture networks on the roof walls of workout areas in mining engineering. Perhaps, it will also be possible to some day predict the failure and fall of the roof of workout areas by means of the fractal dimensions of fracture networks on the roof wall.

14.10 FRACTAL CHARACTER AND MECHANISM OF ROCK BURSTS

Rock bursts are experienced in underground mining at various localities in the world, causing death and injury to underground miners and damage to mine structures (drifts, stopes, etc.). To date, many studies have been conducted to understand the causes of rock bursts and outbursts and to predict their occurrence. The approaches made to detect rock bursts include: the microgravity method, rheological method, rebound method, drilling-yield method, microseismic method, and so on (Haramy & McDonnell, 1988). Although all of these methods have been used, none is completely reliable, and few are useful in the rapidly advancing mining environment. The US Bureau of Mines recognized microseismic technology as a potential tool for rock burst prediction as early as 1939, however, to date only a few successful predictions have been achieved (Jenkins et al., 1990). The causes may be due to two aspects: the first may be that the exact physical process of rock bursts is very complicated and too difficult to measure, and the second may be that the initial data recorded by these approaches are not completely utilized. Even so, the microseismic technique (or acoustic emission technique) is still commonly used in monitoring rock bursts.

Microseismic data analysis, as an initial step, includes both the location of rock noise sources and the rate at which the noises occur. Other analyses may include energy release rates, energy per event, and, at a more advanced level, analyses of the microseismic wave form to determine the failure mechanism and stress conditions within the rock. It has been determined that the microseismic noise rate (number of microseismic events per unit of time) increases as failure approaches, and the near-failure noise rate may be 10 to 100 times the background level or stable noise rate. This increase will then be followed by a dramatic decrease of microseismicity in a broad region surrounding the location of the impending burst. This phenomenon is called the seismicity anomaly. Therefore, the microseismic noise rate is commonly used as the precursor factor of rock bursts. Observational results (Haramy & McDonnell, 1988; Kneisley, 1989; Sato et al., 1986; and Brady, 1978) have also indicated that the occurrence of a rock burst certainly corresponds to seismicity anomaly, however, reversing this case does not hold true. On the other hand, some rock bursts occur in active zones of microseismicity, while others do not. Thus, microseismic noise rate alone is not reliable in the prediction of rock bursts. In this regard, the distribution of the microseismic event locations is often overlooked. In fact, this distribution is a realistic record of the damage evolution process experienced by a rock burst. It may play an important role in the prediction of rock bursts. More recently, Coughlin & Kranz (1991) have studied this distribution by using the correlation exponent method (see Section 14.3) and have found that it has a fractal structure.

In this section, we use the number-radius relationship to examine the distributions of microseismic event locations. We find that these distributions really are fractals, and a very interesting fractal character of the distributions is obtained. By using damage mechanics and fractal geometry, we analyze the fractal and physical mechanisms of rock bursts to obtain a better understanding of rock bursts or earthquakes (Xie & Pariseau, 1992a).

14.10.1 *Fractal measurement methods of microseismicity*

As discussed in Section 14.3, rock, like many other materials, when stressed, produces an impulse generated by microcracking associated with the release of strain energy stored internally within its structure and manifested in the form of elastic waves. These elastic waves travel from the point of origin (hypocenter) within the rock to a boundary. The location of the hypocenter (also called the microseismic event location) can be measured by using acoustic emission. From a physical point of view, the microseismic noise events occurring during the monitoring process of rock bursts characterize the microfracturing process (or damage evolution process) before rock bursting. Studies (see Sections 14.2-14.9, Chapters 13 and 16) have indicated that the damage evolution process from the small-scale (a microcraking, cm) to the large-scale (an earthquake, km) is a fractal. Therefore, rock burst which is a damage and fracture within an intermediate-scale (m) may also be a fractal. The microseismic event locations construct a spatial distribution of a point set in which a point corresponds to a cracking surface or volume element in physical space. Thus, the fractal dimension of the damage evolution process experienced by a rock burst can be directly measured from the distribution of the point set.

A direct method of the fractal measurement is based on the number-radius relationship defined by Equation (2.29) (for details, see Section 2.7). The fractal dimension D obtained in this way is called the clustering dimension which is equal to the slope of $\log M(r) - \log(r)$ plot (Figure 14.41) where $M(r)$ is the total number of microseismic events inside a sphere with radius r over the distribution. It should be pointed out that, in this fractal measurement, the center point of the spheres with different radii r_i should be chosen as the mass center of the distribution (see Figure 2.21).

The fractal dimension of the distribution is also obtained from a correlation integral concept (see Section 2.7). The correlation integrals $C(r)$ for the hypocenter distributions $(p_1, p_2, \ldots, p_N)$ as shown in Figure 14.40 can be obtained from Equation (2.24). The fractal dimension D obtained in this way is called the correlation exponent giving the lower limit of the Hausdorff dimension (Section 2.1). This correlation exponent method can be combined with the determination of epicenters in seismicity by which the coordinates of the pair (p_i, p_j) can be calculated. In this way, the fractal measurement can be performed by the computer software monitoring microseismic or macroseismic events. Thus, like information on daily rock noise rate, the fractal dimension can be daily shown by the micro- or macro-seismic monitoring system.

14.10.2 *Fractal character of rock bursts*

Leighton (1982) reported monitoring microseismic data of a rock-burst-prone pillar in a metallic ore vein in the Galena mine, up to and through failure. The stope was being excavated by overhand cut-and-fill mining. A major rock burst occurred on May 24, 1979. In addition to information on the daily rock noise rate, plots of rock noise locations were compiled by computer. The plots of the rock noise locations for five days prior to the May 24 burst are reproduced in Figure 14.40.

We apply the fractal measurement method to measure the number-radius relationship

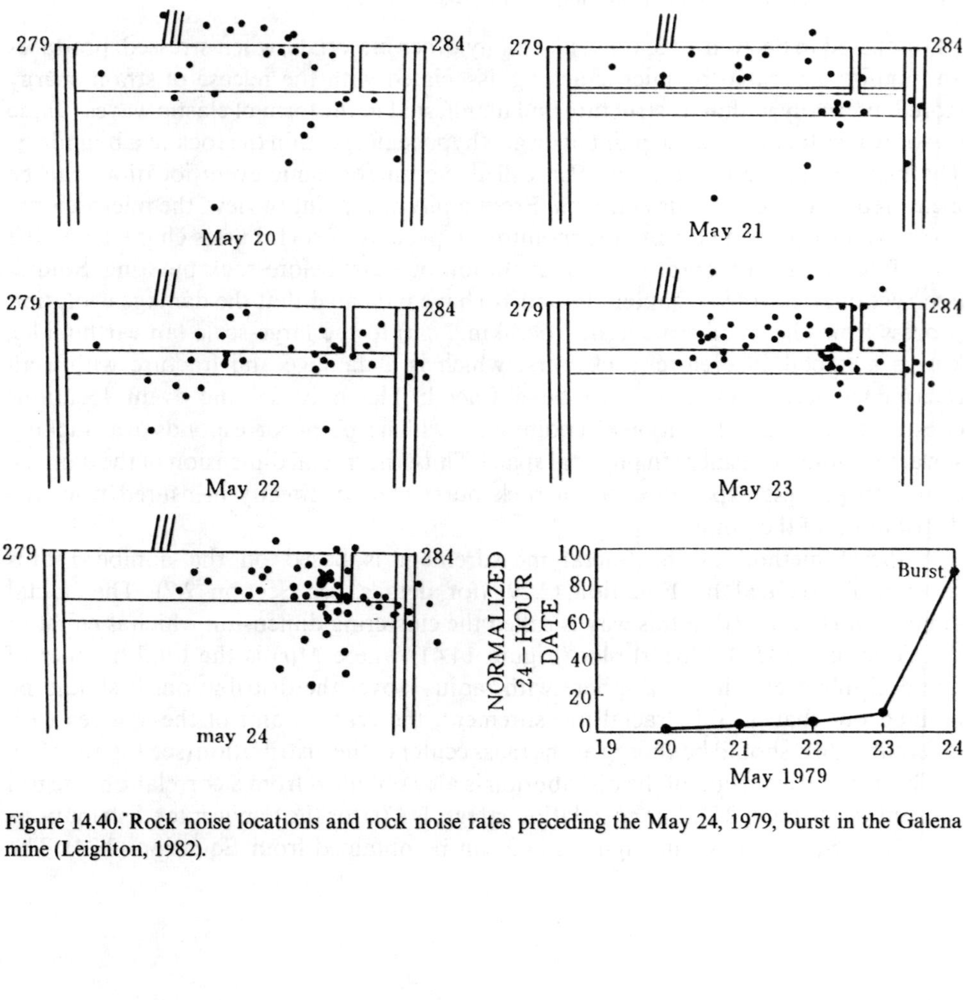

Figure 14.40. Rock noise locations and rock noise rates preceding the May 24, 1979, burst in the Galena mine (Leighton, 1982).

of the distribution of the rock noise locations. The plots of log $M(r)$ vs. log (r) are shown in Figure 14.41. From Figure 14.41, it is seen that there exists a good linear correlationship between log $M(r)$ and log (r). This indicates that the distribution of the rock noise locations is a fractal and has statistical self-similarity. The reliabilities of

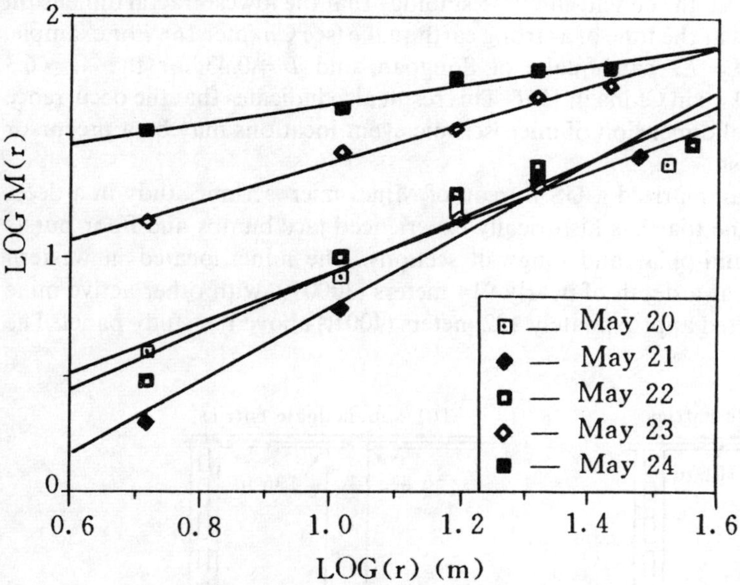

Figure 14.41. The plots of log $M(r)$ vs. log (r) for the distributions of rock noise locations given in Figure 14.40.

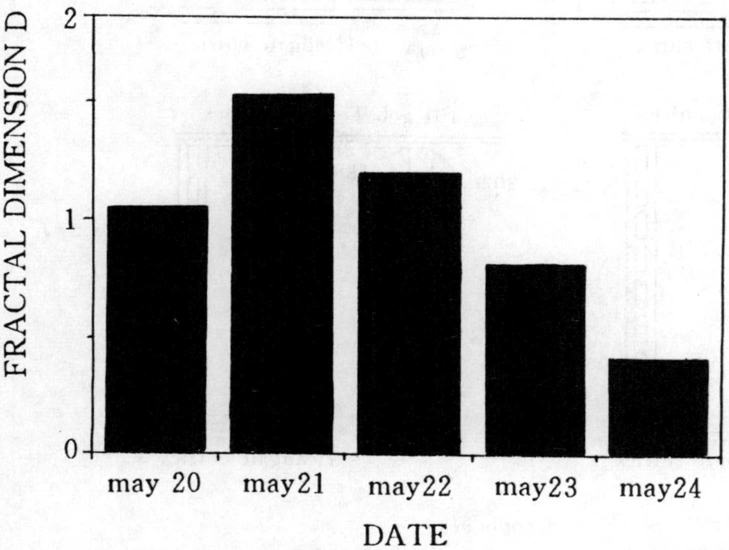

Figure 14.42. The correlation between fractal dimensions and date for the five days prior to the May 24 rock burst.

linear regression are all larger than 0.96. The fractal dimensions for the five days prior to the May 24 burst are directly obtained from the slopes of plots in Figure 14.41. Figure 14.42 shows the relationship between the fractal dimension and the date for these five days. From Figure 14.42, a very interesting result was found; prior to the major rock burst, the lowest fractal dimension ($D = 0.4104$) occurred on May 24. This result is consistent with the conclusion in seismology that the lowest fractal dimension (or b-value) occurs near the time of a strong earthquake (see Chapter 16). For example, $D = 0.45$ for the $M = 7.2$ earthquake of Songpan, and $D = 0.43$ for the $M = 6.3$ earthquake of Dengkou in China in 1976. This result also indicates that the occurrence of a minimum fractal dimension of microseismic event locations may be a precursor to a major rock burst.

Kneisley (1989) summarized a US Bureau of Mines microseismic study in a deep, western US coal mine that has historically experienced face bumps and floor bursts in both the room-and-pillar and longwall sections. The mine, located in western Colorado, operated at a depth of nearly 914 meters (3000 ft), with other active mine workings being located approximately 122 meters (400 ft) above the study panel. The

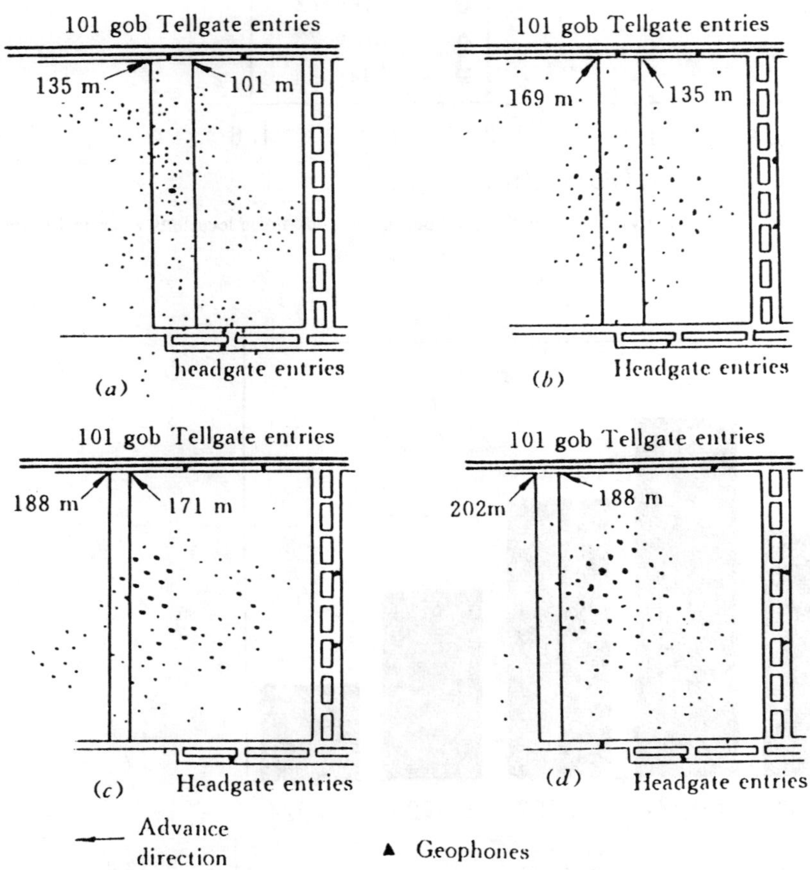

Figure 14.43. Microseismic source locations, February through May. (a) February; (b) March; (c) April; (d) May. (Kneisley, 1989).

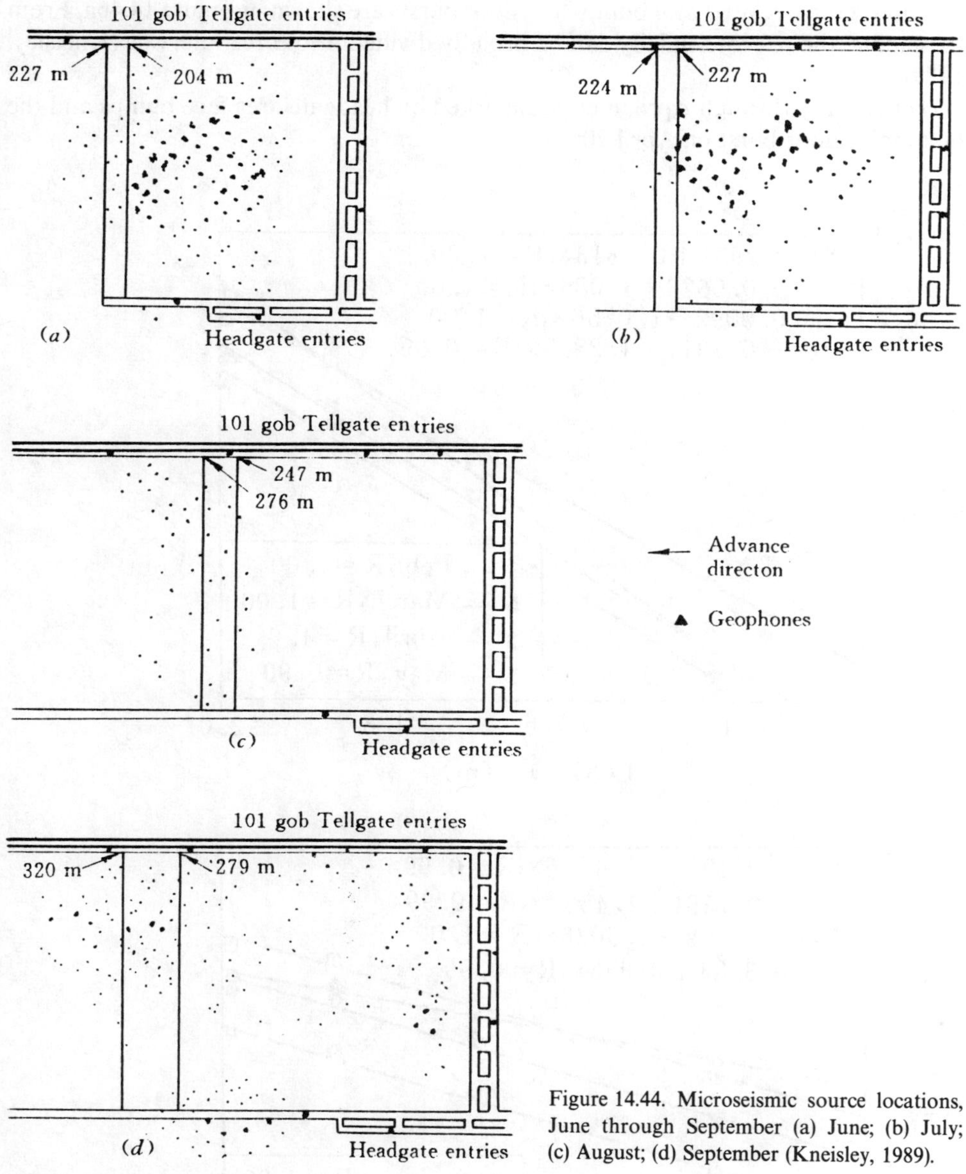

Figure 14.44. Microseismic source locations, June through September (a) June; (b) July; (c) August; (d) September (Kneisley, 1989).

3.048-meter-thick coalbed had been mined using the advancing longwall mining method. Prior to 1983, stress relief by volley firing was practiced to destress the longwall face corners and mining sections at depths exceeding 610 meters (2000 ft). Soon after mining began on the longwall panel, face bumps occurred . From February through September, 1983, several coal bumps and floor bursts occurred as a result of either mining or destressing (Figure 14.46a). General microseismic activity was summarized using both the microseismic event rate and location of these seismic sources. Microseismic source locations for each month of this study (from February to September, 1983) are reproduced in Figures 14.43 and 14.44. The microseismic event rate and several related coal bumps and floor bursts are shown in Figure 14.46a. From these data, the microseismicity can be broadly divided into three intervals (Kneisley, 1989):

(1) February through April, a period marked by headgate area face bumps and the first major floor burst on April 20;

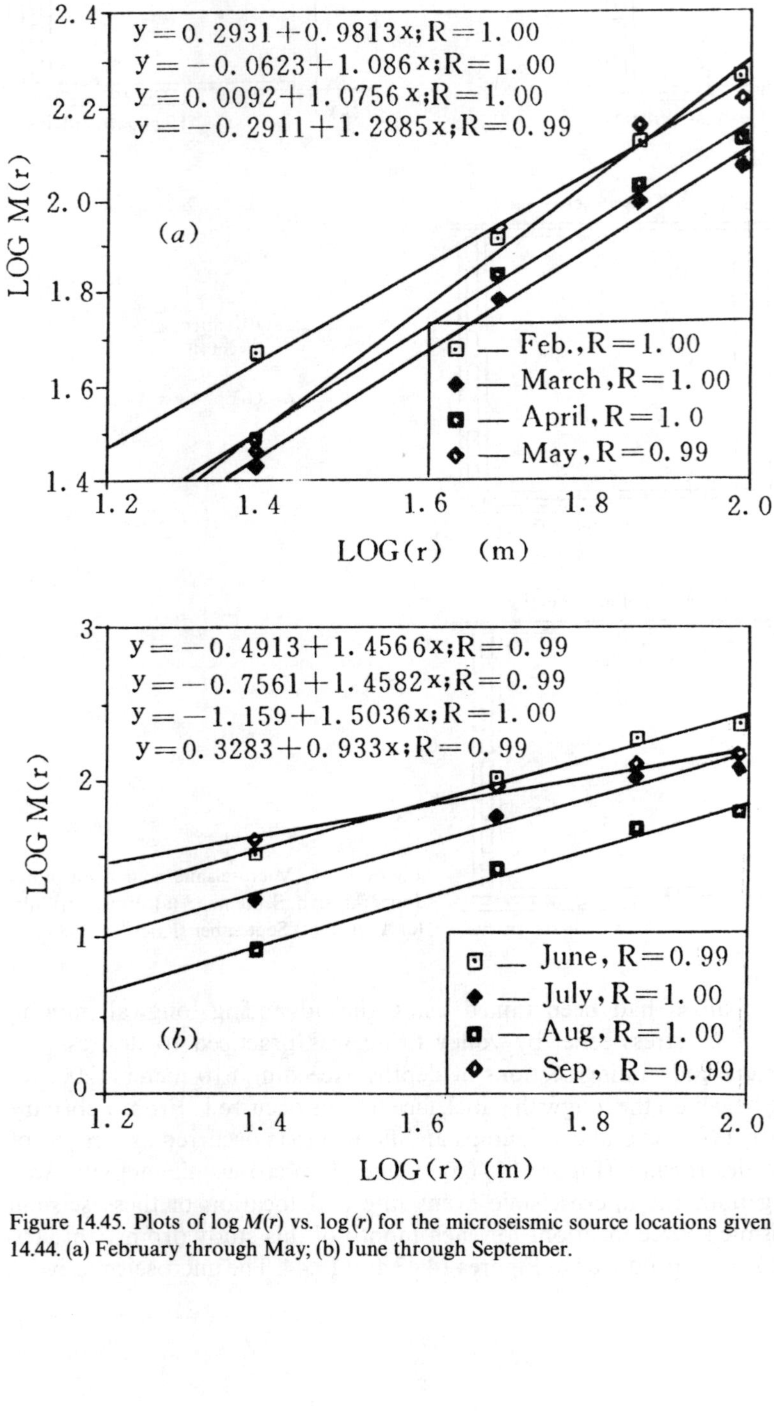

Figure 14.45. Plots of log $M(r)$ vs. log (r) for the microseismic source locations given in Figures 14.43 and 14.44. (a) February through May; (b) June through September.

(2) May through late August, when intensive face and floor destressing began, but no bumps or bursts occurred;

(3) late August through September, a period of gradually increasing microseismic activity, and floor bursts occurred in September.

Note that activity increases of equal or greater magnitude were also recorded in the

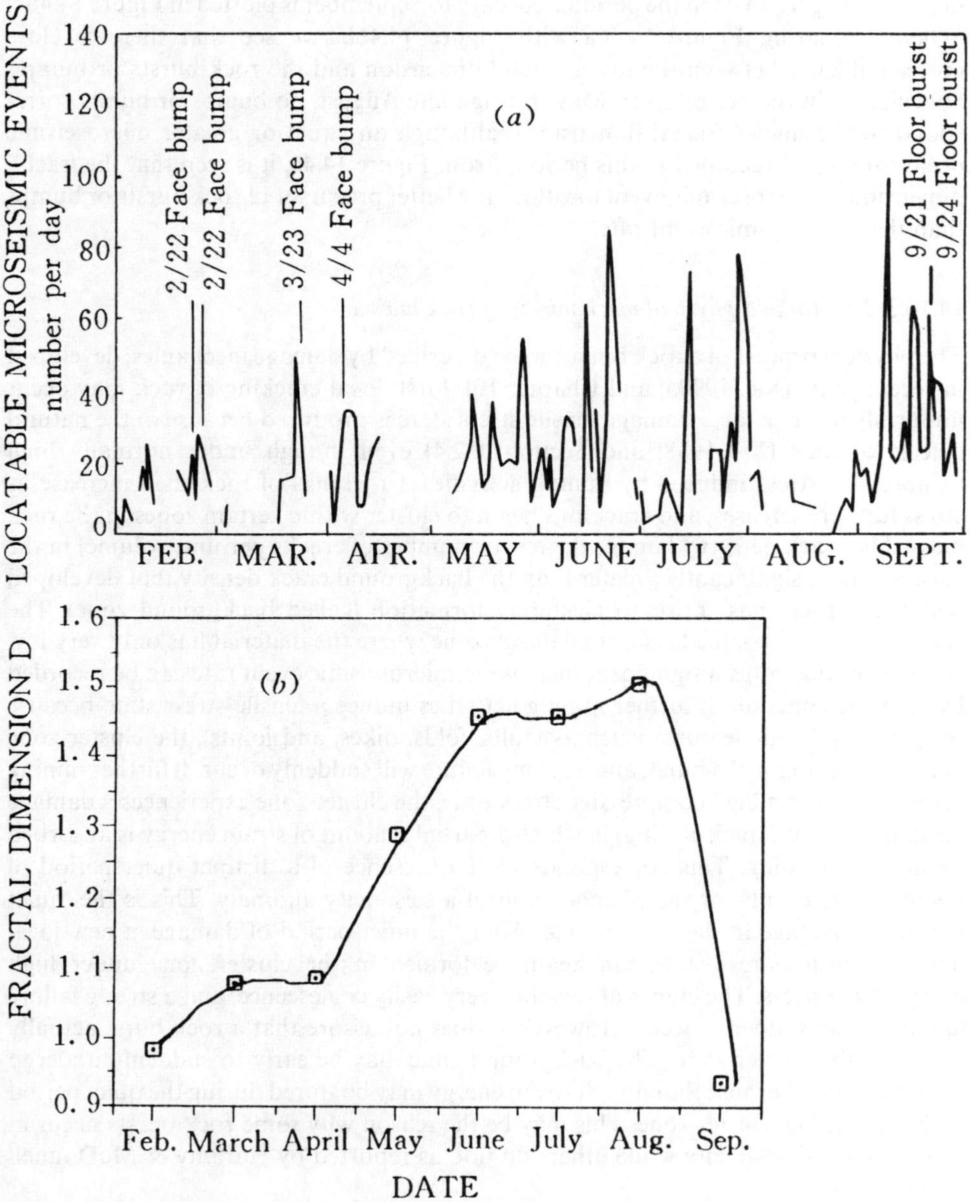

Figure 14.46. The correspondence between the fractal dimension and main rock bursts and bumps: (a) Locatable microseismic events from February through September (Kneisley, 1989); and (b) The change tendency of the fractal dimension within the timeframe of February through September.

second interval. This indicates that the microseismic event rate is really not reliable in the prediction of rock bursts.

By using the fractal measurement method, we measure the distributions of microseismic event locations as shown in Figures 14.43 and 14.44. The measuremental results are illustrated in Figure 14.45. Excellent linear correlations between log $M(r)$ and log (r) actually exist. The change of the fractal dimensions obtained directly from the slopes of plots in Figure 14.45 in the period February to September is plotted in Figure 14.46b. When comparing Figure 14.46a with Figure 14.46b, we see that there is close correspondence between the lower fractal dimension and the rock bursts or bumps. Specifically, in the period from May through late August, no bumps or bursts correspond to the higher fractal dimensions, although an equal or greater microseismic event rate is still recorded in this period. From Figure 14.46, it is seen that the fractal dimension of microseismic event locations is a better precursor of rock bursts or bumps than the microseismic event rate.

14.10.3 *Fractal and physical mechanisms of rock bursts*

The physical process of a rock burst can be described by damage mechanics, developed in recent years (Xie, 1990a; and Chapter 10). First, local cracking of rock may occur uniformly where a local damage tensile stress state is produced because of the natural defects of rock (Xie 1988; and Section 10.2.4) even though under normally high compressive stress induced by mining activities. Crackings of rock then increase as stress further increases, and crackings begin to cluster within certain zones in the rock mass. The crack density (damage degree, or number of cracks per unit volume) in the cluster zone is significantly greater than the background crack density that developed within the rock mass prior to clustering formation (called background zone). The cluster zone is also called a severe damage zone where the material has only very low stiffness. In this zone, a significant increase in microseismic event rate can be recorded by acoustic emission. If further mining activities induce a tensile stress state because of specific geologic features (such as faults, folds, dikes, and joints), the cluster zone becomes the origin of a burst, and a strong failure will suddenly occur. If further mining activities induce a high compressive stress state, the cluster zone experiences a damage 'healing' process (crack closing) in which a partial amount of strain energy is absorbed by damage healing. This corresponds to the presence of a distinct quiet period of microseismic events, or the phenomenon of a seismicity anomaly. This is the 'quiet period' of damage in the cluster zone. After the quiet period of damage, a new local damage tensile stress state can again be formed in the cluster zone under high compressive stress. The cracks at this time very easily coalescence, and a strong failure (a burst) may suddenly occur. However, it does not assure that a rock burst actually occurs in the cluster zone. The background zone may be early to suddenly undergo failure (a burst) because the enough strain energy may be stored during the quiet period of damage in the cluster zone. This may be the reason why some rock bursts occur in active zones of seismicity while others do not, as reported by Haramy & McDonnell (1988).

Although the physical process experienced by a rock burst is very complicated, mathematically, the microseismic events of this process are only simple geometry – fractal clustering geometry. The microseismic events that are almost uniformly distributed in the high stress area in the time period far from rock bursting have a higher fractal dimension. The more the clustering of microseismic events occur near the rock burst, the lower is the fractal dimension that will be produced. This is a fractal geometrical mechanism of rock bursts.

Since each of the microseismic events in the two-dimensional case corresponds to a cracking surface element area (or a crack island area) A_i (note that a crack volume will be considered if in three-dimensional case), the strain energy E_i stored in the rock masses at least needed to be dissipated for the occurrence of this cracking of rock can be expressed as

$$E_i = \gamma_s A_i \tag{14.34}$$

where γ_s is the surface free-energy of rock. Since the measured results given in the previous section show that the distribution of cracking events or microseismic events is a fractal, from the basic definition of fractal geometry (Mandelbrot, 1975a), the number of cracking events (or crack islands) with cracking surface area greater than A_i should thus satisfy the fractal distribution as follows

$$n = n_0 A_i^{-D/2}. \tag{14.35}$$

Note that $n_0 \to 1$, as seismic events (or cracking events) approach a cluster. Since a rock burst is actually equivalent to a fractal cluster of crackings within the rock, the total energy release of rock bursting thus becomes:

$$E_t = \gamma_s A_i n = \gamma_s A_i^{1-D/2}, \tag{14.36}$$

or

$$E_t/\gamma_s = A_i^{1-D/2}, \tag{14.37}$$

where the scale of A_i is (meter)^2 because our analyses is performed in the intermediate-scale of mines, and the fractal dimension D is in the range from 0 to 2.0. Similarly, for the three-dimensional case, Equation (14.37) can be written as

$$E_t \propto V_i^{1-D/3}, \tag{14.38}$$

where V_i is the scale of cracking volume elements, and the fractal dimension ranges from 0 to 3.0.

Based on Equation (14.37), the required energy release ratio E_t/γ_s for different fractal dimensions can be calculated where the scale of A_i is chosen as (2 m^2). The calculated values are listed in Table 14.6. The tendency of E_t/γ_s to vary in relation to the fractal dimension is shown in Figure 14.47. From Figure 14.47, it is seen that the lower the fractal dimension, the more the strain energy needs to be released. This corresponds to the occurrence of a main rock burst. For the three-dimensional case, i.e., $D \in [0, 3]$, the same tendency of energy release E_t to vary in relation to the fractal dimension can be obtained from Equation (14.38).

The theoretical results (Equations (14.37) and (14.38), Figure 14.47) are consistent

Table 14.6. The relationship between strain energy rlease and fractal dimension.

D	0.1	0.3	0.5	0.7	0.9	1.1	1.3	1.5	1.7	1.9
E_t/γ_s	1.93	1.80	1.68	1.57	1.46	1.37	1.28	1.19	1.11	1.04

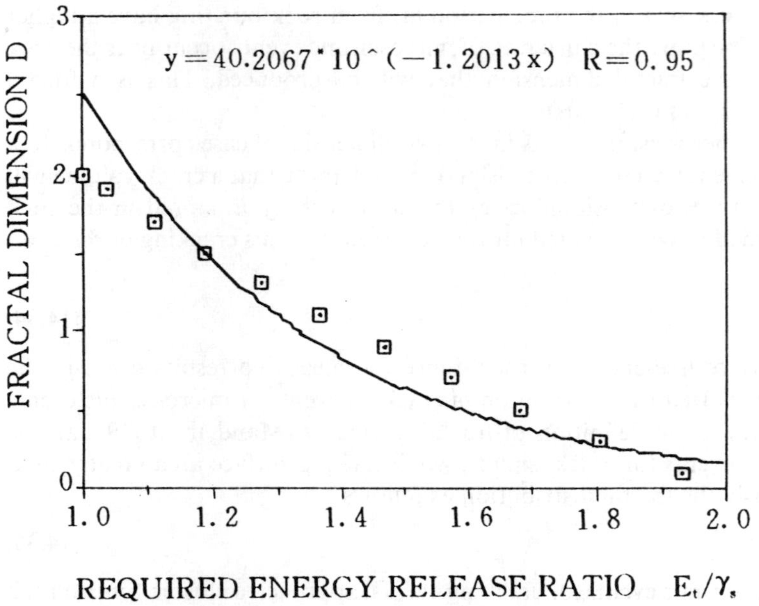

Figure 14.47. The change tendency of the strain energy release with the fractal dimension, obtained by theoretical analyses.

with the fractal description of earthquakes defined by Gutenberg-Richter's relationship (also called the magnitude-frequency relationship) given by

$$\log N = a - bM, \tag{14.39}$$

where a and b are constants for a region, and N is the frequency of magnitude M. Aki (1981) and Turcotte (1986b) have shown that Equation (14.39) has a fractal behavior for earthquakes, and the b-value in Equation (14.39) is strongly related to the fractal dimension of earthquakes in the form (for details, see Section 16.2):

$$D = 3b/C. \tag{14.40}$$

In general, $C = \frac{3}{2}$. Thus, Equation (14.40) can be rewritten as

$$D = 2b. \tag{14.41}$$

Sato et al. (1986) have found that the magnitude-frequency relationship of micro-seismic activity associated with hydraulic mining also follows Gutenberg-Richter's relationship (Equation (14.39)). Their measured results of the b-values and micro-seismic activity at the S3-860L mining panel of the Sunagawa coal mine in Japan are listed in Table 14.7. Here the fractal dimensions are calculated directly from Equation (14.41) based on the b-values given in Table 14.7. The varying tendency of seismic energy release with the fractal dimension is shown in Figure 14.48. From Figure 14.48, it is seen that the measured results obtained by Sato et al. have the same tendency as those of the theoretical analyses given by Equations (14.37) and (14.38). From Figures 14.47 and 14.48, we find that the energy release will exponentially increase with the

Table 14.7. The *b*-values, fractal dimensions, and microseismic activity at the S3-860L mining panel of the Sunagawa coal mine in Japan (* from Sato et al., 1986).

Number of events*	Seismic energy* (KJ)	Seismic energy release rate* (J/m²)	*b*-values*	Fractal dimensions
51	54.8	50.2	0.96	1.92
73	209.7	60.6	0.67	1.34
123	10.7	6.1	1.50	3.00
211	31.3	5.6	1.39	2.78
34	52.4	57.8	0.83	1.66
94	86.9	18.0	1.01	2.02
46	77.2	20.3	1.39	2.78

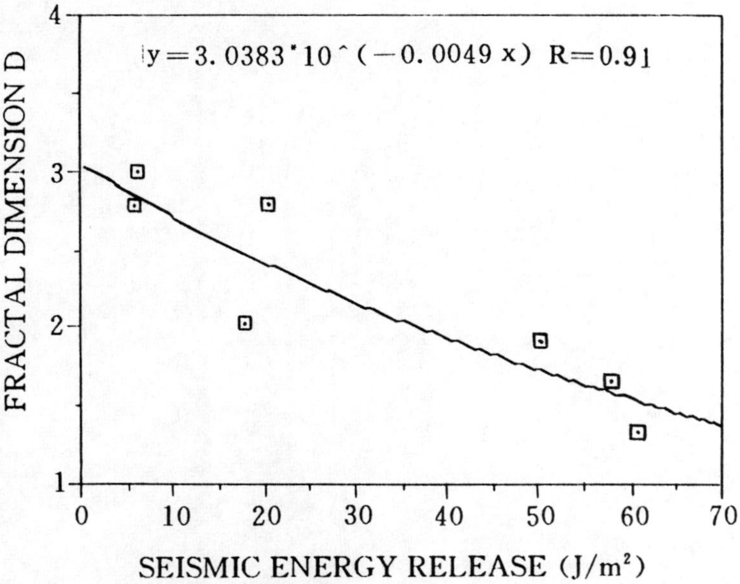

Figure 14.48. The relationship between the fractal dimension and the seismic energy release obtained by microseismic measurements.

decrease of the fractal dimension. After generalizing the regression equations shown in Figures 14.47 and 14.48, we can conclude a general formulation of the energy release to vary in relation to the fractal dimension as follows:

$$D = C_1 * \exp[-C_2 E], \qquad (14.42)$$

where C_1 and C_2 are constants varying with the region and the measurement scale, and the fractal dimension ranges from 0.0 to 3.0.

Equation (14.42) explains, in theory, a fractal and physical mechanism why the lower fractal dimension corresponds to the occurrence of a main rock burst in mines, and a main earthquake in seismology. A lower fractal dimension value indicates the formation of larger new cracking surfaces or volumes within the rock mass, indicating possible rock instability. More strain energy needs to be released, corresponding to the occurrence of a strong failure (a main rock burst or an earthquake). The theoretical analyses here also shows that it is possible to predict the occurrence of main rock bursts by a decrease in the fractal dimension for the distribution of microseismic event locations.

It should be pointed out that the above conclusion is only suitable for clustering fractals. Other fractals may need to have opposite conclusions. For example, as discussed in Section 14.4, the conclusion that the larger fractal dimension is associated with the more serious damage state (a failure) is obtained because another definition of fractals is adopted in which crackings are considered as line events, rather than point events in this section.

CHAPTER 15

Fractal description of the roughness of rock joints

The earliest idea of fractals (Mandelbrot, 1967) originated from the measurement of the roughness of coastlines. So far, terrain, landscapes, faults, and many irregular and rough objects have been described by fractals. One naturally poses the questions: Are joint surfaces or discontinuity profiles in rocks also fractals? What relationships are there between JRC (Joint Roughness Coefficient), flow paths through joints, and fractal dimensions? In this chapter, we shall mainly discuss these topics.

15.1 SURFACE ROUGHNESS, DEFORMATION AND STRENGTH OF ROCK JOINTS

Rock joints have an important influence on the deformational behavior of rock systems. Properties of the joints include orientation, extent, planarity, roughness, and the strength of wall rock asperities. Roughness, which influences the friction angle, dilatancy, and peak shear strength, refers to local departures from planarity at both small and large scales. There have been a considerable number of studies on the influence of surface roughness (Li, 1988). The purpose of these studies is for the eventual estimation or calculation of dilatation and shear strength of jointing rocks.

When a joint is compressed in the normal direction, the contact between the joint walls starts from some point on the lines. During compression, a part of the contacting surface is broken or split, and the contacting area increases. Some measurements indicate that even at a very high compressive stress level, the total contact area is only 40-70 per cent of the entire wall area. The actual process is so complicated that the deformation and strength properties of jointing rocks are usually studied by testing, and the regularity in the test results can be summarized into empirical formulas. One of the empirical formulas is Goodman's formula (Goodman, 1976) as follows.

Assuming that the thickness of a joint is e (as illustrated in Figure 15.1a), the initial stress is σ_0, and the maximum contraction (closing displacement) caused by normal compression, V_{mc}, is less than e. Thus, the hyperbola-type relationship between compressive stress σ and contraction ΔV can be defined by

$$(\sigma - \sigma_0)/\sigma_0 = A[\Delta V/(V_{mc} - \Delta V)]^t, \quad \text{if } \Delta V < V_{mc}. \tag{15.1}$$

Equation (15.1) can be taken as the constitutive equation of normal compressive

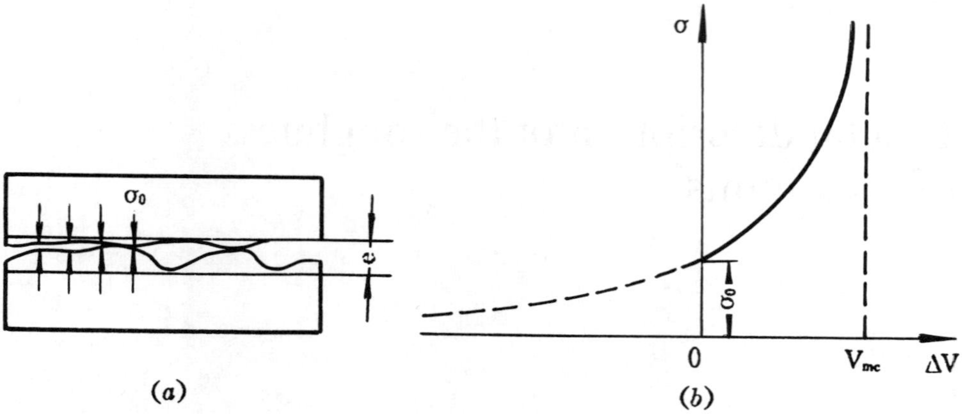

Figure 15.1. (a) Thickness of the joint and initial stress; (b) Compression deformation curve of a joint (Li, 1988).

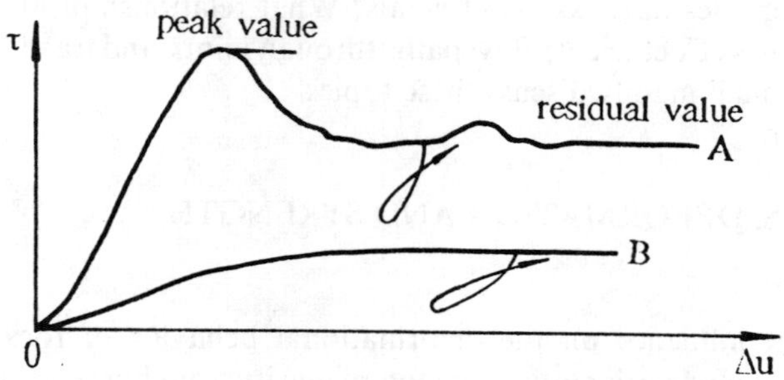

Figure 15.2. Shear stress-displacement curve of shearing tests along a joint (Li, 1988).

contraction of the jointing rock. The slope of the tangent of the curve shown in Figure 15.1b is called the stiffness coefficient of the joint under normal compression.

When shear stress acts along the joints, the shear stress-displacement curve of the joints has a peak and residual value, (curve *A* in Figure 15.2 is called an *A*-type joint). The ratio of the residual strength to the peak strength is about 0.3-0.6, whereas the curve of a filled joint has no peak value occurrence (curve *B* in Figure 15.2, called *B*-type joints).

The shearing test along a joint is commonly performed under a certain amount of normal stress. Thus, each of shear stress-displacement curves ($\tau - \Delta u$) corresponds to the normal stress σ (= constant). The $\tau - \Delta u$ curves of *A*-type joints can be approximately described by sectional linearization as shown in Figure 15.3.

Depending on shearing direction, dilatancy can occur while shearing the toothed joint along one direction, whereas contractancy occurs while shearing in the reverse direction. The so-called dilatancy or contractancy is the expansion or contraction of the joint volume when the jointed rock is sheared along the joint, as indicated in Figure 15.4. The actual toothed surface and the occurrence of a gap in the joint are very complicated. Dilatancy is more frequently produced, whereas contractancy is quite rare.

The strength of a joint depends on the material properties of the rock mass at both

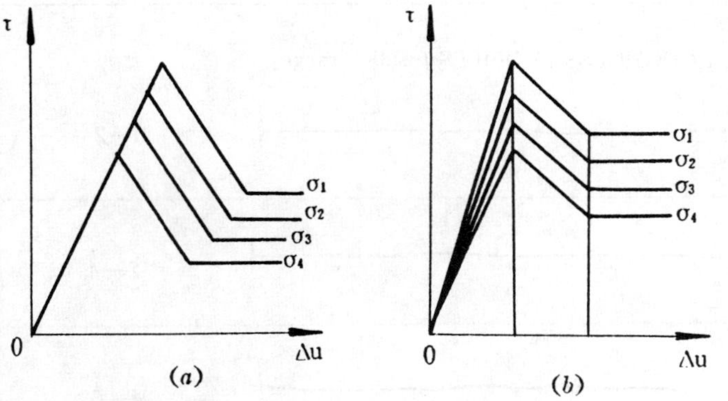

Figure 15.3. (a) and (b) two types of sectional linear models of A-type joints ($\sigma_1 > \sigma_2 > \sigma_3 > \sigma_4$) (Li, 1988).

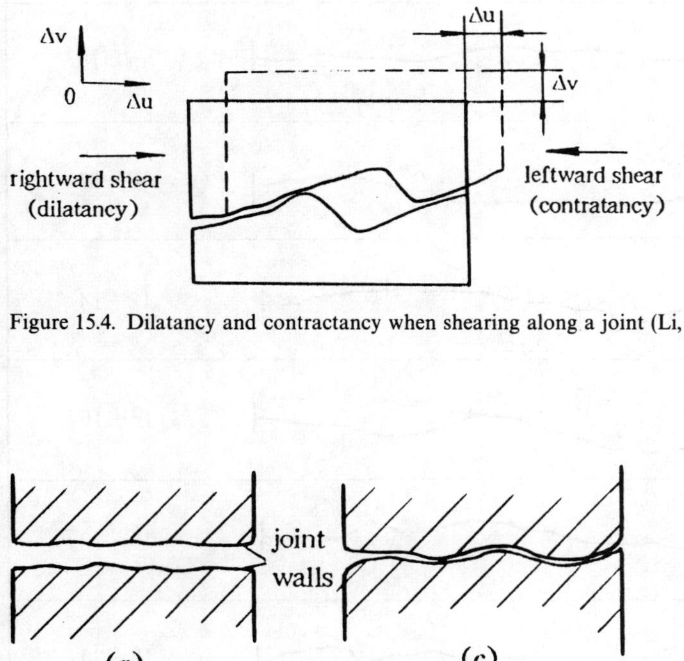

Figure 15.4. Dilatancy and contractancy when shearing along a joint (Li, 1988).

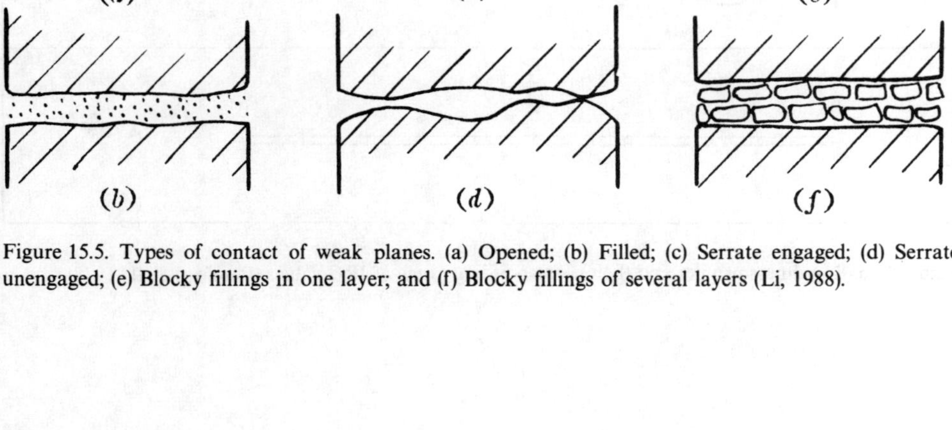

Figure 15.5. Types of contact of weak planes. (a) Opened; (b) Filled; (c) Serrate engaged; (d) Serrate unengaged; (e) Blocky fillings in one layer; and (f) Blocky fillings of several layers (Li, 1988).

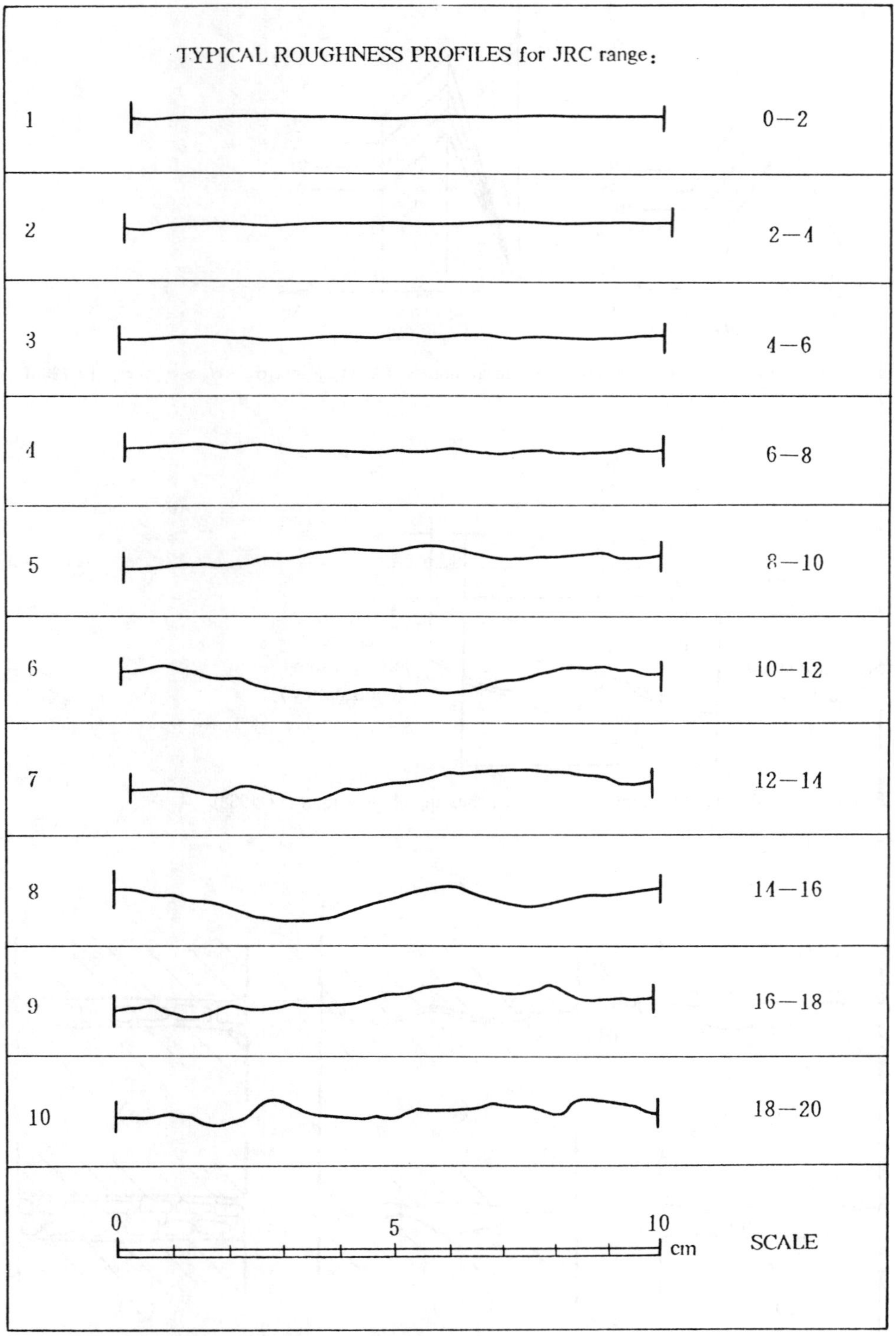

Figure 15.6. Roughness profiles and their corresponding range of JRC values (Bamford et al., 1978).

sides of the joint, the surface roughness of both walls and the type of contact between the two walls as shown in Figure 15.5. Except for joints filled with siliceous and calciferous materials, all joints can be considered as non-resistant to tensile stress. Therefore, the strength of the joint is mainly referred to as the compressive strength perpendicular to the joints and the shear strength along the joints (Li, 1988).

In 1979 the ISRM (International Society for Rock Mechanics) suggested measuring the joint wall compressive strength (JCS) proposed by Schmidt Hammer in the form

$$\log(\mathrm{JCS}) = 0.00088rR + 1.01\,(\mathrm{MPa}), \tag{15.2}$$

where r is the volume weight of rock in $\mathrm{KN/m^3}$, and R is the rebound factor (dimensionless).

The shear strength along joints in rock is closely related to the surface roughness of the joints. An approach to the problem of predicting the shear strength of rough joints was proposed by Barton (1973). Based on careful tests and observations carried out on artificially-produced rough joints in material used for model studies, Barton developed 10 typical profiles of increasing roughness and assigned coefficients ranging from 0 to 20 which form the 10 joint roughness coefficients (JRC) with a range as indicated in Figure 15.6. These standard profiles were later adopted by ISRM in their suggested procedure for measuring the roughness of joints (Bamford et al. 1978). The JRC value can be expressed by means of the test data of compressive shearing tests and joint wall compressive tests as follows:

$$\mathrm{JRC} = [\tan^{-1}(\tau/\sigma) - \phi_b]/\log(\mathrm{JCS}/\sigma), \tag{15.3}$$

where τ and σ are the peak values of shear stress and normal stress, respectively, in the compressive shearing test and ϕ_b is the basic friction angle (friction angle of surface of an unweathered smooth joint). Barton & Choubey (1976) proposed a criterion of shear strength associated with JRC in the form

$$\tau = \sigma\tan[(\mathrm{JRC})\cdot\log(\mathrm{JCS}/\sigma) + \phi_b], \quad \text{or } \tau = \sigma\tan(i + \phi_b). \tag{15.4}$$

where i is joint roughness angle. When JRC = 0, Equation (15.4) becomes

$$\tau = \sigma\tan\phi_b,$$

which means that the formula is transformed into Coulumb criterion for a smooth flat joint (without cohesion).

15.2 RELATIONSHIP BETWEEN THE JRC VALUE AND FRACTAL DIMENSION

When the profile of rock joints is assumed to be self-similar, we can measure the fractal dimensions of rock joints from Equations (2.7) and (2.11) (see Chapter 2). Figure 15.7 shows the measurement of the fractal dimension and how the measured length, $L(r) = r^{1-D} = N\cdot r$, of an irregular curve depends only on the size of the measuring unit (Carr & Warriner, 1987; Turk et al., 1987; Carr, 1989; Lee et al., 1990).

As an example of fractal measurement of rock joints, the works of Lee et al. (1990) will be mainly introduced in this and the next sections. Figure 15.8 shows their measuremental results of a roughness profile (JRC range of 10–12, and $D = 1.0056$).

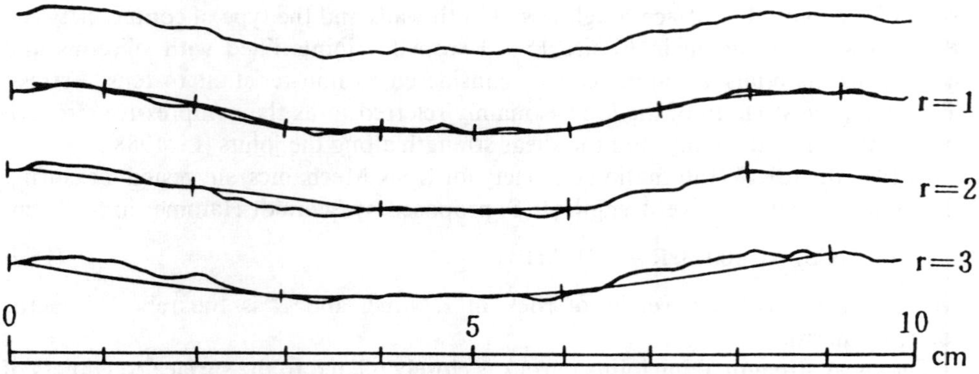

Figure 15.7. Interaction of an irregular curve and the measuring 'ruler length' (Lee et al., 1990).

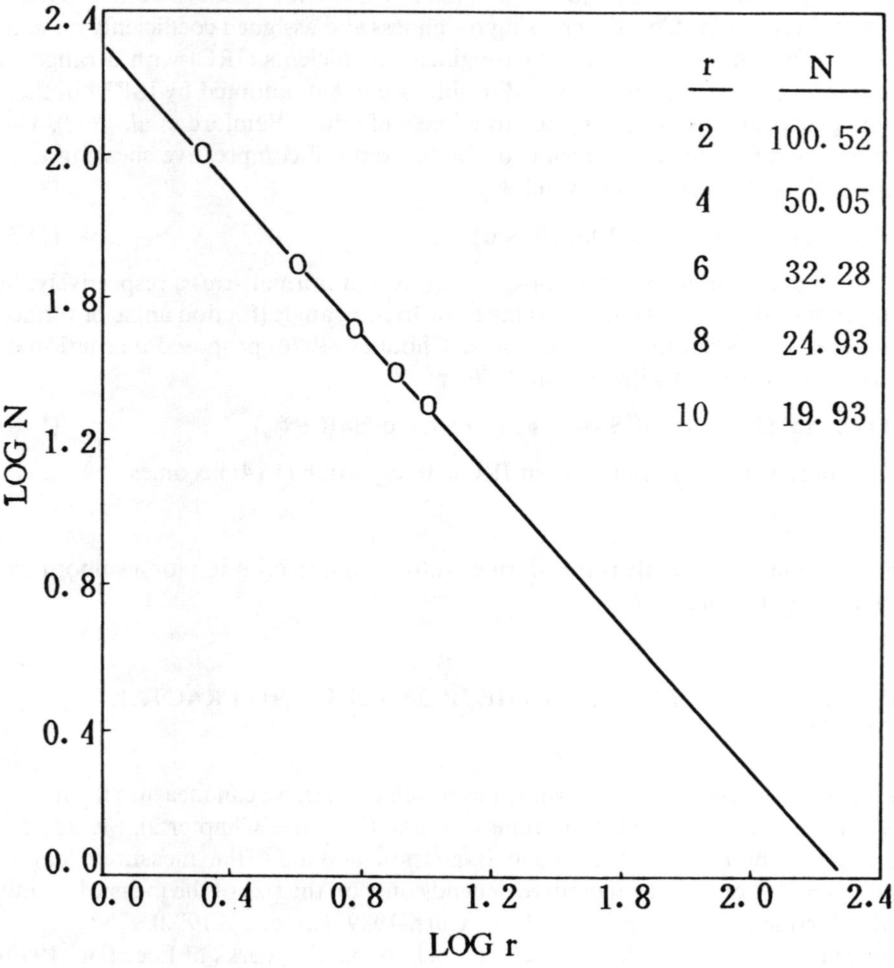

Figure 15.8. Log r vs. log N plot for a roughness profile (JRC range 10-12) with a fractal dimension $D = 1.0056$ (Lee et al., 1990).

The measurement of roughness profile characteristics can be performed with a digitizing table simply by tracing the curve with the cursor. Basic data obtained are the (*x*, *y*) coordinates along the profile at preselected intervals. Thus, measurements of the fractal dimension can be programmed using the computer. Figure 15.9 illustrates a diagrammatic flowchart for calculating the fractal dimension of joint profiles. The

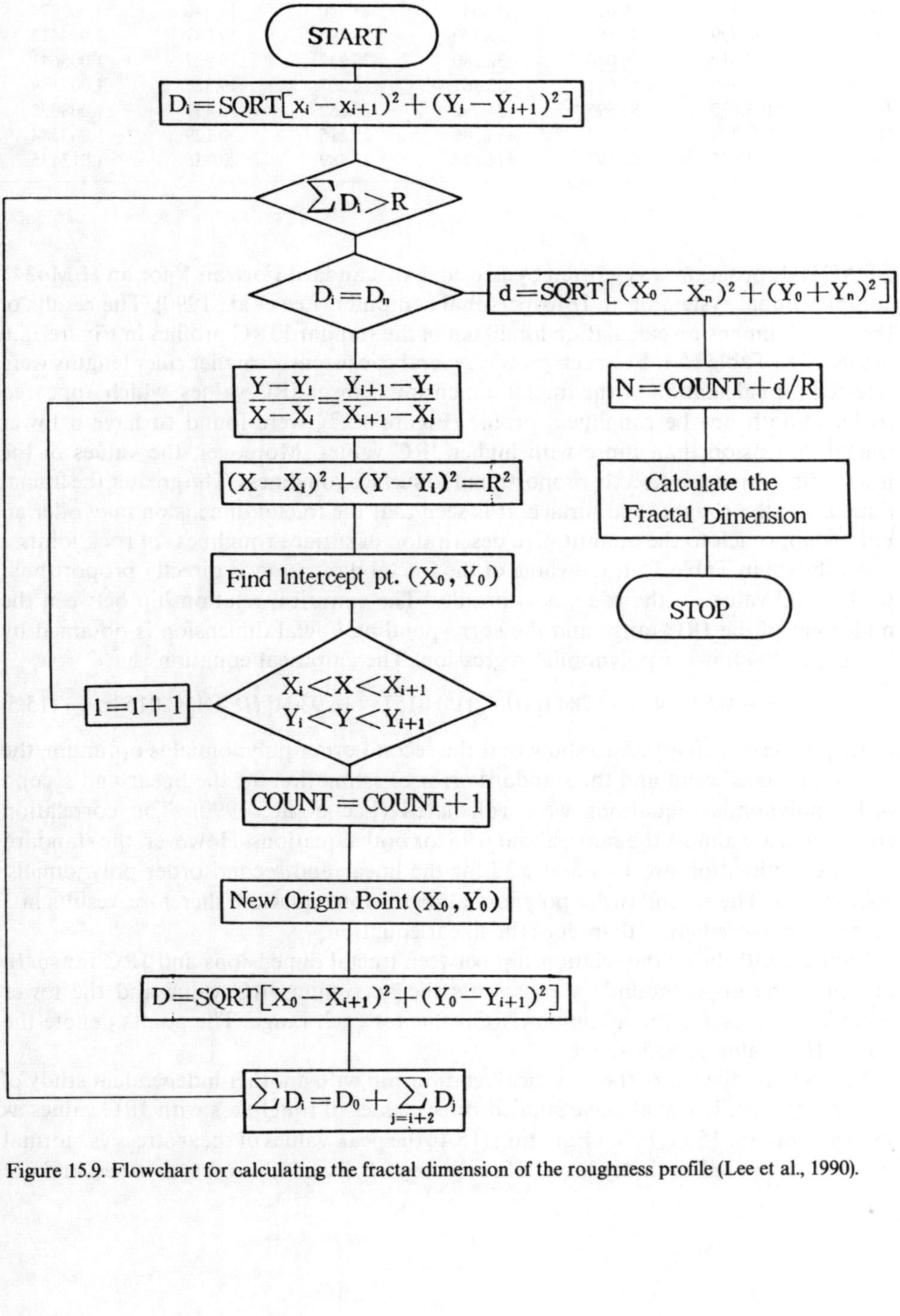

Figure 15.9. Flowchart for calculating the fractal dimension of the roughness profile (Lee et al., 1990).

Table 15.1. Fractal dimension calculation for JRC range (Lee et al., 1990).

JRC range	No. of steps (N)					Fractal dimension
	$r=2$	$r=4$	$r=6$	$r=8$	$r=10$	
0-2	98.098	49.028	32.681	24.509	19.605	1.000446
2-4	98.878	49.849	33.229	24.899	19.922	1.001687
4-6	100.281	50.031	33.297	24.955	19.977	1.002805
6-8	99.427	49.627	33.037	24.750	19.746	1.003974
8-10	98.599	49.182	32.700	24.494	19.587	1.004413
10-12	100.519	50.046	32.280	24.932	19.927	1.005641
12-14	97.799	48.601	33.361	24.207	19.327	1.007109
14-16	103.670	51.595	34.282	25.652	20.459	1.008055
16-18	102.328	50.612	33.786	25.215	20.129	1.009584
18-20	103.866	51.497	34.122	25.489	20.336	1.013435

FRACTAL program was written by Lee et al. in standard Fortran *V* for an IBM4341 computational system or an IBM personal computer (Lee et al., 1990). The results of the fractal dimension calculation for all ten of the standard JRC profiles in Figure 15.6 are listed in Table 15.1. For each profile, several consistently smaller ruler lengths were selected for calculation of the fractal dimension. Lower JRC values, which appeared to be smooth on the roughness profile (Figure 15.7), were found to have a lower fractal dimension than those with higher JRC values. Moreover, the values of the fractal dimension are directly proportional to surface roughness. The greater the fractal dimension, the rougher the surface. It is seen that the fractal dimension may offer an unique approach to the quantitative description of surface roughness of rock joints.

As shown in Table 15.1, the value of the fractal dimension is directly proportional to the JRC value for the roughness profiles. The empirical relationship between the mid-point of the JRC range and the corresponding fractal dimension is obtained by Lee et al. (1990) using polynomial regression. The empirical equation is

$$\mathrm{JRC} = -0.87804 + 37.7844[(D-1)/0.015] - 16.9304[(D-1)/0.015]^2. \qquad (15.5)$$

In this regression analysis, to show that the second order polynomial is optimum, the correlation coefficient and the standard error of estimation for the linear and second order polynomial equations were calculated (Lee et al., 1990). The correlation coefficients are almost the same, about 0.95 for both equations. However, the standard errors of estimation are 1.43 and 1.37 for the linear and second order polynomials, respectively. The second order polynomial regression equation, therefore, results in a more reliable prediction than does the linear equation.

Figure 15.10 shows the relationship between fractal dimensions and JRC range. In this plot, the upper boundary represents the maximum JRC value and the lower boundary represents the minimum JRC value for each range. The points denote the center JRC value of each range.

To compare the use of the empirical relationship with another independent study of rock roughness, Lee et al. have studied three classes of roughness with JRC values as shown in Figure 15.11. From Equation (15.4), the peak values of shear stress vs. normal stress plots related to three classes of roughness *A*, *B* and *C* can be obtained and are

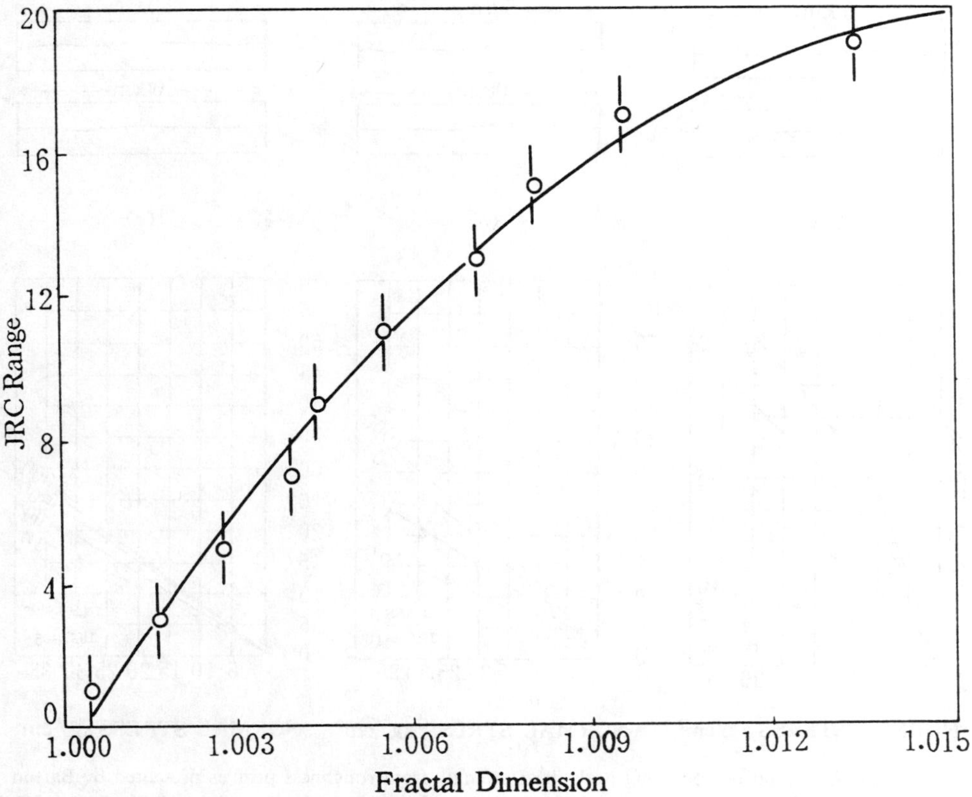

Figure 15.10. Plot of fractal dimension vs. JRC range (Lee et al., 1990).

directly applied to the criterion of shearing failure. In the direct shear testing of unfilled rock joints, the values of σ and ϕ_b in Equation (15.4) can be measured or accurately estimated, but the JRC value is really unknown and must be estimated using engineering and experience judgement. However, the fractal dimension of joints provides an approach to quantitatively estimate JRC value from Equation (15.5). Table 15.2 gives the JRC values for the three profiles determined by using Equation (15.5) and a comparison of JRC values calculated from the fractal dimension with those quoted by Barton in Figure 15.11. The agreement is excellent, indicating that the fractal dimension is ideally suited for the description of joint roughness, and correlated well with JRC values.

Another interesting study is fractal estimation of joint roughness angle proposed by Turk et al. (1987). Turk & Dearman (1985) showed that the joint roughness angle (i) of a joint surface can be determined by comparing the trace length (L_t) and direct length (L_d) of its profile as:

$$\cos(i) = L_d/L_t.$$

This is a very simple method of joint roughness angle determination and assumes that the overall profile traces of asperities can be represented by an equilateral triangle.

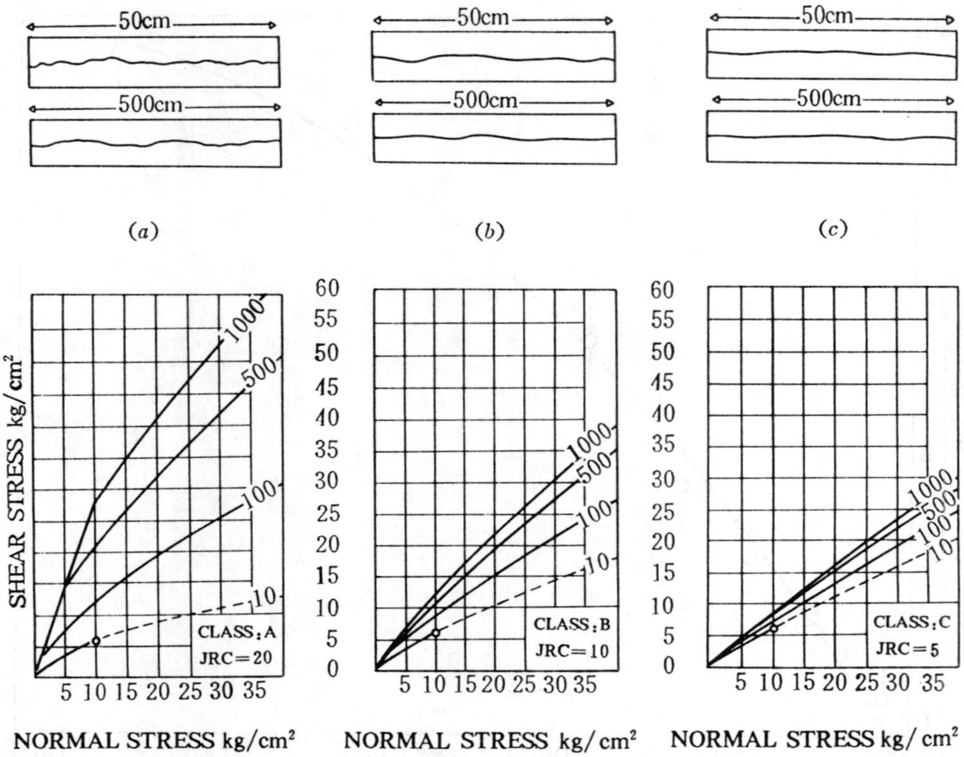

Figure 15.11. A method of estimating peak shear strength from roughness profiles presented by Barton (1973) from Equation (15.4). (a) JRC = 20, $\phi_b = 30°$; (b) JRC = 10, $\phi_b = 30°$; and (c) JRC = 5, $\phi_b = 30°$ (Lee et al., 1990).

Table 15.2. Comparison of JRC values calculated from fractal dimension and Barton's empirical results (Lee et al., 1990).

Profile	Fractal dimension	JRC range	Figure 15.11
A	1.022239	20.00	20.00
B	1.004553	10.03	10.00
C	1.002149	5.19	5.00

This method gives an average roughness angle of a rock joint surface without the need to measure the individual asperity angle. Additionally this approach takes into account both the first and second order asperities of rock joints. When the trace length is used as a measuring step length equal to unity, both the trace and direct length of the joint profiles satisfies power-law relationship in the form:

$$L_t = L_d^D.$$

Since, $\cos(i) = L_d/L_t$, and L_t is replaced by L_d^D

$$\cos(i) = L_d^{(1-D)}.$$

However, for the general case, it can be shown that:

$$\cos(i) = (CL_d)^{1-D},$$

where C is a constant. This is the fundamental relation for estimating the roughness angle of rock from the fractal dimension. However, the roughness angles estimated by this approach is not in good agreement with those determined by other methods. For detailed discussion see the paper of Turk et al. (1987).

15.3 FRACTAL MEASUREMENT OF ROCK JOINTS IN THE LABORATORY AND FIELD

Haas et al. (1975) reported their shear tests in an investigation of the shear resistance of bolted and unbolted rock discontinuities utilizing rock blocks having a fracture for the shear surface. The fracture was created by splitting 0.61 m cubes of Salem limestone with wedged feathers as shown in Figure 15.12. Four wedges were used to control the direction of the fracture. The fracture plane was chosen parallel to the bedding. Lee et al. have measured these fracture profiles by a device constructed as indicated in Figure 15.13, along grid lines spaced 5.1 cm apart. In each case the profiles were measured on the rigidly held block. The long bar was supported by a pivot point at its midpoint. The pivot point was fixed to a bracket on a tool post which, in turn, slid against a guide bar and on the ground surface of a small steel table. As the scribe moved up the fracture surface, the pen moved down the same amount and recorded

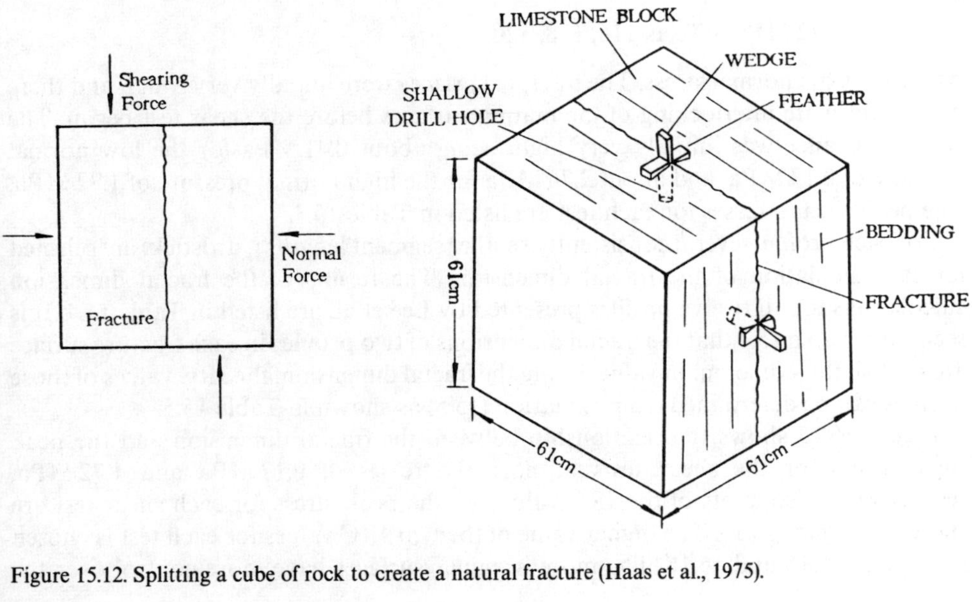

Figure 15.12. Splitting a cube of rock to create a natural fracture (Haas et al., 1975).

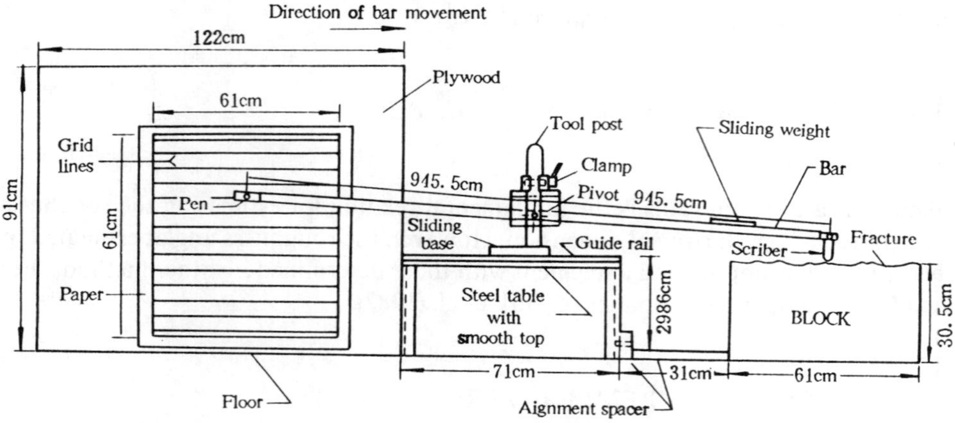

Figure 15.13. Device to record fracture surface profiles (Haas et al., 1975, after Lee et al., 1990).

a trace of the fracture profile. Such profiles, spaced 5.1 cm apart, were recorded on six blocks which were then used in shear tests with or without a bolt.

Lee et al. (1990) used this device to record all fracture profiles (of the six blocks) which were parallel to the direction of shear displacement as shown in Figure 15.14. Each pair of blocks, after being recorded, were carefully put back together as they were placed in the shear loading frame so that the fracture surfaces mated well. Each pair of blocks were sheared only once, because the surface characteristics were significantly changed during the shear. The six shearing tests without bolts were performed by Lee et al. (1990) under conditions as follows:

$$\sigma_n = 0.17\,\text{MPa—Tests 107, 108, 109;}$$

and

$$\sigma_n = 1.72\,\text{MPa—Tests 116, 118, 120;}$$

where σ_n is the normal stress. The fracture surfaces were initially very rough and there was an intimate interlocking of the mating surfaces before the shear test began. The shear resistance was initially very high, being about 0.41 MPa for the low normal pressure of 0.17 MPa, and about 2.76 MPa for the high normal pressure of 1.72 MPa. The peak shear stresses for each test are listed in Table 15.3.

For each profile, several consistently smaller segment lengths (yardstick) are selected for the calculation of the fractal dimension. The results of the fractal dimension calculations for all twelve profiles presented by Lee et al. are listed in Table 15.4. It is seen from Table 15.4 that the fractal dimensions of two profiles in each fracture surface are only slightly different in value. Using the fractal dimension, the JRC values of those profiles can be determined from Equation (15.5) as shown in Table 15.5.

Figure 15.15 shows the relationship between the fractal dimension and the peak shear stress for the shear tests at normal stresses of 0.17 MPa and 1.72 MPa, respectively. Also, plots of the JRC values vs. the peak stress for each shear test are shown in Figure 15.16. The higher value of the two JRC values for each test is chosen in Figures 15.15 and 15.16. The more irregular surfaces have more control over the

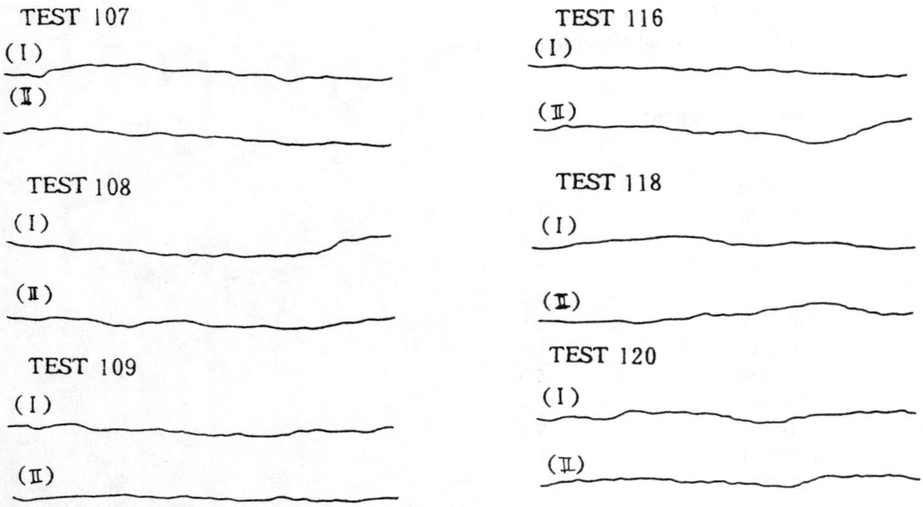

Figure 15.14. Roughness profiles of a fracture surface for a shear test (Lee et al., 1990).

Table 15.3. Peak shear stress of six shear tests obtained by Lee et al. (1990).

Normal stress (MPa)	Test	Peak shear stress (MPa)
0.17	107	0.39
0.17	108	0.46
0.17	109	0.42
1.72	116	2.91
1.72	118	2.71
1.72	120	3.01

Table 15.4. Fractal dimension calculation for the fracture surface of the shear test (Lee et al., 1990).

Test no.	No. of steps (*N*)					Fractal dimension
	$r = 0.2$	$r = 0.4$	$r = 0.6$	$r = 0.8$	$r = 1.0$	
107-1	118.785	59.305	39.525	29.581	23.622	1.003252
107-2	118.386	59.028	39.315	29.450	23.552	1.003330
108-1	119.158	59.416	39.530	29.599	23.625	1.005178
108-2	119.618	59.652	39.668	29.704	23.731	1.005064
109-1	118.201	58.970	39.191	29.370	24.510	1.003953
109-2	118.664	59.181	39.394	29.527	23.583	1.003739
116-1	117.892	58.826	39.181	29.366	23.470	1.002748
116-2	118.585	59.182	39.406	29.526	23.612	1.002842
118-1	119.896	59.869	39.865	29.880	23.910	1.002004
118-2	117.112	58.457	38.930	29.177	23.332	1.002449
120-1	117.044	58.336	38.836	29.084	23.237	1.004465
120-2	115.644	57.677	38.377	28.775	22.970	1.004014

Table 15.5. JRC range of the fracture surfaces of the shear test determined by the fractal dimension (Lee et al., 1990).

Profile	Fractal dimension	JRC range
107-1	1.003252	7.5
107-2	1.003330	7.6
108-1	1.005178	11.2
108-2	1.005064	11.0
109-1	1.003953	8.9
109-2	1.003739	8.5
116-1	1.002748	6.5
116-2	1.002842	6.7
118-1	1.002004	5.0
118-2	1.002449	5.8
120-1	1.004465	9.9
120-2	1.004014	9.1

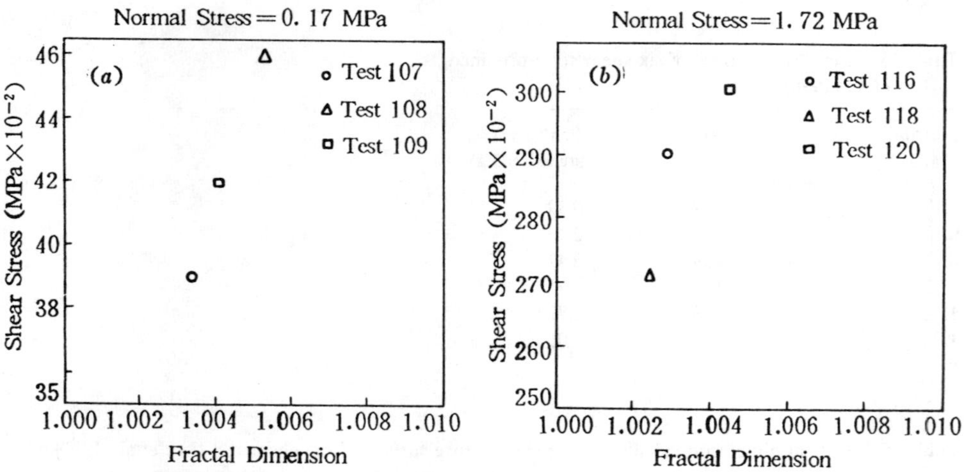

Figure 15.15. Plot of the fractal dimension vs. the peak shear stress for shear tests. (a) Normal stress = 0.17 MPa; (b) Normal stress = 1.72 MPa. (Lee et al., 1990).

peak shear stress. It is seen that the peak shear stress increases with an increment of the JRC value or the fractal dimension value (Lee et al., 1990).

Fractal measurements of rock jointing surfaces discussed above can be applied to a quantitative assessment of the roughness of the rock slide surface in the field. For example, the rock slide around the field data area occurred at Libby Dam, Montana on January 31, 1971 (Hamel 1974). This slide consisted of a rock wedge which slid towards the dam abutment. The volume of this rock was approximately 46 000 cubic meters. The rock wedge was formed and bounded by the intersection of a bedding

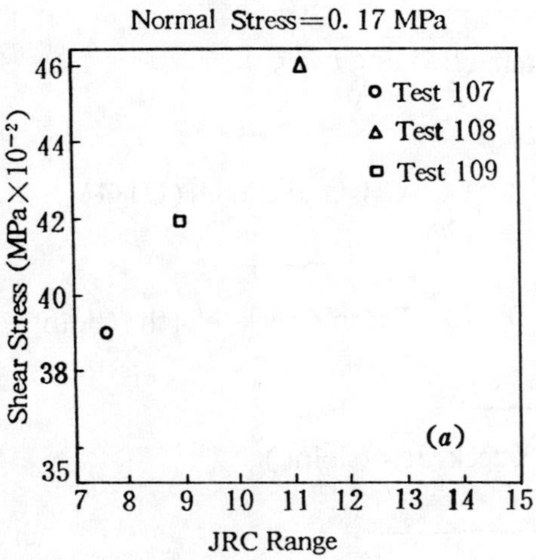

Figure 15.16. Plot of the JRC range vs. the peak shear stress for shear tests. (a) Normal stress = 0.17 MPa; (b) Normal stress = 1.72 MPa. (Lee et al., 1990).

plane and a joint. Clearly, this description requires the quantitative characterization of spatial variation in the roughness across the planar surfaces of these joints with the help of fractal measurements.

Carr (1989) has measured 16 profiles of the joints and the bedding planes in the field as illustrated in Figure 15.17. The fractal dimensions are computed for each of the 16 profiles by the fractal computer program (Lee et al., 1990). The range of the fractal dimension of these joints is 1.004 to 1.005. This means that the profiles appear to be almost planar, except for the Old Notch #1 (joint).

Similarly, the JRC values of the field data can be determined from fractal dimension (Carr, 1989; Lee et al., 1990). These results are listed in Table 15.6. The JRC values for

Table 15.6. JRC ranges of the field data determined from fractal dimension (Lee et al., 1990).

Profile	Fractal dimension	JRC range
Island Notch (bedding)	1.001409	3.5
Island Notch (joint)	1.000107	0.4
Island Notch (UTG)	1.003104	7.2
Dunn Creek Notch (bedding)	1.008608	16.2
Dunn Creek Notch (joint)	1.004735	10.4
Wolf Creek Jct (joint)	1.005641	11.9
Backes Notch #1 (bedding)	1.001059	2.7
Backes Notch #1 (joint)	1.004233	9.4
Backes Creek Notch #1 (bedding)	1.000891	2.3
Old Notch #1 (joint)	1.021994	19.1
DS+122 (C joint)	1.003045	7.1
DS+122 (minor joint)	1.003369	7.8
DS−122 (joint #2)	1.003469	8.0
914 Rib (joint)	1.003334	7.7
930 Joint (upper end)	1.004516	9.7

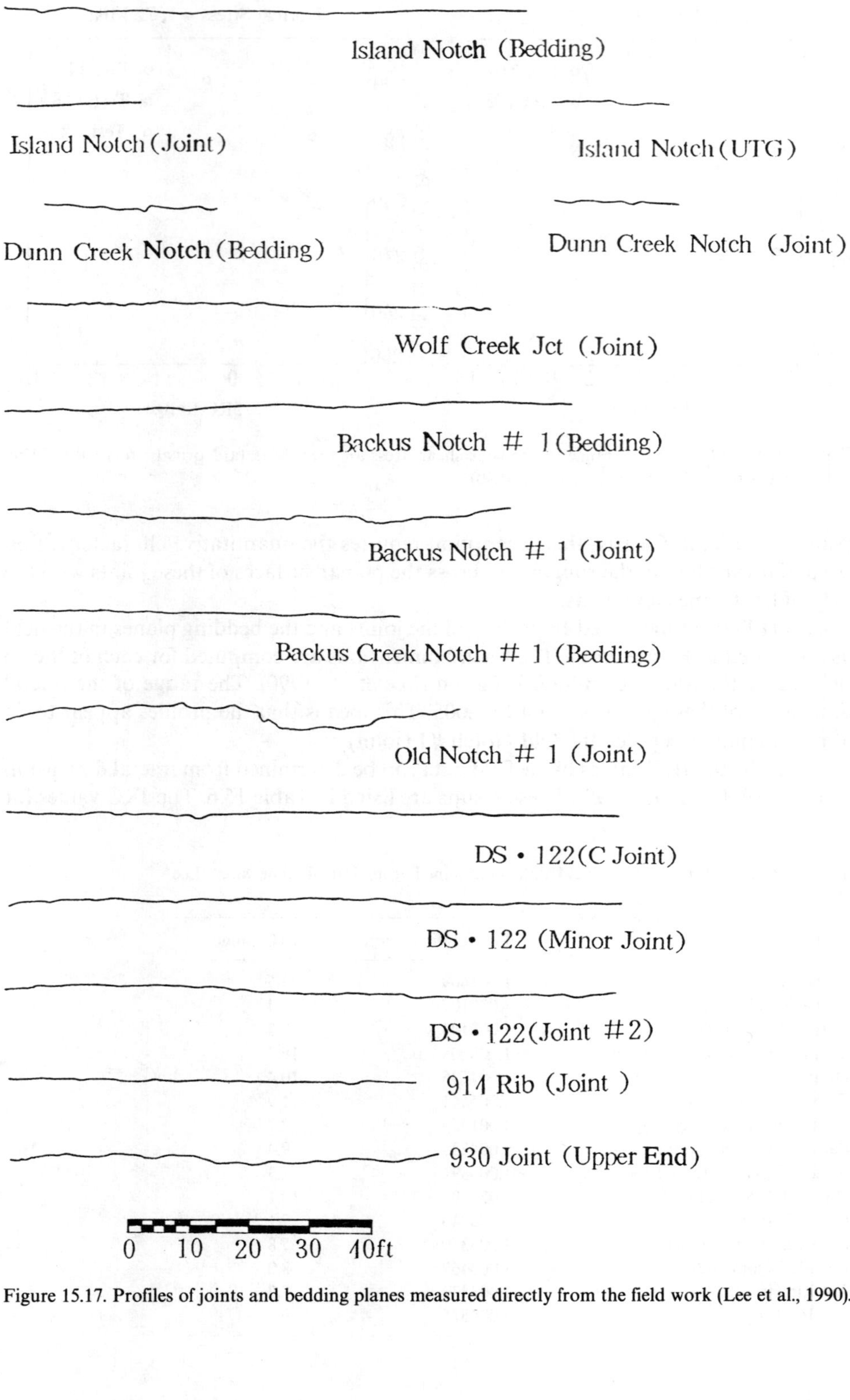

Figure 15.17. Profiles of joints and bedding planes measured directly from the field work (Lee et al., 1990).

many of the field profiles are low, i.e. less than 10, and the surfaces appear to be nearly smooth planar upon field examination. The fractal dimension has therefore been found to be useful for the analysis of the texture of joint or bedding surfaces from field profile measurements.

15.4 BROAD BANDWIDTH STUDY OF THE TOPOGRAPHY OF JOINT SURFACES IN ROCKS

The concept of fractals provides a useful framework to study surface topography as discussed in Chapter 3. In this section, we will introduce the fractal studies of Brown & Scholz (1985) on the topography of natural rock surfaces. Unlike the discussions in the previous sections, they indirectly measured fractal dimensions of rough surfaces of rock joints by means of the power spectra from the profiles.

The power spectral methods rely on time series analyses, a long-standing and well-established part of applied mathematics. In this approach, power spectra are computed from the profile data. The longest and shortest wavelengths for which spectral methods provide information are equal to the profile length L and a length corresponding to double the sampling interval, respectively. Generally, spectra calculated from self-similar or self-affine surfaces will have power spectral density functions $G(w)$ of the form:

$$G(w) \propto w^s.$$

As discussed in Sections 2.5 and 2.6, two fractal functions, the fractional Brownian function and the Weierstrass-Mandelbrot function can serve as models of surface profiles. Both have the power spectral density $G(w)$ given by

$$G(w) \propto w^s = w^{-(5-2D)}, \tag{15.6}$$

where $w = 2\pi/\lambda$ is the angular frequency. Using this relationship, an estimate of the fractal dimension of a surface profile can be made from slope (s) of the power spectrum on a log-log plot. Thus, the measurement of fractal dimensions of surface topography is transformed into the measurement of power spectrums (for surface fractals, $s = -(7-2D)$ in Equation (15.6)).

Using two digital surface profiles, a laboratory instrument and a field instrument, Brown & Scholz studied the topography of several natural rock surfaces from wavelengths of 1 m to 10 μm. The field instrument was directly bolted to the rock outcrop; small cores of the surface were taken for the laboratory measurements. Five field profiles and eight laboratory profiles were taken at about 20 cm spacing in the same direction on each surface. Two perpendicular sets of profiles were taken on those surfaces and they showed obvious anisotropy. The surfaces studied by them were naturally occurring rock surfaces from roadside outcrops in New York, which included fresh natural joints in both crystalline and sedimentary rock, a frictional surface was formed by glaciation, and a bedding plane surface (Table 15.7). A typical set of profiles is indicated in Figure 15.18.

A commercially available stylus profilometer capable of recording wavelengths from 1 cm to 10 μm and heights from 2 mm to 0.1 μm was used as the laboratory instrument. The field instrument was a stylus profilometer capable of recording wavelengths from

Table 15.7. Rock surfaces studied by Brown & Scholz (1985).

Key	Description
A	Palisades 1: natural joint in Palisades Diabase, Palisades, New York. Large columnar joints (cooling fractures).
B	Palisades 2: natural joint in Palisades Diabase, Palisades, New York. Large columnar joints (cooling fractures).
C	Hamilton 1: natural joint in siltstone (Hamilton Group), Kingston, New York. Fracture perpendicular to bedding.
D	Hamilton 2: natural joint in siltstone (Hamilton Group), Kingston, New York. Fracture perpendicular to bedding. Anisotropic with major axis of anisotropy parallel to trace of bedding planes, nearly sinusoidal morphology perpendicular to bedding D1, parallel to bedding D2, perpendicular to bedding.
E	Glacial: glacially striated bedding plane in siltstone (Hamilton Group), Kingston, New York. Anisotropic with frictional wear grooves on all scales. El, parallel to wear grooves. E2, perpendicular to wear grooves.
F	C-Plume: natural joint in siltstone, Watkins Glen, New York. Surface morphology: rhythmic c-type described in the work of Bahat & Engelder in 1984.
G	S-Plume: natural joint in siltstone, Watkins Glen, New York. Surface morphology: S-type described in the work of Bahat & Engelder in 1984. Profiles taken parallel to bedding near the center of plume structure.
H	Bedding: bedding plane in siltstone, Watkins Glen, New York.

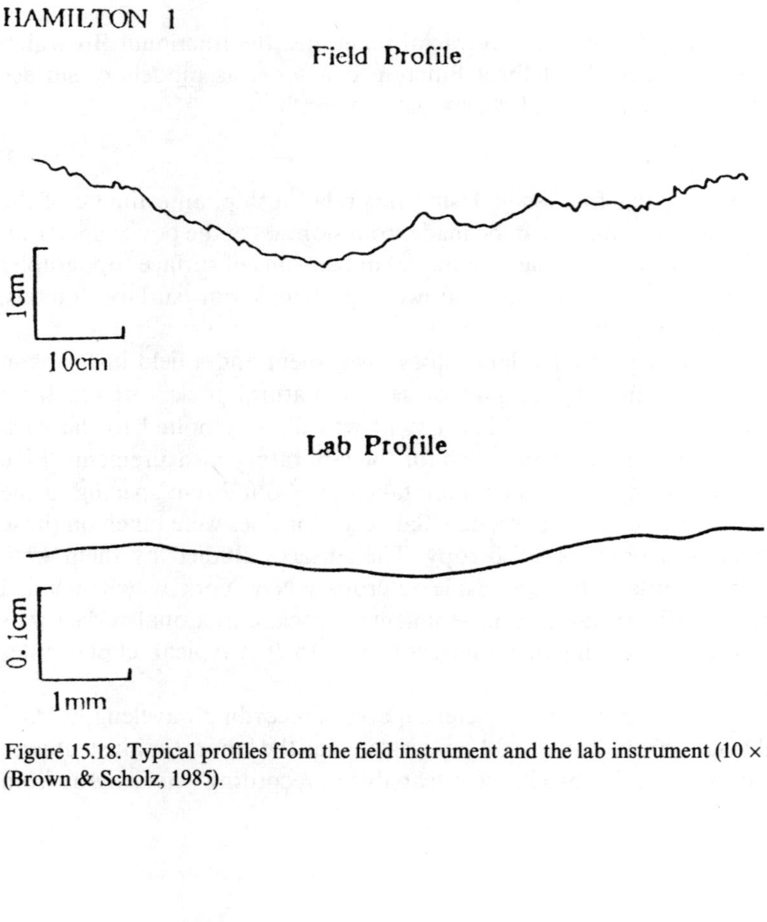

Figure 15.18. Typical profiles from the field instrument and the lab instrument (10 × vertical exaggeration) (Brown & Scholz, 1985).

1 m to 1 mm and heights from 5 cm to 10 μm (Brown, 1984, Brown & Scholz, 1985). The largest wavelength recorded depends on both the sample interval and the dimension of the profilometer styles. The smallest wavelength possible is twice the sample interval (Nyquist cutoff).

Power spectra from each profile were computed by Brown & Scholz using fast Fourier transform calculations introduced by Bendat & Piersol (1971). Raw estimates of a power spectral density function obtained in this way should be smoothened to further reduce the spurious variations in power between adjacent frequencies resulting from the calculation procedures. For this purpose, ensemble averages of the spectra from both the field and laboratory instruments were taken. An ensemble average was computed by stacking the individual spectra for each surface (i.e., by averaging the power at each frequency). Ensemble averaging of many spectra from a single uniform surface is equivalent to averaging the spectra from many short segments of a single long profile. This has the desired effects of reducing the variability in the power from one frequency to the next while maintaining a small resolution bandwidth. In all cases, the averaged spectra from the field instrument agreed well with those from the laboratory instrument within the 1 cm to 1 mm band of overlap. Together these spectra span nearly five decades in spatial frequency. The spectra from each surface obtained by Brown & Scholz are reproduced in Figures 15.19 and 15.20.

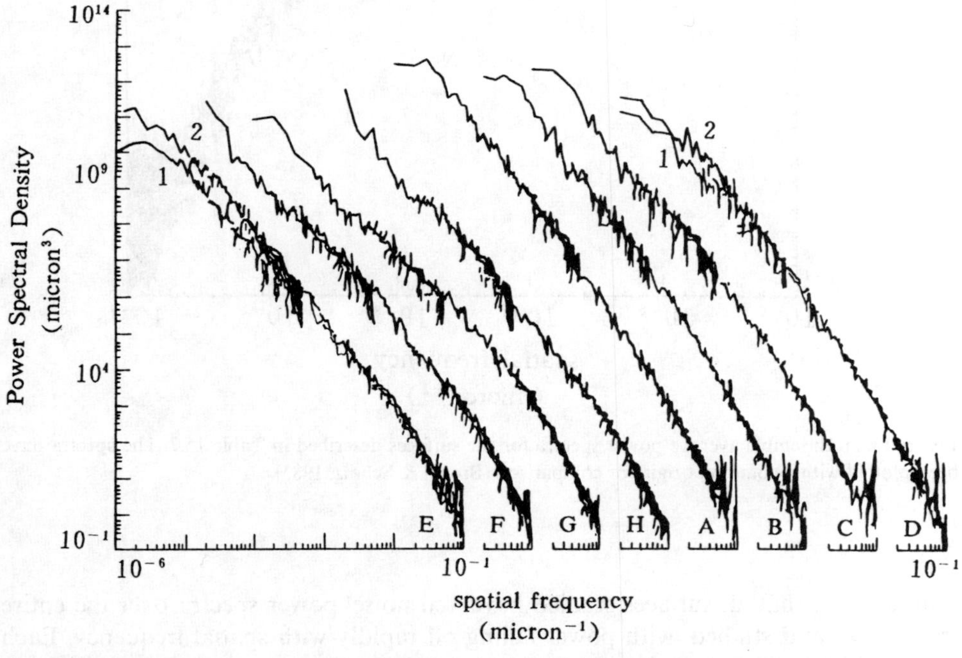

Figure 15.19. Ensemble average power spectra for the surfaces described in Table 15.7. The spectra have been offset in frequency by 1 decade for clarity. Spectra from the anisotropic surfaces D1-D2 and E1-E2 have been overlaid for comparison (the directions of the profiles are indicated by the Labels 1 and 2; refer to Table 15.7) (Brown & Scholz, 1985).

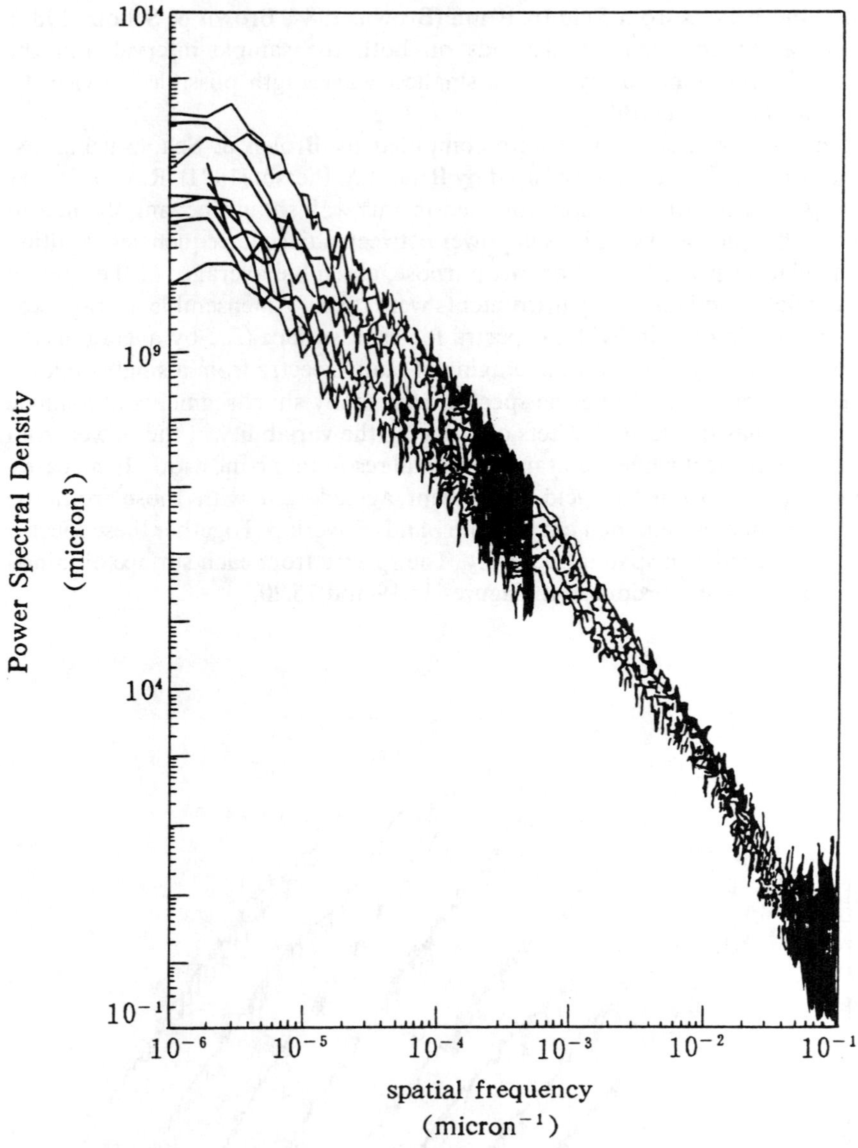

Figure 15.20. Ensemble average power spectra for the surfaces described in Table 15.7. The spectra have been plotted with a common origin for comparison (Brown & Scholz, 1985).

It is noted that all surfaces studied have 'red noise' power spectra over the entire frequency band studied, with power falling off rapidly with spatial frequency. Each spectrum is approximately linear in log-log plots with slopes ranging between -2 and -3. Slopes (s) of the spectra and the corresponding fractal dimensions for three frequency bands determined from Equation (15.6) are listed in Table 15.8 (Brown & Scholz). The slopes are computed by least squares fitting of a straight line to the

Table 15.8. Least squares slope S and fractal dimension D for arbitrary frequency bands (refer to Figures 15.19 and 15.20) (Brown & Scholz, 1985).

Surface (refer to Table 15.7)	10^{-5}-10^{-4} μm^{-1}		10^{-4}-10^{-3} μm^{-1}		10^{-3}-10^{-2} μm^{-1}	
	S	*D*	*S*	*D*	*S*	*D*
A, Palisades 1	−2.48	1.26	−2.60	1.20	−3.07	0.97
	(±0.19)	(±0.10)	(±0.10)	(±0.05)	(±0.04)	(±0.04)
B, Palisades 2	−2.64	1.18	−2.62	1.19	−3.35	0.82
	(±0.23)	(±0.11)	(±0.10)	(±0.05)	(±0.09)	(±0.04)
C, Hamilton 1	−1.99	1.50	−2.61	1.20	−2.90	1.05
	(±0.20)	(±0.10)	(±0.11)	(±0.06)	(±0.08)	(±0.04)
D1, Hamilton 2	−1.82	1.59	−2.54	1.23	−3.08	0.96
	(±0.19)	(±0.10)	(±0.10)	(±0.05)	(±0.09)	(±0.04)
D2, Hamilton 2	−2.01	1.50	−2.60	1.20	−3.05	0.97
	(±0.28)	(±0.14)	(±0.13)	(±0.07)	(±0.11)	(±0.05)
E1, Glacial	−2.27	1.37	−2.38	1.31	−2.14	1.43
	(±0.21)	(±0.11)	(±0.11)	(±0.05)	(±0.07)	(±0.04)
E2, Glacial	−2.27	1.37	−2.58	1.21	−2.25	1.37
	(±0.21)	(±0.11)	(±0.10)	(±0.05)	(±0.07)	(±0.04)
F, C-Plume	−1.79	1.60	−2.18	1.41	−2.36	1.32
	(±0.19)	(±0.09)	(±0.10)	(±0.05)	(±0.10)	(±0.05)
G, S-Plume	−1.63	1.68	−2.96	1.52	−2.32	1.34
	(±0.18)	(±0.09)	(±0.10)	(±0.05)	(±0.09)	(±0.05)
H, Bedding	−1.81	1.60	−2.54	1.23	−2.40	1.30
	(±0.19)	(±0.09)	(±0.10)	(±0.05)	(±0.09)	(±0.04)

spectrum in log-log space. The frequency bands are chosen to avoid the influence of electrical noise at the highest frequencies and the roll off at the lowest frequencies. From Figure 15.19 and Table 15.8 it is seen that the slope is not constant for a given surface and tends to gradually decrease as one moves from high to low frequency. On many spectra two somewhat subjective breaks in slope can be seen. Above a frequency corresponding approximately to the grain size observed in hand specimen ($\sim 10^2$-10^3 μm), the slope rapidly increases and often becomes steeper than -3 ($s < -3$). This may mean that the profilometer stylus traverses smooth crystal cleavage surfaces. The other change in slope is observed at a frequency corresponding to lengths of 10's of centimeters ($\sim 10^5$ μm) below which the slope again rapidly increases. This region of high power is an expression of the outcrop scale shape, which is often observed in profiles as a long wavelength high amplitude arc (Figure 15.18). The observed variation in spectral slope with frequency means these surfaces are fractally inhomogeneous with no single fractal dimension able to describe the entire surface.

The fractal dimension can be considered constant only over a limited range of spatial frequencies; thus, these surfaces are not self-similar on all scales, and no simple scaling laws exist. Therefore, for fractals to be quantitatively useful in the study of natural rock surfaces, one should explicitly include the variation of the fractal dimension with spatial frequency, i.e., one should adopt the concept of non-uniform fractals (Xie, 1989c) and multifractals discussed in Chapter 6.

From the above Brown & Scholz's researches, we can see that the concept of fractals is a significant advance in the study of surface topography. Unlike earlier descriptions of topography, the scaling of surface roughness is a fundamental part of the fractal model. The fractal dimensions can potentially tie the scaling of surface topography to that of other natural phenomena such as damage and fracture (Chapter 14), and earthquake processes (Chapter 16). Also, it leads to convenient methods for the numerical generation of realistic model surfaces (see Chapter 3).

15.5 A FRACTAL MODEL OF JOINT PROFILES

In Sections 15.2 and 15.3, we introduced Lee et al.'s studies on fractal measurements of joint profiles. The same studies have also been performed by Turk et al.(1987), Carr (1989), and Muralha & Charrua-Graca (1990). However, most of these studies have too small of a measured fractal dimension ($D \simeq 1.00$) for joint profiles. This may arise from the measurement method of the fractal dimension. In fractal geometry, the length of a fractal curve is defined by Equation (2.7a), and is the limit value of the yardstick scale, $\delta \to 0$. This means that a true value, D, can be obtained only when the yardstick scale, δ, is small enough in the measurement of the fractal dimension. There is a critical value of the yardstick scale, $\delta_{\min}$, only when the yardstick scale, $\delta \leq \delta_{\min}$, is used to measure the fractal object, the measured fractal dimension approaches its true value (see discussions in Section 3.4). In fact, fractal dimensions with different orders of magnitude for a given fractal object are obtained by different researchers because they use different yardstick scales. The smaller the yardstick scale, the more are the details of the fractal object that can be measured, and the more accurate the measured fractal dimension that can be obtained (Xie & Pariseau, 1992b). Therefore, the measurement of a fractal dimension for a joint profile is a hard work.

So far, there are four methods which can be used to measure the fractal dimensions of the roughness profiles. These four methods include divider and box methods as discussed in Chapter 2, the spectral method as used in Sections 15.4 and 16.5, and modified divider method (Brown, 1987a). The modified divider method proposed by Brown consists of using horizontal divider spans (x-increments) rather than walking the dividers along the profile. Thus, the incremental lengths (i.e., segments) along the profile from x_i to x_{i+1} are summed to obtain the total estimated profile length for each

Table 15.9. Summary of roughness measures for a set of joint profiles (Miller, 1990).

Profile	D_s	D_b	D_d	D_m
A	1.258	1.099	1.161	1.129
B	1.246	1.158	1.192	1.123
C	1.202	1.118	1.193	1.157
D	1.191	1.159	1.094	1.103
E	1.383	1.143	1.261	1.199
F	1.309	1.122	1.155	1.128
G	1.263	1.136	1.182	1.200
H	1.274	1.146	1.183	1.103
J	1.124	1.041	1.084	1.058

given x-increment, which is denoted as δ for the case. The linear $\log(\delta)$-$\log(L)$ relationship holds where $\delta = x$-increment, L = estimated profile length, and $D = 1 - s$, where s is the slope of the log-log plot. Actually, the modified divider method is the one-dimensional projection of the box method. Miller (1990) measured a set of profiles by using these four methods. His measured results are listed in Table 15.9 where the fractal dimensions were calculated by different methods: D_b, by the box method; D_d, by the divider method; D_m, by the modified divider method; D_s, by the spectral method. From Table 15.9, it is seen that the four methods do not give consistent estimations of D for a given profile. However, the values of the fractal dimensions for the profiles measured by Miller (1990) are much larger than those obtained by other investigators.

In this section, we mainly introduce our recent studies on joint profiles (Xie & Pariseau, 1992b), in which the roughness of joint profiles is considered as a fractal structure with finite generations. Based on the triadic Koch curve, a theoretical fractal model has been established to measure the fractal dimension of joint profiles, making the fractal dimension measurement easier and simpler, and this theoretical fractal model can then lead to a fractal estimation of joint roughness coefficients.

15.5.1 *A fractal model of joint profile roughness*

A rock joint profile has more roughness and can be regarded as a natural fractal. From the physical point of view, a joint is experienced during the rock fracturing process from the microscopic to the macroscopic level, which behaves as a fractal structure (for details, see Chapters 13 and 14). The joint roughness at the microscopic level is caused by intergranular and transgranular brittle fractures of rock (see Chapters 10 and 13). This roughness scale is grain size. The macroscopic scale of the joint roughness may be in meters or kilometers (see Chapter 16). Figure 15.21 shows a random joint profile at the macroscopic scale. It is statistically similar to the triadic Koch curve (Figure 2.6). The difference is only that the former occurs with finite generations ranging from grain size to kilometer size, and the latter with infinite generations. Based on this idea, we assume that joint profiles in nature have a similar construction to the Koch curve. In order to simulate random joint profiles in nature, we generalize the generation rule of the Koch curve as shown in Figure 15.22. Unlike the generator of the triadic Koch curve ($\theta = 60°$), the generalized generator has $\theta \in [0°, 90°]$. In this fractal model, we have

$$N = 4, \quad 1/r = 2[1 + \cos\tan^{-1} 2h/L]. \tag{15.7}$$

Thus, the fractal dimension is defined by

$$D = \log 4/\log[2(1 + \cos\tan^{-1} 2h/L)], \tag{15.8}$$

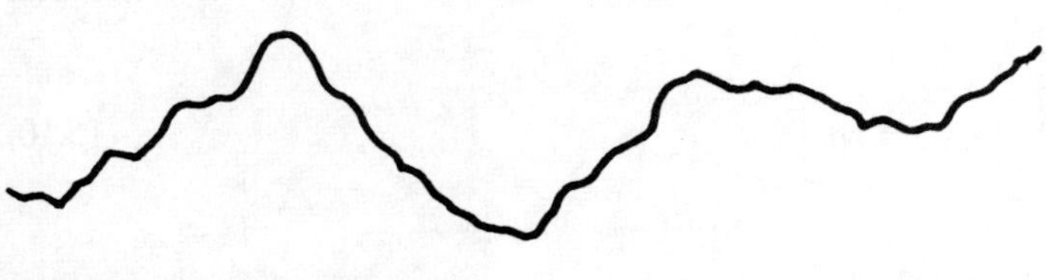

Figure 15.21. A random joint profile.

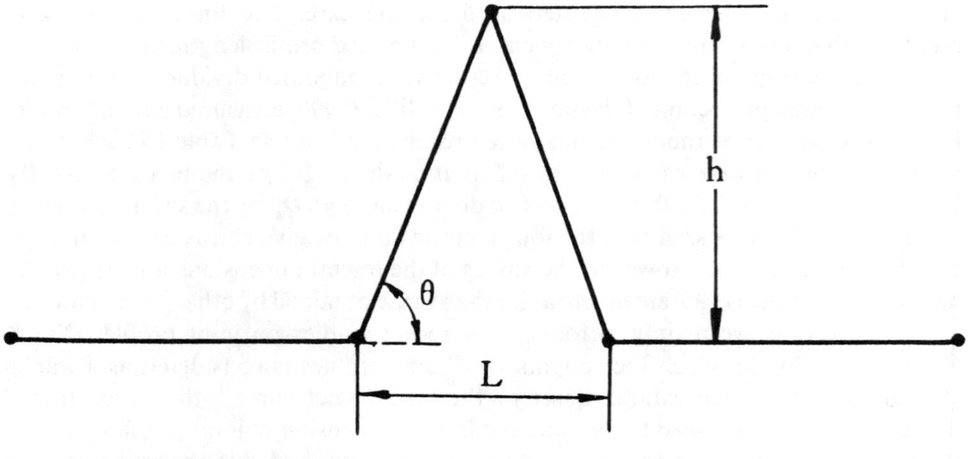

Figure 15.22. The generator of generalized Koch curve used to simulate the roughness of rock joint profiles.

where h and L are, respectively, the statistical height and base length of higher-order asperities of a rock joint. It is seen from Equation (15.7) that the fractal dimension of the joint profile ranges from 1 (corresponding to $h=0$) to 2 (corresponding to $L=0$).

Equation (15.7) is tested by computer-generated fractal Brownian motions as shown in Figure 15.23. It should be pointed out here that for actual joint profiles (natural fractals), h and L in Equation (15.7) are replaced by h^* and L^*, the average height and average base length, respectively, measured in higher-order asperities of the profiles as shown in Figure 15.24:

$$h^* = 1/M \sum_{i}^{M} h_i; \quad L^* = 1/M \sum_{i}^{M} L_i. \tag{15.9}$$

Table 15.10 lists the measured results of h^*, L^* and the fractal dimension estimated by Equation (15.8) for computer-generated fractal Brownian motions. The fractal dimension determined by Equation (15.8) is in good agreement with its theoretical values. It seems that the theoretical fractal model established in this section can be used to simulate natural fractal curves such as rock joint profiles.

15.5.2 *Fractal JRC approximation*

The quantification of joint roughness has always been a problem in joint shear modelling. At present, one commonly adopts Tse & Cruden's empirical statistical relationship (Tse & Cruden,1979) to calculate typical JRC values. This functional relationship is defined by

$$Z_2 = \left[1/(MDx^2) \sum_{i=1}^{M} (y_{i+1} - y_i) \right]^{1/2}, \tag{15.10}$$

and

$$\text{JRC} = 32.20 + 32.47 \log Z_2, \tag{15.11}$$

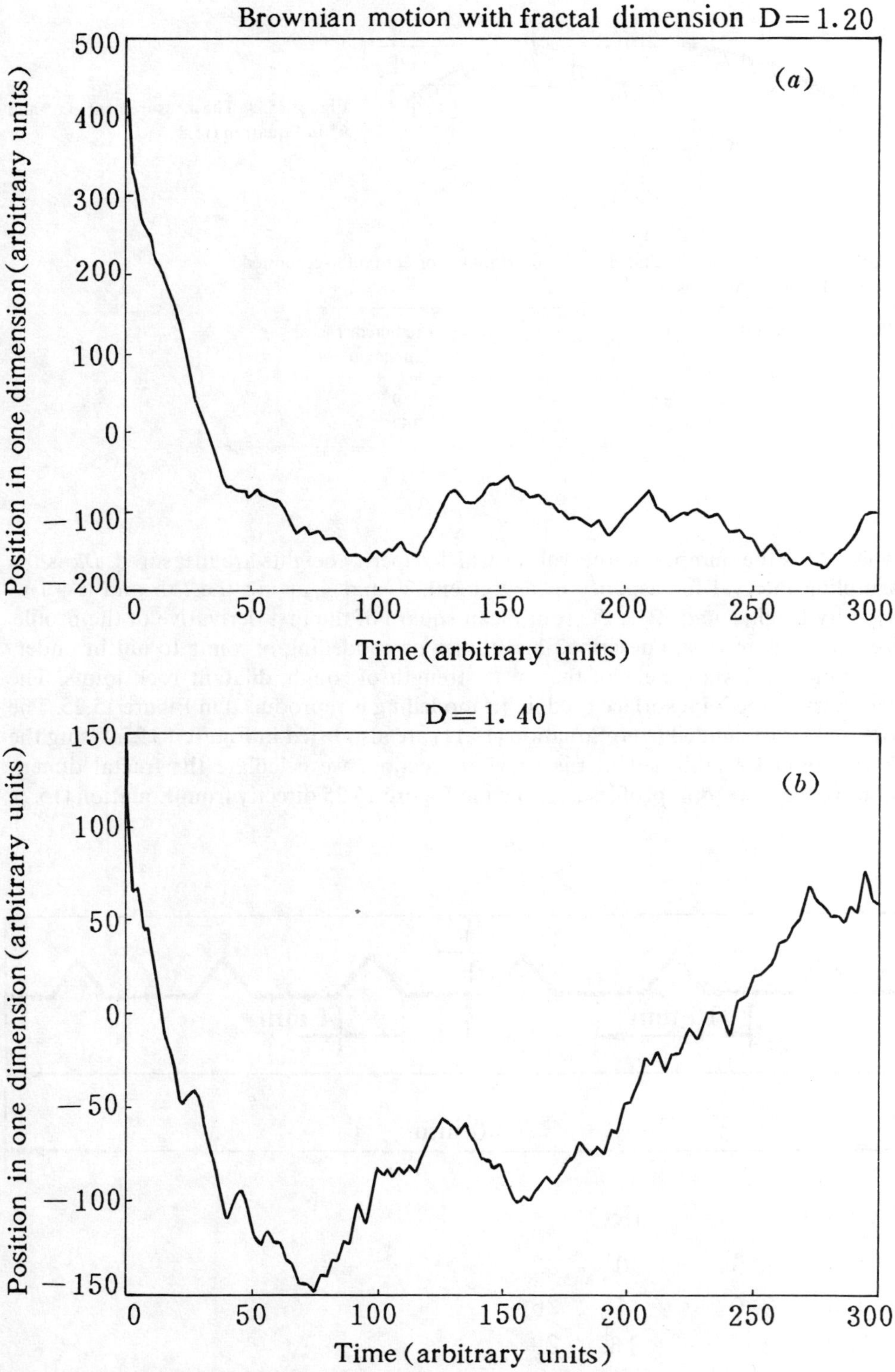

Figure 15.23. Computer-generated fractal Brownian motions. (a) $D = 1.20$; (b) $D = 1.40$.

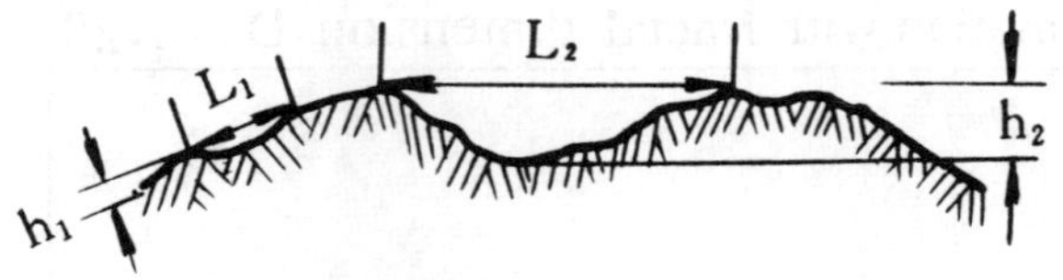

Figure 15.24. The measurement of L^* and h^* in Equation (15.8).

Table 15.10. h^*, L^* and fractal dimension estimates of computer-generated fractal Brownian motions.

Theoretical fractal dimension	L^*	h^*	Predicted fractal dimension
1.20	6.0	3.9714	1.119
1.40	6.0	8.550	1.369

where M is the number of intervals at which asperity heights are measured, Dx is the sampling interval for asperity measurement, y_i and y_{i+1} are the i-th and $(i+1)$th asperity heights, and Z_2 is the root mean square of the first derivative of the profile. Venkatachalam (1985) described his studies on modelling of joints to aid in understanding the development of the shear strength of rough, dilatant rock joints. The geometry of the joint surface used in his modelling is reproduced in Figure 15.25. The JRC values calculated from Equation (15.11) are also listed in Figure 15.25. Using the fractal model established in the previous section, we calculate the fractal dimensions of these six joint profiles as shown in Figure 15.25 directly from Equation (15.7).

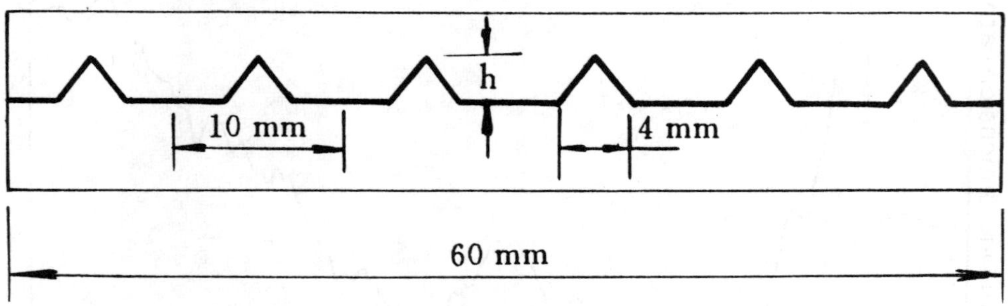

h (mm)		JRC	
0.0	1.5	0	22
0.5	2.0	6	26
1.0	2.5	16	29

Figure 15.25. Joint profiles used by Venkatachalam in the modelling of joint shear strength.

Table 15.11. Fractal dimensions of six joint profiles used by Venkatachalam in the modelling of shear strength.

h(mm) with $L = 4$ mm	JRC	Fractal dimension
0.0	0	1.0
0.5	6	1.010969
1.0	16	1.0407116
1.5	22	1.0822527
2.0	26	1.12895
2.5	29	1.17635

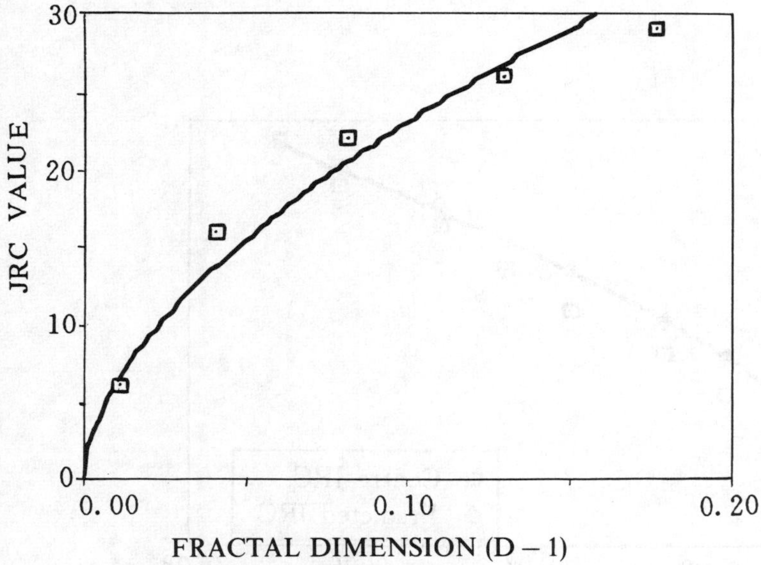

Figure 15.26. The relationship between the fractal dimension and JRC value.

The results are listed in Table 15.11. Then, using computer regression, we obtained the relational curve between the fractal dimension and the JRC value (as shown in Figure 15.26). An empirical relationship between the fractal dimension and the JRC value can be written by

$$\mathrm{JRC} = 85.2671(D-1)^{0.5679}; \quad (R = 0.99). \tag{15.12}$$

It means that the fractal dimension, D, is really a measurement of the joint roughness coefficient. The larger the D is, the larger the JRC value.

15.5.3 *Fractal estimation of JRC*

The typical JRC range curves proposed by Barton as shown in Figure 15.6 are used to test the fractal JRC approximation defined by Equation (15.12). h^* and L^* are measured

Table 15.12. Fractal estimation of JRC for typical ten JRC ranges.

JRC range	L^*	h^*	D (from Eq. (15.7))	Predicted JRC values
0-2	84.0	4.5	1.0020596	2.5
2-4	25.0	2.0	1.0045686	3.995
4-6	19.2	2.4	1.010969	6.573
6-8	18.5	3.0	1.0181337	8.7453
8-10	15.2	2.92	1.0250134	10.4979
10-12	11.5	2.683	1.0359008	12.88918
12-14	14.5	3.75	1.0432832	14.3333
14-16	13.0	3.666	1.0505344	15.6512
16-18	14.666	4.666	1.0623758	17.6388
18-20	15.5	5.25	1.069403257	18.74

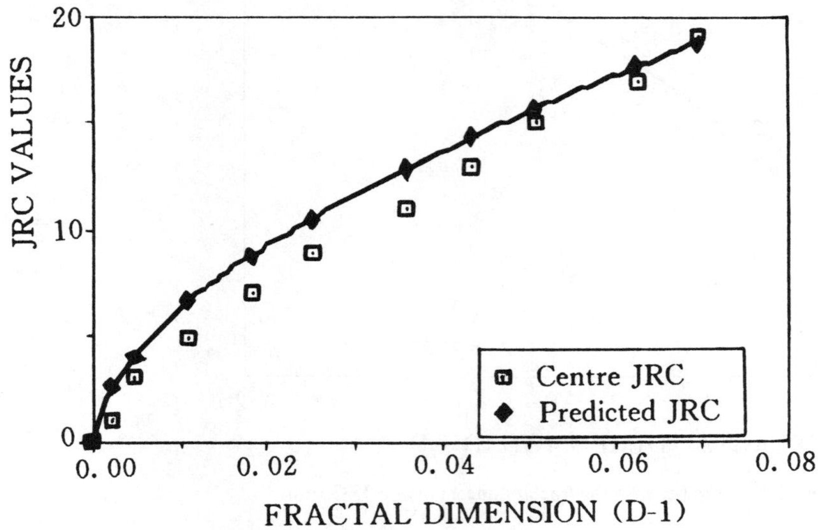

Figure 15.27. Plot of predicted JRC values and typical JRC values vs. fractal dimensions.

on higher-order asperities of the profiles, and fractal dimensions are obtained from Equation (15.7), as listed in Table 15.12. From Table 15.12, it is seen that the fractal dimensions are all larger than those obtained by Lee et al. (1990) from the yardstick method of fractal measurement. However, the fractal dimension ($D = 1.069403257$) of the JRC range (18-20) is almost the same as the fractal dimension value ($D = 1.0632$) determined by Muralha & Charrua-Graca (1990) from the box covering method of fractal measurements.

From the fractal dimension determined by Equation (15.7), we predict the JRC values for ten typical JRC ranges. The comparison between center JRC values of ten typical JRC ranges and predicted JRC values from Equation (15.12) are illustrated in Figure 15.27.

To further assure that fractal JRC approximation (Equation (15.12)) is appreciated, we have collected six joint profiles with typical JRC values from Barton (1973) and Barla et al. (1985) as indicated in Figure 15.28. h^*, L^* and the fractal dimension are obtained from the theoretical fractal model. The JRC values are then predicted by Equation (15.12) as listed in Table 15.13. From Table 15.13, it is calculated that the maximum error of typical JRC values relating to predicted JRC values is about 1.4%, and the minimum error is about 0.34%. Figure 15.29 shows a comparison of predicted JRC values with typical JRC values.

It is seen from Figures 15.27 and 15.29 that the fractal JRC approximation proposed in this section is useful in estimating the JRC value for the rock joint profile. For a given joint profile, only the two statistical parameters, L^* and h^*, the average base

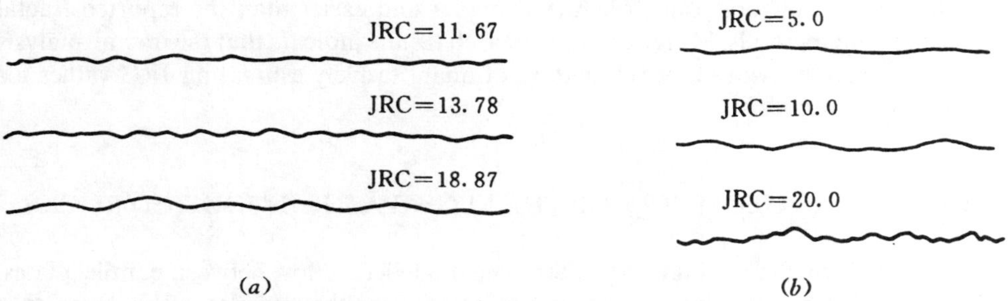

Figure 15.28. Joint profiles with typical JRC values. (a) From Barla et al. (1985); (b) From Barton (1973).

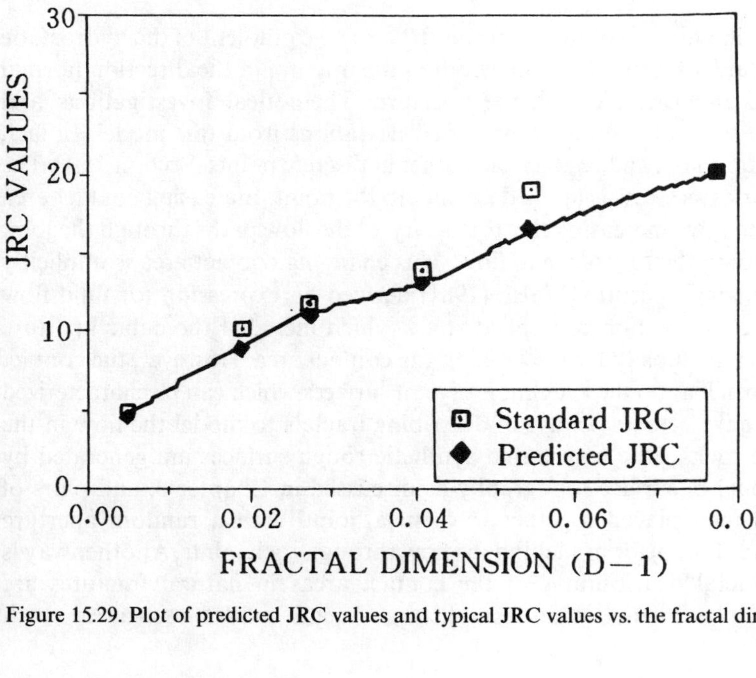

Figure 15.29. Plot of predicted JRC values and typical JRC values vs. the fractal dimension.

Table 15.13. Fractal estimation of JRC values for six joint profiles from Barton (1973) and Barla et al. (1985).

Typical JRC values	L^*	h^*	Fractal dimension D	Predicted JRC values
5	26.0	2.4	1.004772273	4.6908
10	18.0	3.0	1.019108466	9.00927
20	7.75	2.8	1.077341775	19.9302
11.67	10.333	2.0833	1.0273865	11.05245
13.78	9.5	2.3833	1.040970907	13.893333
18.87	21.4	5.3	1.05449547	16.336558

length and average height of asperities of the joint profile, respectively, are necessary for estimating the fractal dimension and JRC value of the joint profile. This makes the fractal analysis of rock joint roughness simpler and easier than the reported fractal measurement methods. Moreover, the research results indicate that the fractal analysis proposed here provides a novel method of quantitatively estimating JRC values for rock discontinuities.

15.6 FRACTAL GEOMETRY OF THE FLOW PATHS IN JOINTED ROCKS

Fluid flow through rock joints has often been modelled as flow between parallel plates. The dependence of the flow on the separation between the two plates (this separation is called the aperture) can be analytically expressed for which the volume flow rate varies as the cube of the joint aperture,

$$Q = -L_y(b^3/12\mu)\cdot(\mathrm{d}P/\mathrm{d}x), \tag{15.13}$$

where Q is the flow in units of volume per time, $\mathrm{d}P/\mathrm{d}x$ is the gradient of the hydrostatic pressure, μ is the fluid viscosity, L_y is the width of the fracture in the direction normal to fluid flow, and the variable b is the aperture. Theoretical investigations and experimental evidence, however, have predicted deviations from this model. In fact, real joint surfaces are rough and contact each other at discrete points (Section 15.1). The contact area increases as stress is applied normal to the point. Increasing contact area produces barriers to flow, increasing the 'tortuosity' of the flow paths through the joint and thereby increasing the resistance to flow. This changing contact area is implicitly related to the changing aperture. Walsh (1981) derived an expression for fluid flow through fractures as a function of applied stress which included the cubic aperture term (b^3) as well as a tortuosity term related to the contact area. However, the contact area is strongly dependent on the roughness of joint surfaces which can be characterized by fractals. There have been two ways of describing fractals to model the flow in the natural fracture in rocks. One way is that synthetic rough surfaces are generated by using a fractal model of surface topography as discussed in Chapter 3, and pairs of these surfaces are then placed together to form a 'joint' with a random aperture distribution (Brown, 1987b) for modelling the flow through rock joints. Another way is to say that the fractal distributions of the contact areas in natural fractures are determined from experiments, then establish a fractal model associated with continuum percolation to quantitatively reproduce the flow path geometry (Nolte et al., 1989).

From the discussions in Section 15.4, it is known that profiles of natural rock fractures have spectra with slopes between -2 and -3 corresponding to a surface fractal dimension ranging from 2.5 to 2.0 (from $s = -(8 - 2D)$ for surface fractals), respectively. Brown (1987b) generated some surfaces whose profiles have fractal dimensions within this range for use in the flow simulations. Examples of the two extreme cases are reproduced in Figure 15.30. Both the surfaces have the same standard deviation of height σ, but have different autocorrelation functions. Nearby points on the lower surface (Figure 15.30b) are more highly correlated with each other than corresponding points on the upper surface (Figure 15.30a). This correlation is described by either the spectral slope or the fractal dimension. To form a fracture, two surfaces with the same fractal dimension, but comprised of different sets of random numbers, are placed together at some fixed distance. For all fractures, the surfaces are held

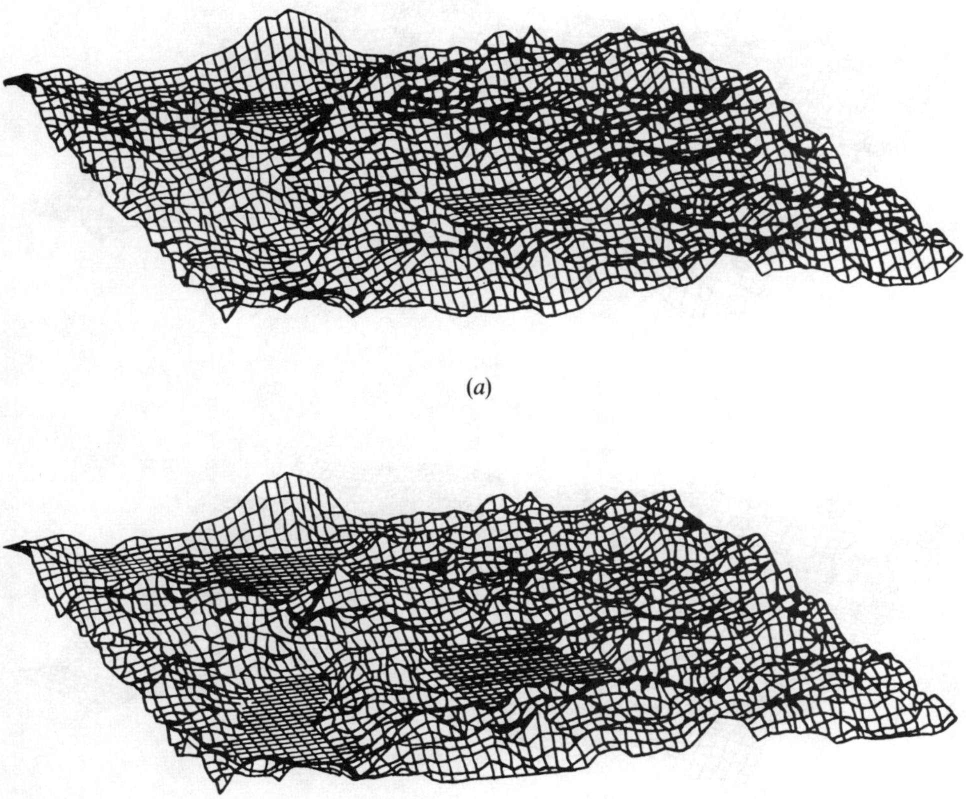

(*a*)

(*b*)

Figure 15.30. Examples of fractal surfaces generated for flow simulations. Both surfaces have the same standard deviation of height σ. Both were generated with the same set of random numbers, but have a different fractal dimension D; (a) $D = 2.5$, and (b) $D = 2.0$ (Brown, 1987b).

parallel and the mean separation between them is referred to as the mechanical aperture b_m. The aperture distribution $b(x, y)$ used in the flow simulations is taken as the local distance between the surfaces. Real rough surfaces have an absolute maximum and minimum height and, thus, have truncated height distributions. The model surfaces used by Brown have Gaussian height distributions with a standard deviation σ. The distributions are truncated for surface heights greater or less than 3σ from the mean level. Following a well-known result of statistics for the sum of two random functions, two uncorrelated Gaussian surfaces with a standard deviation of height σ will just touch when $b_m \propto 3\sqrt{(2\sigma)}$, assuming that the highest peaks on the surfaces are about 3σ from the mean plane. Thus, the surfaces are separated at approximately $b_m > 4.24\sigma$ and touching at smaller values. For $b_m < 4.24\sigma$, the numerical surfaces overlap or interpenetrate at some places. Figure 15.31 shows examples of the local aperture distributions for a single fracture at two different mechanical apertures

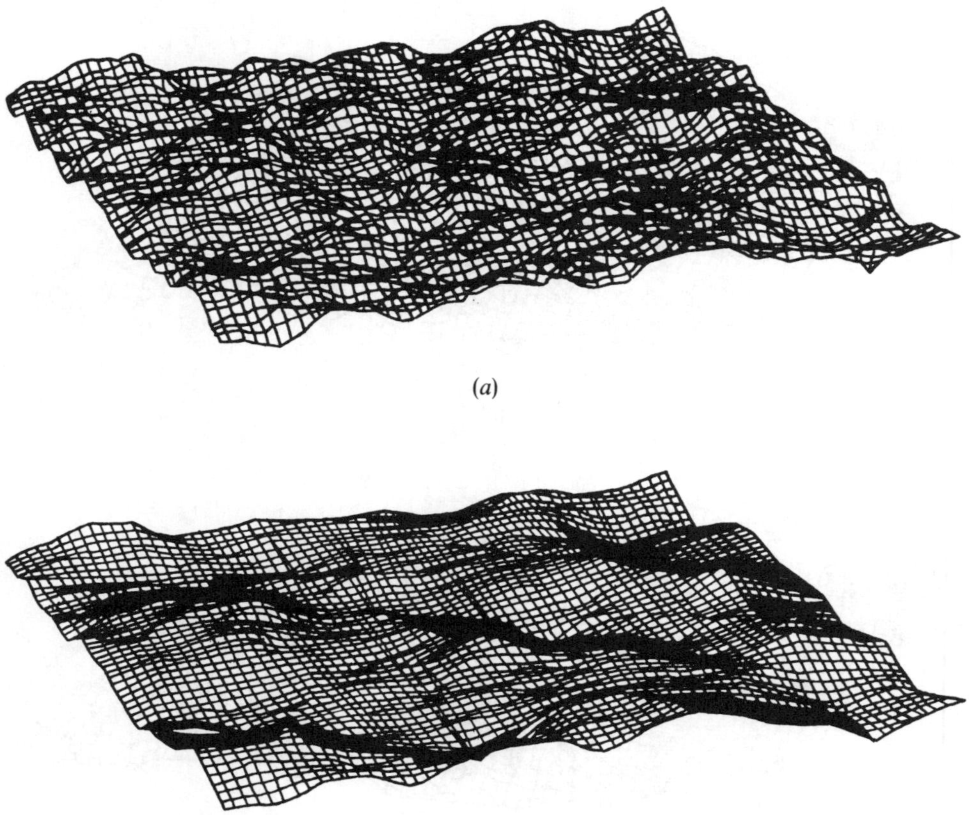

(*a*)

(*b*)

Figure 15.31. Aperture distribution $b(x, y)$ formed by placing two surfaces ($D = 2.5$) together at separations of (a) $b_m = 2\sigma$, and (b) $b_m = 1\sigma$. Flat areas are regions of zero aperture or 'contacts', defined by the overlap of the two surfaces (Brown, 1987b).

(Brown, 1987b). In addition to the mechanical aperture, another relevant measure in terms of the fluid flow is the mean aperture (Brown, 1987b):

$$\langle b \rangle = [1/(L_x L_y)] \int_{x=0}^{L_x} \int_{y=0}^{L_y} b(x, y)\,\mathrm{d}x\,\mathrm{d}y,$$

where L_x and L_y are the dimensions of the fracture. The mean aperture more closely represents the aperture available to flow. For example, one may imagine a situation where $b_m = 0$, but $\langle b \rangle \neq 0$. This means that there are still open fluid pathways. Brown used three pairs of generating surfaces with fractal dimensions 2.0, 2.25 and 2.5, respectively, in which each pair of surfaces has the same fractal dimension for flow simulation. The surfaces were placed together at various mechanical apertures to form a rough-walled fracture. Ten values of b_m were considered, ranging from 10σ (complete separation) to 1σ (intimate contact). In the computer simulation performed by Brown, Reynolds equation for steady state flow of an incompressible fluid between stationary surfaces $[\nabla(b^3/(12\mu)\nabla P) = 0]$ was written in dimensionless form, with all dependent and independent variables scaled to appropriate characteristic values. This dimensionless equation was then written in finite-difference form and solved for the pressure field $P(x, y)$ by standard matrix inversion techniques (Lapidus & Pinder, 1982).

Figure 15.32. Flow field resulting from the finite difference solution of Reynolds equation for a surface separation of $b_m = 10\sigma$. Surfaces are completely separated. Left and right sides of the fracture are the constant pressure boundaries, with the left side having the highest pressure. No fluid is allowed to flow out from the top and bottom. Vectors represent the relative magnitude and direction of the local volume flow rate. The total volume flow rate through the fracture is given by the sum of the x components (components parallel to the noflow boundaries) of the local flow vectors along any line parallel to the constant pressure boundaries (Brown, 1987b).

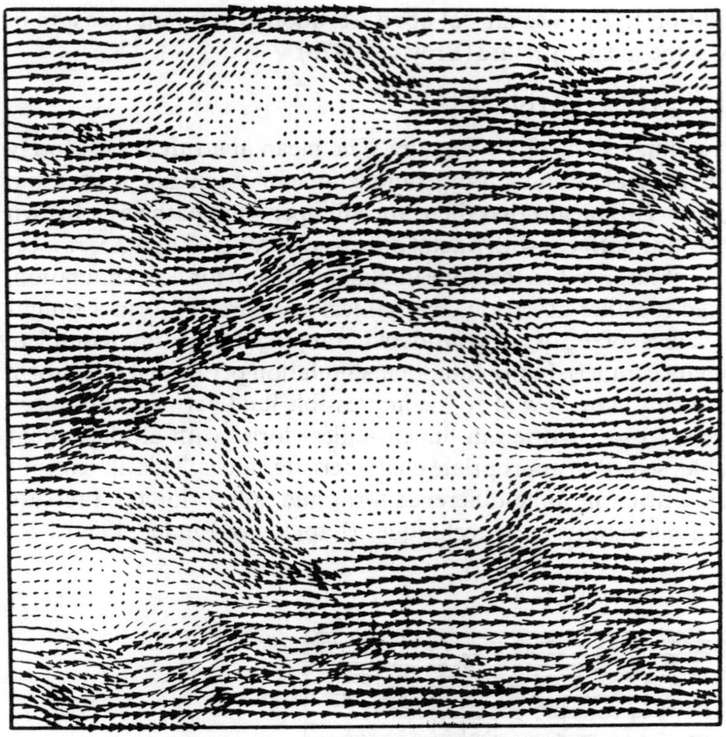

Figure 15.33. Flow field for the same fracture as in Figure 15.32, but with a surface separation of $b_m = 4\sigma$. Surfaces are just touching with a fractional contact area of about 0.4%. Contacts are shown as blank patches. Notice that the flow is through high aperture channels between the contacts (Brown, 1987b).

Figures 15.32 and 15.33 show the flow simulation results obtained by Brown (1987b). Figure 15.33 illustrates that when the surfaces touch, the flow is tortuous and tends to be channelled through high-aperture regions around the contacts. When the surfaces are completely separated, the effect of surface roughness is still apparent (Figure 15.32). This is in contrast to the parallel plate model, where all flow vectors would be parallel and of equal length. In Brown's work, the total volume flow rate and the macroscopic pressure gradient were used in the cubic law (Equation (15.13)) to obtain an effective parallel plate aperture or 'hydraulic' aperture b_h for the rough-walled fracture. For various surface roughnesses (fractal dimensions) the hydraulic aperture was compared to the mean separation of the surfaces. At large separations, the surface topography has little effect. At small separations, the flow is tortuous, tending to be channelled through the high-aperture regions. The parameter most affecting fluid flow through rough joints is the ratio of the mean separation between the surfaces to the root-mean-square surface height. This parameter describes the distance the surface asperities protrude into the fluid and accounts for most of the disagreement with the parallel plate model. Variations in the fractal dimension produce only a second-order effect on the fluid flow. For the range of joint closures expected during elastic deformation, these results show that the actual flow rate between rough surfaces is about 70-90% of that predicted by the parallel plate model (Brown, 1987b).

Noite et al. (1989) measured the contact area distribution of the core specimens labelled E30, E32, and E35 (stripe granite) by using the technique of injecting molten wooden metal into fractures which are initially evacuated. During the injection, stress was maintained on the fracture through a servo-controlled, stiff- testing machine. These contact area distributions are quite similar to the rock slit islands shown in Photo 14.1 and behave as fractals. Thus, a random Sierpinski carpet, shown in Figure 15.34, was chosen by Noite et al. (1989) to measure fractal dimensions of these experimentally obtained contact area distributions. A random Sierpinski carpet is constructed iteratively by successively removing small squares (or tremas) from the original black square (initiator). In the carpet shown, eight out of nine sub-squares, of scale $b = \frac{1}{3}$, remain at each level. This gives the approximate fractal dimension, $D = \log 8/\log 3 = 1.89$, of the carpet. In principle, the fractal dimension of a Sierpinski carpet can be measured by counting the number of tremas (corresponding to single areas of contact) with area, a larger than some set value, A. In practice, counting is carried out by superposing grids with successively smaller spacing and counting the number of grid squares at each level which are occluded more than 50% by a single trema. Thus, the fractal dimensions are derived from the slope of $\log(b^2 - N)$ vs. $\log b$, where b is the scale size of the grid, and N is the number of squares occluded for that scale b. The quantity $(b^2 - N)$ is the number of grid squares which remain uncut by tremas at this scale size. Table 15.14 gives the measuring results for the specimens E30 and E32 obtained by Noite et al. in which scale $b = 1$ corresponds to 32 mm. Since the contact

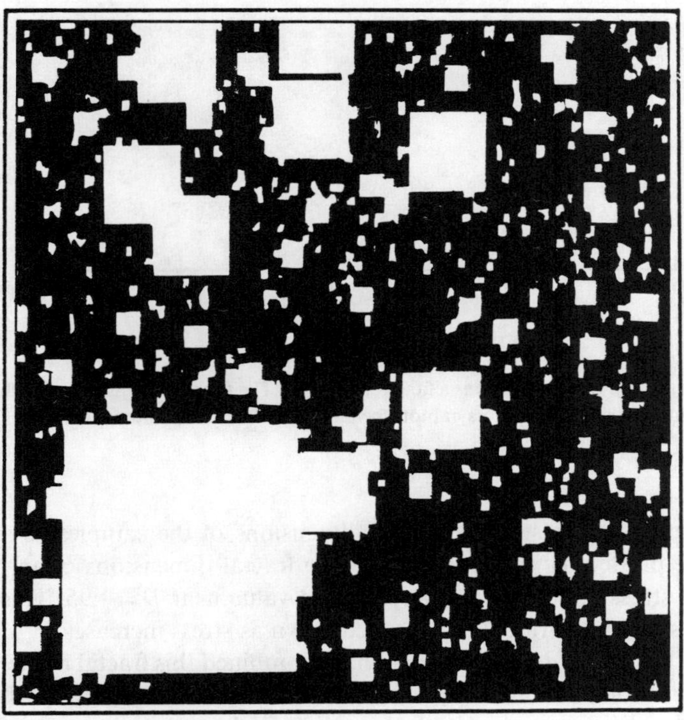

Figure 15.34. Model of a random Sierpinski carpet.

Table 15.14. Area fraction ($b^2 - N$) as a function of scale size (Noite et al., 1989).

Scalar 1/b	Specimen E30 3 MPa	33 MPa	85 MPa	Specimen E32 3 MPa	33 MPa	85 MPa
Size						
1	1	1	1	1	1	1
1/2	4	4	4	4	4	4
1/4	6	14	14	16	16	16
1/8	64	59	59	64	58	60
1/16	253	233	225	255	234	229
1/32	1016	922	894	1014	916	925
Fractal dimension D	1.996	1.975	1.964	1.997	1.995	1.995 ± 0.11

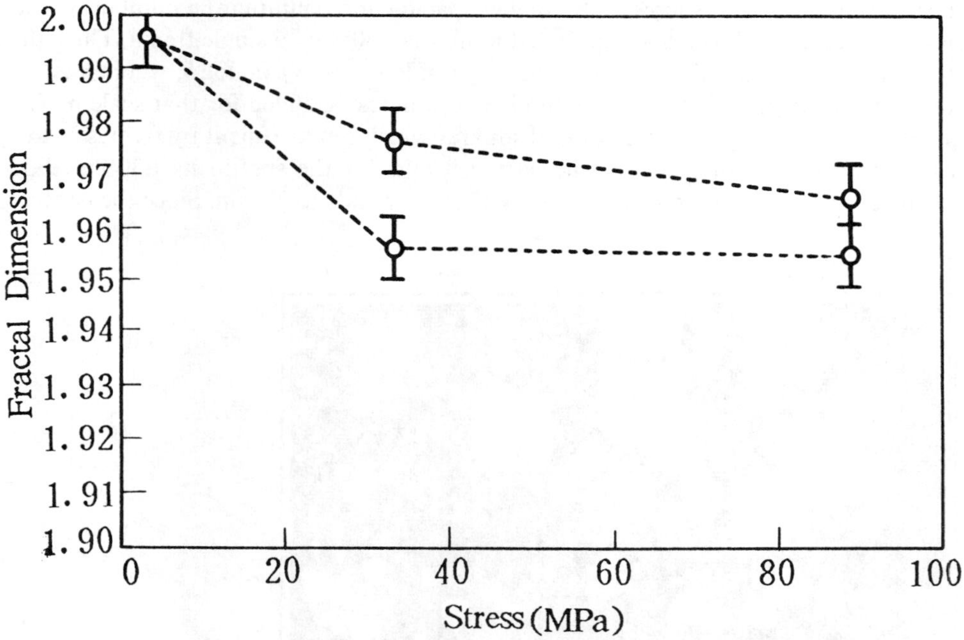

Figure 15.35. The fractal dimensions of the flow paths as a function of stress. The fractal dimensions of both fractures non-linearly depend on stress and appear to asymptote to values near $D = 1.95$ (Noite et al., 1989).

area increases as the stress is increased, the fractal dimensions of the samples as a function of stress can be plotted as in Figure 15.35. The fractal dimensions clearly decrease with increasing stress and appear to approach a value near $D = 1.95$. This means that the roughness of joint surfaces is smoothed down as stress increases.

A more interesting work is that of Noite et al. who have combined this fractal model with the continuum percolation theory to quantitatively reproduce the flow path geometries. The so-called continuum percolation theory is a fairly recent extension of the standard lattice percolation models discussed in Chapter 4. One of the possible models on the flow paths in the fractures involves randomly distributed circles in the plane. The occupied area fraction in this model is (Pike & Seager, 1974)

$$AF(N) = 1 - \exp[-(\pi r^2/\text{Area})N], \tag{15.14}$$

where r is the radius of the circles, and Area is the area in which N points are placed. The quantity in the exponential is simply the mean number of centers that fall within an area of radius r. The critical area fraction for the onset of percolation has been determined to be $AF_c = 0.688$ (Pike & Seager, 1974) which corresponds to the mean value of 1.14 points per circle of radius r. A percolation plot near the percolation limit with $AF = 0.7$ is shown in Figure 15.36 (Noite et al., 1989). Because continuum percolation is assumed to be in the same universality class as lattice percolation, the mean size of the percolation clusters (correlation length) is expected to be the same if the occupation probability of lattice percolation is replaced by the area fraction (Elam et al., 1984). However, the conductivity exponent, t, defined by $\sigma \propto (P - P_c)^t$, is a different matter. The conductivity exponent in lattice percolation is the same for both electrical conductivity and fluid conductivity because all the bonds have equal apertures, whereas in continuum percolation the flow paths have different apertures, which influence the flow of electrical and fluid currents differently. Noite et al. proposed an improved continuum percolation model which is constructed by considering the

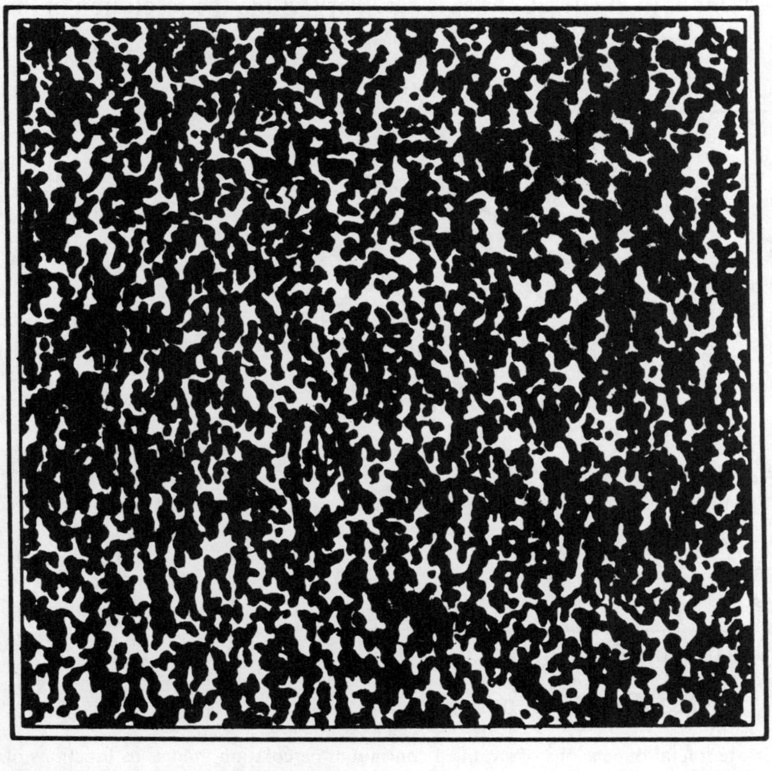

Figure 15.36. A random percolation plot near the percolation limit with area fraction $AF = 0.70$ and fractal dimension $D = 2.0$ (Noite et al., 1989).

auto-correlation function of the percolation plots, and is a hybrid between a Sierpinski carpet and the random percolation plot. For random percolation, if a given site is occupied, the probability that a site with a distance R away is also occupied is independent of R. However, it is apparent that sites of contact have a high probability of being immediately surrounded by other sites of contact from the examination of the contact areas of the joints. This means that the contact area is self- or auto-correlated. Thus, the auto-correlation of a distribution $f(r)$ is defined by Noite et al. in the form

$$F(R) = \int f(r) f(r+R) \mathrm{d}r. \tag{15.15}$$

It is clear that if $f(r)$ represents a power-law probability distribution, then the distribution is a fractal, and its auto-correlation function varies as $F(R) \propto R^{d_E - D}$ (d_E is the Euclidean dimension). For the random continuum model, $F(R)$ is a constant, which gives a fractal dimension $D = 2.00$. Therefore, a better model for the flow paths through the joints should be a continuum percolation model that possesses correlations. For this purpose, Noite et al. chose an approach which produced percolation plots that scaled with adjustable fractal dimensions, but which preserved some properties of the standard continuum percolation model. This percolation model proposed by Noite et al. can be iteratively constructed. Start by randomly distributing N sites within a given square region called a tier. Each site defines the center of a new sub-tier which is reduced from the preceding tier by a scale factor b. Within each of these new sub-tiers, N points are randomly distributed to define the centers of new sub-subtiers. This procedure continues for as many iterations as is necessary within the resolution of the

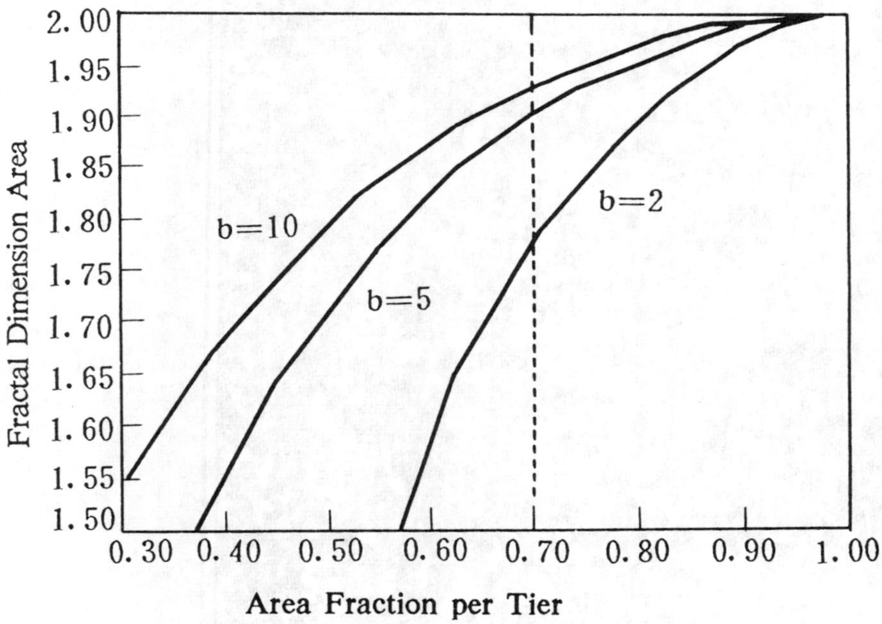

Figure 15.37. Approximate fractal dimensions of stratified continuum percolation models as functions of area fraction per tier ($AFPT$) for three scale values: $b = 2$, 5, and 10. The percolation threshold occurs near $AFPT = 0.7$, independent of resolution (Noite et al., 1989).

graphics. This model was called the stratified continuum percolation model (Noite et al., 1989). If the initial area fraction per tier $AFPT$ is given by $AFPT = AF(N)$ (defined in Equation (15.14)), then the fractal dimension of the pattern is approximately estimated as

$$D = 2 + \log[AFPT + (1 - AFPT)(AFPT)^2]/\log(b). \qquad (15.16)$$

In the absence of overlap, the fractal dimension would be $D = 2 + \log[AFPT]/\log(b)$ because the overlap fraction per tier is $AFO = AF(N) \times AF(N-1) = (AFPT)^2$. It is

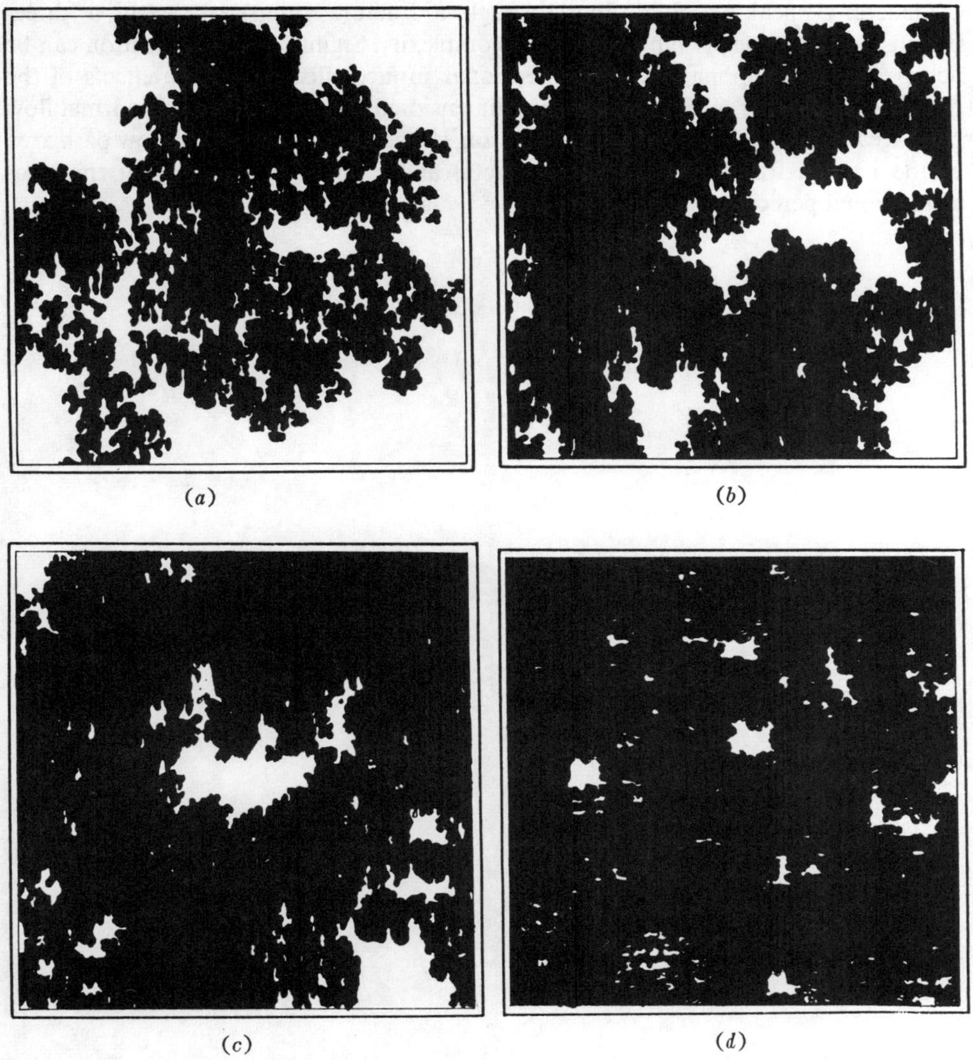

Figure 15.38. Stratified continuum percolation plots for a succession of fractal dimensions (5-tiers): (a) $b = 2.7$, 7 points per tier, $AFPT = 0.7$, $D = 1.80$; (b) $b = 2.4$, 9 points per tier, $AFPT = 0.79$, $D = 1.94$; (c) $b = 2.4$, 10 points per tier, $AFPT = 0.89$, $D = 1.97$; and (d) 2-tiers, $b = 9$, 250 points per tier, $AFPT = 0.96$, $D = 1.89$ (Noite et al., 1989).

noted that the total area fraction of the stratified percolation plots is not uniquely defined for a given scale b and $AFPT$. Instead, it depends on the resolution and upper cut-off of the pattern which is a fundamental property of fractals. For the n-th iteration, the total area fraction can be expressed as

$$AF_n = AFPT[AFPT + (1 - AFPT)(AFPT)^2]^{n-1}. \tag{15.17}$$

Figure 15.37 shows the results from Equation (15.16) obtained by Noite et al. (1989).

Figure 15.38 indicates the flow path patterns simulated by Noite et al. using their stratified percolation model. It is seen that these fractal stratified percolation plots are in good agreement with the real data obtained from experiments.

Generally speaking, fluid flow through a fracture (joint) is clearly a three-dimensional process of high geometrical complexity. Significant simplification can be achieved by decoupling the two-dimensional surface effects from the effects of the third, perpendicular direction. This two-dimensional projection defines a normal flow path area. From the discussion in this section, it is seen that this normal flow path area of the natural fractures (joints) is a fractal and can be analyzed using the two-dimensional percolation theory.

CHAPTER 16

Fractal nature of cluster systems

Fractal nature has been widely found in geology and seismology under a variety of circumstances. Examples include the distribution of islands, earthquakes, faults, and geographic phenomena. King (1983) discussed in detail the self-similarity of a faulting system and proposed a fractal approach to tectonics. Kagan & Knopoff (1981) showed that earthquakes, at least for magnitudes less than 6, are fractally clustered in time. Turcotte (1986b, 1989) gave a fractal model for crustal deformation and discussed fractal applications to complex crustal problems. It is recognized, both from the distribution of earthquakes and geographic phenomena, that displacements on a major fault zone, clustering earthquakes and surface geology are fractally distributed over a region. It is possible to use fractal concepts to better understand these distributions.

16.1 FRACTAL GEOMETRY IN THE SAN ANDREAS FAULT SYSTEM

A specific example of fractal nature in cluster systems is the San Andreas fault system in central California (Figure 16.1). This fault system is recognized as a major boundary between the Pacific and North American plates. It has been noted that the spatial distribution of earthquakes in the San Andreas fault system is related to the complexity of the fault system. The observed offset along the San Andreas fault is being achieved by strain release mechanisms in addition to the occurrence of great earthquakes like the 1857 Fort Tejon earthquake or the 1906 San Francisco earthquake, and the strong motion from the 1966 Parkfield earthquake. A number of studies have suggested that it is not only the long-term seismic behavior but also the detailed behavior of an earthquake fault, from nucleation of the dynamic instability through the propagation to the ultimate arrest of the rupture, which is heavily influenced by geometric fault complexity. This is just as Segall & Pollard (1980) state: a knowledge of the detailed fault geometry is fundamental to an understanding of fault mechanics. Okubo & Aki (1983, 1987) and Aviles et al. (1987) have applied fractal geometry to analyze the geometric fault complexity of the San Andreas fault system in different size ranges. As typical examples of fractal measurement of faults, Okubo & Aki's (1987) and Aviles et al.'s (1987) results will be mainly introduced in this section.

Okubo & Aki illustrated a fractal measurement method of fault length as shown in Figure 16.2. In this figure, two parallel fault segments of equal length, a, are

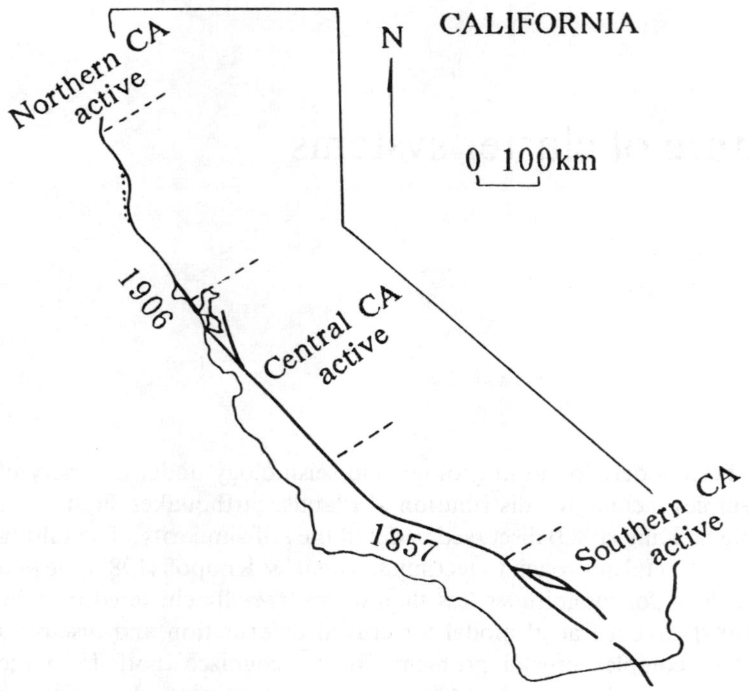

Figure 16.1. Map of California showing three active regions and sections of the San Andreas Fault that broke during the great earthquakes during 1857 and 1906 (after Allen, 1968, in Okubo & Aki, 1987).

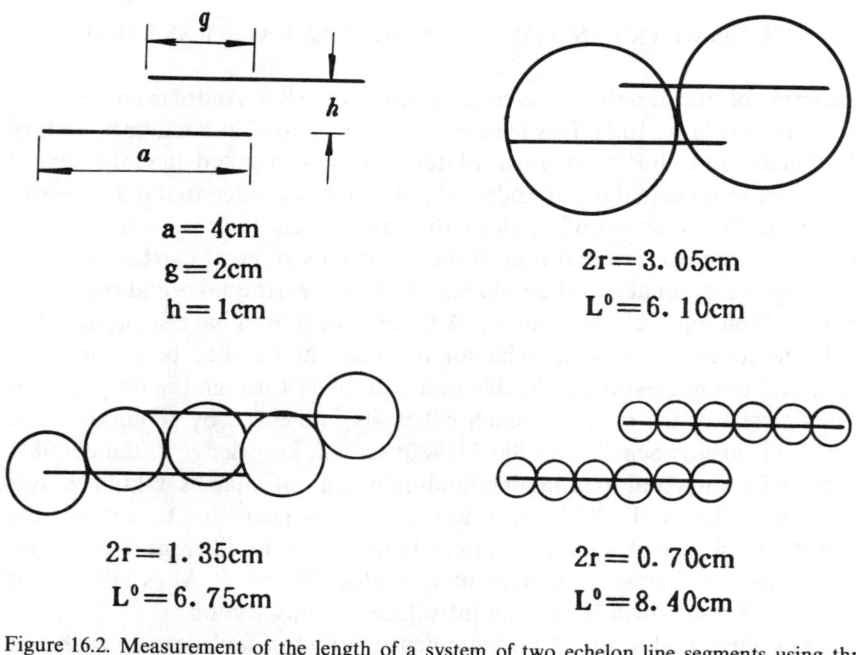

Figure 16.2. Measurement of the length of a system of two echelon line segments using three different 'yardstick' lengths (Okubo & Aki, 1987).

separated in the x and y directions by spacings g and h, respectively, where, as an example, $a = 4$ cm, $g = 2$ cm, and $h = 1$ cm. With a covering circle of radius $r = 1.52$ cm, these fault segment traces can be covered by two circles; $L^0(r) = 6.10$ cm. Five circles of diameter 1.35 cm are required to cover these same traces, and the measured fault length increases to $L^0(r) = 6.75$ cm. When the circle diameter is reduced to less than the spacing h between the fault segments, then lengths of the segments are measured separately, the same as the measurement of coastlines (Chapter 2). With $r = 2.3$ cm, the measured length increases to $L^0(r) = 8.28$. Obviously, the measured fault lengths will tend towards its true value, $L^0(r) = 8$ cm, as r decreases. Thus for nonuniform spacings and fault segment lengths, the fault length measurements will vary over a wider range of r which is determined by the total extent of the fault segments and the distributions of segment length, a, and spacings g and h.

One of the important factors necessary for considering measurement of natural fractal curves such as fault-traces is how to chose fractal cutoffs of yardstick lengths; that is, how to determine the upper and lower bounds of the fractal existing over this range. The cutoffs are selected, in principle, on the basis of specific physical processes being considered. In fractal analysis of the San Andreas fault system, Okubo & Aki selected an upper bound of 15 km to be associated with the thickness of the seismogenic crust in California, and the lower bound is of the order of 1 km. Figure 16.3 is a reduction of the 1:750 000 CDMG (the California Division of Mines and Geology) map with the 30-km-wide 'San Andreas fault' sketched in. The fault length measurements in the studies of Okubo & Aki include the portions of faults like the San Jacinto or the Garlock faults which intersect the San Andreas lying within this band. Where the San Andreas branches into northern and southern reaches southeast of San Bernardino, Okubo & Aki focused only the fault band about a median line between these two reaches and subsets of the San Andreas fault. In Figure 16.3, subset P ranges from Bitterwater to Cholame where fault creep and small-to-moderate earthquakes are observed. This subset includes that part of the San Andreas which ruptured in the 1966 Parkfield earthquake. Subsets Q, R and S include the southern locked portion of the San Andreas fault from Cholame to Wrightwood. Subset R ranges from the southern end of the Carrizo plain to Lake Hughes. This part of the fault contains the 'big bend' portion where the San Andreas is intersected by the Big Pine and Garlock faults. Subsets T and U extend through the southern active portions of the fault from Wrightwood to the Salton Sea near Durmid. Using the method of fault length measurement shown in Figure 16.2, Okubo & Aki measured the length of these subsets of the San Andreas fault system. Examples of the measurements, subsets Q and R, are given in Figure 16.4. The fault length and measuring radius or yardstick length data obtained by Okubo & Aki are replotted on log-log plots in Figure 16.5. Fractal dimension D is estimated from the slopes of straight-line least squares fits to these data. It is seen from Figure 16.5 that fractal dimension D for the entire length of this part of the San Andreas fault is 1.31 ± 0.02; from the northwest to the southeast, D increases from 1.12 to 1.43. We can also see that the fractal character exhibited over the range of the measurement is uniform. More important is that the variations in fractal dimension measured by Okubo & Aki reflect the differences in fault complexity between different segments of the San Andreas fault system discussed by Allen (1968). At the scale of 1:750 000, subset P, the northernmost segment from Bitterwater to Cholame, is relatively linear in mapped fault detail except at the edges of the set and

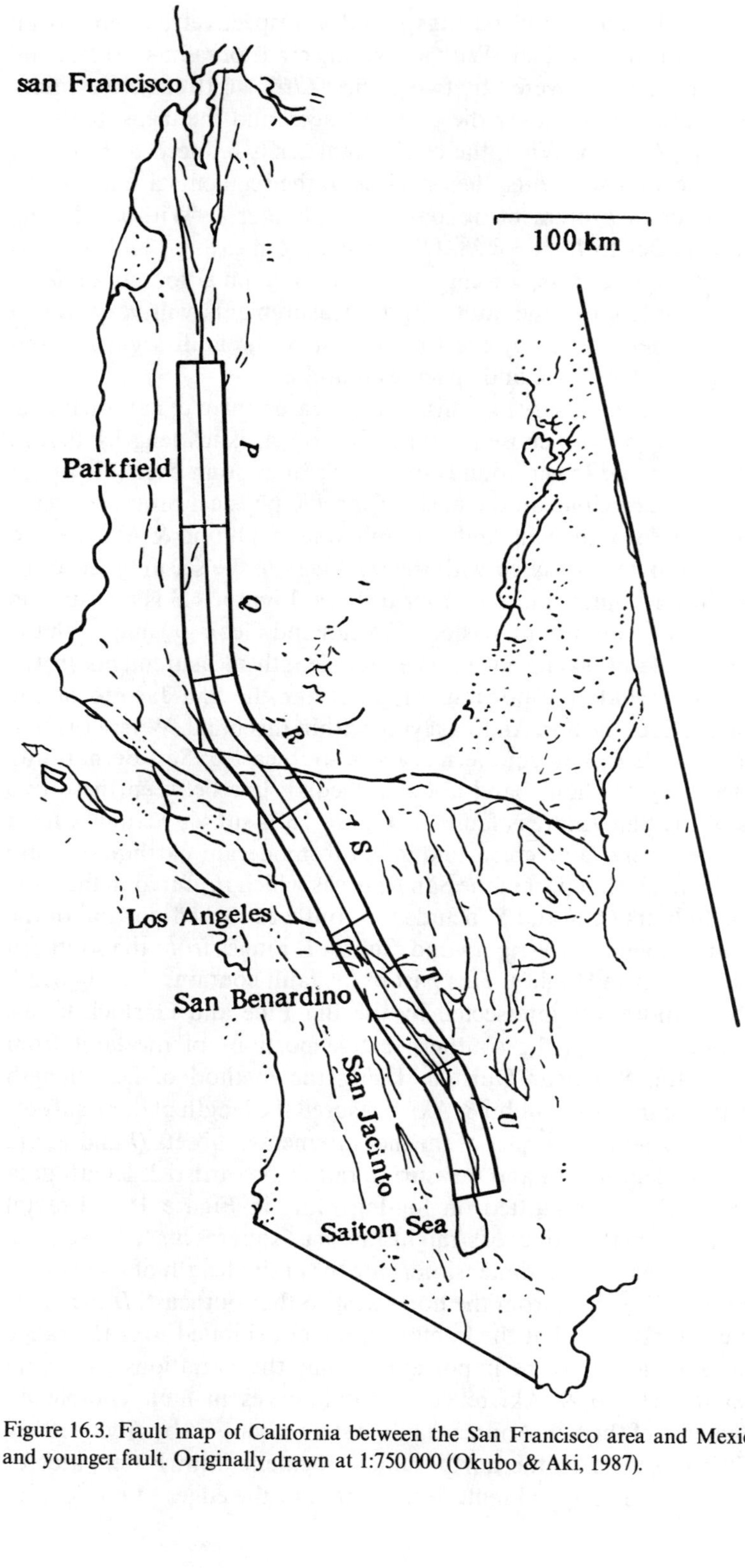

Figure 16.3. Fault map of California between the San Francisco area and Mexico, showing quaternary and younger fault. Originally drawn at 1:750 000 (Okubo & Aki, 1987).

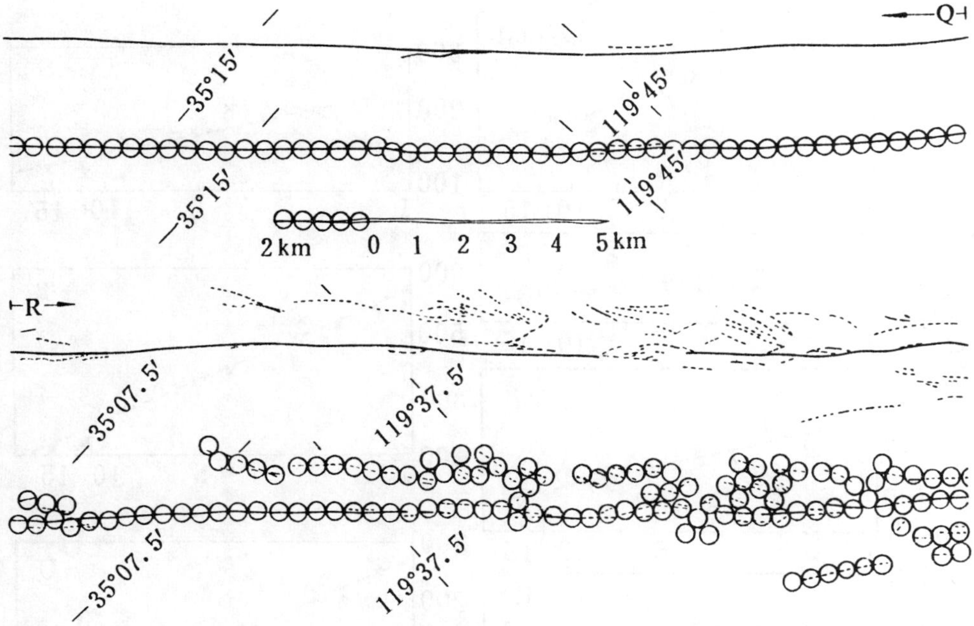

Figure 16.4. Adjoining southern end of subset Q and northern end of subset R showing most recent fault breaks mapped at 1:24 000. Method of covering fault traces by circles of radius $r = 0.2$ km is also shown for each section (Okubo & Aki, 1987).

is characterized by fractal dimension $D = 1.12 \pm 0.02$. To the southeast in set Q, R, and S, values of D are 1.20 ± 0.02, 1.25 ± 0.03 and 1.21 ± 0.03, respectively. The presence of other faults within the 30-km-wide fault band contributes to the measured fault lengths with the larger degree of complexity associated with the 'big bend' portion of the fault where the San Andreas is intersected by the Big Pine and the Garlock Faults in subset R. Farther to the southeast, sets T and U from Wrightwood to the Salton Sea are characterized by the highest estimated values of D; $D = 1.42 \pm 0.02$ in T and $D = 1.43 \pm 0.02$ in U, reflecting the branching of the San Andreas and the proximity of the Banning, Cucamonga, and San Jacinto faults to the San Andreas. The southern locked portion of the fault which last ruptured during the 1857 earthquake is associated with a lower fractal dimension. The southern California active region exhibits a higher fractal dimension.

Okubo & Aki (1987) also measured the fault-trace lengths of the most recently active breaks along the San Andreas mapped at scales of 1:62 500 in subset P and 1:24 000 in subset Q and R as part of the US Geological Survey (USGS) 'strip map' series. They subdivided the fault in the same manner as above, but it is important to note that these trip maps were intended to delineate only those most recently active fault breaks presumed to be associated with the 1966 Parkfield and 1857 Fort Tejon earthquakes on the San Andreas and indicated only a 4-km-wide band centered about the main fault trace rather than the 30-km-wide band used in the preceding discussion. In that case, other major faults in addition to the San Andreas contributed to the measured fault lengths. Since the finer map scales are examined here, the inner cutoff is reduced

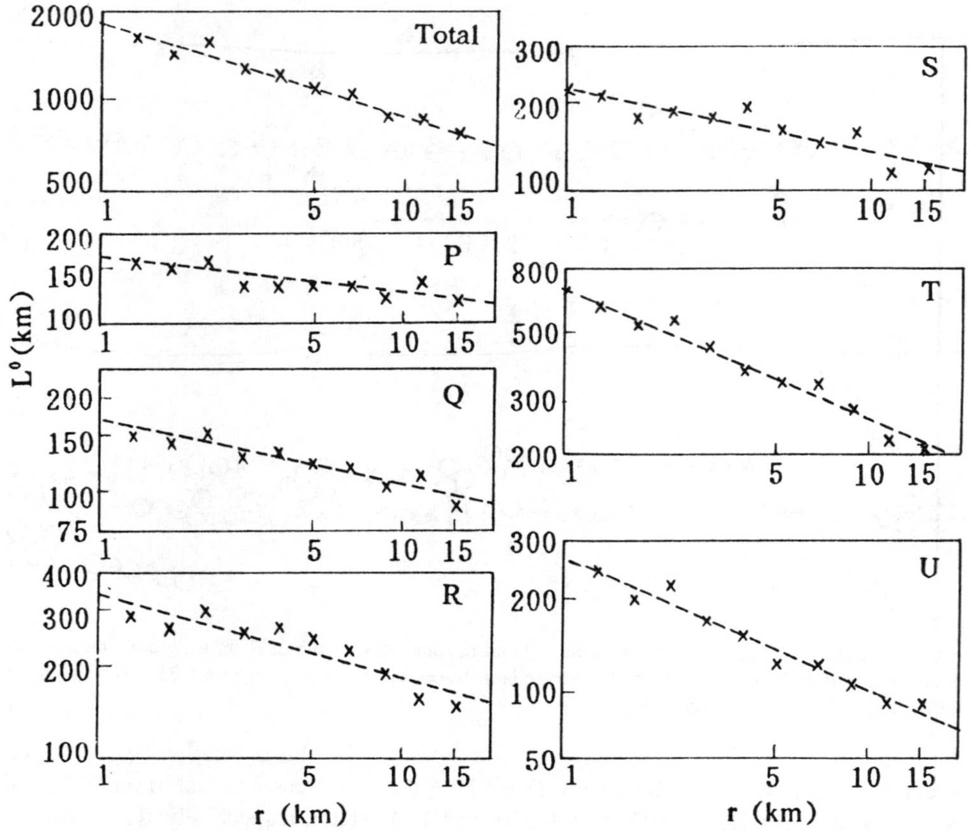

Figure 16.5. Fault length $L^0(r)$ vs. measuring radius data for San Andreas fault and subsets measured at 1:750 000. Dashed lines are best least squares fits to data (Okubo & Aki, 1987).

to less than 100 m. The fault length and measuring radius data obtained by Okubo & Aki are replotted in Figure 16.6. In contrast to Figure 16.5, the measurements on the strip map renderings of the fault breaks show changes in the rates of change of fault length L^0 at a scale length less than 1 km. Thus, the fractal character for the mapped fault breaks is exhibited only over a portion of the range of length between the cutoffs. There is a critical radius or length r_c, so that the mapped fault trace geometry is not a fractal (or self-similar) when the measuring radii are larger than r_c. The value of r_c estimated from graphically best fit straight-line segments to L^0 vs. r data are $r_c \sim 500$ m in subset P, 300-400 m in subset Q and ~ 1 km in subset R.

Aviles et al. (1987) also examined the fractal characteristics of the San Andreas Fault system, but only centered the irregularity of the main fault trace. Unlike the method of fault length measurement given by Okubo & Aki, Aviles et al. used the ruler method ($L(\delta) = L_0\delta^{1-D}$ as discussed in Chapter 2) for measuring the fault length. They divided the San Andreas Fault into five segments, delineated in Figure 16.7 as solid at their ends. Within any segment, the amount of coseismic slip has varied along strike. A subdivision of the 1906 break at Black Mountain and of the 1857 break at the southern

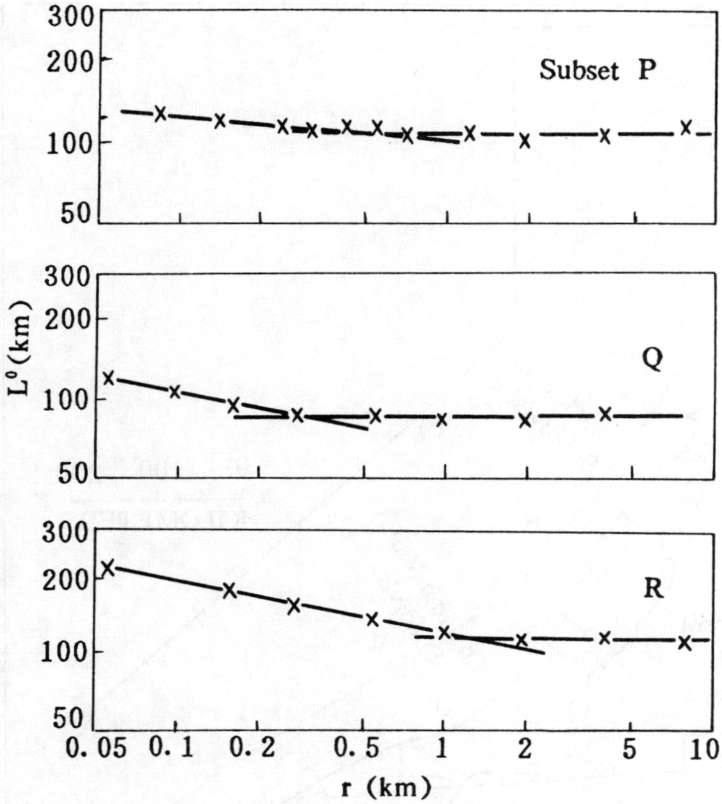

Figure 16.6. L^0 vs. r plots for subsets P, Q, and R mapped at scales of USGR strip maps (1:24 000 and 1:62 500). Dashed lines that are graphically estimated best fits to data (Okubo & Aki, 1987).

end of Cholame Valley has been suggested on the basis of observed slip, seismicity patterns, historic events, and geological complexity indicated by the dashed bars in Figure 16.7. Because the most active trace south of San Bernadino is difficult to identify, three alternatives are illustrated in the local enlargement in Figure 16.7.

The fault length data as a function of measuring ruler length for the entire San Andreas Fault and the major and minor subdivisions are replotted in Figures 16.8-16.10. Plots of length vs. measuring ruler for the four longest San Andreas Fault segments are shown in Figure 16.8, whereas the plots for the four intermediate- and short-length segments are shown in Figures 16.9 and 16.10, respectively. It is seen that the fractal characteristics are nonuniform in the entire band (0.5-100 km) shown, but a single straight line fits the observations very well in the range from point *a* to *b*.

The fractal dimensions calculated by Aviles et al. for each fault segment are reproduced in Figure 16.11 for the entire band and in Figures 16.12 and 16.13 for the long- and short-length bands, respectively. The fractal dimensions given in Figures 16.11-16.13 are only slightly larger than 1.0, and they reflect the smoothness of the main San Andreas Fault trace. Although the fractal dimensions calculated differ from 1.0 only slightly, the difference is significantly greater than the error. This means that

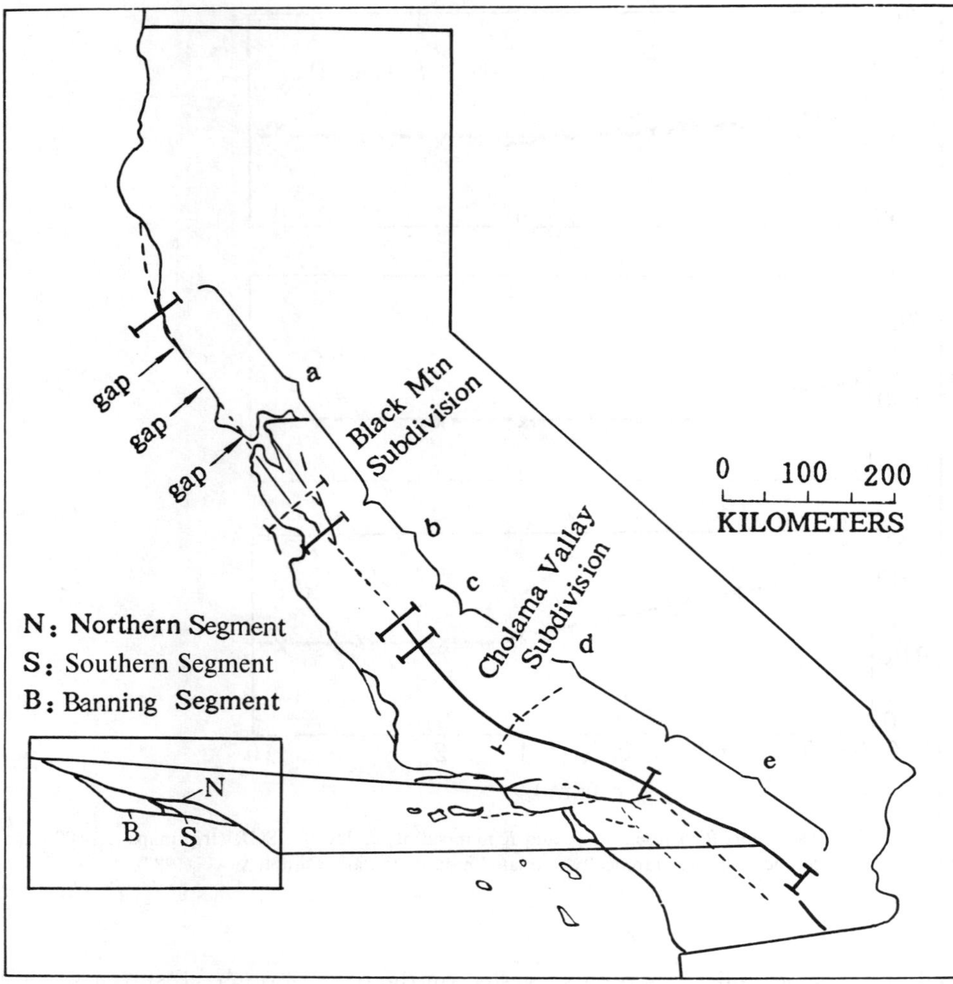

Figure 16.7. San Andreas fault divided into five major sections (solid bars) based on characteristic earthquake behavior. Subdivisions of 1906 and 1857 sections shown with dashed bars. Enlargement gives northern (*N*), southern (*S*), and Banning (*B*) alternatives for the southern sections (Aviles, et al., 1987).

all major segments of the fault really have detectable fractal behavior. Furthermore, there are significant variations in the fractal dimension along strike from the 1906 segment to the southern segments, with the southern segments having the largest *D* values. The three alternative segments for the southern segment exhibit comparable *D* values. Subdivisions indicate that the southern part of the 1906 segment is more jagged than the northern part over all length scales represented.

It is concluded, from the studies of Okubo & Aki and Aviles et al. on fractal characteristics of the San Andreas Fault system, that fractal concepts can be applied in quantitative analyses of fault geometry. The variation of fractal dimension depends on the degree of complexity and irregularity of the fault system. Small *D* values indicate

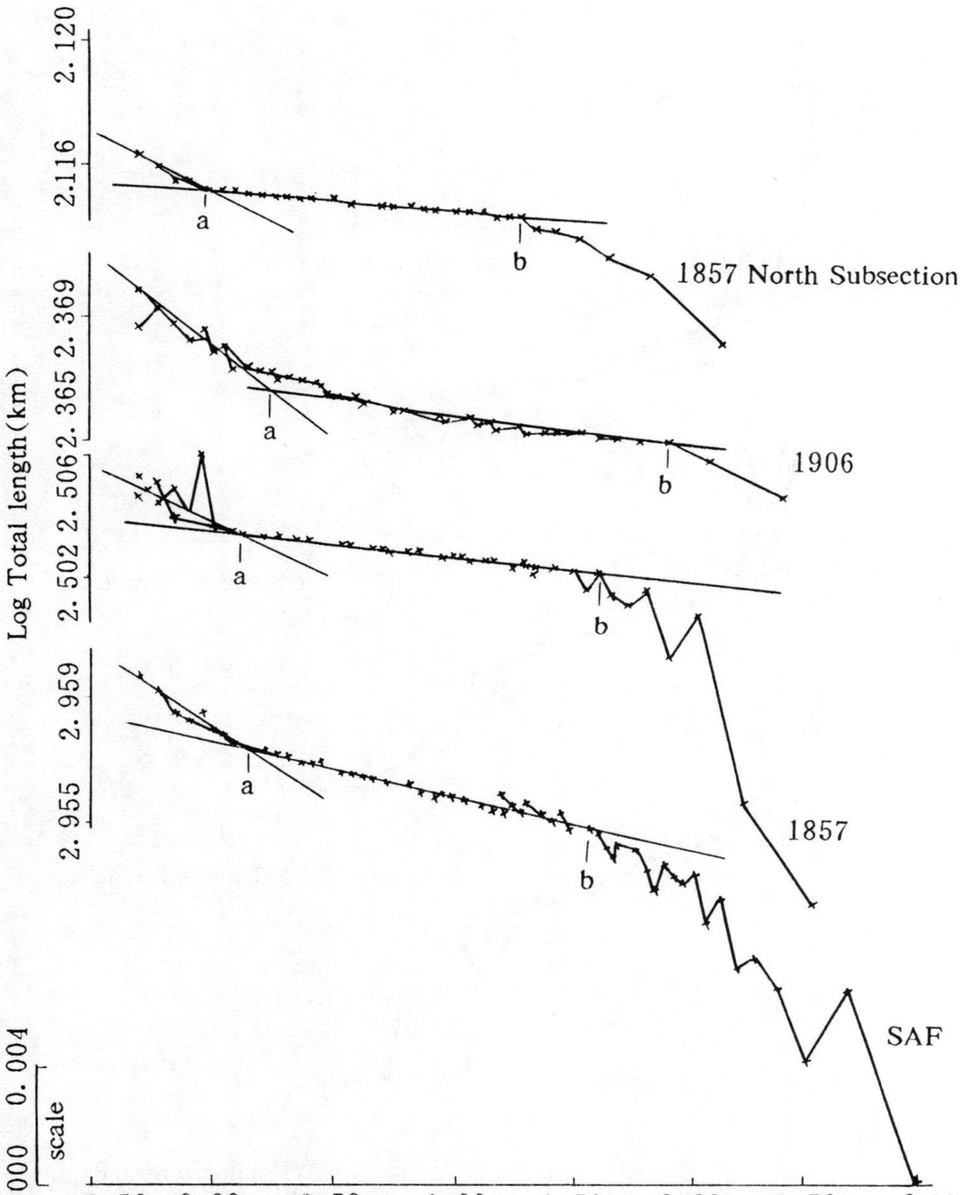

Figure 16.8. Logarithm of total length as a function of the logarithm of the ruler for the four longest fault sections. Separate length measurements starting from the north (diamonds) and south (crosses) ends of the sections are shown. Label *a* is the demarcation point between the short- and long-length trend. Label *b* marks the onset of end effects in the measurement (Aviles, et al., 1987).

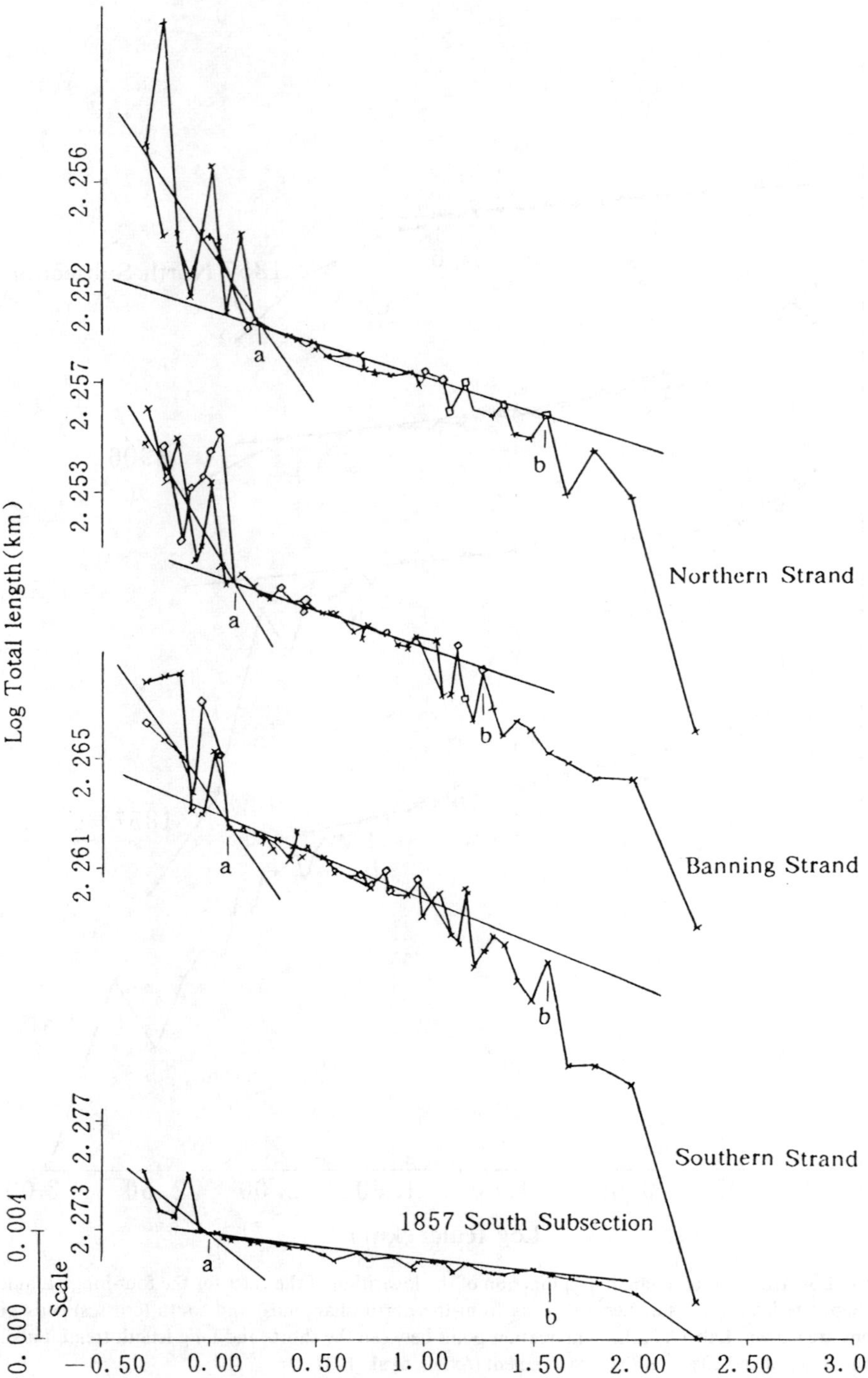

Figure 16.9. Log length vs. log ruler length for the four intermediate-length fault sections. Refer to Figure 16.8 for symbol definition (Aviles, et al., 1987).

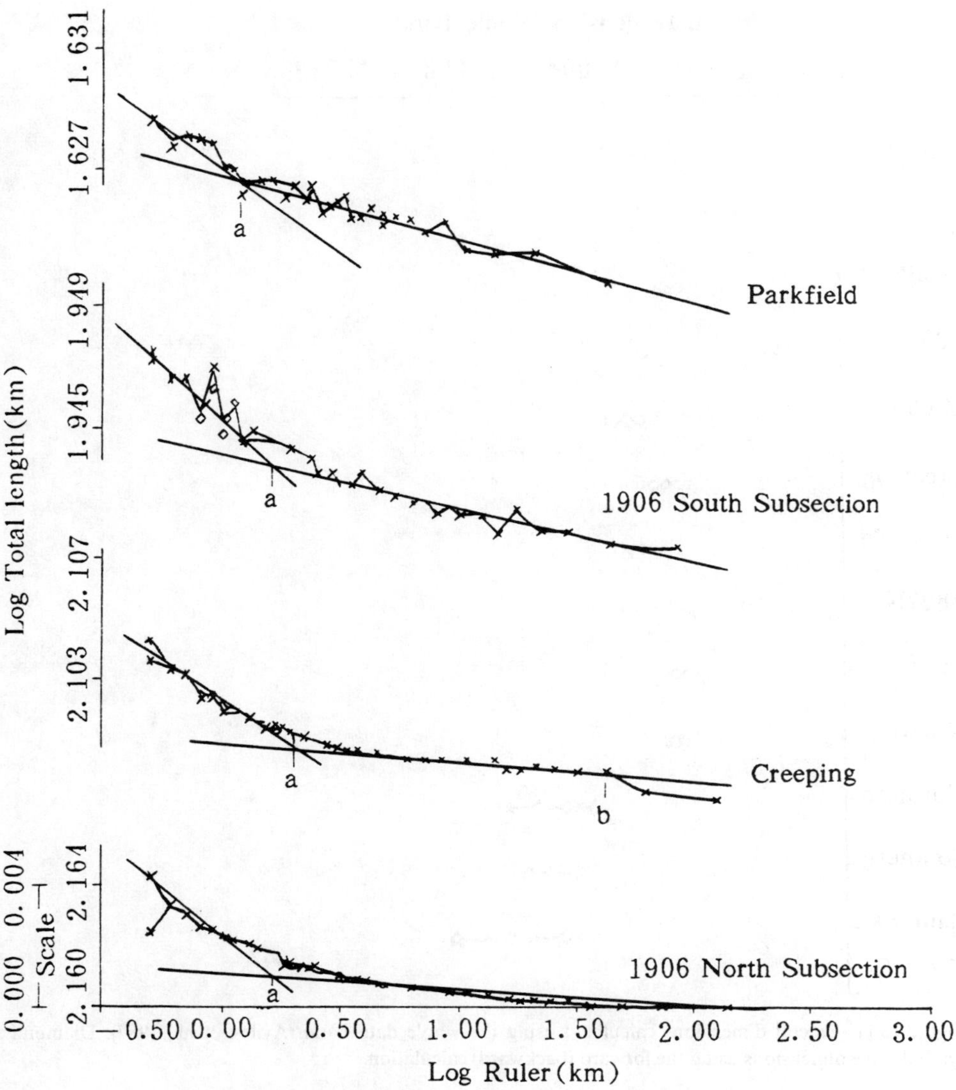

Figure 16.10. Log length vs. log ruler length for the four shortest fault sections. Refer to Figure 16.8 for symbol definition (Aviles, et al., 1987).

relative smoothness of the fault traces. It should be noted that, however, values of the fractal dimension of the fault system are also dependent on a different selection of cutoffs, the different manner of defining the relevant fault traces examined, and the scale range used in fractal measurements. In fact, the fault system is a nonuniform self-similar fractal. For example, Figures 16.8 and 16.9 show three different self-similarity ranges, that is, three different fractal dimension values which are dependent on the scale range.

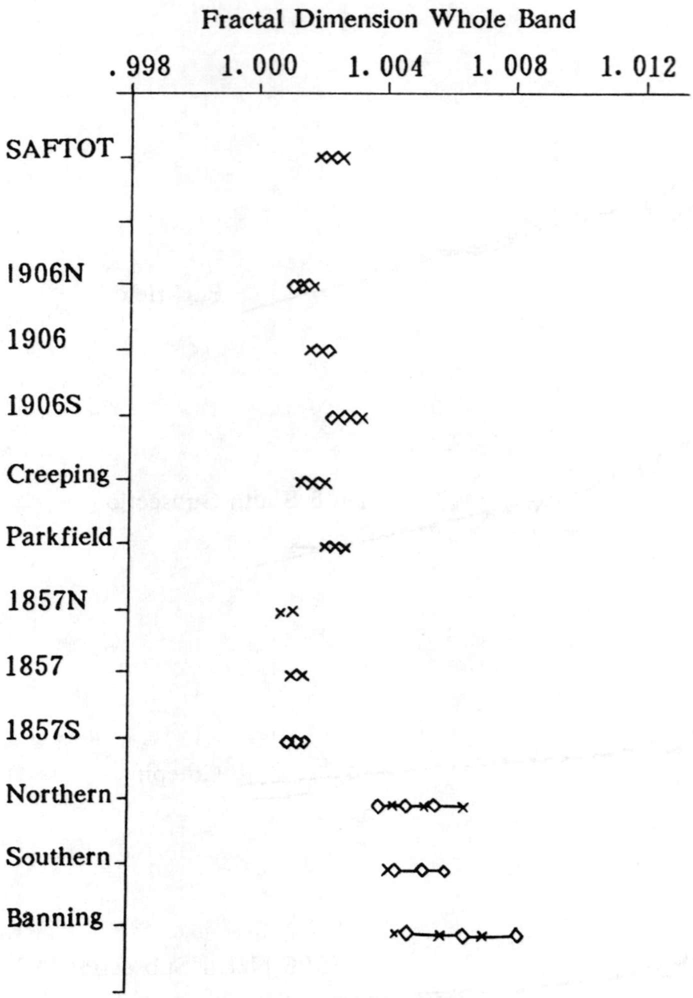

Figure 16.11. Fractal dimensions calculated using the whole data band (Aviles, et al., 1987). Diamond symbols give dimensions using the forward (backward) calculation.

16.2 FRACTAL NATURE IN SEISMOLOGY

Many studies have been performed on the distribution of earthquakes in time and space in order to better understand the earthquake generation process and possibly assist in earthquake prediction. The location, timing, and magnitude of earthquakes can be accurately determined from seismic arrays which exhibit a well-defined statistical fractal distribution. Under many circumstances the number of earthquakes N with a magnitude greater than m satisfies the empirical relationship given by Gutenbery & Richter (1956):

$$\log N = -bm + a, \qquad (16.1)$$

where a and b are constants. The b-value varies from region to region but generally

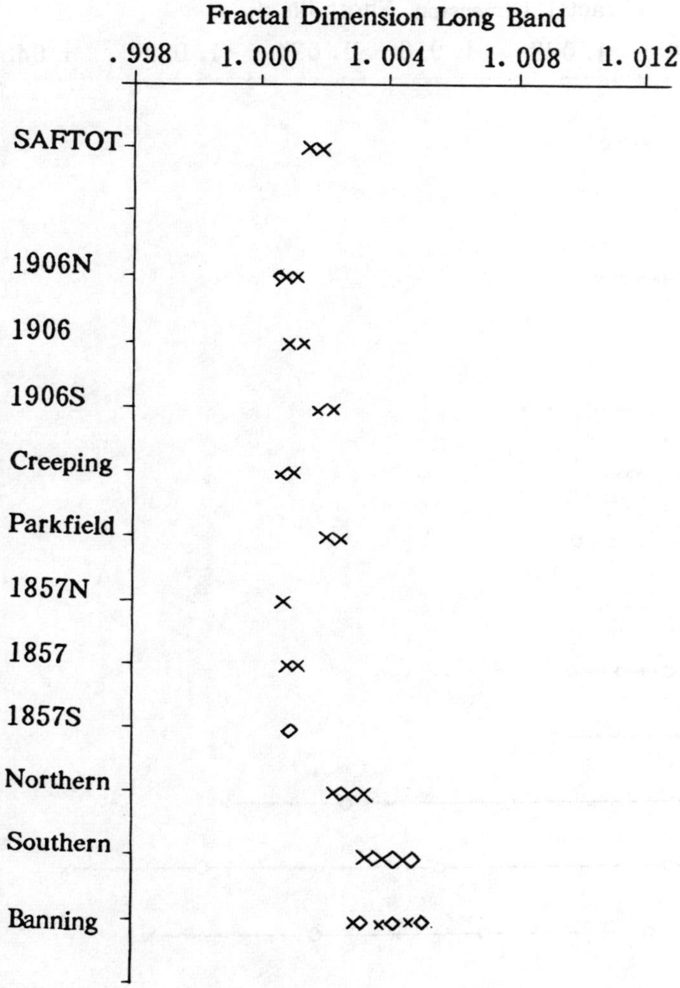

Figure 16.12. Fractal dimension D calculated over only the long-period band (Aviles, et al., 1987).

lies in the range $0.5 \leq b \leq 1.5$. Aki (1981) and Turcotte (1986a) have shown that the empirical relation is consistent with a fractal behavior for earthquakes. The moment M of an earthquake is defined by:

$$M = \mu \delta A, \tag{16.2}$$

where μ is the shear modules, A, the area of the fault break, and δ, the mean displacement on the fault break. The moment of an earthquake can be related to its magnitude by

$$\log M = cm + d, \tag{16.3}$$

where c and d are also constants. Kanamori and Anderson (1975) have established a

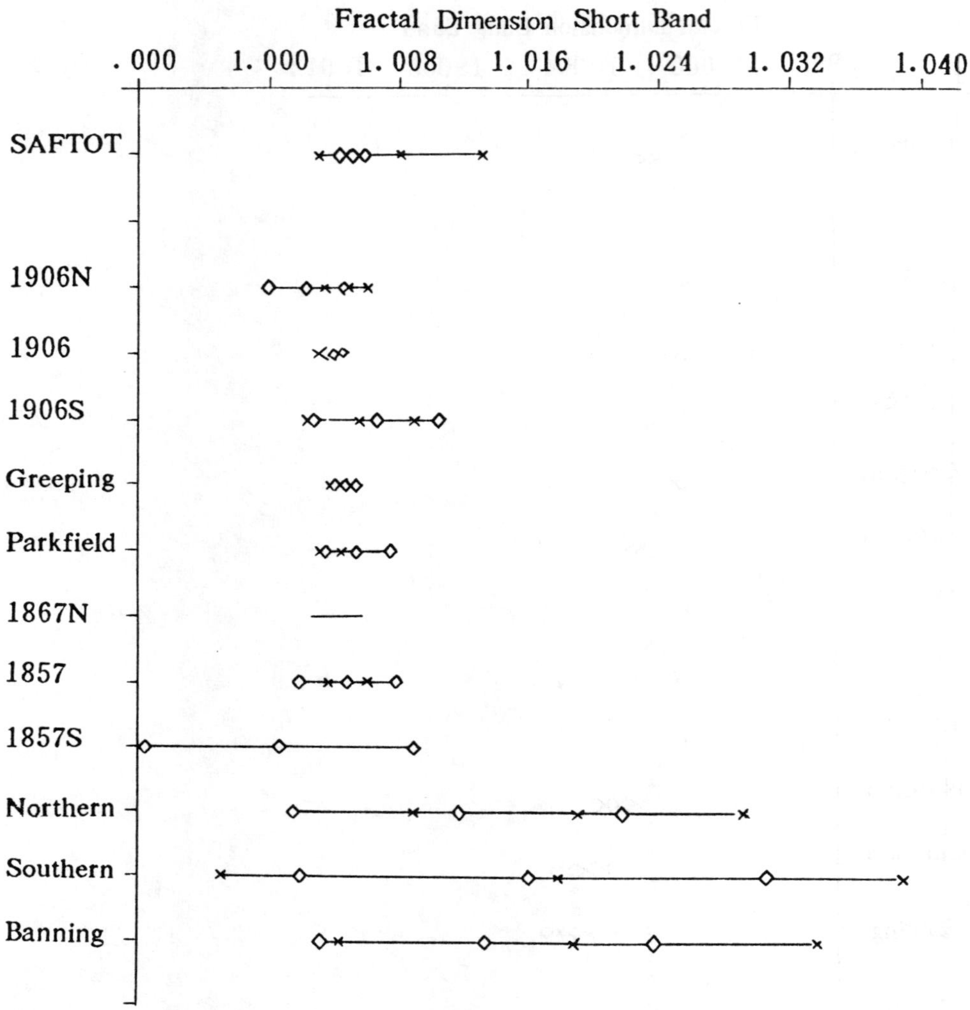

Figure 16.13. Fractal dimension D calculated over only the short period band (Aviles, et al., 1987).

theoretical basis for taking $c = 1.5$. They have also shown that it is a good approximation to take

$$M = \alpha r^3, \tag{16.4}$$

where $r = A^{1/2}$ is the linear dimension of an earthquake. Combining Equations (16.1), (16.3) and (16.4) yields

$$\log N = (-3b/c)\log r + \log \beta, \tag{16.5}$$

where

$$\log \beta = (bd/c) + a - (b/c)\log \alpha. \tag{16.6}$$

Thus, Equation (16.5) can be rewritten as

$$N = \beta r^{-3b/c}. \tag{16.7}$$

This is a fractal distribution with fractal dimension (Turcotte, 1986a)

$$D = 3b/c. \tag{16.8}$$

Recently, Niu (1991) derived a relationship similar to Equation (16.8) in terms of the statistical physics theory of fractal micro-fractures. This indicates that the regional frequency-magnitude distribution of earthquakes is indeed a fractal distribution. As an example of a regional frequency-magnitude distribution, Turcotte (1989) considered seismicity in southern California. Based on data from 1932 to 1972, the number of earthquakes per year, N, with magnitude greater than m is given as a function of m (Figure 16.14). In the magnitude range $4.25 < m < 6.5$ the data are in excellent agreement with Equation (16.1) taking $b = 0.89$ or $D = 1.78$. With $c = 1.5$, $d = 16$ and $\alpha = 3.27 \times 10^7$ dyne/cm^2 the corresponding range for r from Equations (16.4) and (16.3) is $0.9\,\text{km} < r < 12\,\text{km}$. Data for smaller amplitude earthquakes; $m < 4.25$, would be expected to fit this relationship, but in broad regions instrumented coverage is inadequate.

Also included in Figure 16.14 is the N associated with great earthquakes on the southern section of the San Andreas fault as given by Sieh (1978): $m = 8.05$ and $N = 0.006\text{a}^{-1}$ (a repeat time of 163 years). An extrapolation of the fractal relationship for regional seismicity appears to make a reasonable prediction of great earthquakes on this section of the San Andreas fault (Turcotte, 1989).

It is found that b-value (also D value) varies with time and space. In particular, b-value has a strong variation before the occurrence of a large earthquake. Figure 16.15 shows the b-value variation before the occurrence of the Tangshan earthquake with

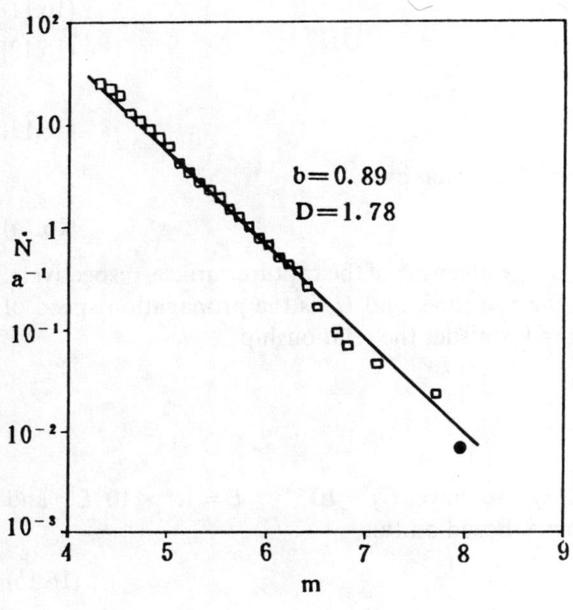

Figure 16.14. Number of earthquakes occurring per year, N, with a surface wave magnitude greater than m. The open squares are data for southern California from 1932-1972, and the solid circle is the rate of occurrence of great earthquakes in southern California from Equation (16.1) (Turcotte, 1989).

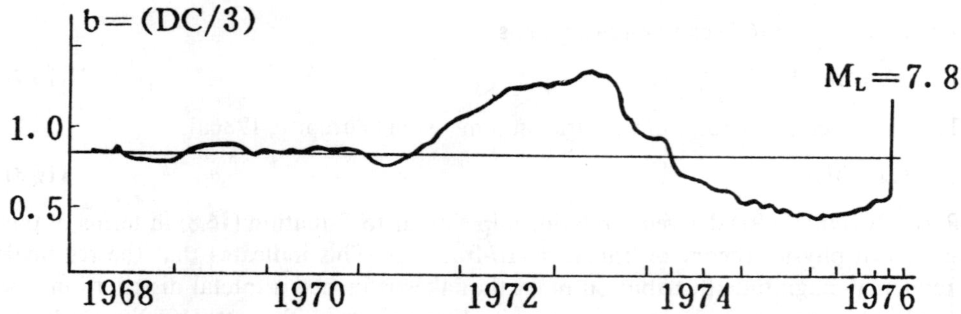

Figure 16.15. The *b*-value ($D = 3b/c$) variation before the occurrence of large earthquake with magnitude $m = 7.8$ in Tangshan (in China)

$m = 7.8$. The occurrence of a minimum *b*-value (or *D*-value) may be a precursor of a large earthquake. The fractal and physical mechanism of this phenomenon has been discussed in Section 14.10.

Based on the Equation (16.1), Li (1990) has also studied earthquake self-similar phenomena. The geometric size of an earthquake can be defined by 'source length, *L*':

$$m = \alpha \log(L) + \beta. \tag{16.9}$$

For real data of different temporal-spatial domain, the parameters $\alpha = 1.9 - 2.1$, $\beta = 3.3 - 3.8$ of which most of $\alpha = 2.0$, and $\beta = 3.6$. Thus Equation (16.9) is generalized to

$$m = 2.0 \log(L) + 3.6. \tag{16.10}$$

Some researchers have made several presumptions for source parameters. These presumptions are

$$C_1 = d/L \quad \text{geometric similar;} \tag{16.11}$$

$$C_2 = U/L \quad \text{strain drop fixed;} \tag{16.12}$$

and

$$C_3 = U_r T/L \quad \text{dynamic similar.} \tag{16.13}$$

Taking $C_4 = W/L$, the source volume V is given by

$$V = L \cdot d \cdot W = C_1 \cdot C_4 L^3, \tag{16.14}$$

where d, W, and L are the height, width, and length of the rupture surface, respectively. U is the dislocation distance, T is the rise time, and U_r is the propagation speed of rupture. The total earthquake energy E satisfies the relationship:

$$E = (8/3\pi)\mu U^2 L,$$

with

$$\log U = 0.5m - 1.1.$$

From Equations (16.10) and (16.12), we have $C_2 = 10^{-4.3}$, $E = 1.1 \times 10^3 L^3$, and $E_s = 10^{2.2} L^3$ for $\mu = 5 \times 10^{11}$ dyne/cm^2. Based on this,

$$\eta = E_s/E = 0.15. \tag{16.15}$$

Mei (1987) used deformation data to reverse and obtained $C_4 = 0.2$. C_1 can be calculated by

$$C_1 = 10^{2.2}/(C_4 \eta e), \tag{16.16}$$

where $e = \mu\varepsilon^2/2$. It is seen that the macroscopic (Equations (16.9) and (16.10)) or microscopic formula (Equations (16.11)-(16.14)) is used to show that the earthquake intensity can be expressed by a scale quantity L. One of the basic physics concepts of tectonic seismics is that rock mass ruptures or rapidly slips when elastic shear deformation has exceeded shear intensity by which rock can be restrained. This rupture mechanism is generally available for different earthquakes ranging ten orders of scale length. This is one of the physical fundamental of earthquake self-similarity with different magnitudes.

On the other hand, a large number of facts show that time-sequence of seismicity generally exhibits inhomogeneous distribution in different stratification. This is usually called 'cluster-quiet' phenomena. These phenomena also show the characteristic of multiple stratification with a self-similar mosaic structure. Inhomogeneous measurement of seismic temporal-spatial distribution can be realized by measuring its fractal dimensions (the fractal dimension in temporal distribution is denoted by D_t, and that in spatial distribution is denoted by D_s). The research results (Li, 1990) indicate that smaller earthquakes still remain same time structure and fractal dimension in smaller time scales. The main behaviors of fractal dimensions can be summarized as follows (here fractal dimension D_t is measured by a random Cantor set model which will be discussed in the next section):

(1) Fractal dimension reduction phenomenon occurs before a strong earthquake with a magnitude greater than 6 (also see Figure 16.15).

During the last ten years, three strong earthquakes $m > 6$ occurred in the Gang-Ning network, China. The D_t value decreased before the Menyuan (China) earthquake in 1986 ($m = 6.4$). One year before the earthquake occurred, $D_t = 0.42$. The D_t value also dropped obviously below 0.5 before two strong earthquakes occurred in 1976 (see Table 16.1).

(2) The D_t values of aftershock, earthquake swarm, and foreshock all have their interval, which is 0.75-0.84, 0.63-0.69, and 0.41-0.50, respectively, as listed in Tables 16.2-16.4.

Table 16.1. D_t values of three years before Shongpan (in China) Earthquake (1976) $m = 7.2$ and Dengkou Earthquake $m = 6.3$ (Li, 1990).

Earthquake	Statistical region	m	1973	1974	1975
Aug. 16, 1976 Shongpan 7.2 (China)	32-33°20′N 103°20′-104°30′E	$m > 2.0$	0.97 ± 0.04	0.41 ± 0.31	0.45 ± 0.03
Sept. 23, 1976 Dengkou 6.3 (China)	39-40°30′N 105°40′-106°50′E	$m > 1.0$	0.62 ± 0.03	0.50 ± 0.05	0.43 ± 0.07

Table 16.2. D_t values of aftershock of some major earthquakes (Li, 1990).

Mainshock		Fractal dimension of aftershock time			
Time	region	m	Δt (day)	N	D_t
March 8, 1966	Xingtai	7.2	14	503	0.78 ± 0.04
Feb. 4, 1975	Haicheng	7.3	3	772	0.78 ± 0.05
May 29, 1976	Longling	7.6	2	129	0.84 ± 0.09
July 28, 1976	Tangshan	7.8	2	984	0.82 ± 0.06
Aug. 16, 1976	Shongpan	7.2	6	296	0.78 ± 0.06

Table 16.3. D_t values of some earthquake swarm (Li, 1990).

Time	Earthquake swarm area	Time-fractal m_{max}	N	D_t
Oct. 14, 1967-Oct. 22, 1967	Zhuoluo of Herbei	4.2	158	0.66 ± 0.01
Oct. 18, 1971-Nov. 20, 1971	Gucheng of Hubei	4.0	321	0.67 ± 0.07
Jan. 9, 1972-Feb. 29, 1972	Xiaohongsan of Wuchang	2.2	173	0.63 ± 0.04
Dec. 21, 1974-Jan. 31, 1975	Qinwaq of Liaoning	5.2	185	0.64 ± 0.07
Feb. 5, 1976-March 9, 1976	Neixiang of Henan	4.0	211	0.69 ± 0.07
Feb. 4, 1976-April 5, 1976	Changdao of Shandong	4.1	277	0.69 ± 0.03
March 6, 1979-March 29, 1979	Wujiahe of Neimeng	1.2	73	0.67 ± 0.03
Oct. 28, 1987-Nov. 1, 1987	Maqin of Qinghai	2.1	139	0.63 ± 0.05

Table 16.4. D_t values of some foreshock (Li, 1990).

Foreshock time	Site	N	D_t
Jan. 2, 1966-March 8, 1966	Xingtai	47	0.49 ± 0.03
Feb. 3, 1975-April, 1975	Haicheng	36	0.41 ± 0.04
Oct. 5, 1975-May 29, 1976	Longling	431	0.49 ± 0.05

If an epicenter is regarded as a 'point source', the epicenter distribution is similar to the Sierpinski carpet. D_s values of epicenter distribution of earthquakes $m \geq 2.0$ during 1970-1986 for the Gan-Ning-Qing (in China) region and its eastern subregion are listed in Table 16.5. It is seen that, from Table 16.5, the D_s value (about 1.5) is very close to that in different time intervals. The Shongpan earthquake $m = 7.2$ occurred in this region in 1976. In that year, the D_s values were 0.58 (large region) and 0.52 (subregion). They were clearly lower than normal value. Three years ago (1973-1975) before the Shongpan (China) earthquake occurred, the D_s value of the large region was obviously lower ($D_s = 0.66$).

Now let us return to the discussion of the regional magnitude-frequency relationship (Equation (16.1)). The results obtained from various regions suggested that the *b*-value is likely to remain nearly constant at the same tectonic region and in the case of the statistical time field longer. But the *a*-value in Equation (16.1) varies with statistical time field and *a* is an implicit function of time and not a constant. Hence, Equation (16.1) is not ideal for reflecting the similarity of the earthquake quantity law of distribution. Using the concept of annual rate of earthquake occurrence and taking statistical time field Δt, Li (1990) has obtained a better expression. He defined

$$\bar{A} = a - \log \Delta t;$$

and

$$\bar{N}(m) = N(m)/\Delta t.$$

Then the magnitude-average annual relational expression of frequency can be written by

$$\log \bar{N} = \bar{A} - bm. \tag{16.17}$$

The study of the Gan-Ning-Qing region (in China) has shown that $\bar{A}$ and *b* parameters are constants in the same seismic belt, and not relative to the time field determined. For example, the statistical *b* and $\bar{A}$ values of the northern part of the Southnorthern seismic belt and Hexi corridor seismic belt are identical for historical time (1220 years and 345 years) and recent earthquakes (17 years) of a different level, but the *a*-value has large differences as listed in Table 16.6. This means that Equation (16.17) can reflect the magnitude-frequency correlation well for different statistical time fields. This is also an alternative embodiment of the fractal nature on the earthquake quantity law of distribution in a seismic belt.

Table 16.5. D_s values of different range field in Gan-Ning-Qing (China) large region and its eastern subregion (Li, 1990).

Time ($m \geq 2.0$)	D_s (L97°-107°E)	D_s (S103°-107°E)
1970-1972 (3 years)	1.50 ± 0.027	1.60 ± 0.01
1973-1975 (3 years)	0.66 ± 0.05	1.44 ± 0.15
1976 (1 year)	0.58 ± 0.08	0.52 ± 0.10
1970-1975 (6 years)	1.30 ± 0.10	1.59 ± 0.12
1976-1981 (6 years)	1.35 ± 0.10	1.62 ± 0.11
1982-1986 (5 years)	1.45 ± 0.15	1.53 ± 0.09
1970-1986 (17 years)	1.30 ± 0.06	1.55 ± 0.01

Table 16.6. Magnitude-frequency statistical results (Li, 1990).

Seismic zone	Statistical magnitude	Time range	Δt (year)	b	$\bar{A}$	a
Northern segment	$m \geq 6.5$	734-1954	1220	0.77	3.45	6.54
of S-N seismic zone	$m < 6.5$	1970-1986	17	0.77	3.49	4.72
Seismic zone of	$m \geq 6.5$	1609-1954	345	0.74	3.22	5.76
Hexi (in China)	$m < 6.5$	1970-1986	17	0.74	3.17	4.40

16.3 A FRACTAL APPROACH TO THE CLUSTERING OF EARTHQUAKES

If the occurrence of each earthquake was totally uncorrelated with other earthquakes, then the distributions of earthquakes would be a random process. Thus, the random fractals discussed in Chapter 4 could be used to model these clustering earthquakes. Smalley, et al. (1987) proposed a random fractal approach to quantify the temporal and spatial clustering of earthquakes. Earthquake epicenters can be considered to be point events in space and time. To study fractal clustering, the distribution of events over a wide range of scales should be examined. Thus, fractal temporal clustering of earthquakes is more representative since timing of events and the total length of time provide quite a wide range of scales for analyses. In this section, therefore, we focus on introducing the temporal fractal clustering of earthquakes. For spatial fractal clustering of earthquakes occurring at random, a similar analytical procedure can be adopted. Some regional seismicity in which the distribution described as nets, swarms, and clusters are often observed, but the distributions are correlated to the main shock; that is, they are not random, can be simulated by normal fractal concepts (uniform fractals).

The basic idea of a fractal approach to clustering earthquakes proposed by Smalley et al. is that if the seismic events are taken as a 'point process' on the time axis, then the seismicity can be simulated by a random Cantor set. Consider the temporal distribution of N earthquakes occurring in a time interval τ_0. The number of earthquakes is determined by the area of the region considered and the magnitude cut-off. For practical considerations, the magnitude cut-off is made to insure that all earthquakes in the region are detected by the regional network. The interval τ_0 is the interval over which data are collected and the natural period for the data is τ_0/N. The interval τ_0 can be divided into a series of similar intervals (subintervals), τ, defined in the form

$$\tau = \tau_0/n, \quad n = 2, 3, 4, \ldots \tag{16.18}$$

The measure of clustering will be the fraction $x(\tau)$ of the intervals in which an earthquake occurs as a function of the interval length τ.

As a specific example, consider a uniform (equally spaced in time) series of earthquakes. For this example,

$$x = \begin{cases} \tau N/\tau_0 & \text{if} \quad \tau < \tau_0/N \\ 1 & \text{if} \quad \tau > \tau_0/N \end{cases}. \tag{16.19}$$

If the number of earthquakes is greater than the number of intervals, then an earthquake occurs in every interval and $x = 1$. If the number of earthquakes, N, is less than the number of intervals, n, then $x = N/n$.

If the fraction x of the intervals of length τ containing earthquakes is related to the time interval by $x \sim \tau^{1-D}$, then fractal clustering occurs with fractal dimension $D(0 < D < 1)$.

The Cantor set discussed in Chapter 2 is scale-invariant, highly ordered, and deterministic. A scale-invariant random set as discussed in Chapter 4 is easily generated by randomly removing the first, middle, or last third of each closed interval instead of the middle third in the construction of the classical Cantor set, as illustrated in Figure 16.16. The result is scale-invariant fractal clustering that is also nondeterministic. For

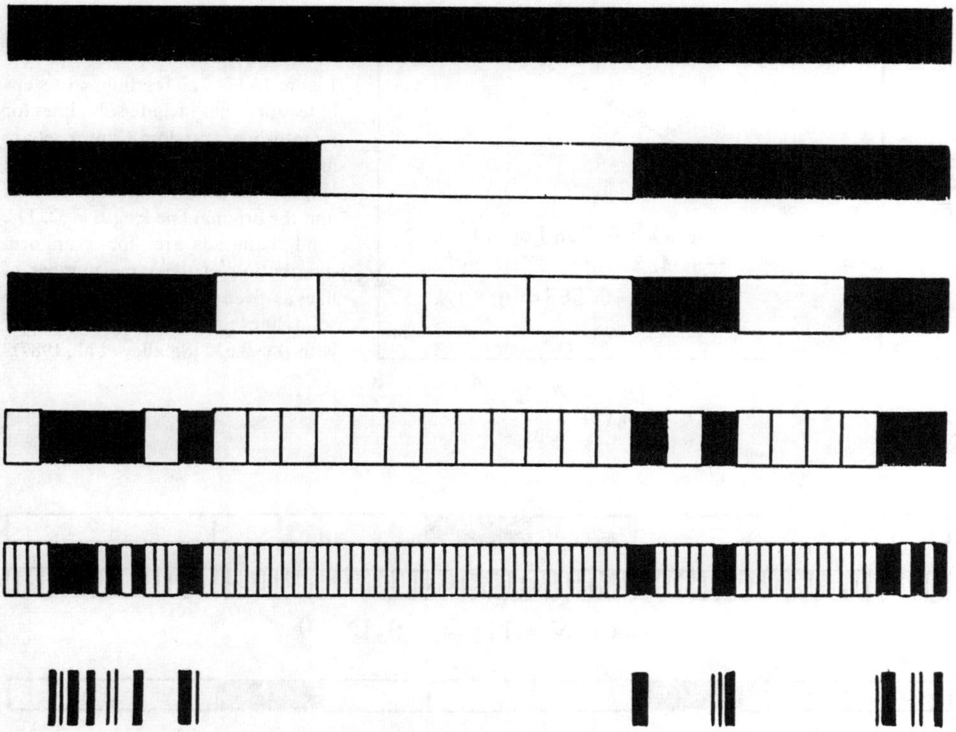

Figure 16.16. Illustration of the first six orders of a random Cantor set. At each step, a random third of each solid line is removed (Smalley et al., 1987).

actual applications of clustering, the number of events is finite. This means the order of generation is not required to approach infinity. The number of events associated with the order m of construction is 2^m. The approach to the fractal analyses of clustering proposed by Smalley et al. is replotted in Figure 16.17, showing the results for a random Cantor set of order 9. First, rescaling the length, so that the width of an element of the truncated random Cantor set is unity, yields a total length of 3^9, and contains 2^9 elements. To measure the fractal dimension by the box method (see Chapter 2), intervals of length $r = 2^n$ are taken, and the fraction x that includes at least one element is determined. This fraction is given as a function of the interval length by the open squares in Figure 16.17. The best-fit straight line has a slope of 0.368 so that $x \propto r^{0.368}$ and $D = 0.632$. If the same number of lines is uniformly distributed across the interval (no clustering), the probability of finding a line within an interval from Equation (16.19) is given by the solid diamonds in Figure 16.17. In this case, the slope is unity for $r < (\frac{3}{2})^9$ and zero for $r > (\frac{3}{2})^9$. Thus, $D = 0$ for $r < (\frac{3}{2})^9$, i.e., a set of isolated points, and $D = 1$ for $r > (\frac{3}{2})^9$, i.e., a line. To illustrate the relationship between clustering and the fractal dimension D, the first two iterations in the construction of a series of Cantor-like sets with increasing D are indicated in Figure 16.18. In this series, the closed unit interval [0,1] is divided into nine equal intervals of length, $r = \frac{1}{9}$, and at each iteration in the construction of the set, M of these subintervals is removed (note

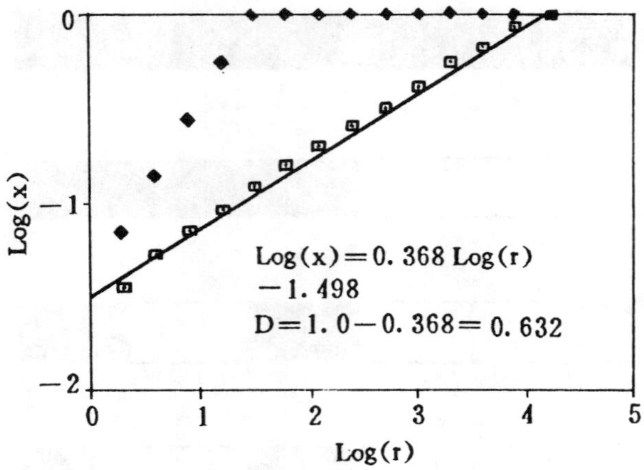

Figure 16.17. The fraction x of steps of length r that include solid lines for a ninth-order random Cantor set are given by the open squares. The unit length is the length of the shortest line, and the original line length is 3^9. The solid diamonds are for a uniform distribution of the same number of lines as given by Equation (16.19). The solid line corresponds to $x_i \propto r_i^{1-D}$ with $D = 0.632$ (Smalley et al., 1987).

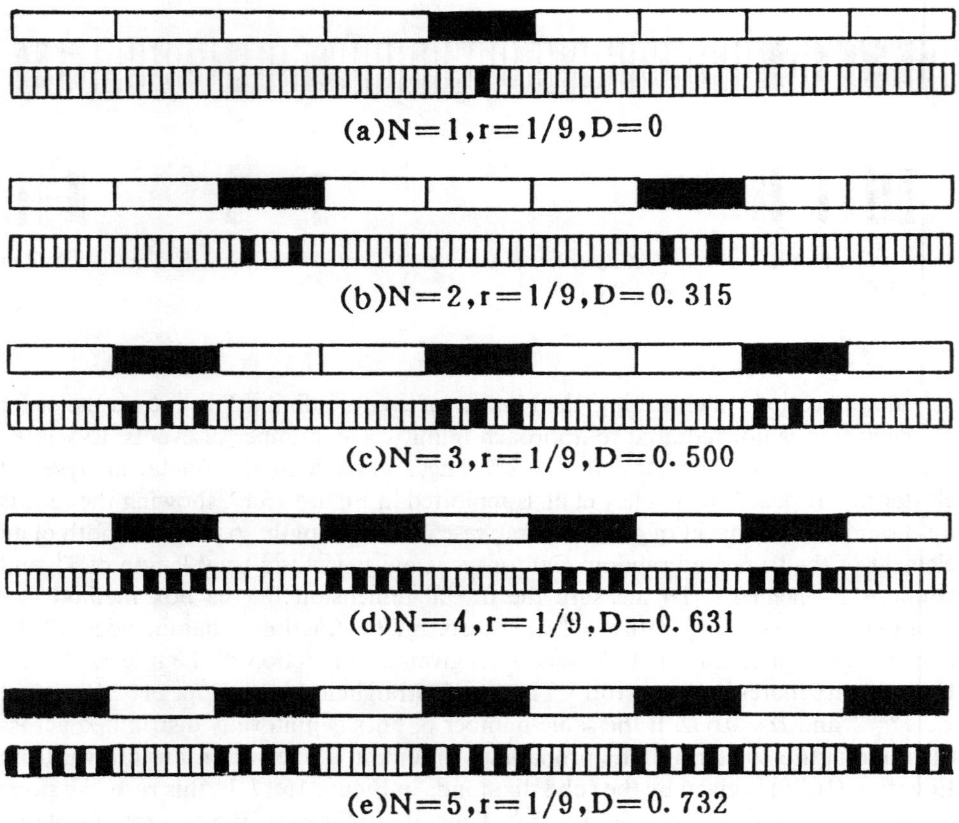

Figure 16.18. Illustration showing the relationship between clustering and fractal dimension. Clustered fractal sets with different values of D are constructed in an iterative process similar to that used for the Cantor set. The original solid line is divided into nine parts, and from one (a) to five (e) closed segments are retained at each iteration (e.g., in the first iteration in (a), the intervals $[0,\frac{1}{9}]$, $[\frac{1}{9},\frac{2}{9}]$, $[\frac{2}{9},\frac{3}{9}]$, $[\frac{3}{9},\frac{4}{9}]$, $[\frac{5}{9},\frac{6}{9}]$, $[\frac{6}{9},\frac{7}{9}]$, $[\frac{7}{9},\frac{8}{9}]$, and $[\frac{8}{9},1]$ are removed, leaving the closed interval $[\frac{4}{9},\frac{5}{9}]$). The corresponding values of D from Equation (2.28) are also given (Smalley et al., 1987).

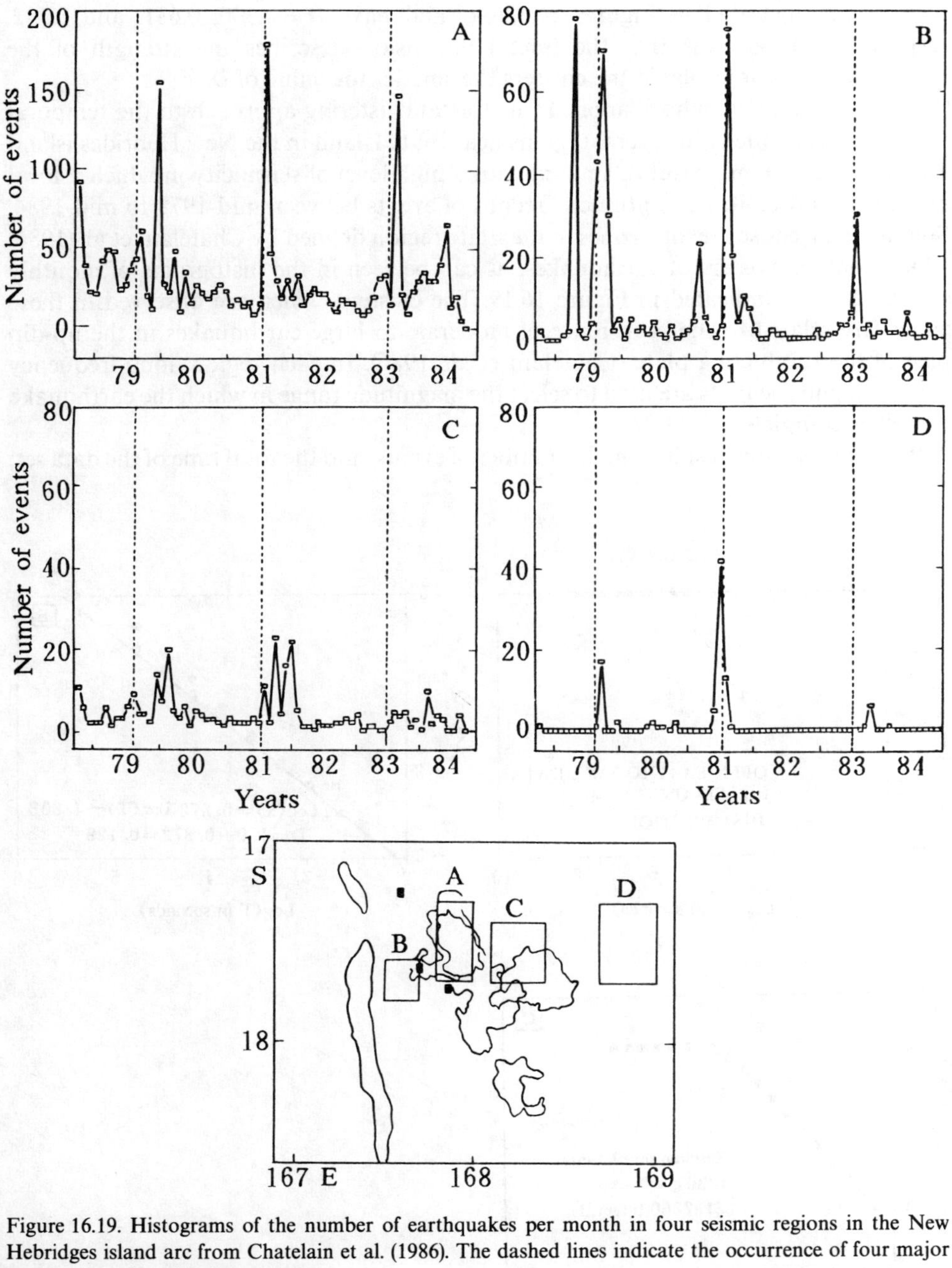

Figure 16.19. Histograms of the number of earthquakes per month in four seismic regions in the New Hebridges island arc from Chatelain et al. (1986). The dashed lines indicate the occurrence of four major earthquakes (Smaller et al., 1987).

to carefully make the choice of open or closed ends on the intervals). In Figure 16.18a, $M = 8$ intervals are removed, $N = 1 = 9 - M$ corresponding to a point, and from Equation (2.28), $D = 0$. In Figure 16.18b, $M = 7$ intervals are removed, $N = 2$, and from Equation (2.28), $D = 0.315$. Sets with $M = 6$, 5, and 4 intervals removed, or $N = 3$, 4, and 5, are illustrated in Figures 16.18(c-e) and have $D = 0.500$, 0.631, and 0.732, respectively. It is clear that the fractal dimension describes the strength of the clustering; the more isolated the clusters the smaller the value of D.

Smalley et al. (1987) have applied this fractal clustering approach to the temporal variation of seismicity in several regions near Efate Island in the New Hebrides island arc (mapped in Figure 16.19), an area with a high level of seismicity in which a local network has provided a continuous record of events between mid-1978 to mid-1984. Smalley et al. chose the four zones in the Efate region defined by Chatelain et al. (1986) which contain clusters of earthquakes, as can be seen in the histograms of monthly seismic activity indicated in Figure 16.19. The clusters, which are observed in these zones, are related to the occurrence of moderate to large earthquakes in the up-dip part of the subducting plate (Chatelain et al., 1986). In each region, long frequency versus magnitude plots are used to select the magnitude range in which the earthquake sample is complete.

Based on the time resolution, the number of events, and the total time of the data set,

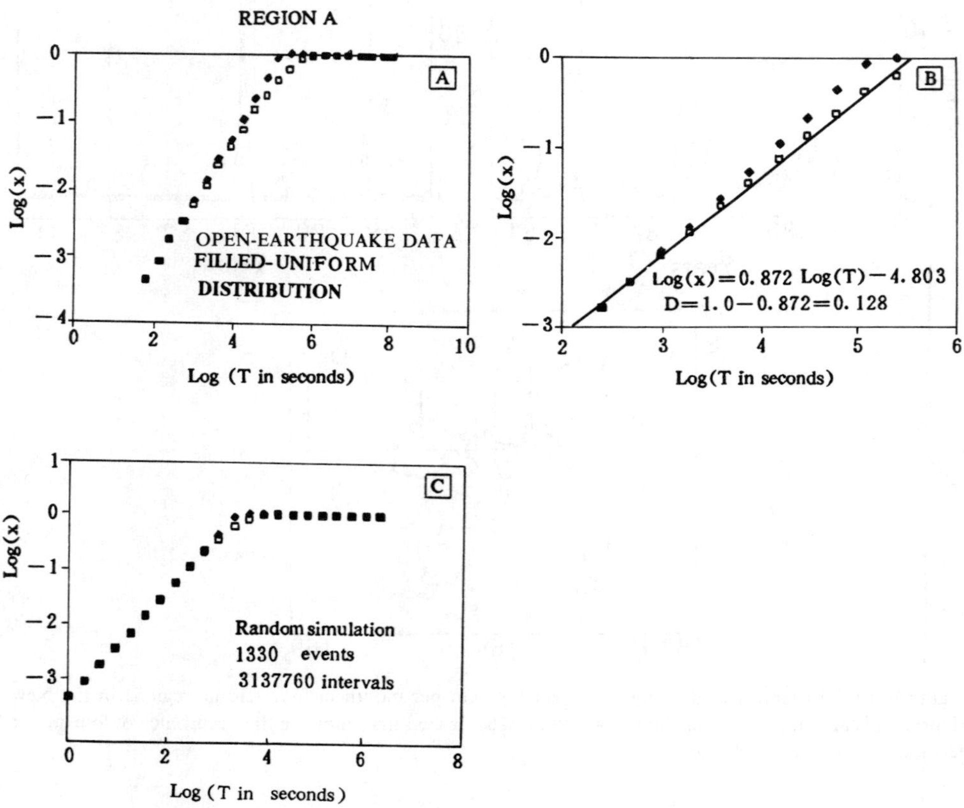

Figure 16.20. (A) The open squares are the fraction x of time intervals of length t that contain earthquakes in region A. The solid diamonds are for a uniform distribution of 1.33 earthquakes as given by Equation (16.19). (B) The best fit of the data from (A) with $x \propto \tau^{1-D}$ in the range $4 < \tau < 4096$ min. The derived fractal dimension is $D = 0.128$. (C) Open squares are a random simulation of 1330 events for the Poisson process (Smalley et al., 1987).

Smalley et al. (1987) have chosen the smallest time interval to be 1 min ($\tau = 60$ sec.). This interval of time is increased by factors of 2 up to $2^{21} \sim 2 \times 10^6$ min. The total time interval for which data is available is about 3×10^6 min. For each zone, the fraction of intervals that include an earthquake, as a function of the interval size, is plotted, and the fractal dimension is determined from $D = 1 - s$ (s is the slope of the plot of $\log x$ vs. $\log T$) (Smalley et al., 1987) as shown in Figures 16.20-16.23.

For region *A* shown in Figure 16.19, 1330 events occurred in the magnitude window. The results of the fractal analysis of these earthquake data are indicated in Figure 16.20A. For $\tau \leq 4$ min, only single event occur in each time interval, and no clustering is observed. For $\tau \geq 32\,768$ minutes $= 22.8$ days, every interval contains an earthquake. For values of τ between these limits, a fractal clustering is obtained which gives a fractal dimension of $D = 0.128$, corresponding to a slope of 0.872 (Figure 16.20B, given by Smalley et al.). Figure 16.20C gives simulation results of 1330 events in the time interval studied by Smalley et al. It is seen that the simulation is significantly different from the earthquake data and close to the uniform distribution. The fractal analyses results for regions *B*, *C*, and *D* obtained by Smalley et al. are illustrated in Figures 16.21A-16.23A. Figures 16.21B-16.23B show the results for time intervals with scale-invariant clustering, 16 min $\leq \tau \leq 16\,786$ min for region *B*, 16 min $\leq \tau \leq 32\,768$ min for region *C*,

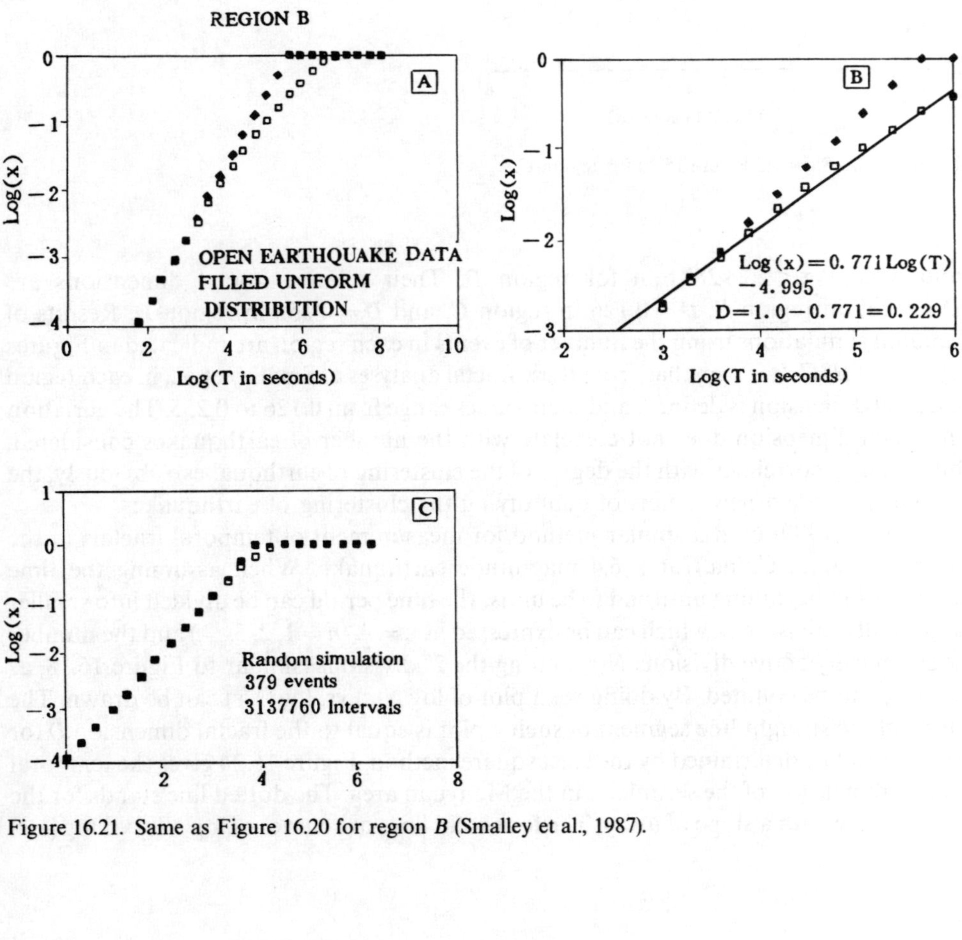

Figure 16.21. Same as Figure 16.20 for region *B* (Smalley et al., 1987).

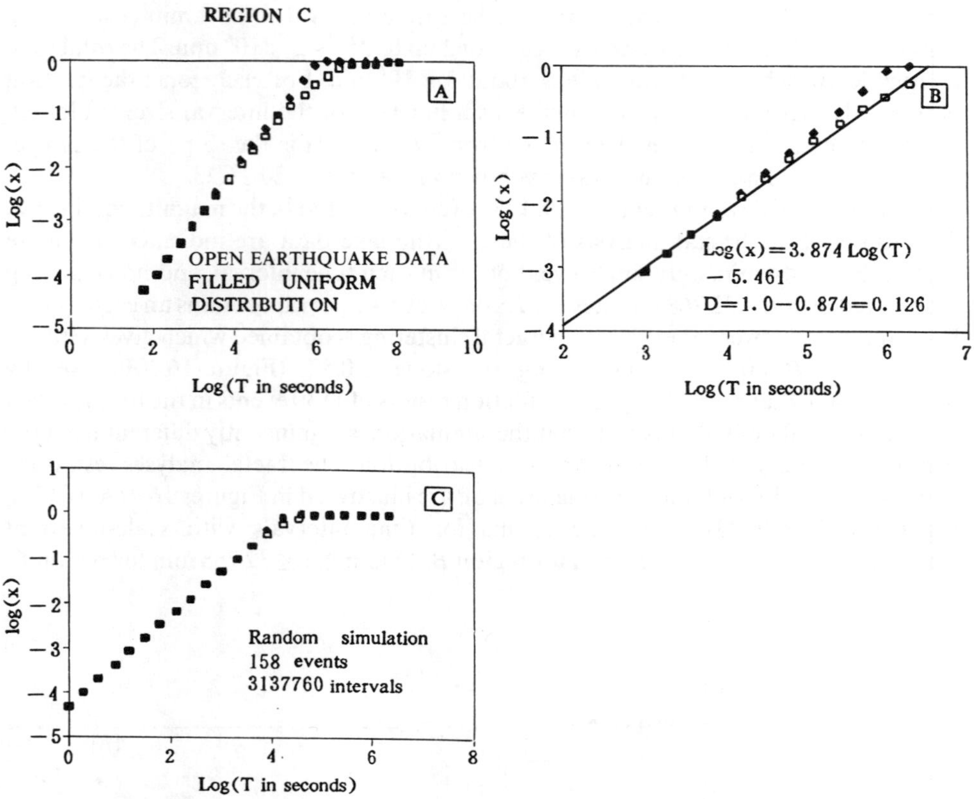

Figure 16.22. Same as Figure 16.20 for region *C*.

and $8\ \text{min} \leq \tau \leq 524288\ \text{min}$ for region *D*. Their relative fractal dimensions are $D = 0.239$ in region *B*, $D = 0.126$ in region *C*, and $D = 0.255$ in region *D*. Results of random simulations using the number of events in each region are indicated in Figures 16.21C-16.23C. It is seen that, from these fractal analyses of Smalley et al., in each region a fractal dimension is defined, and their values range from 0.126 to 0.255. The variation in fractal dimension does not correlate with the number of earthquakes considered, but strongly correlates with the degree of the clustering of earthquakes. Obviously, the fractals provide a new means of quantifying the clustering of earthquakes.

Li et al. (1989) used a similar method for measurement of temporal fractals in the Menyuan area, China, for a 6.4 magnitude earthquake. When assuming the time chosen from beginning until end to be unity, the time period can be divided into smaller and smaller divisions ε which can be expressed as $\varepsilon = \frac{1}{2^n}$ $(n = 1, 2, 3, \ldots)$ and the number of seismically active divisions $N(\varepsilon)$ among the 2^n divisions (similar to Figure 16.18, as $2^n \to 3^n$) can be counted. By doing so, a plot of $\log N(\varepsilon)$ vs. $\log(1/\varepsilon)$ can be drawn. The slope of the straight line segment of such a plot is equal to the fractal dimension D (or D_t), and can be determined by the least square method. Figure 16.24 gives the temporal fractal dimension of the seismicity in the Menyuan area. The dotted line stands for the straight line with a slope of unity. The four solid lines are curves of $\log N(\varepsilon)$ vs. $\log(1/\varepsilon)$

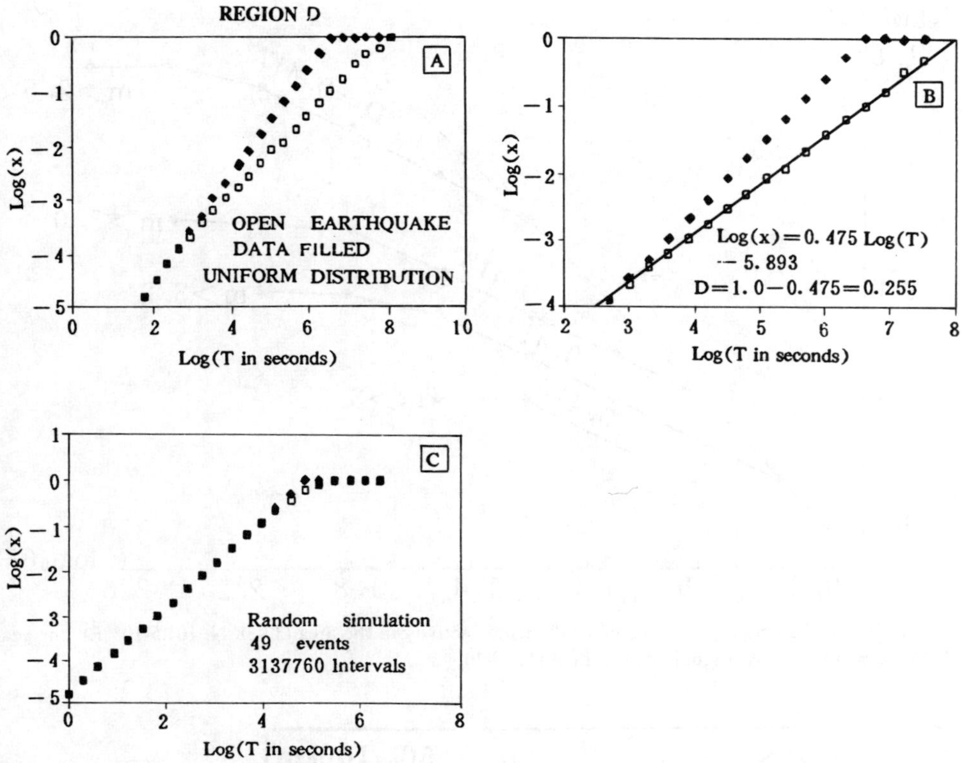

Figure 16.23. Same as Figure 16.20 for region *D* (Smalley et al., 1987).

with different lower limits of magnitude in the same temporal and spatial ranges. From this figure, it is seen that each of the four curves can be divided into three segments: two straight segments with their slopes equal to 1 and to a fraction, respectively, and a curved segment with a slope approaching zero. Among them, the first represents that $N(\varepsilon)$ and (ε) are in a strict reciprocal relationship. The last shows that $N(\varepsilon)$ changes little with (ε), and that N is tending to be saturated. The second straight segment with the slope equal to a fraction is a segment where the fractal exists. Figure 16.25 illustrates the distribution of seismic stations in the Menyuan area, China, which is the area of lat.37-38°N and long. 101.5-103°E, including Gulang and Jiutiaoling. Li et al. (1989) focused on the study of 16 years seismic data from 1970 to 1985 (one year before a main earthquake of magnitude 6.4). The fractal dimension values are calculated for the four time divisions of 1 year, 4 years, 8 years, and 16 years as shown in Figure 16.26.

A total of 442 earthquakes with $m \geq 2.0$ which occurred in the Menyuan area during 1970-1985 is plotted in Figure 16.26a. There is a fractal segment with a D value of 0.580 (correlation coefficient $r = 0.98$). The 16 year period is divided into two 8-year divisions (1970-1977 and 1978-1985) as shown in Figures 16.26b and c. In the first division (1970-1977) the temporal distribution of the seismicity has no fractal structure (see Figure 16.26b). In the second division (1978-1985) a D value of 0.540 exists ($r = 0.99$). This indicates that although the temporal distribution of seismicity of the entire 16-year

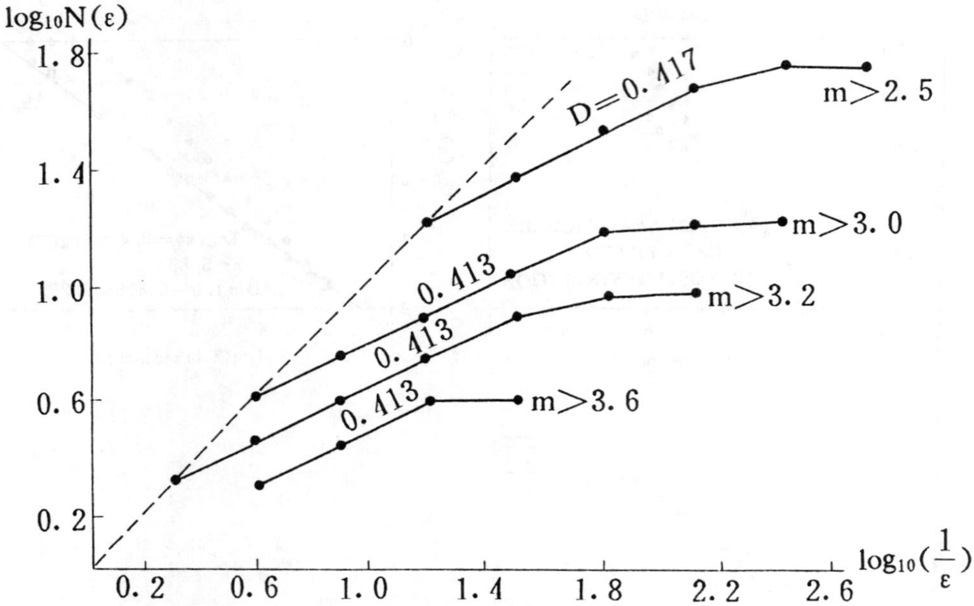

Figure 16.24. The temporal fractals of small seismic activity in the area (37-38°N, 101.5-103°E), one year before the Menyuan earthquake of 6.4 (Li et al., 1989).

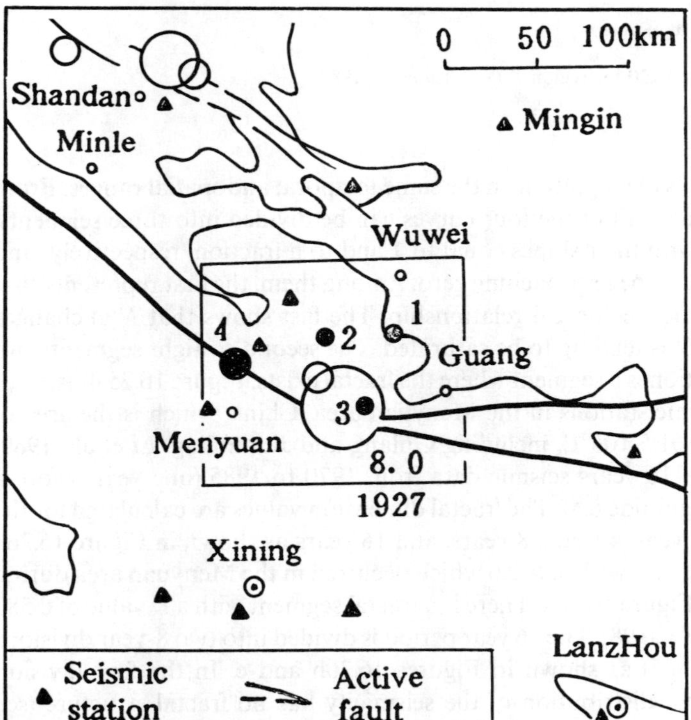

Figure 16.25. The distribution of seismic stations in the Menyuan area and its vicinity (Li et al., 1989).

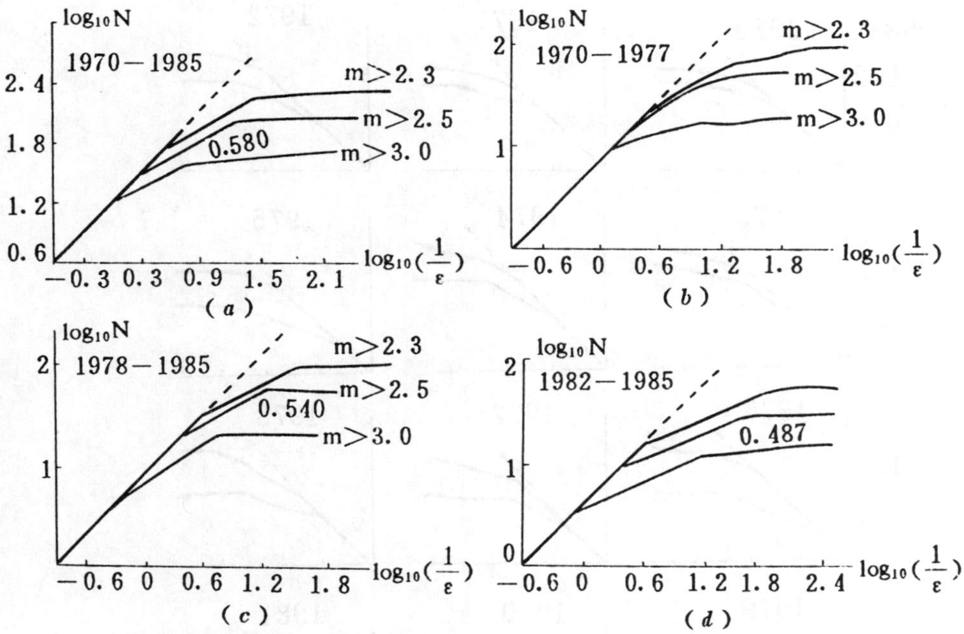

Figure 16.26. The temporal distribution structure in the Menyuan area in different phases from 1970 to 1985. (a) 1970-1985; (b) 1970-1977; (c) 1978-1985; (d) 1982-1985 (Li et al., 1989).

period may have a fractal dimensional structure, a part of it may not. Inspection of the four 4-year divisions shows that a D value exists only in the last division (1982-1985) as indicated in Figure 16.26d. Its fractal dimension is about 0.487 ($r = 0.996$). The first three divisions have no fractal dimensional structures. Figure 16.27 shows the year-by-year curves from 1970 to 1984. It is seen that, from this figure, although certain individual years (e.g., 1981, 1983, and 1984) show some tendency of having a D-value, it is not satisfactory on the whole. Furthermore, identical results are obtained by regarding the second half of a year and first half of the next year as a one-year time division. However, in the division nearest to the strong earthquake, i.e., from July 1984 to 1985, a D-value of 0.420 ($r = 0.99$) exists (see Figure 16.24).

From the results obtained by Li et al., it is concluded that since seismic events are a point process discontinuous everywhere on the time axis, the D-value has a value close to unity over a certain period of time only when the temporal distribution of seismicity is not a clustering, but is homogeneous or symmetrical. Therefore, if a D-value less than one exists for a reliable earthquake sequence, the temporal distribution of this sequence must be clustering. The D-values of the above four sets, Viz., 16-year, 8-year, 4-year and 1-year sets, are 0.580, 0.540, 0.487, and 0.420, respectively, which shows a gradually decreasing trend. This indicates that the more clustering or more heterogeneous the temporal structure of seismicity in the area becomes, the closer it is to the main earthquake. It can also be seen from these results that with the nearing of the occurrence of a strong earthquake, the D-value not only exists but also shows a tendency to gradually decrease with time. This conclusion is

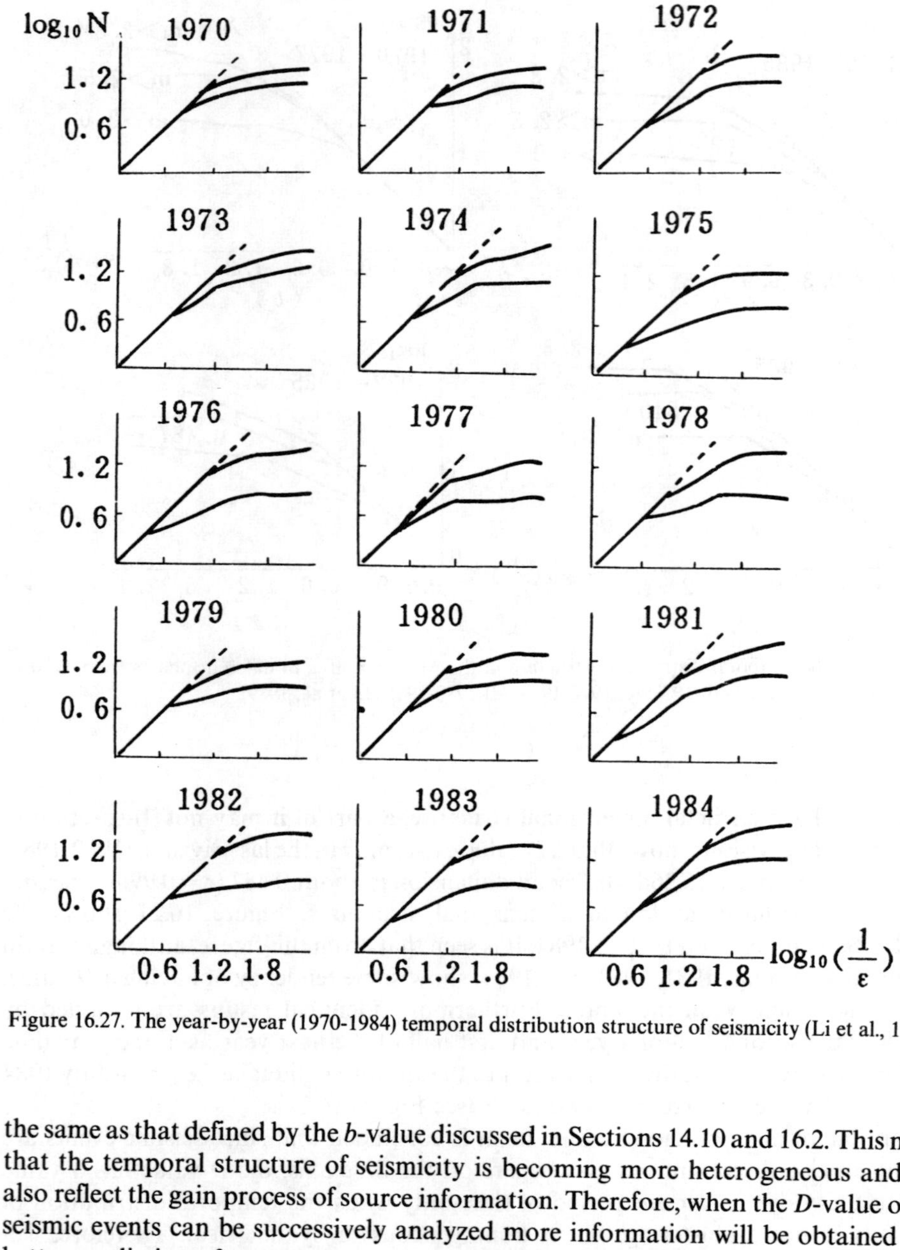

Figure 16.27. The year-by-year (1970-1984) temporal distribution structure of seismicity (Li et al., 1989).

the same as that defined by the *b*-value discussed in Sections 14.10 and 16.2. This means that the temporal structure of seismicity is becoming more heterogeneous and may also reflect the gain process of source information. Therefore, when the *D*-value of new seismic events can be successively analyzed more information will be obtained for a better prediction of a strong earthquake. This 'step-by-step' or 'follow-up' method of analyses should be adopted in the prediction of earthquakes.

16.4 A FRACTAL MODEL FOR CRUSTAL DEFORMATION

King (1983) has applied the fractal concepts to triple junctions where three strike-slip faults intersect. He indicated that a fractal distribution of faults yielded the magnitude-frequency relationship given in Equation (16.1). Turcotte (1986b) extended this work and formulated a fractal model for the continuum deformation of the crust.

Turcotte hypothesized a scale-invariant matrix of vertical strike-slip faults and assumed that all deformations were in the horizontal plane. The basic fault pattern proposed by Turcotte is a symmetric triple junction between $N_{f1} = 3$ first-order, strike slip faults of length h_1 as shown in Figure 16.28a. The depth of each first-order fault is also taken to be h_1. This triple junction forms the interior of a hexagon as shown; the hexagon is basic building block (or generator). The relative velocity across each of the first-order faults is V_1. If all the faults are right lateral (or left lateral), the displacements are instantaneously compatible but will result in a deformation at the triple junction. After assuming complete symmetry, i.e. an infinite set of similar hexagons, the distance between parallel faults is $W_1 = 3^{1/2} h_1$. Neglecting complex two-dimensional interactions, the earthquake displacement, δ_1, on a first-order fault is related to the stress drop σ_0 by

$$\delta_1 = \sigma_0 W_1/\mu = 3^{1/2}\sigma_0 h_1/\mu. \tag{16.20}$$

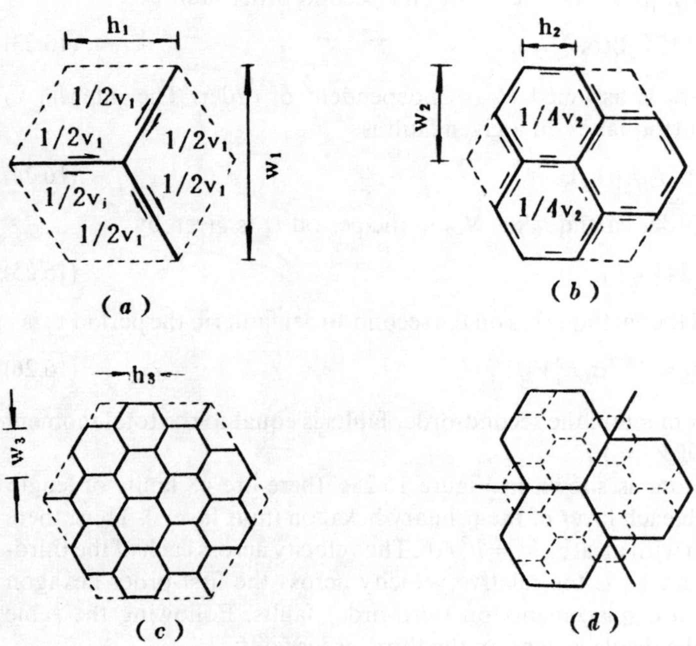

Figure 16.28. Illustration of the fractal model for crustal deformation. (a) Three first-order, strike-slip faults of length h_1 (solid lines) form a symmetric triple junction. The relative velocity across each fault is V_1. (b) Twelve second-order, strike-slip faults of length $h_2 = h_1/2$ (solid lines) form a series of symmetric triple junctions. There are actually two layers of faults, each with a depth $h_1/2$ so that the total number of second-order faults is 24. The relative velocity across each fault is $V_2/2$. (c) Forty-eight third-order, strike-slip faults of length $h_3 = h_1/4$ (solid lines) form a series of symmetric triple junctions. There are actually four layers of faults, each with a depth $h_1/4$ so that the total number of third-order faults is 192. The relative velocity across each faults is $V_3/4$. (d) The pattern of first, second, and third-order faults is illustrated (Turcotte, 1986b).

Thus the time interval τ_1 between first-order earthquakes is

$$\tau_1 = \delta_1/V_1 = 3^{1/2}\sigma_0 h_1/(\mu V_1), \tag{16.21}$$

and there will be $N_{e1} = 3$ earthquakes in this interval. The total moment of the three earthquakes is:

$$M_1 = 3\mu h_1^2 \delta_1 = 3^{3/2}\sigma_0 h_1^3. \tag{16.22}$$

In this idealized model given by Turcotte, the three earthquakes on the first-order faults are assumed to occur simultaneously in order to maintain geometrical compatibility. As a fractal construction, the set of second-order hexagons with side length $h_2 = h_1/2$ is indicated in Figure 16.28b. Thus, there are 12 faults of length h_2. In order to maintain scale-invariance, the second-order hexagons with a depth h_2 are assumed so that there are two layers of second-order hexagons. Thus, there are $N_{f2} = 24$ faults each with an area $h_2^2 = h_1^2/4$. Note that the assumption of $h_2 = h_1/2$ appears to be arbitrary and the results are basically independent of this choice. The relative velocity across each of the second-order faults is $V_2/2$, where V_2 is the relative velocity across the first-order hexagon that is attributed to displacements on the second-order faults. When each fault is taken to be right lateral, following the same procedure as for the first-order faults, the earthquake displacement on a second-order fault is:

$$\delta_2 = 3^{1/2}\sigma_0 h_2/\mu = (3^{1/2}/2)(\sigma_0 h_1/\mu), \tag{16.23}$$

where the stress drop σ_0 is assumed to be independent of order. The interval τ_2 between second-order earthquakes on a given fault is:

$$\tau_2 = 2\delta_2/V_2 = 3^{1/2}\sigma_0 h_1/(\mu V_2). \tag{16.24}$$

The number of second-order earthquakes N_{e2} in the period τ_1 is given by

$$N_{e2} = N_{f2}\tau_1/\tau_2 = 24 V_2/V_1. \tag{16.25}$$

Similarly, the moment of the earthquakes on the second-order faults in the period τ_1 is

$$M_2 = 24\mu h_2^2 \delta_2 \tau_1/\tau_2 = 3^{3/2}\sigma_0 h_1^3 V_2/V_1. \tag{16.26}$$

As expected, the total moment on the second-order faults is equal to the total moment on the first-order faults if $V_2 = V_1$.

For the third generation, as shown in Figure 16.28c, there are 48 faults of length $h_3 = h_1/4$ associated with each layer of the primary hexagon (four layers). Thus, there are $N_{f3} = 192$ faults each with an area $h_3^2 = h_1^2/16$. The velocity across each of the third-order faults is $V_3/4$, where V_3 is the relative velocity across the first-order hexagon that can be attributed to displacements on third-order faults. Following the same procedure, the earthquake displacement on the third-order fault:

$$\delta_3 = 3^{1/2}\sigma_0 h_3/\mu = (3^{1/2}/4)(\sigma_0 h_1/\mu). \tag{16.27}$$

The interval between third-order earthquakes is

$$\tau_3 = 4\delta_3/V_3 = 3^{1/2}\sigma_0 h_1/(\mu V_3). \tag{16.28}$$

The number of third-order earthquakes in the period τ_1 is given by

$$N_{e3} = N_{f3}\tau_1/\tau_3 = 192 V_3/V_1. \tag{16.29}$$

and the total moment of the earthquakes on the third-order faults in the period τ_1 becomes

$$M_3 = 192\mu h_3^2 \delta_3 \tau_1/\tau_3 = 3^{3/2}\sigma_0 h_1^3 V_3/V_1. \tag{16.30}$$

Using such recursive procedures, the above results are easily generalized to the nth-order. There are $N_{fn} = 3 \times 2^{3(n-1)}$ faults each with an area $h_n^2 = h_1^2/2^{2(n-1)}$. The velocity across each of the nth-order faults is $(V_n/2^{(n-1)})$, where V_n is the relative velocity across the first-order hexagon that can be attributed to displacements on the *n*th-order faults. Thus, the earthquake displacements on *n*th-order faults are

$$\delta_n = 3^{1/2}\sigma_0 h_n/\mu = (3^{1/2}/2^{n-1})(\sigma_0 h_1/\mu). \tag{16.31}$$

The interval between *n*th-order earthquakes is

$$\tau_n = 2^{n-1}\delta_n/V_n = 3^{1/2}\sigma_0 h_1/(\mu V_n). \tag{16.32}$$

The number of *n*th-order earthquakes in the period τ_1 is given by

$$N_{en} = N_{fn}\tau_1/\tau_n = 3 \times 2^{3(n-1)} V_n/V_1, \tag{16.33}$$

and the total moment of the earthquakes on the *n*th-order faults in the period τ_1 is

$$M_n = N_{fn}\mu h_n^2 \delta_n \tau_1/\tau_n = 3^{3/2}\sigma_0 h_1^3 V_n/V_1. \tag{16.34}$$

If the number of earthquakes has a fractal distribution (as discussed in Section 16.2), letting $r = 1/2^{n-1}$, then

$$N_{en}/N_{e1} = (1/r)^D = (2^{n-1})^D = 2^{D(n-1)}. \tag{16.35}$$

A comparison of Equations (16.33) and (16.35) yields

$$V_n/V_1 = 2^{-(3-D)(n-1)}. \tag{16.36}$$

The total velocity V across the primary hexagon is the sum of the velocities at all orders and is defined by

$$V = \sum_{n=1}^{\infty} V_n = V_1 \sum_{n=1}^{\infty} 2^{-(3-D)(n-1)} = V_1/[1 - 2^{-(3-D)}]. \tag{16.37}$$

Thus,

$$V_n/V = (1 - 2^{-(3-D)})2^{-(3-D)(n-1)}. \tag{16.38}$$

The fraction of the total velocity V that is accommodated on the primary (first-order) fault is

$$V_1/V = 1 - 2^{-(3-D)}. \tag{16.39}$$

Clearly, if $D = 2$, then $V_1/V = \frac{1}{2}$ and one half of the total velocity on the fault system is on the primary faults. As D approaches 3 virtually all the displacements occur on small faults.

Turcotte (1986b) also applied this fractal approach to the San Andreas fault system in central California. From the *b*-value of this area, fractal dimension is determined to be $D = 1.90$. After assuming that the full relative velocity between the Pacific and North American Plates occurs on this system, the total velocity is taken to be $V = 5.5$ cm/yr. Thus, from Equation (16.39), the predicted relative velocity across the primary fault is $V_1 = 2.93$ cm/yr. The remainder of the relative motion is taken up on higher-order faults. In fact, the average slip-rate on the San Andreas fault at Wallace Creek on the Carrizo Plain of central California has been determined by Sieh & Jahns (1984). For the past 3700 years they give a value of 3.39 ± 0.29 cm/yr., which is in reasonably good agreement with the value of 2.93 cm/yr. predicted by Turcotte, using his fractal approach. As interval and magnitude prediction of earthquakes, Turcotte chose the June 27, 1966 Parkfield earthquake as an example. This earthquake occurred on the primary strand of the San Andreas so that it corresponds to an earthquake on a first-order fault in Turcotte's model. Trifunac & Udwadia (1974) have studied the Parkfield earthquake in considerable detail and given a mean fault displacement $\delta_1 = 120$ cm, a moment $M_1 = 4.4 \times 10^{25}$ dyne cm, and a shear modulus $\mu = 3 \times 10^{11}$ dyne/cm^2. From Equation (16.2) the fault area is $A_1 = 122$ km^2 so that $h_1 = 11$ km. Thus, from Equation (16.20), the corresponding stress drop is $\sigma_0 = 20 \times 10^6$ dyne/cm^2 which is quite a reasonable value. Taking $\delta_1 = 120$ cm and $V_1 = 3.39$ cm/yr, the time interval between earthquakes predicted from Equation (16.21) is $\tau_1 = 35.4$ (years). Bakun & McEvilly (1984) argue that previous earthquakes on the Parkfield section of the San Andreas fault occurred in 1857, 1881, 1901, 1922, and 1934. Thus, the actual interval between 1934 and 1966 is 32 years, which is in quite good agreement with the predicted interval 35.4 years. Using Equation (16.3) with $c = 1.5$ and $d = 16$, then $m = 6.43$, which is consistent with magnitude estimates for the 1966 earthquakes ranging from 5.3 to 6.5 ($m_b = 5.3$ USGS, $m_L = 5.5$ Berkeley, $m_L = 5.8$ Pasadena, $m_\sigma = 6.5$ Palisades). Turcotte also applied his fractal approach to the distribution of seismicity of faults of various orders. The studied results indicate that fractal dimension specifies the fraction of the total displacement that occurs on faults with a characteristic size greater than h; once the fractal dimension has been specified, the number of faults with a characteristic size greater than h is specified. Also, the number of earthquakes and the mean interval between earthquakes are specified.

A more interesting application of this fractal approach given by Turcotte is to dilatancy. As discussed above, the deformation of the hexagons is instantaneously compatible for infinitesimal deformations. For large deformations, however, the corners of the hexagons become offset. This can result either in the removal of material or in the creation of void space. For shallow levels in the crust, the creation of void

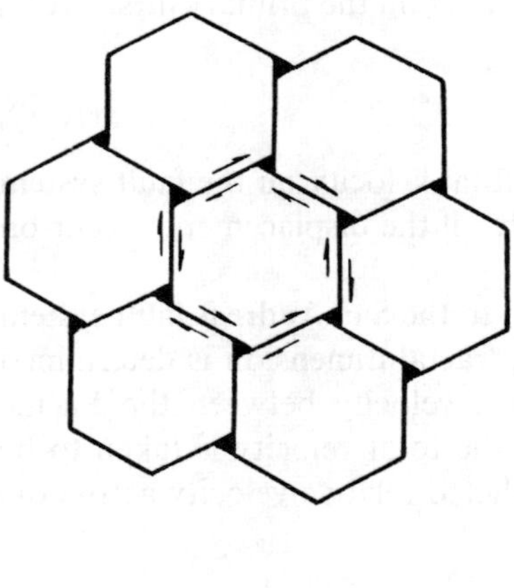

Figure 16.29. Illustration of the creation of void space (dilatancy) for right lateral strain with $\varepsilon_{xy} = 0.05$ (Turcotte, 1986b).

space many result in dilatancy. The creation of void space is illustrated in Figure 16.29 for right lateral strain with $\varepsilon_{xy} = 0.05$. A triangular void with an area $\delta^2/2$ is created at each corner of the hexagons. Thus, the volume dilatancy is

$$\Delta = 2(\delta^2/2)/(3^{1/2}/2 \cdot W^2) = (\tfrac{2}{3})^{1/2}/(\delta/W)^2. \tag{16.40}$$

This result is valid at all orders of deformation so that the dilatancy can be related to the total strain by

$$\Delta = (\tfrac{4}{3})^{1/2} \cdot \varepsilon_{xy}. \tag{16.41}$$

16.5 FRACTAL TOPOGRAPHY

Topography is often the result of displacements on faults. As discussed in previous sections, the distribution of earthquakes and faults is a fractal. Thus, it should not be surprising that the resulting topography is also fractal. In Chapters 2 and 3, some examples of fractal topography have been given.

Fractal dimensions for topography can be easily obtained in several ways. The first is, to directly use the measurement of fractal curve length discussed in Chapter 2 from topographic maps. Several examples are given in Figure 16.30. The length along specified contours P is obtained using dividers of different length r. The results generally satisfy Equation (2.7a, b), and a fractal dimension can be obtained. The second is the method of the Weierstrass-Mandelbrot fractal function. Several authors have shown that the Weierstrass-Mandelbrot fractal function discussed in Chapter 2 (Equation (2.18)) has a power spectrum $p(w)$ that varies approximately as $w^{-(5-2D)}$ in one-dimensional fractal topography and a variance of increments $V(t) = \langle [W(t_0) -$

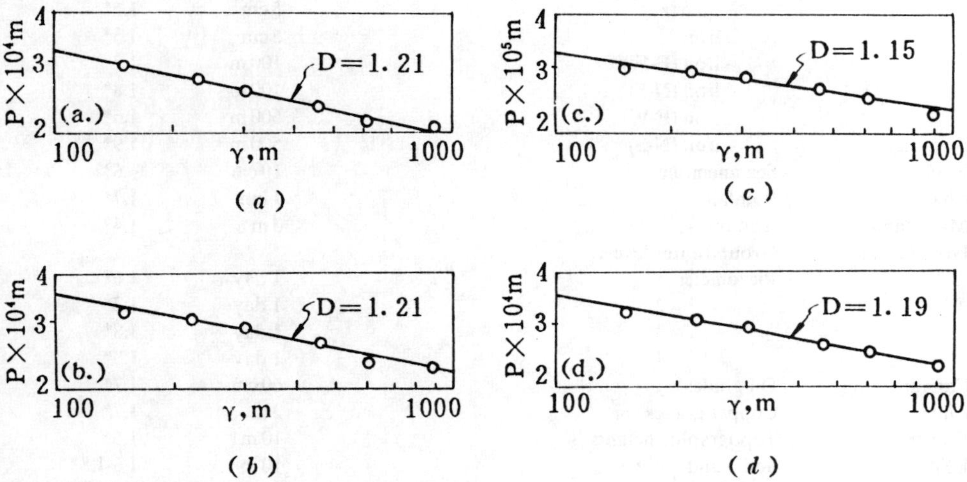

Figure 16.30. Fractal dimensions for specified topographic contours in several mountain belts: (a) 3000 ft. contour of the Cobblestone Mountain quadrangle, Transverse Ranges, California; (b) 5400 ft. contour of the Tatooh Buttes quadrangle Cascade Mountains, Washington; (c) 10000 ft. contour of the Byers Peak quadrangle, Rocky Mountains, Colorado; and (d) 1000 ft. contour of the Silver Bay quadrangle, Adirondack Mountains, New York (Turcotte, 1989).

$W(t_0 + t)]^2\rangle$ that varies as t^{4-2D} at the origin. These results allow us to estimate the fractal dimension D of a real series, either from the slope of the log-log plot of the power spectrum as $t \to 0$ (see Section 15.4, and William & Terry, 1991), or from

$$d \log V(t)/d \log t = 4 - 2D \quad (t \to 0).$$

Table 16.7. Estimated D values for various environmental series (Burrough, 1981).

Location	Property	Lag	D as lag$\to 0$
Wales	Soil – sodium content	15.2 m	1.7-1.9*
	– stone content	15.2 m	1.1-1.8*
	(both over four directions)		
England	Soil – thickness of cover loam	20 m	1.6*
England	Soil – electrical resistivity (4 directions)	1 m	1.4-1.6*
England	Surface of airport runway	30 cm	1.5+
Deserts in Africa and America	Soil – mean cone index	~1 km	1.9#
	– silt + clay in 0-15 cm layer	~1 km	1.8#
	– mean diameter of surface stones	~1 km	1.8#
	– coarse sand fraction in 0.15 cm layer	~1 km	1.8#
	Vegetation cover	~1 km	1.6#
Sourth Arica	Gold	Various	1.9+
Australia	Soil – phosphorus level	5 m	2.0#
	– pH	5 m	1.5#
	– potassium level	5 m	1.6#
	– bulk density	5 m	1.5#
	– 0.1 bar water	5 m	1.5#
France	Iron ore in rocks		
	– chlorite	15 μm	1.6*
	– quartz	15 μm	1.9*
	– quartz	5 cm	1.6*
	– iron	5 cm	1.5*
	– iron (E-W)	100 m	1.7*
	– iron (N-S)	100 m	1.8*
	– iron (E-W)	500 m	1.6*
	– iron (N-S)	500 m	1.9*
France	Sea anemone	10 cm	1.6$
Chad	Rainfall	1 km	1.7*
Mauritania	Iron ore	3 m	1.4*
Ivory Coast	Groundwater levels		
	Piezometer 1	1 day	1.6*
	2	1 day	1.7*
	3	1 day	1.8*
	4	1 day	1.3*
Canada	Oil grades	60 cm	1.7*
Chile	Copper grades	2 m	1.7*
France	Topographic heights	10 m	1.5*
USA	Soil – and content	10 m	1.6-1.8*
	– pH	10 m	2.0*
Worldwide	Crop yields	1-1,000 m	1.6-1.8#
India	Water table depth	250 m	1.6*

Note: *Estimated from variogram. #Estimated from block variance. +Estimated from power spectrum. $Estimated from covariance.

The variance of increments is often used in geostatistical studies where, computed over distances, it is referred to as the variogram. Computing the variogram is the first step in the interpolation procedure known as kriging, which is used to assist in the estimation of mineral reserves, contouring groundwater surfaces, and so on (Burrough, 1981). Table 16.7 shows the fractal dimensions of landscapes and other environmental data obtained from this method. Recently Jones et al. (1989) extended this method and studied fractal properties of computer-generated and natural geophysical data.

The second method is based on the fact that, after one- or two-dimensional Fourier transforms, the amplitudes for the Earth's topography generally have a power law dependence on wave number in the form of Equation (15.6) (Brown & Scholz, 1985, Mandelbrot, 1986, Huang & Turcotte, 1989, Goff, 1990, William & Terry, 1991; also see Section 15.4). Using this method, Huang & Turcotte (1989) digitized topography for the state of Arizona (7 points/km) to obtain maps of fractal dimension and roughness amplitude. Figure 16.31 gives their results. For Arizona, the mean fractal dimension for two-dimensional Fourier spectral analyses is $D = 2.59$; for one-dimensional Fourier spectral analyses, the mean fractal dimension is $D = 1.52$, close to the Brown noise value $D = 1.5$. Meantime, Turcotte (1989) introduced synthetic profiles and topography generated by using this method. Examples of synthetically generated fractional Brownian motion are given in Figure 16.32. The method used to generate random time series was as follows:

(1) The time interval was divided into a large number of increments. Each increment was given a random number based on a Gaussian probability distribution. This is then a Gaussian white noise sequence;

(2) The Fourier transform of the sequence was taken;

(3) The resulting Fourier coefficients were filtered by multiplying by $f^{-S/2}$;

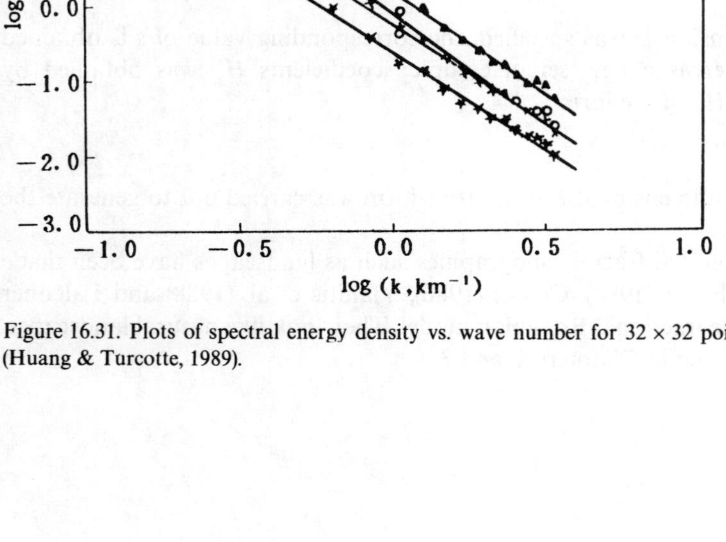

Figure 16.31. Plots of spectral energy density vs. wave number for 32 × 32 point subregions in Arizona (Huang & Turcotte, 1989).

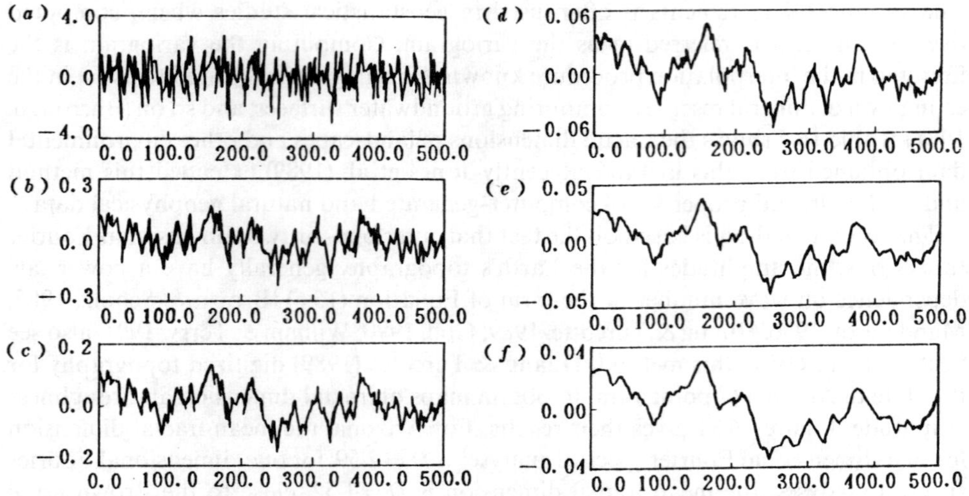

Figure 16.32. Synthetically generated fractional Brownian motion. (a) Gaussian white noise; (b) $s = 1.2$, $D = 1.9$; (c) $s = 1.6$, $D = 1.7$; (d) $s = 2.1$, $D = 1.5$; and (f) $s = 2.8$, $D = 1.1$ (Turcotte, 1989).

(4) An inverse Fourier transform was taken using the filtered coefficients;
(5) In order to remove edge effects (periodicities), only the central portion of the time series was retained.

The above procedure is extended for the generation of synthetic topography (Turcotte, 1989). Two-dimensional white noise was generated on a $N \times N$ grid using a Gaussian distribution of values. Then, a fractal filtering technique was applied to the synthetic white noise using the two-dimensional Fourier spectra approach. By carrying out a two-dimensionally described Fourier transform, a $N \times N$ array of complex Fourier transform coefficients H_{st} was obtained using the relationships:

$$H_{st} = \sum_{n=0}^{N-1} \sum_{n=0}^{N-1} h_{nm} \exp[-(2\pi i/N)(sn + tm)].$$

Each pair of indices (s, t) in the transform domain was assigned an equivalent radial index K_{st} according to the relationship:

$$K_{st} = (s^2 + t^2)^{1/2}.$$

When a fractal dimension D was specified, the corresponding value of s is obtained by $D = (7 - s)/2$, whereas a new set of complex coefficients H'_{st} was obtained by filtering the original H_{st} in the form

$$H'_{st} = H_{st}/(K_{st})^{s/2}.$$

Thus, an inverse two-dimensional Fourier transform was carried out to generate the synthetic topography.

The computer-generated fractal topographies such as landscapes have been made by Voss (1985a, b), Feder (1988), Clarke (1986), Yfanitis et al. (1988) and Falconer (1990) by means of the fractional Brownian model. They look like natural landscapes. Some examples are given in Chapters 2, and 3.

References

Adams, L. H. & Williamson, E. D. 1923. The compressibility of minerals and rocks at high pressure. *J. Franklin Inst.* 195: 475–525.

Aharony, A. 1986. Percolation in direction. In G. Grinstein & G. Mazenko (eds.), *Condensed matter physics.* Singapore: World Scientific Publishers: 1–50.

Aharony, A. & Stauffer, D. 1987. Percolation. *Encyclopedia of Physical Science and Technology* 10, 226–244.

Aki, A. 1981. A probabilistic synthesis of precursory phenomena. In D. W. Simpson & P. G. Richards (eds.), *Earthquake Prediction,* American Geophysical Union. Washington D.C.: 556–574.

Albano, E. V. & Martin, H. O. 1989. Adsorption isotherms on fractal substrata. *Phys. Rev.* A39: 6003–6009.

Allegre, J., Le Mouel, J. L. & Provost, A. 1982. Scaling rules in rock fracture and possible implications for earthquake prediction. *Nature* 297: 47–49.

Allen, C. R. 1968. The tectonic environments of seismically active and inactive areas along the San Andreas fault system. *Proc. of Conf. on Geologic Prob. of the San Andreas Fault System. Stanford Univ. Publ. Geo. Sci.* 11: 70–82.

Ambegaokar, V., Halperin, B. I. & Langer, J. S. 1971. Hopping conductivity in disordered systems. *Phys. Rev. B.* 4: 2612–2615.

Amitrano, C., Coniglio, A. & Di Liberto, F. 1986. Growth probability distribution in kinetic aggregation processes. *Phys. Rev. Lett.* 57: 1016–1019.

Anscombe, F. J. 1973. Graphs in statistical analysis. *Am. Statist.* 27: 17–21.

Arya, L. M. & Paris, J. F. 1981. A physicoempirical model to predict the soil moisture characteristic from particle-size distribution and bulk density data. *Soil Sci. Am. J.* 45: 1023–1030.

Aviles, C. A., Scholz, C. H. & Boatwright, J. 1987. Fractal analysis applied to characteristic segments of the San Andreas Fault. *J. of Geophysical Research* 92 B1: 331–344.

Avnir, D., Farin, D. & Pfeifer, P. 1983. Chemistry in noninteger dimensions between two and three. II fractal surfaces of adsorbents. *J. Chem. Phys.* 79: 3566–3571.

Avnir, D., Farin, D. & Pfeifer, P. 1984. Molecular fractal surface. *Nature* 308 (15): 261–263.

Avnir, D., Farin, D. & Pfeifer, P. 1985. Surface geometric irregularity of particulate materials. *J. of Colloid and Interface Science* 103(1): 112–123.

Badii, R. & Politi, A. 1984. Hausdorff dimension and uniformity of strange attractors. *Phys. Rev. Lett.* 52: 1661–1664.

Bakun, W. H. & McEvilly, T. V. 1984. Recurrence models and Parkfield, California, earthquakes. *J. Geophys. Res.* 89: 3051–3058.

Bamford, W. E., et al. 1978. Suggested methods for the quantitative description of discontinuities in rock masses. *Int. J. Rock Mech. Min. Sci. & Geomech. Abstr.* 15: 319–369.

Banaver, J. R., Kopik, J. & Winklen, K. W. (eds.) 1987. *Physics and chemistry of porous media II.* New York., America Institute of Physics.

Barla, G., Forlati, F. & Zaninetti, A. 1985. Shear behavior of filled discontinuities. *Proc. of Int. Symp. on Fundamentals of Rock Joints,* Rotterdam: Balkema, Bjorkliden: 163–172.

Barnsley, M. 1988. *Fractal everywhere.* Academic Press, Inc.

Barnsley, M. & Denko, S. 1986. Iterated function systems and the global construction of fractals. *Proc. of the Royal Soc. London* A399: 243–275.

Barton, C. & Larsen, E. 1985. Fractal geometry of two dimensional fractal networks at Yucca mountain, Southwestern Nevada. *Proc. of Int. Symp. of Fundamentals of Rock Joints*, Bjorklichen. Rotterdam: Balkema.

Barton, N. R. 1973. Review of a new shear strength criterion for rock joints. *Engineering Geology* 7: 287–332.

Barton, N. R. & Choubey, V. 1976. The shear strength of rock joints in theory and practice. *Rock Mech.* 10: 1–54.

Beauchamp, E. K. & Purdy, B. A. 1986. Decrease in fracture toughness of chert by heat treatment. *J. Mater. Sci.* 21: 1963–1969.

Bendat, J. S. & Piersol, A. G. 1971. Random data: *Analysis and measurement procedures.* New York: Interscience: 407–418.

Bennett, J. G. 1936. Broken coal. *J. Inst. Fuel* 10: 22–39.

Bles, J. L. & Feuga, B. (eds.) 1986. *The fracture of rocks.* England: North Oxford Academic Publishers Ltd.

Blumenfeld, R., Meir, Y., Aharony, A. & Harris, A. B. 1987. Resistance fluctuations in randomly diluted networks. *Phys. Rev.* B35: 3524–3535.

Botsis, J. & Kunin, B. 1987. On self-similarity of crack layer. *Int. J. of Fracture* 35(3): R51–R56.

Boyle, J. T. & Spence, J. 1983. *Stress analysis for creep.* England: Butterworth & Co. Ltd: 40–43.

Brady, B. T. 1978. Prediction of failures in mines – An overview. *RI8285, U.S. Bureau of Mines.*

Brien, H. K. 1989. *A random walk through fractal dimension.* VCH Publishers.

Broadbent, S. R. & Hammersley, J. M. 1957. Percolation processes I. Crystals and mazes. *Proc. Cambridge Philos. Soc.* 53: 629–641.

Broberg, H. 1975. Creep damage and rupture. Diss. Chalmers University of Tech. Gothenburg.

Broek, D. 1974. *Elementary engineering fracture mechanics.* Leyden: Noordhoff.

Brown, S. R. 1984. Fundamental study of the closure property of joints. Ph.D. thesis, Columbia Univ., New York.

Brown, S. R. 1987a. A note on the description of surface roughness using fractal dimension. *Geophys. Research Lett.* 14: 1095–1098.

Brown, S. R. 1987b. Fluid through rock joints: The effect of surface roughness. *J. of Geophy. Res.* 92 (B2): 1337–1347.

Brown, S. R. & Scholz, C. H. 1985. Broad bandwidth study of the topography of natural rock surfaces. *J. of Geophys. Res.* 90 (B14): 12575–12582.

Brown, W. K., Karpp, R. R. & Grady, D. E. 1983. Fragmentation of the universe. *Astrophys. Space Sci.* 94: 401–412.

Bui-Quoc, T., et al. 1971. Cumulative fatigue damage under stress controlled conditions. *J. of Basic Engineering Trans. ASME*: 691–698.

Burrough, P. A. 1981. Fractal dimensions of landscapes and other environmental data. *Nature* 294(19): 240–242.

Carr, J. R. 1989. Fractal characterization of joint surface roughness in welded tuff at Yucca mountain, Nevada. *In Rock Mechanics as a Guide for Efficient Utilization of Natural Resources.* Rotterdam: Balkema: 193–200.

Carr, J. R. & Warriner, J. B. 1987. Rock mass classification using fractal dimension. *28th U.S. Symp. on Rock Mechanics, Tucson.* Rotterdam: Balkema: 73–80.

Chatelain, J. L., Isacks, B. L., Cardwell, R. K., Prevot, R. & Bevis, M. 1986. Patterns of seismicity associated with asperities in the central New Hebrides island arc. *J. Geophys. Res.* 91: 12497–12519.

Che, Y. T., et al. 1989. The test study on deformation failure and pore pressure of the saturated sandstone under the conditions of various temperature and confining pressure. *Acta seism. Sinica* 5(1): 9–15.

Chen, Z. D. 1979. Geometric field theory of finite deformation in continuum mechanics. *Acta Mechanica Sinica* 2: 107–117 (in Chinese).

Chen, Z. D. 1983. *Lecture on rational mechanics.* Beijing Graduate School, China University of Mining & Technology Press (in Chinese).

Chen, Z. D. 1986. On the representation of finite rotation of nonlinear field theory for continuum mechanics. *Applied Math. and Mech.* 7: 959–968.

Chen, Z. D. 1988. *Rational mechanics.* China University of Mining & Technology Press (in Chinese).

Cherbit, G. 1991. *Fractals: non-integral dimensions and applications.* John Wiley & Sons.

Chiles, J. P. 1988. Fractal and geostatistical methods for modelling of a fracture network. *Mathematical Geology* 20(6): 631–654.

Chudnovsky, A. 1984. Crack layer theory. *NASA contractor Rep.* 174634.

Chuoke, R. L., et al. 1959. The instability of slow, immiscible, viscous liquid-liquid displacements in permeable media. *Trans. Metall. Soc. of AIME* 216: 188–194.

Clark, N. N. 1986. Three techniques for implementing fractal analysis of particle shape. *Powder Tech.* 46: 45–52.

Clarke, K. C. 1986. Computation of the fractal dimension of topographic surfaces using the triangular prism surface area method. *Comput. Geosci.* 12(5): 713–722.

Cohen, M. N. & Anderson, M. P. 1985. The chemistry and physics of porous media. In M. Tomkiewicz & P. N. Sen (eds.), *The Electrochemical Society*. New Jersey: Pennington.

Cohen, W. H. & Knight, B. H. 1974. *J. Soc. Chem. Ind.* 66: 357–366.

Coster, M. & Cherment, J. L. 1983. Recent developments in quantitative fractograpny. *Int. Met. Rev.* 28(4): 228–250.

Costin, I. S. 1983. A microcrack model for the deformation and failure of brittle rock. *J. Geophys. Res.* 88(B11): 9485–9492.

Coughlin, J. & Kranz, R. 1991. New approaches to studying rock burst-associated seismicity in mines. *Proc. 32nd U.S. Symp. on Rock Mechanics.* Rotterdam: Balkema: 491–500.

Davison, L. & Stevens, A. L. 1973. Thermomechanical constitution of spalling elastic bodies. *J. Applied Phys.* 44(2): 668–674.

Dohnanyi, J. S. 1969. Collisional model of asteroids and their debris. *J. Geophys. Res.* 74: 2531–2554.

Dohnanyi, J. S. 1972. Interplanetary objects in review: statistics of their masses and dynamics. *Icarus* 17: 1–48.

Donnison, J. R. & Sugden, R. A. 1984. The distribution of asteroid diameters. *Mon. Not. R. Astron. Soc.* 210: 673–682.

Dougill, J. W., Lau, J. C. & Burt, N. J. 1976. Mechanics in engineering. *AS CE. EMD*: 333–355.

Dragon, A. & Mroz, Z. 1979. A continuum model for plastic-brittle behavior of rock and concrete. *Int. J. Engineering Sciences* 17: 121–137.

Drake, J. M. & Klafter, J. 1990. Dynamics of confined molecular systems. *Phys. Today* 43: 46–55.

Edgar, G. A. 1990. *Measure, topology and fractal geometry.* Berlin: Springer-Verlag.

Elam, W. T., et al. 1984. Critical properties of the void percolation problem for spheres. *Phys. Rev. Lett.* 17: 1516–1519.

Engelberts, W. F. & Klikenberg, L. J. 1951. Laboratory experiments on the displacement of oil by water from packs of granular material. *Petr. Congr. Proc. Third World*: 544–554.

Ewalds, H. L. & Wanhill, R. J. H. 1984. *Fracture mechanics.* The Netherlands: Martinus Nijhoff Publishers.

Falconer, K. J. 1985. *The geometry of fractal sets.* Cambridge: Cambridge University Press.

Falconer, K. J. 1990. *Fractal geometry: mathematical foundations and applications.* England: John Wiley & Sons.

Feder, J. 1988. *Fractals.* New York: Plenum Press.

Fisher, M. E. 1974. The renormalization group in the theory of critical behavior. *Rev. of Modern Physics.* 46(4): 597–616.

Fisher, P. & Smith, W. R. 1985. *Chaos, fractals and dynamics.* Marcel Dekker Inc.

Fonseka, G. U. & Krajcinovic, D. 1981. The continuous damage theory of brittle materials-II: Uniaxial and plane response modes. *J. Appl. Mech.* 48: 816–824.

Frantziskonis, G. & Desai, C. S. 1987. Constitutive model with strain softening. *Int. J. Solid Structures* 23(6): 733–768.

Frieson, W. I. & Mikula, R. J. 1987. Fractal dimensions of coal particles. *J. of Colloid and Interface Science* 20(1): 263–271.

Fripiat, J. J., Gatineau, L. & Van Damme, H. 1986. Multilayer physical adsorption on fractal surfaces. *Langmuir* 2: 562–567.

Fujiwara, A., et al. 1977. Destruction of basaltic bodies by high-velocity impact. *Icarus* 31: 277–288.

Gefen, Y., et al. 1980. Critical phenomena of fractal lattices. *Phys. Rev. Lett.* 45: 855–859.

Goff, J. A. 1990. Comment on 'Fractal mapping of digitized images: application to the topography of Arizona and comparison with synthetic images'. *J. of Geophys. Res.* 95(B4): 5159–5161.

Goodchild, M. F. 1988. Lakes on fractal surfaces: a null hypothesis for lake-rich landscapes. *Mathematical Geology* 20(6): 615–629.

Goodchild, M. F. & Mark, D. M. 1987. The fractal nature of geographic phenomena. *Annals of the Association of American Geographers* 77(2): 265–279.

Goodman. R. E. 1976. *Methods of geological engineering.* West publishing Co.: 472–485.

Gordon, J. M., et al. 1986. Superconducting-normal phase boundary of a fractal network in a magnetic field. *Phys. Rev. Lett.* 56: 2280–2283.

Grady, D. E. & Kipp, M. E. 1980. Continuum modelling of explosive fracture in oil shale. *Int. J. Rock Mech. Min. Sci. & Geomech. Abstr.* 17: 147–157.

Grassberger, P. 1983. Generalized dimensions of strange attractors. *Phys. Lett.* A 97: 227–230.

Grassberger, P. & Procaccia, I. 1983. Measuring the strangeness of strange attractors. *Physica* 9: 189–208.

Griffith, A. A. 1920. The phenomena of rupture and flow in solids. *Phil. Trans. Roy. Soc.* A 221: 163–198.

Griffith, A. A. 1924. The theory of rupture. *Proc. First Int. Congr. Appl. Mech.* Also in J. Waltman (ed.), *Technische Boekhandel en Drukkerij*, Delft (1925): 55–63.

Gutenberg, B. & Richter, C. F. 1956. Earthquake magnitude, intensity, energy acceleration. *Bull. Seism. Soc. Am.* 46: 105–145.

Haas, C. J., et al. 1975. An investigation of the interaction of rock and types of rock bolts for selected loading conditions. *Second Annual Report. U.S. Bureau of Mines Contract*: H0122110.

Hack, J. T. 1957. Studies of longitudinal stream profile in Viginia and Maryland. *U.S. Geol. Surv. Prof. Pap.* 294–B: 45–97.

Hamel, J. V. 1974. Rock strength from failure cases: Lefe slope stability study, Libby dam and lake Koocanusa, Montana. *Tech. Report MRD-1-74, Missouri River Div. Corps of Engineers.* Omana: Nebraska.

Han, S., et al. 1986. A dynamic analysis of jointed rock mass. *Proc. of Int. Symp. on Eng. in Complex Rock Formations. Beijing, China.* Science Press: 338–344.

Hao, Q. W. 1985. Modelling on substance of mixture of rocks. Master's thesis. Beijing Graduate School, China University of Mining and Technology.

Haramy, K. Y. & McDonnell, J. P. 1988. Causes and control of coal mine bumps. *RI9225, U.S Bureau of Mines.*

Hartmann, W. K. 1969. Terrestrial, lunar and interplanetary rock fragmentation. *Icarus* 10: 201–213.

Hausdorff, F. 1919. Dimension und äusseres mass math. *Annalen* 79: 157–179.

Haverkamp, R. & Parlange, J. Y. 1986. Predicting the water vetention curve from particle-size distribution: Study soils without organic matter. *Soil Sci.* 142: 325–399.

Hawkins, G. S. 1960. Asteroidal fragments. *Astrophys J.* 65: 318–322.

Hele-Shaw, H. S. 1898. The flow of water. *Nature* 58: 34–36.

Hellyer, B. 1971. The fragmentation of asteroids II. *Mon. Not. R. Astron. Soc.* 154: 279–291.

Hentschel, H. G. E. & Procaccia, I. 1984. Building blocks of percolation clusters: volatile fractals. *Phys. Rev. Lett.* 53: 1121–1124.

Herrmann, H. J. & Roux, S. 1990. *Statistical models for the fracture of disordered media.* Elsevier Science Publishers B.V.

Hirata, T., et al. 1987. Fractal structure of spatial distribution of microfracturing in rock. *Geophys. J. R. Astr. Soc.* 90: 369–374.

Hjelmfelt, A. T. 1988. Fractals and the river-length catchment-area ratio. *Water Resources Bulletin* 24(2): 455–459.

Hoek, E. 1967. Brittle failure of rock. In K. G. Stagg (ed.), *Rock Mechanics in Engineering Practice.* London: Wiley.

Hoek, E. & Bieniawski, Z. T. 1984. Brittle fracture propagation in rock under compression. *Int. J. of Fracture* 26: 276–294.

Homsy, G. M. 1987. Viscous figuring in porous media. *Ann. Rev. Fluid Mech.* 19: 271–311.

Hong, S. H. & Hong, S. M. 1988. A study of fractal in geoscience: drainages, earthquakes and other. *Exploration of Nature* 7(2): 33–40 (in Chinese).

Hornbogen, E. 1989. Fractals in microstructure of metals. *Int. Materials Rev.* 34(6): 277–296.

Horton, R. E. 1945. Erosional development of streams and their drainage basins: hydrophysical approach to quantitative morphology. *Geal. Soc. Am. Bull.* 56: 275–370.

Huang, J. & Turcotte, D. L. 1989. Fractal mapping of digitized images: application to the topography of Arizoda and comarisions with synthetic images. *J. of Geophys. Res.* 94(B6): 7491–7495.

Hurd, A. J., et al. 1989. Surface areas of fractally rough particles studied by scattering. *Phys. Rev.* B39: 9742–9745.

Irwin, G. R. 1964. Structural aspects of brittle fracture. *Int. J. Properties and Testing Engng. Mater.* 3(2): 65–81.

Jaeger, J. C. & Cook, N. C. W. 1979. Fundamentals of rock mechanics. *3rd Edn*: Chapman & Hall: 593–601.

Janson, J. & Hult, J. 1977. Fracture mechanics and damage mechanics – a combined approach. *J. de Mecanique Appliquee* 1: 69–84.

Jenkins, F. M., Williams, T. J. & Wideman, C. J. 1990. Rock burst mechanism studies at the Lucky Friday mine. *Proc. 31st U.S. Symp. on Rock Mechanics*, WV Univ., Rotterdam: Balkema: 955–962.

Jensen, M. H., et al. 1987. Effect of gravity on the Saffman-Talor meniscus: theory and experiment. *Phys. Rev.* A35: 2221–2227.

Jonas, J. J. & Baudelet, B. 1977. Effect of crack and cavity generation on tensile stability. *Acta Melatturgica* 25: 43–50.

Jones, J. G., et al. 1989. Fractal properties of computer-generated and natural geophysical data. *Computer & Geosciences* 15(2): 227–235.

Jullien, R. & Rotet, R. 1987. *Aggregation and fractal aggregates.* Singapore: World Scientific.

Kachanov, L. M. 1958. On the time to failure under creep conditions. *Izv. Akad. Nauk. SSR-Otd. Tekhn.* N.8: 26–31.

Kachanov, M. L. 1980. Microcrack model for rock inelasticity. Ph.D. Dissertation, Rutgers University, Piscataway. NJ: 08854.

Kagan, Y. Y. & Knopoff, L. 1981. Stochastic synthesis of earthquake catalogs. *J. Geophys. Res.* 86: 2853–2862.

Kanamori, H. & Anderson, D. L. 1975. Theoretical basis of some empirical relations in seismology. *Bull. Seismol. Soc. Am.* 65: 1073–1096.

Kaplan, J. & Yorke, J. A. 1979. Chaotic behavior of multidimensional difference equations. In Springer Lecture Notes. Berlin: Springer: 228–231.

Kawakami, K., et al. 1990. Size-frequency distribution of fragments under impacting. In *Seismography of Mathematics and Physics* (Chinese edition). Seima Press: 79–83.

Kawamoto, T., et al. 1988. Deformation and fracturing behavior of discontinuous rock mass and damage mechanics theory. *Int. J. for Numerical and Analy. Methods in Geomech.* 12: 1–30.

Katz, A. J. & Thompson, A. H. 1985. Fractal sandstone pores: implications for conductivity and pore formation. *Phys. Rev. Lett.* 54(12): 1325–1328.

Katz, A. J. & Thompson, A. H. 1986. Quantitative prediction of permeability in porous rock. *Phys. Rev. B.* 34(11): 8179–8181.

Kay, B. H. 1986. The description of two-dimensional rugged boundaries in fine particle science by means of fractal dimensions. *Powder Tech.* 46: 245–254.

Kiefer, B. V. & Hilton, P. D. 1985. Combined viscous and plastic deformations in two-dimensional large strain finite element analysis. *J. of Eng. Materials and Technology* 13: 107–118.

King, G. 1983. The accommodation of large strains in the upper lithosphere of the earth and other solids by self-similar fault systems: the geometrical origin of b-value. *Pure Appl. Geophys.* 121: 761–815.

Kinlock, A. J. & Young, R. J. 1983. *Fracture behavior of polymers.* Applied Sci. Publishers LTD.

Kirkpatrik, S. 1979. The geometry of the percolation threshold. In R. Balian, et al. (eds.), *I11-Condensed matter*. Amsterdam: North-Holland: 99–116.

Kneisley, R. O. 1989. Microseismic data analysis of failure occurrence in a deep western U.S. coal mine: a case study. *RI 9228, U.S. Bureau of Mines.*

Knott, J. F. 1973. *Fundamentals of fracture mechanics.* London: Butterworths.

Krajcinovic, D. 1979. Distributed damage theory of beams in pure bending. *J. Applied Mech.* 46: 592–596.

Krajcinovic, D. 1989. Damage mechanics. *Mechanics of materials* 8: 117–197.

Krajcinovic, D. & Fonseka, G. U. 1981. The continuous damage theory of brittle materials. Part I: General theory. *J. of Applied Mech.* 48: 809–815.

Krajcinovic, D. & Silva, M. A. G. 1982. Statistical aspects of the continuous damage theory. *Int. J. Solids Structures* 18(7): 551–562.

Krohn, C. E. 1988a. Sandstone fractal and Euclidean pore volume distributions. *J. Geophys. Res.* 93(B4): 3286–3296.

Krohn, C. E. 1988b. Fractal measurements of sandstones, shales and carbonates. *J. Geophys. Res.* 93(B4): 3297–3305.

Krohn, C. E. & Thompson, A. H. 1986. Fractal sandstone pores: automated measurement using scanning-electron-microscope images. *Phys. Rev.* B 33(9): 6366–6374.

Langbein, W. B., et al. 1947. Topographic characteristics of drainage basins. *U.S. Geol. Surv. Water-Supply Pap.* 968-C: 125–257.

Lange, M. A., Ahrens, T. J. & Boslough, M. B. 1984. Impact cratering and spall failure of gabbro. *Icarus* 58: 383–395.

Lapidus, L. & Pinder, G. 1982. *Numerical solution of particle differential equations in science and engineering.* New York: John Wiley.

La-Point, P. R. 1988. A method to characterize fracture density and connectivity through fractal geometry. *Int. J. Rock Mech. Min. Sci. & Geomech. Abstr.* 25(6): 421–429.

La-Point, P. R. & Hudson, J. A. 1985. Characterization and interpretation of rock mass joint patterns. *Special paper 199, Geological Society of America. Boulder Co. 37.*

Leckie, F. A. & Onat, E. T. 1980. Tensorial nature of damage measuring internal variables. In J. Hult & J. Lemaitre (eds.), *Proc. IUTAM Symp. on Physical Non-Linearities in Structural Analysis*: Springer-Verlag.

Lee, Y. H., Carr, J. R., Bars, D. J. & Haas, C. J. 1990. The fractal dimension as a measure of the roughness of rock discontinuity profiles. *Int. J. Rock Mech. Min. Sci. & Gemech. Abstr.* 27: 453–464.

Legendre, D. & Mazars, J. 1984. Damage and fracture mechanics for concrete (a combined approach). *ICF-6* 4: 2841–2848.

Leighton, F. 1982. A case history of a major rock burst. *RI 8701, U.S. Bureau of Mines.*

Lemaitre, J. 1984. How to use damage mechanics. *Nuclear Engineering and Design Journal* 80: 215–226.

Lemaitre, J. & Chaboche, J. L. 1974. A nonlinear model of creepfatigue damage cumulation and interaction. *Proc. IUTAM Symp. of Mech. of Visco-elastic Media and Bodies.* Gothenburg: Springer-Verlag.

Li, H. 1986. Damage mechanics. Lecture Notes of Middle China University of Technology.

Li, H. H. 1990. A preliminary study on earthquake self-similar phenomena. *Acta Seismological Sinica* 11: 1–10.

Li, H. H., Zhang, W. X., Zhang, Y. L., Ma, W. J. & Guo, Y. P. 1989. The temporal fractal structure of seismicity before the M6.4 Menyuan earthquake. *Earthquake Res. in China* 3(1): 55–63.

Li, J. 1990. Size, shape and chemical characteristics of mining-generated and laboratory-generated coal dust. M.S. thesis, the Pennsylvania State University.

Li, S. P. 1988. *A concise course of rock mechanics.* China Coal Ind. Publishers.

Louis, E., Guinea, F. & Flores, F. 1986. The fractal nature of fracture. In L. Pietronero & E. Tosatti (eds.), *Fractals in Physics.* Amsterdam: North-Holland: 177–180.

Lovejoy, S. 1982. Area-perimeter relation for rain and cloud areas. *Science* 216: 185–187.

Lovejoy, S. & Schertzer, D. 1985. Generalized scale invariance in the atmosphere and fractal models of rain. *Water Resour. Res.* 21: 1233–1250.

Lung, C. W. 1985. Fractals and fracture of cracked metals. *Preprint of the Int. Center for Theoretical Physics.* Trieste, Italy: IC/85/123.

Lung, C. W. & Mu, Z. Q. 1988. Fractal dimension measured with perimeter-area relation and toughness of materials. *Physical Rev. B* 38(16): 11781–11784.

Lung, C. W. & Zhang, S. Z. 1989. Fractal dimension of the fractured surface of materials. *Physica D* 38: 242–245.

Ma, X. Y., et al. 1981. *Songshan structural deformation.* Geological Publishing House (in Chinese).

Madden, T. R. 1983. Microcrack connectivity in rocks: A renormalization group approach to the critical phenomena of conduction and failure in crystalline rocks. *J. Geophys. Res.* 88: 585–592.

Maløy, K. J., Boger, F., Feder, J. & Jøssang, T. 1987. Dynamics and structure of viscous fingers in porous media. In R. Pynn & T. Riste (eds.), *Time-dependent effects in disordered materials.* Plenum Press: 111–138.

Maløy, K. J., Feder, J. & Jøssang, T. 1985a. Viscous fingering fractals in porous media. *Phys. Rev. Lett.* 55: 2688–2691.

Maløy, K. J., Feder, J. & Jøssang, T. 1985b. Radial viscous fingering in a Hele-Shaw cell. *Report Series, Cooperative Phenomena Project, Dept. of Physics, Univ. of Oslo* 85–9: 1–15.

Mandelbrot, B. B. 1967. How long is the cost of Britain? statistical self-similarity and fractal dimension. *Science* 155: 636–638.

Mandelbrot, B. B. 1974. Intermittent turbulence in self-similar cascades: Divergence of high moments and dimension of the carrier. *J. Fluid Mech.* 62: 331–358.

Mandelbrot, B. B. 1975a. Stochastic models for the Earth's relief, the shape and the fractal dimension of the coastlines, and the number-area rule for islands. *Proc. Natl. Acad. Sci. U.S.A.* 72: 3825–3828.

Mandelbrot, B. B. 1975b. *Les objects fractals: forme, hasard et dimension.* Paris: Flammarion.

Mandelbrot, B. B. 1977. *Fractals: form, chance and dimension.* San Francisco: W.H. Freeman.

Mandelbrot, B. B. 1982. *The fractal geometry of Nature.* New York: W.H. Freeman (Page numbers correspond to the 1983 edition).

Mandelbrot, B. B. 1983. Fractals in physics: squing clusters diffusions, fractal measures and the unicity of fractal dimensionality. *J. Stat. Phys.* 34: 895–930.

Mandelbrot, B. B. 1986. Self-affine fractal sets. In L. Pietronero & E. Tosatti (eds.), *Fractals in physics.* Amsterdam: North-Holland: 3–28.

Mandelbrot, B. B. 1987. Fractals. *Encyclopedia of Physical Science and Technology* 5: 579–593.

Mandelbrot, B. B., et al. 1984. Fractal character of fracture surfaces of metals. *Nature* 308: 721–723.

Mandelbrot, B. B. & Van Ness, J. W. 1968. Fractional Brownian motions, fractional noises and applications. *SIAM Rev.* 10: 422–437.

Mathis, J. S. 1979. The size distribution of interstellar particles II: polarization, *Astrophys. J.* 232: 747–753.

McCrosky, R. E. 1968. Distributions of large meteoric bodies. Spec. Rep. 280. *Smithson, Astrophys. Observ. Washington D.C.*

Mecholsky, J. J., et al. 1986. Fractal aspects of materials. *Mat. Res. Soc.* 5: 207–217.

Mecholsky, J. J. & Mackin, J. J. 1988. Fractal analysis of fracture in Ocala chert. *J. of Materials Science Lett.* 7: 1145–1147.

Mei, Shiromg 1987. *Earthquake prediction research.* Center of Analysis and Prediction, State Seismological Bureau. Beijing 1: 8–12 (in Chinese).

Miller, S. M. 1990. Ambiguities in estimating fractal dimensions of rock fracture surfaces. In W. A. Hustrulid & G. A. Johnson (eds.), *Rock Mech. Contributions and Challenges: Proc. of 31st U.S. Symp. on Rock Mech.* Rotterdam: Balkema: 471–478.

Mu, Z. Q. & Lung, C. W. 1988. Studies on the fractal dimension and fracture toughness of steel. *J. phys. D: Appl. Phys.* 21: 848–850.

Murakami, S. & Ohno, N. 1981. *A continuum theory of creep and creep damage, creep in structures.* Berlin: Springer: 422–444.

Muralha, J. & Charrua-Grace, J. G. 1990. An experience on the application of fractal theory to basic shear strength studies. In N. R. Barton, et al. (eds.), *Rock Joints.* Rotterdam: Balkema: 667–671.

Niemeyer, L., Pietronero, L. & Wiesmann, H. J. 1984. Fractal dimension of dielectric breakdown. *Phys. Rev. Lett.* 52: 1033–1037.

Nittmann, J., et al. 1987. Experimental evidence for multifractality. *Phys. Rev. Lett.* 58: 619–622.

Niu, Zhiren 1991. Statistical theory of fractal fracture of rock. *Chinese J. of Geophys.* (Preprint).

Noite, D. D., Pyrak-Noite, L. T. & Cook, N. G. 1989. The fractal geometry of the flow paths in natural fractures in rock and the approach to percolation. In C. H. Scholz & B. B. Mandelbrot (eds.), *Fractals in Geophysics.* Boston: Birkhauser.

Nolen-Hoeksema, R. C. & Gordon, R. B. 1987. Optical detection of crack patterns in the opening-mode fracture of marble. *Int. J. Rock Mech. Min. Sci. & Geomech. Abstr.* 24(2): 135–144.

Okeefe, J. D. & Ahrens, T. J. 1985. Impact and explosion crater ejecta, fragmental size and velocity. *Icarus* 62: 328–338.

Okubo, P. G. & Aki, K. 1983. Fractal geometry of the San Andreas fault system. *Eos Trans. AGU* 64: 766–776.

Okubo, P. G. & Aki, K. 1987. Fractal geometry in the San Andreas fault system. *J. Geophys. Res.* 92(B1): 345–355.

Orford, J. O. & Whalley, W. B. 1983. The use of the fractal dimension to quantify the morphology of irregular-shaped particles. *Sedimentology* 30: 655–668.

Ostoja-Starzewski, M. 1989. Damage in a random microstructure: Size effects, fractals, and entropy maximization. *Appl. Mech. Rev.* 42(11): 5202–5212.

Pande, C. S., et al. 1987. Fractal characterization of fracture surfaces. *Acta Metall.* 35(7): 1633–1637.

Panagiotopoulos, P. D. 1981. Non-convex superpotentials in the sense of F. H. Clarke and applications. *Mech. Res. Comm.* 8: 335–347.

Panagiotopoulos, P. D. 1983. Non-convex energy functions, hemivariationals inequalities and substationarity principles. *Acta Mechanica.* 42: 363–375.

Peitgen, H. O. & Richter, P. H. (eds.) 1986. *The beauty of fractal.* Berlin: Springer-Verlag.

Peitgen, H. O. & Saupe, D. (eds.) 1988. *The science of fractal images.* Berlin: Springer-Verlag.

Pfeifer, P. & Avnir, D. 1983. Chemistry in noninteger dimensions between two and three I. fractal theory of heterogeneous surfaces. *J. Chem. Phys.* 79: 3558–3565.

Pfeifer, P., Wu, Y. J., Cole, M. W. & Krim, J. 1989. Multilayer adsorption on a fractally rough surface. *Phys. Rev. Lett.* 62: 1997–2000.

Pietronero, L., Evertsz, C. & Siebesma, A. P. 1986. Fractal and multifractal structure in kinetic critical phenomena. In S. Albeverio, et al. (eds.), *Stochastic Processes of Physics and Engineering*: Boston.

Pietronero, L., Erzan, A. & Evertsz, C. 1988. Theory of fractal growth. *Physica A* 151: 207–218.

Pietronero, L. & Tosatti, E. (eds.) 1986. *Fractals in Physics.* Amsterdam: North-Holland.

Pike, G. E. & Seager, C. H. 1974. Percolation and conductivity: a computer study. *Phys. Rev.* B10: 1421–1434.

Porod, G. 1982. General theory. In O. Glatter & O. Kratky (eds.), *Small angle X-ray scattering.* New York: Academic Press. 17–51.

Priest, S. D. & Samaniego, A. 1984. A model for the analysis of discontinuity characteristics in two dimensions. *Proc. 5th Congr. ISRM*: F199–F207.

Rabotnov, Y. N. 1969. Creep rupture. *Proc. XII Int. Congr. Appl. Mech.* Berlin: Springer.

Robert, A. & Roy, A. G. 1990. On the fractal interpretation of the mainstream length-drainage area relationship. *Water Resources Research* 26(5): 839–842.

Robert, L. D. & Linda, K. 1989. Chaos and fractals: The mathematics behind the computer graphics. *Proc. of Symp. in Applied Math*: 39–51.

Roberts, J. N. 1986. Comment about fractal sandstone pores. *Phys. Rev. Lett.* 56: 2111–2121.

Rockafeller, R. T. 1980. Generalized directional derivatives and subgradients of nonconvex functions. *Can. J. Math. XXXII*: 257–270.

Rosso, R., et al. 1991. Fractal relation of mainstream length to catchment area in river networks. *Water Resources Research* 27(3): 381–387.

Rys, F. S. & Waldvogel, A. 1986. Fractal shape of hail clouds. *Phys. Rev. Lett.* 56: 784–787.

Saffman, P. G. & Tayloy, G. I. 1958. The penetration of a fluid into a medium or Hele-Shaw cell containing a more viscous liquid. *Proc. R. Soc. Lond.* 245: 312–329.

Santarelli, F. T. & Brown, E. T. 1989. Failure of three sedimentary rocks in triaxial and hollow cylinder compression tests. *Int. J. Rock Mech. Min. Sci. & Geomech. Abstr.* 26(5): 401–413.

Sato, K., Isobe, T., Mori, N. & Goto, T. 1986. Microseismic activity associated with hydraulic mining. *Int. J. of Rock Mech. Min. Sci. & Geomech. Abstr.* 23(1): 85–94.

Satpathy, S. 1986. Dielectric breakdown in three dimensions. In L. Pietronero & E. Tosatti (eds.), *Fractals in Physics.* Elsevier Science Publishers B. V.: 173–176.

Scholz, C. H. & Mandelbrot, B. B. (eds.) 1989. *Fractals in geophysics.* Basel: Birkhauser Verlag.

Schoutens, J. E. 1979. Empirical analysis of nuclear and high-explosive cratering and ejecta. *Nuclear Geoplosics Sourcebook 55 Part 2, Section 4, Rep. DNA 65 01H-4-2, Def. Nucl. Agency*: Bethsda Md.

Seaman, L., Curran, D. R. & Murri, W. J. 1985. A continuum model for dynamic tensile microfracture and fragmentation. *J. of Applied Mech.* 52: 593–600.

Segall, P. & Pollard, D. D. 1980. Mechanics of discontinuous faults. *J. Geophys. Res.* 85: 4337–4350.

Serra, J. 1982. *Image analysis and mathematical morphology.* Academic Press: 127–161.

Shante, V. K. S. 1977. Introduction to percolation theory. *Phys. Rev.* B16: 2597–2562.

Sieh, K. E. 1978. Slip along the San Andreas fault associated with the great 1857 earthquake. *Bull. Seismol. Soc. Am.* 68: 1421–1448.

Sieh, K. E. & Jahns, R. H. 1984. Holocene activity of the San Andreas fault at Wallace creak, California. *Geol. Soc. Am. Bull.* 95: 883–896.

Sih, G. G. 1977. *Elastodynamic crack problems.* 4 Noord Hoff Int. Publishing: XVII–XLVII.

Smalley, R. F., et al. 1987. A fractal approach to the clustering of earthquakes applications to the seismicity of the new Hebrides. *Bull. of Seismol. Soc. Am.* 77(4): 1368–1381.

Smalley, R. F., Turcotte, D. L. & Solla, S. A. 1985. A renormalization group approach to the sick-slip behavior of faults. *J. Geophys. Res.* 90: 1894–1900.

Smith, E. 1986. The effect of crack front of irregularity on the fracture toughness of brittle materials. *Arch. Mech.* 38(1–2): 185–190.

Sokolov, I. M. 1986. Dimensionalities and other geometric critical exponents in percolation theory. *Sov. Phys. Usp.* 29(10): 924–945.

Sondergeld, C. H. & Estey, L. H. 1981. Acoustic emission study of microfracturing during the cyclic loading of westerly granite. *J. Geophys.* 86: 2915–2924.

Sprunt, E. S. & Brace, W. F. 1974. Direct observations of microcavities in crystalline rocks. *Int. J. Rock Mech. Min. Sci.* 11: 139–150.

Stanley, H. E. & Ostrowsky, N. 1986. *On growth and form.* Dordrecht: Martinus Nijhoff.

Stauffer, D. 1985. *Introduction to percolation theory.* London: Taylor & Francis.

Stevens, Roger T. 1989. *Fractal Programming in C*. M & T Publishing Inc.

Stevens, Roger T. 1990. *Fractal programming in Turbo Pascal*. M & T Publishing Inc.

Su, H., Zhang, Y. & Yan, Z. Q. 1989. Fractal analysis of impact fracture of metal. *Acta Metall. Sinica* 25(6): A466–468 (in Chinese).

Synge, J. L. & Chien, W. Z. 1941. The intrinsic theory of shells and plates. *Ka'rmall Anniversary Volume*: 111–122.

Takayasu, H. 1982. Differential fractal dimension of random walk and its applications to physical systems. *J. Phys. Soc. Japan*. 51(9): 3057–3064.

Takayasu, H. 1985. Simulation of electric breakdown and resulting variant of percolation fracture. *Phys. Rev. Lett.* 54: 1099–1103.

Takayasu, H. 1986. Pattern formation of dendritic fractals in fracture and electric breakdown, In L. Pietronero & E. Tosatti (eds.), *Fractals in Physics*. Elsevier Science Publishers B.V.: 181–184.

Tao, Z. Y., et al. 1989. The damage model of jointed rock mass. *Proc. of Int. Conf. on Advance in Constitutive Laws for Engineering Materials*. Int. Academic Publishers: 655–658.

Taylor, L. M., et al. 1986. Damage accumulation due to microcracking in brittle rock under dynamic loading. *ICF–6*: 95–104.

Termonia, Y. & Meakin, P. 1986. Formulation of fractal cracks in a kinetic fracture model. *Nature* 320(3): 429–431.

Termonia, Y., Meakin, P. & Smith, P. 1985. Theoretical study of the influence of the molecular weight on the maximum tensile strength of polymer fibers. *Macromolecules* 18: 2246–2252.

Thompson, A. H. 1991. Fractals in rock physics. *Annual Rev. of Earth and Planetary Sciences* (to be published).

Thompson, A. H., Katz, A. J. & Krohn, C. E. 1987. The microgeometry and transport properties of sedimentary rock. *Advances in Physics*. 36(5): 625–694.

Trifunac, M. D. & Udwadia, F. E. 1974. Parkfield, California, earthquake of June 27, 1966: A three-dimensional moving dislocation, *Bull. Seismol. Soc. Am.* 64: 511–533.

Tse, R. & Cruden, D. M. 1979. Estimating joint roughness coefficients. *Int. J. Rock Mech. Min. Sci. & Geomech. Abstr.* 16: 303–307.

Turcotte, D. L. 1986a. Fractal and fragmentation. *J. Geophys. Res.* 91, 132: 1921–1926.

Turcotte, D. L. 1986b. A fractal model for crustal deformation. *Tectonophysics* 132: 261–269.

Turcotte, D. L. 1989. Fractal application to complex crustal problems. In P. Germain, M. Piau & D. Caillerie (eds.), *Theoretical and Applied Mechanics, IUTAM*. Elsevier Science Publishers B.V. 337–349.

Turk, N. & Dearman, W. R. 1985. Investigation of some rock joint properties: roughness angle determination and joint closure. *Proc. Int. Symp. on Fundamentals of Rock Joints, Bjorhliden*. Balkema: Rotterdam: 197–204.

Turk, N., et al. 1987. Characterization of rock joint surfaces by fractal dimension. *28th U.S. Symp. on Rock Mechanics, Tucson*. Balkema: Rotterdam: 1223–1236.

Tyler, S. W. & Wheatcraft, S. W. 1989. Application of fractal mathematics to soil water retention estimation. *Soil Sci. Soc. Am. J.* 53: 987–996.

Underwood, E. E. & Banerji, K. 1986. Fractals in fractography. *Materials Science and Engineering* 80: 1–14.

Van Genuchten, M. T. 1980. A closed form equation for predicting the hydraulic conductivity of unsaturated soils. *Soil Sci. Soc. Am. J.* 44: 892–898.

Velde, B., et al. 1990. Fractal analysis of fractures in rocks: the Cantor's dust method. *Tectonophysics* 179(N3/4): 345–352.

Velde, B., et al. 1991. Fractal patterns of fractures in granites. *Earth and Planetary Sci. Letters, Bjorhliden.* 104: 25–35.

Venkatachalam, G. 1985. Modelling of rock joints. *Proc. of Int. Symp. on Fundamentals of Rock Joints:* Rotterdam: Balkema: 453–459.

Vicsek, T. 1989. *Fractal growth phenomena*. World Scientific Co.

Voss, R. F. 1985a. Random fractals: characterization and measurement. In R. Pynn & A. Skjeltorp (eds.), *Scaling phenomena in disordered systems*. New York: Plenum Press: 1–11.

Voss, R. F. 1985b. Random fractal forgeries. In R. A. Earnshaw (ed.), *Fundamental Algorithms in Computer Graphics*. Berlin: Springer-Verlag: 805–835.

Walsh, J. B. 1981. Effect of pore pressure and confining pressure on fracture permeability. *Int. J. Rock Mech. Min. Sci. & Geomech. Abstr.* 18: 429–435.

Wang, X. W., Dong, L. K. & Lung, C. W. 1990. The computer simulation of fractal dimensionality with perimeter-area relation. *Advance on Material Science* 4(3): 58–64.

Watanabe, K. 1986. Stochastic evaluation of the two dimensional continuity of fractures in a rock masses. *Int. J. Rock Mech. Min. Sci.* 23(6): 431–437.

William, L. P. & Terry, E. T. 1991. Euclidean and fractal models for the description of rock surface roughness. *J. of Geophys. Res.* 96(B1): 415–424.

Williford, R. E. 1988a. Multifractal fracture. *Scripta Metallurgica* 22: 1749–1754.

Williford, R. E. 1988b. Scaling similarities between fracture surfaces, energies and a structure parameter. *Scripta Metallurgica* 22: 197–200.

Wilson, K. G. & Kogut, J. 1974. The renormalization group and the ε-expansion. *Phys. Rep.* 12C: 75–200.

Witten, T. A. & Sander, L. M. 1983. Diffusion-limited aggregation. *Phys. Rev.* B27: 5686–5697.

Wong, Po-zen 1985. Scattering by inhomogeneous systems with rough internal surface: porous solids and random-field systems. *Phys. Rev.* B 32: 7417–7424.

Wong, Po-zen 1987. Fractal surfaces in porous media: physics and chemistry of porous media II. In J. Banavar, J. Koplik & K. Winkler (eds.), *AIP conference Proc.* 154. American Institute of Physics, New York: 304–318.

Wong, Po-zen 1988. The statistical physics of sedimentary rock. *Physics Today* 41: 24–32.

Wong, Po-zen, Howard, J. & Lin, J. S. 1986. Surface roughening and the fractal nature of rocks. *Phys. Rev. Lett.* 57: 637–640.

Wong, Po-zen & Lin, J. S. 1988. Studying fractal geometry on submicron length scales by small-angle scattering. *Mathematical Geology* 20(6): 656–665.

Xia, X. L., et al. 1985. Mixed model(I,II) fracture test of rock subjected to compressive loading. *J. of Rock and Soil Engineering* 7(2): 66–74 (in Chinese).

Xie, H. 1984. Nonlinear analysis of large deformation. M.D. thesis. Beijing Graduate School, China University of Mining & Technology.

Xie, H. 1985. Finite element analysis of large deformation for continuum in gravity field. *Proc. of Int. Conf. on Nonlinear Mech.* China: 1337–1343.

Xie, H. 1987a. Analysis of rock creep damage for large deformation and its fractal models of micro-fractures. Ph.D. Dissertation. China University of Mining & Technology.

Xie, H. 1987b. Fractal geometry and its application to analysis of fracture surfaces. *Proc. MMM-2.* Shanghai: China Mechanics Society.

Xie, H. 1988. Tensile failure of local damage of rock materials. *Chinese J. of rock Mech. and Eng.* 7(2): 147–154.

Xie, H. 1989a. Studies on fractal model of the microfracture of marble. *Chinese Science Bulletin* 34(15): 1292–1296.

Xie, H. 1989b. Fractal characteristics in geological material mechanics. *Proc. of Int. Conf. of Applied Mech.* Beijing: China. also in *Mech, & Practice* 12(4): 1–9(1990).

Xie, H. 1989c. *Linear fractals, non-uniform fractals and self-inverse fractals.* Lecture Notes of Spring College on Fractal Aspects of Materials. Held by Int. Center of Material Physics of China Academy of Sciences.

Xie, H. 1989d. Fractal effect of irregularity of crack branching on the fracture toughness of brittle materials. *Int. J. of Fracture* 40(4): 267–274.

Xie, H. 1990a. *Damage mechanics of rock and concrete materials.* CUMT Press (in Chinese).

Xie, H. 1990b. Fractals-Rock mechanics. Lecture Notes given at the University of Utah.

Xie, H. 1991. Fractal nature on damage evolution of rock materials. *Proc. 2nd Int. Symp. on Min. Sci. & Tech.* (China University of Mining & Technology) Vol. 2: 697–704.

Xie, H., Bhaskar, R. & Li, J. 1993. Generation of fractal models for characterization of pulverized materials. *Minerals & Metallurgical Processing Joural* 30: 36–42.

Xie, H. & Chen, Z. D. 1988a. Analysis of non-smooth deformation of rock masses. *Proc. of Int. Symp. on Modern Min. Sci. and Tech.* China: 275–282.

Xie, H. & Chen, Z. D. 1988b. Research on the model of continuous damage mechanics of rock. *J. of China Coal Society* 13(1): 56–68.

Xie, H. & Chen, Z. D. 1988c. Fractal geometry and rock fracture. *Acta Mechanica Sinica* 4(3): 264–274.

Xie, H. & Chen, Z. D. 1989. Micromechanisms of rock fracture. *J. of China Coal Society* 14(2): 82–95.

Xie, H. & Chen, Z. D. 1990. Analysis of creep damage for large deformation caused by extravation in jointed rock masses. *J. of China Coal Society* 15(4): 23–34.

Xie, H. & Gao, F. 1991. Fractal studies on rock damage evolution. *Chinese J. of Rock Mech. and Engineering* 10(1): 65–73.

Xie, H. & Pariseau, W. G. 1992a. Fractal character and mechanism of rock bursts. *33rd U.S. Symp. on Rock Mech., Santa Fe.* Rotterdam: Balkema.

Xie, H. & Pariseau, W. G. 1992b. Fractal estimation of joint roughness coefficients. *J. of Chinese Sciences* (to be published) and Int. Conf. on Fractured and Jointed Rock Masses, Berkeley.

Xie, H. & Zhang, Y. P. 1989. Self-affine fractal geometry. *Nature Journal* 12(9): 655–665.

Xu, J., et al. 1986. Experimental study on whole microcracking process of sandstone under uniaxial compression. *Mech. & Practice* (in Chinese) 4: 16–20.

Xu, Z. B. 1985. The study of Wu zhi-ling fault structural features in Song mountain region. Master's thesis: Beijing Graduate School, China University of Mining & Technology.

Yangidani, T., et al. 1985. Localization of dilatancy in Ohshima granite under constant uniaxial stress. *J. Geophys. Res.* 90: 6840–6858.

Yfanitis, E. A., et al. 1988. Simulation of geological surfaces using fractals. *Mathematical Geology* 20(6): 667–672.

Zhang, Y. P. & Xie, H. 1989. Fuzzy fractals. *Proc. of First National Symp. on Fractal Theory and Its Application,* China. Also in Natural Journal 13(2) (1990) (in Chinese).

Zhang, Y. P. & Xie, H. 1990. Fractal geometry derived from geometric inversion. *Appl. Math. and Mech.* 11(11): 1075–1079.

Zhang, Y. P. & Xie, H. 1991. Study on estimating fractal dimension of fracture surfaces. *Mechanics & Practice* 13(1): 43–47 (in Chinese).

Zhao, Z. Y., et al. 1990. Fractal analysis applied to cataclastic rocks. *Tectonophysics* 178 (2/4): 373–377.

Subject index

Author index